Light Scattering Technology for Food Property, Quality and Safety Assessment

Contemporary Food Engineering

Series Editor
Professor Da-Wen Sun, Director
Food Refrigeration & Computerized Food Technology
National University of Ireland, Dublin
(University College Dublin)
Dublin, Ireland
http://www.ucd.ie/sun/

Light Scattering Technology for Food Property, Quality and Safety Assessment, *edited by Renfu Lu* (2016)

Advances in Heat Transfer Unit Operations: Baking and Freezing in Bread Making, *edited by Georgina Calderon-Dominguez, Gustavo F. Gutierrez-Lopez, and Keshavan Niranjan* (2016)

Innovative Processing Technologies for Foods with Bioactive Compounds, *edited by Jorge J. Moreno* (2016)

Edible Food Packaging: Materials and Processing Technologies, *edited by Miquel Angelo Parente Ribeiro Cerqueira, Ricardo Nuno Correia Pereira, Oscar Leandro da Silva Ramos, Jose Antonio Couto Teixeira, and Antonio Augusto Vicente* (2016)

Handbook of Food Processing: Food Preservation, *edited by Theodoros Varzakas and Constantina Tzia* (2015)

Handbook of Food Processing: Food Safety, Quality, and Manufacturing Processes, *edited by Theodoros Varzakas and Constantina Tzia* (2015)

Edible Food Packaging: Materials and Processing Technologies,
edited by Miquel Angelo Parente Ribeiro Cerqueira, Ricardo Nuno Correia Pereira, Oscar Leandro da Silva Ramos, Jose Antonio Couto Teixeira, and Antonio Augusto Vicente (2015)

Advances in Postharvest Fruit and Vegetable Technology,
edited by Ron B.H. Wills and John Golding (2015)

Engineering Aspects of Food Emulsification and Homogenization,
edited by Marilyn Rayner and Petr Dejmek (2015)

Handbook of Food Processing and Engineering, Volume II: Food Process Engineering, *edited by Theodoros Varzakas and Constantina Tzia* (2014)

Handbook of Food Processing and Engineering, Volume I: Food Engineering Fundamentals, *edited by Theodoros Varzakas and Constantina Tzia* (2014)

Juice Processing: Quality, Safety and Value-Added Opportunities, *edited by Víctor Falguera and Albert Ibarz* (2014)

Engineering Aspects of Food Biotechnology, *edited by José A. Teixeira and António A. Vicente* (2013)

Engineering Aspects of Cereal and Cereal-Based Products, *edited by Raquel de Pinho Ferreira Guiné and Paula Maria dos Reis Correia* (2013)

Fermentation Processes Engineering in the Food Industry, *edited by Carlos Ricardo Soccol, Ashok Pandey, and Christian Larroche* (2013)

Modified Atmosphere and Active Packaging Technologies, *edited by Ioannis Arvanitoyannis* (2012)

Advances in Fruit Processing Technologies, *edited by Sueli Rodrigues and Fabiano Andre Narciso Fernandes* (2012)

Biopolymer Engineering in Food Processing, *edited by Vânia Regina Nicoletti Telis* (2012)

Operations in Food Refrigeration, *edited by Rodolfo H. Mascheroni* (2012)

Thermal Food Processing: New Technologies and Quality Issues, Second Edition, *edited by Da-Wen Sun* (2012)

Physical Properties of Foods: Novel Measurement Techniques and Applications, *edited by Ignacio Arana* (2012)

Handbook of Frozen Food Processing and Packaging, Second Edition, *edited by Da-Wen Sun* (2011)

Advances in Food Extrusion Technology, *edited by Medeni Maskan and Aylin Altan* (2011)

Enhancing Extraction Processes in the Food Industry, *edited by Nikolai Lebovka, Eugene Vorobiev, and Farid Chemat* (2011)

Emerging Technologies for Food Quality and Food Safety Evaluation, *edited by Yong-Jin Cho and Sukwon Kang* (2011)

Food Process Engineering Operations, *edited by George D. Saravacos and Zacharias B. Maroulis* (2011)

Biosensors in Food Processing, Safety, and Quality Control, *edited by Mehmet Mutlu* (2011)

Physicochemical Aspects of Food Engineering and Processing, *edited by Sakamon Devahastin* (2010)

Infrared Heating for Food and Agricultural Processing, *edited by Zhongli Pan and Griffiths Gregory Atungulu* (2010)

Mathematical Modeling of Food Processing, *edited by Mohammed M. Farid* (2009)

Engineering Aspects of Milk and Dairy Products, *edited by Jane Sélia dos Reis Coimbra and José A. Teixeira* (2009)

Innovation in Food Engineering: New Techniques and Products, *edited by Maria Laura Passos and Claudio P. Ribeiro* (2009)

Processing Effects on Safety and Quality of Foods, *edited by Enrique Ortega-Rivas* (2009)

Engineering Aspects of Thermal Food Processing, *edited by Ricardo Simpson* (2009)

Ultraviolet Light in Food Technology: Principles and Applications, *Tatiana N. Koutchma, Larry J. Forney, and Carmen I. Moraru* (2009)

Advances in Deep-Fat Frying of Foods, *edited by Serpil Sahin and Servet Gülüm Sumnu* (2009)

Extracting Bioactive Compounds for Food Products: Theory and Applications, *edited by M. Angela A. Meireles* (2009)

Advances in Food Dehydration, *edited by Cristina Ratti* (2009)

Optimization in Food Engineering, *edited by Ferruh Erdoğdu* (2009)

Optical Monitoring of Fresh and Processed Agricultural Crops, *edited by Manuela Zude* (2009)

Food Engineering Aspects of Baking Sweet Goods, *edited by Servet Gülüm Sumnu and Serpil Sahin* (2008)

Computational Fluid Dynamics in Food Processing, *edited by Da-Wen Sun* (2007)

Light Scattering Technology for Food Property, Quality and Safety Assessment

EDITED BY RENFU LU

CRC Press
Taylor & Francis Group
Boca Raton London New York

CRC Press is an imprint of the
Taylor & Francis Group, an **informa** business

CRC Press
Taylor & Francis Group
6000 Broken Sound Parkway NW, Suite 300
Boca Raton, FL 33487-2742

© 2016 by Taylor & Francis Group, LLC
CRC Press is an imprint of Taylor & Francis Group, an Informa business

No claim to original U.S. Government works

Printed on acid-free paper
Version Date: 20160317

International Standard Book Number-13: 978-1-4822-6334-3 (Hardback)

This book contains information obtained from authentic and highly regarded sources. Reasonable efforts have been made to publish reliable data and information, but the author and publisher cannot assume responsibility for the validity of all materials or the consequences of their use. The authors and publishers have attempted to trace the copyright holders of all material reproduced in this publication and apologize to copyright holders if permission to publish in this form has not been obtained. If any copyright material has not been acknowledged please write and let us know so we may rectify in any future reprint.

Except as permitted under U.S. Copyright Law, no part of this book may be reprinted, reproduced, transmitted, or utilized in any form by any electronic, mechanical, or other means, now known or hereafter invented, including photocopying, microfilming, and recording, or in any information storage or retrieval system, without written permission from the publishers.

For permission to photocopy or use material electronically from this work, please access www.copyright.com (http://www.copyright.com/) or contact the Copyright Clearance Center, Inc. (CCC), 222 Rosewood Drive, Danvers, MA 01923, 978-750-8400. CCC is a not-for-profit organization that provides licenses and registration for a variety of users. For organizations that have been granted a photocopy license by the CCC, a separate system of payment has been arranged.

Trademark Notice: Product or corporate names may be trademarks or registered trademarks, and are used only for identification and explanation without intent to infringe.

Visit the Taylor & Francis Web site at
http://www.taylorandfrancis.com

and the CRC Press Web site at
http://www.crcpress.com

Contents

Series Preface ix
Preface xi
Nomenclature xv
Series Editor xvii
Editor xix
Contributors xxi

Chapter 1 Introduction to Light and Optical Theories 1

Renfu Lu

Chapter 2 Overview of Light Interaction with Food and Biological Materials 19

Renfu Lu

Chapter 3 Theory of Light Transfer in Food and Biological Materials 43

Renfu Lu

Chapter 4 Monte Carlo Modeling of Light Transfer in Food 79

Rodrigo Watté, Ben Aernouts, and Wouter Saeys

Chapter 5 Parameter Estimation Methods for Determining Optical Properties of Foods 111

Kirk David Dolan and Haiyan Cen

Chapter 6 Basic Techniques for Measuring Optical Absorption and Scattering Properties of Food 133

Changying Li and Weilin Wang

Chapter 7 Spatially Resolved Spectroscopic Technique for Measuring Optical Properties of Food 159

Haiyan Cen, Renfu Lu, Nghia Nguyen-Do-Trong, and Wouter Saeys

Chapter 8 Time-Resolved Technique for Measuring Optical Properties and Quality of Food 187

Anna Rizzolo and Maristella Vanoli

Chapter 9 Spectral Scattering for Assessing the Quality of Fruits and Vegetables 225

Yibin Ying, Lijuan Xie, and Xiaping Fu

Chapter 10 Light Propagation in Meat and Meat Analog: Theory and Applications 251

Gang Yao

Chapter 11 Spectral Scattering for Assessing Quality and Safety of Meat 283

Yankun Peng

Chapter 12 Light Scattering Applications in Milk and Dairy Processing 319

Czarena Crofcheck

Chapter 13 Dynamic Light Scattering for Measuring Microstructure and Rheological Properties of Food 331

Fernando Mendoza and Renfu Lu

Chapter 14 Biospeckle Technique for Assessing Quality of Fruits and Vegetables 361

Artur Zdunek, Piotr Mariusz Pieczywek, and Andrzej Kurenda

Chapter 15 Raman Scattering for Food Quality and Safety Assessment 387

Jianwei Qin, Kuanglin Chao, and Moon S. Kim

Chapter 16 Light Scattering–Based Detection of Food Pathogens 429

Pei-Shih Liang, Tu San Park, and Jeong-Yeol Yoon

Index 445

Series Preface

CONTEMPORARY FOOD ENGINEERING

Food engineering is a multidisciplinary field of applied physical sciences combined with the knowledge of product properties. Food engineers provide the technological knowledge transfer essential to the cost-effective production and commercialization of food products and services. In particular, food engineers develop and design processes and equipment to convert raw agricultural materials and ingredients into safe, convenient, and nutritious consumer food products. Food engineering topics are continuously undergoing changes to meet diverse consumer demands, and the subject is being rapidly developed to reflect market needs.

In the development of food engineering, one of the many challenges is to employ modern tools and knowledge, such as computational materials science and nanotechnology, to develop new products and processes. Simultaneously, improving food quality, safety, and security continues to be a critical issue in food engineering studies. New packaging materials and teclmiques are being developed to provide more protection to foods, and novel preservation technologies are emerging to enhance food security and defense. Additionally, process control and automation are among the top priorities identified in food engineering. Advanced monitoring and control systems are developed to facilitate automation and flexible food manufacturing. Furthermore, energy saving and minimization of environmental problems continue to be important food engineering issues, and significant progress is being made in waste management, efficient utilization of energy, and reduction of effluents and emissions in food production.

The Contemporary Food Engineering Series addresses some of the recent developments in food engineering. The series covers advances in classical unit operations in engineering applied to food manufacturing as well as such topics as progress in the transport and storage of liquid and solid foods; heating, chilling, and freezing of foods; mass transfer in foods; chemical and biochemical aspects of food engineering and the use of kinetic analysis; dehydration, thermal processing, nonthermal processing, extrusion, liquid food concentration, membrane processes, and applications of membranes in food processing; shelf life and electronic indicators in inventory management; sustainable technologies in food processing; and packaging, cleaning, and sanitation. The books in this series are aimed at professional food scientists, academics researching food engineering problems, and graduate-level students.

The editors of these books are leading engineers and scientists from different parts of the world. All the editors were asked to present their books to address the market's needs and pinpoint cutting-edge technologies in food engineering.

All chapters have been contributed by internationally renowned experts who have both academic and professional credentials. All authors have attempted to provide critical, comprehensive, and readily accessible information on the art and science of a relevant topic in each chapter, with reference lists for further information. Therefore, each book can serve as an essential reference source to students and researchers in universities and research institutions.

Da-Wen Sun
Series Editor

Preface

Light scattering is a phenomenon about the change of light traveling direction in a medium. It takes place when light is incident on a rough surface, when it travels in an optically inhomogeneous medium with a varying refractive index, or from one optically homogeneous medium into another homogeneous medium of different refractive indexes, or when photons encounter scattering particles in the medium. Most food and biological materials are heterogeneous in structure and composition, and the cellular structures (i.e., the organelles and cellular membrane) act as scatterers. Light would thus go through multiple scattering events before it exits from the tissue or is being absorbed. Light scattering in food and biological materials is often accompanied or coupled with the absorption of photons by molecules or atoms, which are then converted into another form of energy (e.g., heat, chemical reaction, fluorescence, etc.). Absorption is related to the chemical compositions of food, whereas scattering (or elastic scattering) is primarily a physical phenomenon that is associated with the structural characteristics of food. Light scattering is dependent on factors such as density, compositions, and cellular structures (size, shape, and spatial distribution). Because absorption and scattering are intertwined and also wavelength dependent, their measurement and quantification can provide a powerful means for determining the structural, rheological/mechanical, chemical, and sensory properties of food products.

In this book, light scattering techniques are broadly defined as those techniques that are developed based on the principles and theories of light transfer, or the utilization of light scattering phenomena, for achieving various application purposes. According to this definition, we would exclude conventional near-infrared spectroscopic technique from consideration in the book, because its commonly used measurement configurations are neither based on the theory of light transfer, nor directly utilize light scattering phenomena. However, the demarcation line here is not always clear, as optical fiber-based visible and near-infrared spectroscopy is often used for measuring the optical absorption and scattering properties of turbid food and biological materials in spatially resolved techniques (Chapters 5 and 7).

Over the past 15 years, considerable research activities have been reported on the development and application of various light scattering–based techniques for assessing structural, rheological, and sensory properties, quality attributes, and the safety of food and agricultural products. Scientific publications about light scattering are dispersed over many different food and agricultural disciplines. Having recognized the increased interest in light scattering technology, the editor organized the first special technical session titled "Spectral Scattering Technology for Food Quality and Safety Evaluation" at the 2010 annual meeting of the American Society of Agricultural and Biological Engineers (ASABE), with invited speakers from around the world. The technical session received considerable attention from researchers, and ASABE has since held a regular technical session on this topic at its annual meetings. Despite the increasing interest and research activities in light scattering technology for food and agriculture, no single book has been dedicated to the subject. In addition, while many books have been published on optical techniques for food and agricultural applications, few of them offer a comprehensive description of the fundamental principles, theories, and modeling of light transfer in food and agricultural products. This book, for the first time, gives the reader an overview of the principles and theory of light transfer in food and biological materials and a comprehensive review of the latest advances in light scattering technology in food and agriculture applications.

This book provides a balanced, comprehensive coverage about light scattering technology in food and agriculture. Each chapter of the book is written by expert(s) in their respective fields from around the world. The first four chapters cover basic concepts, principles, theories, and modeling of light transfer in food and biological materials. Chapters 5 and 6 describe parameter estimation methods and basic (invasive or *ex vivo*) techniques for determining optical absorption and scattering properties of food products. Chapter 7 provides an overview of the spatially resolved measurement technique for determining the optical properties of food and biological materials, whereas Chapter 8 is focused on the time-resolved spectroscopic technique for measuring optical properties and quality or maturity of horticultural products. Chapter 9 gives a broad coverage of practical light scattering techniques for nondestructive quality assessment of fruits and vegetables. Chapter 10 presents the theory of light transfer in meat muscle and the measurement of optical properties for determining the postmortem condition and textural properties of muscle foods and meat analogs. Chapter 11 gives an extensive coverage of the applications of light scattering techniques for assessing the quality and safety of animal products. Chapter 12 summarizes past and recent research in the application of light scattering for milk and dairy processing. Chapter 13 provides an overview of the concepts and principles of dynamic light scattering and its applications for measuring the microstructure and rheological properties of food. Chapter 14 shows the applications of the biospeckle technique, a special form of dynamic light scattering, for assessing the quality and condition of fruits and vegetables. Chapter 15 introduces the concepts and principles of Raman scattering and provides a detailed description of Raman scattering spectroscopic and imaging techniques in food quality and safety assessment. Chapter 16, the final chapter of the book, is focused on applications of light scattering techniques for the detection of food-borne pathogens.

This book is written for graduate students, researchers, and practitioners who are interested in learning the basic concepts, principles, and theories of light transfer and gaining an in-depth knowledge of light scattering technology for measurement and characterization of food and agricultural products. For those readers who do not have an advanced mathematical background, they can skip the first four chapters and directly go to the remaining chapters on specific application topics. It is hoped that this book will stimulate new interest to both incoming and veteran researchers and practitioners in exploring light scattering technology to address a broad range of issues in food property, quality, and safety assessment.

Finally, I thank the U.S. Department of Agriculture, Agricultural Research Service (USDA/ARS) for allowing me to take on this book project. I also thank all chapter contributors for their excellent contributions to the book. Without their enthusiastic participation and excellent cooperation, this book project would have not been possible. I am especially grateful to my wife Shexing for her understanding and support, so that I could spend many extra hours working on this book. I also want to acknowledge my two PhD students, Aichen Wang and Yuzhen Lu, for proofreading Chapters 1 through 3, and for helping me prepare several figures in Chapter 3.

MATLAB® and Simulink are registered trademarks of The MathWorks, Inc.
For product information, please contact

The MathWorks, Inc.
3 Apple Hill Drive
Natick, MA 01760-2098 USA
Tel: 508 647 7000
Fax: 508-647-7001
E-mail: info@mathworks.com
Web: www.mathworks.com

Nomenclature

$a = \mu_s/(\mu_a + \mu_s)$ = Transport albedo
A = Area, m^2
c_0 = Speed of light in vacuum, m/s
c = Speed of light in the tissue or medium, m/s
d = Distance or thickness of the sample, or diameter of a particle, m
$D = 1/[3(\mu_a + \mu'_s)]$ = Diffusion coefficient, cm or m
E = Energy, J
E_0, E = Irradiance at the surface, W/m^2
$E(r_1, r_2)$ = Irradiance at point r_1, r_2, W/m^2
f = Frequency, 1/s
g = Anisotropy factor
h = Planck's constant ($=6.62618 \times 10^{-34}$ J)
I = Radiant intensity, W/m^2
I_0 = Incident radiant intensity, W/m^2
$J(\bar{r}, t)$ = Flux, W/m^2
k = Wave number, 1/m
$L(\bar{r}, \bar{s}, t)$ = Radiance, W/sr-m^2
$l = 1/\mu_t$ = Mean free path, cm or m
n = Refractive index
N = Number of photons or number of light scattering or absorbing particles
$p(\bar{s}' \cdot \bar{s}), p(\theta)$ = Phase function of single scattering, 1/sr
P = Radiant power, W
Q = Radiant energy, J
r = Radius or distance from the origin in the polar coordinate system, m
$\bar{r}$ = Vector position (x, y, z), m
R = Remittance, backscattering, or reflectance with appropriate subscription
s = Distance, m
$\bar{s}$ = Vector position (x, y, z), m
t = Time, s
T = Transmission with appropriate subscription
T_c = Collimated transmission
T_d = Diffuse transmission
$U(r, t)$ = Electric field at the vector position r and time t, V/m
v = Velocity, m/s
V = Volume, m^3
W = Radiant energy density, J/m^3
x, y, z = Cartesian coordinates, m
$\delta = 1/\mu_t$ = Penetration depth of collimated light (mean free path for the attenuation event), m
φ = Azimuthal angle
Φ = Fluence rate, W/m^2
θ = Deflection angle or polar angle

λ = Wavelength, nm
μ_a = Absorption coefficient, 1/cm or 1/m
μ_s = Scattering coefficient, 1/cm or 1/m
$\mu'_s = \mu_s(1 - g)$ = Reduced scattering coefficient, 1/cm or 1/m
$\mu_t = \mu_a + \mu_s$ = Total attenuation coefficient, 1/cm or 1/m
$\mu'_t = \mu_a + \mu'_s$ = Reduced total attenuation coefficient, 1/cm or 1/m
$\mu_{eff} = [3\,\mu_a(\mu_a + \mu_s)]^{1/2}$ = Effective attenuation coefficient, 1/cm or 1/m
υ = Frequency, cycles per second, 1/s
ρ = Density, kg/m^3
ω = Solid angle, sr

Series Editor

Born in southern China, Dr. Da-Wen Sun is a world authority in food engineering research and education; he is a member of the Royal Irish Academy (RIA), which is the highest academic honor in Ireland; he is also a member of Academia Europaea (The Academy of Europe) and a fellow of the International Academy of Food Science and Technology. His main research activities include cooling, drying, and refrigeration processes and systems, quality and safety of food products, bioprocess simulation and optimization, and computer vision/image processing and hyperspectral imaging technologies. His many scholarly works have become standard reference materials for researchers in the areas of computer vision, computational fluid dynamics modeling, vacuum cooling, among others. Results of his work have been published in over 800 papers, including more than 400 peer-reviewed journal papers (Web of Science h-index = 71), among them, 31 papers have been selected by Thomson Reuters' *Essential Science Indicators^{SM}* as highly-cited papers, ranking him No. 1 in the world in Agricultural Sciences (December 2015). He has also edited 14 authoritative books. According to Thomson Reuters *Essential Science Indicators^{SM}* based on data derived over a period of 10 years from the ISI Web of Science, there are about 4500 scientists who are among the top 1% of the most cited scientists in the category of agriculture sciences. For many years, Dr. Sun has consistently been ranked among the top 50 scientists in the world (he was at the 20th position in December 2015), and has recently been named Highly Cited Researcher 2015 by Thomson Reuters.

He earned a first class BSc Honours and MSc in mechanical engineering, and a PhD in chemical engineering in China before working in various universities in Europe. He became the first Chinese national to be permanently employed in an Irish university when he was an appointed college lecturer at the National University of Ireland, Dublin (University College Dublin [UCD]), in 1995, and was then continuously promoted in the shortest possible time to senior lecturer, associate professor, and full professor. Dr. Sun is now the professor of Food and Biosystems Engineering and the director of UCD Food Refrigeration and Computerised Food Technology.

As a leading educator in food engineering, Dr. Sun has significantly contributed to the field of food engineering. He has trained many PhD students who have made their own contributions to the industry and academia. He has also delivered lectures on advances in food engineering on a regular basis in academic institutions internationally and delivered keynote speeches at international conferences. As a recognized authority in food engineering, he has been conferred adjunct/visiting/consulting professorships from 10 top universities in China, including Zhejiang University, Shanghai Jiaotong University, Harbin Institute of Technology, China Agricultural University, South China University of Technology, and Jiangnan University. In recognition of his significant contribution to food engineering worldwide and for his outstanding leadership in the field, the International Commission of Agricultural and Biosystems Engineering (CIGR) awarded him the CIGR Merit Award twice in 2000 and in 2006, the Institution of Mechanical Engineers based in the United Kingdom named him Food Engineer of the Year 2004. In 2008, he was awarded the CIGR Recognition Award in honor of his distinguished achievements as the top 1% of agricultural engineering scientists in the world. In 2007, he was presented with the only AFST(I) Fellow Award by the Association of Food Scientists and Technologists (India), and in 2010, he was presented with the CIGR Fellow Award; the title of Fellow is the highest honor in CIGR and is conferred to individuals who have made sustained, outstanding contributions worldwide. In March 2013, he was presented with the You Bring Charm to the World award by Hong Kong-based Phoenix Satellite Television with other award recipients including Mr. Mo Yan—the 2012 Nobel Laureate in Literature and the Chinese Astronaut Team for Shenzhou IX Spaceship. In July 2013, he received the Frozen Food Foundation Freezing Research Award from the International Association for Food Protection for his significant contributions to enhancing the field of food freezing technologies. This is the first time that this prestigious award was presented to a scientist outside the United States, and in June 2015 he was presented with the IAEF Lifetime Achievement Award. This IAEF (International Association of Engineering and Food) award, highlights the lifetime contribution of a prominent engineer in the field of food.

He is a fellow of the Institution of Agricultural Engineers and a fellow of Engineers Ireland (the Institution of Engineers of Ireland). He is also the editor-in-chief of *Food and Bioprocess Technology—An International Journal* (2012 impact factor = 4.115), former editor of *Journal of Food Engineering* (Elsevier), and editorial board member for a number of international journals, including the *Journal of Food Process Engineering, Journal of Food Measurement and Characterization*, and *Polish Journal of Food and Nutritional Sciences*. He is also a chartered engineer.

At the 51st CIGR General Assembly held during the CIGR World Congress in Quebec City, Canada, on June 13–17, 2010, he was elected incoming president of CIGR, became CIGR President in 2013–2014, and is now a CIGR Past President. CIGR is the world's largest organization in the field of agricultural and biosystems engineering.

Editor

Renfu Lu is a supervisory research agricultural engineer and research leader of the Sugarbeet and Bean Research Unit within the USDA/ARS in East Lansing, Michigan. He also holds an adjunct professor appointment with the Department of Biosystems and Agricultural Engineering at Michigan State University. Dr. Lu earned his PhD and MS degrees in agricultural engineering from Pennsylvania State University and Cornell University, respectively, and his BS degree in engineering from Zhejiang Agricultural University (now Zhejiang University) in China. As a research leader, he leads and supervises the unit's three federal research programs in engineering, genetics, breeding, and pathology for dry beans, sugar beet, and fruits and vegetables. Dr. Lu's research is primarily focused on sensing technologies for quality measurement and grading of horticultural and food products.

During the 30+ years of his professional career, he has made many significant, original contributions to the development and application of optical imaging and spectroscopic techniques for quality evaluation of horticultural and food products. His research has been documented in more than 220 publications, including 95 refereed journal articles and 16 book chapters. Dr. Lu has held numerous leadership positions for the ASABE, including chair of the ASABE Food and Process Engineering Division, Refereed Publications Committee, and several technical committees. He has served as an editor for *Transactions of the ASABE* and *Applied Engineering in Agriculture,* for 6 years, and is on the editorial board of *Journal of Food Measurement and Characterization* and *Postharvest Biology and Technology.* Among his many awards and honors are election as a Fellow of the ASABE (2013), an Outstanding Alumni Award from the College of Agricultural Sciences at Pennsylvania State University (2011), and a Federal Laboratory Consortium Technology Transfer Award (2009).

Contributors

Ben Aernouts
Department of Biosystems
KU Leuven—University of Leuven
Leuven, Belgium

Haiyan Cen
College of Biosystems Engineering
   and Food Science
Zhejiang University
Zhejiang, China

Kuanglin Chao
U.S. Department of Agriculture
   Agricultural Research Service
   (USDA/ARS)
Beltsville Agricultural Research Center
Environmental Microbial and Food
   Safety Laboratory
Beltsville, Maryland

Czarena Crofcheck
Department of Biosystems and
   Agricultural Engineering
University of Kentucky
Lexington, Kentucky

Kirk David Dolan
Department of Food Science and
   Human Nutrition
Department of Biosystems and
   Agricultural Engineering
Michigan State University
East Lansing, Michigan

Xiaping Fu
College of Biosystems Engineering
   and Food Science
Zhejiang University
Zhejiang, China

Moon S. Kim
U.S. Department of Agriculture
   Agricultural Research Service
   (USDA/ARS)
Beltsville Agricultural Research
   Center
Environmental Microbial and Food
   Safety Laboratory
Beltsville, Maryland

Andrzej Kurenda
Institute of Agrophysics
Polish Academy of Sciences
Lublin, Poland

Changying Li
College of Engineering
University of Georgia
Athens, Georgia

Pei-Shih Liang
U.S. Department of Agriculture
   Agricultural Research Service
   (USDA/ARS)
Western Regional Research
   Center
Albany, California

Renfu Lu
U.S. Department of Agriculture
   Agricultural Research Service
   (USDA/ARS)
Sugarbeet and Bean Research
   Unit
East Lansing, Michigan

Fernando Mendoza
Department of Plant, Soil and Microbial Sciences
Michigan State University
East Lansing, Michigan

Nghia Nguyen-Do-Trong
Department of Biosystems
KU Leuven—University of Leuven
Leuven, Belgium

Tu San Park
Department of Agricultural and Biosystems Engineering
The University of Arizona
Tucson, Arizona

Yankun Peng
College of Engineering
China Agricultural University
Beijing, China

Piotr Mariusz Pieczywek
Institute of Agrophysics
Polish Academy of Sciences
Lublin, Poland

Jianwei Qin
U.S. Department of Agriculture
Agricultural Research Service (USDA/ARS)
Beltsville Agricultural Research Center
Environmental Microbial and Food Safety Laboratory
Beltsville, Maryland

Anna Rizzolo
Unità di ricerca per I processi dell'industria agroalimentare
Consiglio per la ricerca in agricoltura e l'analisi dell'economia agraria (CREA-IAA)
Milan, Italy

Wouter Saeys
Department of Biosystems
KU Leuven—University of Leuven
Leuven, Belgium

Maristella Vanoli
Unità di ricerca per I processi dell'industria agroalimentare
Consiglio per la ricerca in agricoltura e l'analisi dell'economia agraria (CREA-IAA)
Milan, Italy

and

Istituto di Fotonica e Nanotecnologie
Consiglio Nazionale delle Ricerche (IFN-CNR)
Milan, Italy

Weilin Wang
Monsanto Company
St. Louis, Missouri

Rodrigo Watté
Department of Biosystems
KU Leuven—University of Leuven
Leuven, Belgium

Lijuan Xie
College of Biosystems Engineering and Food Science
Zhejiang University
Zhejiang, China

Gang Yao
Department of Bioengineering
University of Missouri
Columbia, Missouri

Yibin Ying
College of Biosystems Engineering and Food Science
Zhejiang University
Zhejiang, China

Jeong-Yeol Yoon
Department of Agricultural and
 Biosystems Engineering
The University of Arizona
Tucson, Arizona

Artur Zdunek
Institute of Agrophysics
Polish Academy of Sciences
Lublin, Poland

1 Introduction to Light and Optical Theories

Renfu Lu

CONTENTS

1.1 Basics of Light 1
    1.1.1 Gamma Rays and X-Rays 2
    1.1.2 UV Light 3
    1.1.3 Visible and Infrared Light 4
    1.1.4 Microwaves 4
    1.1.5 Radio Waves 5
1.2 Optical Theories 5
    1.2.1 Ray Theory 7
    1.2.2 Wave Theory 9
    1.2.3 Electromagnetic Theory 10
    1.2.4 Quantum Theory 13
1.3 Light Scattering and Its Applications in Food and Agriculture 15
1.4 Summary 18
References 18

1.1 BASICS OF LIGHT

Light is ubiquitous in our daily life. Without light, there would be no life on the earth. For thousands of years, humans have been fascinated by light phenomena and tried to understand and harness light for better serving their needs. The study of light and its interactions with matter is known as *optics*, which is an important research field in modern physics, engineering, and life science. Over the past 100 years, many important discoveries about light phenomena and developments in optical theories and technologies have taken place. With the advent and rapid advances of the computer, Internet, and wireless technologies over the past 30 years, we have witnessed the emergence of many new optical technologies and their expanding applications in detection or diagnosis, manufacturing, product processing, communications, energy generation, etc.

This chapter presents an introduction to the basic characteristics of light and optical theories, followed by a brief overview of light scattering technology and its applications to food and agricultural products. In our daily life, the term *light* often refers to visible light, which only covers a very narrow section of the electromagnetic spectrum. In this book, most of our discussion about light scattering technology is focused on the visible and near-infrared region because light in this spectral region can undergo multiple scattering and long-distance propagation in food and biological materials.

In classical electromagnetic theory, light is treated as electromagnetic radiation, consisting of both electric and magnetic vector waves. The wave characteristics are described by frequency (υ), wavelength (λ), and velocity (c). Light travels in vacuum at a speed of 300,000 km/s, regardless of its frequency or wavelength. However, in a medium, light travels slower than in vacuum; its actual speed depends on the type of waves or wavelengths and the property of the medium. As the speed of light changes from one medium to another, the light no longer travels in a straight direction. Instead, the light changes its traveling direction after it enters the second medium. This important phenomenon is called *refraction*, which contributes to light scattering and propagation in food and biological materials and will be further discussed in a later section and also in Chapter 2.

Classical electromagnetic theory adequately describes the wave characteristics and most phenomena of light and its interactions with matter, except for a few special situations. Modern quantum theory, on the other hand, considers light to be composed of photons of different frequencies, or small packets of energy. A photon carries electromagnetic energy, but has zero mass at rest. The energy of a photon, E, is proportional to frequency υ (or inversely proportional to wavelength λ), which can be expressed by the following equation:

$$E = h\upsilon = \frac{hc}{\lambda}, \tag{1.1}$$

where E has the unit of joule or J, $h = 6.63 \times 10^{-34}$ J/s is Planck's constant, $\upsilon = c/\lambda$ is the frequency in 1/s, and c is the velocity of light in the medium. This wave–photon duality property provides the most complete explanation for virtually all light phenomena that are now known to us, and are widely used in studying the interaction of light with matter, including biological and food materials.

Since the energy of photons is determined by wavelength, photons of different wavelengths carry different levels of energy and thus behave differently when they interact with matter. The electromagnetic radiation emitting from the sun or a black body consists of a spectrum covering a broad range of frequencies (or wavelengths). The electromagnetic spectrum is normally divided into different regions according to wavelengths or frequencies, in the decreasing order of energy: gamma rays, x-rays, ultraviolet (UV) light, visible light, infrared light, microwaves, and radio waves (Figure 1.1). The major characteristics of these radiation regions and their typical applications are briefly discussed in the following subsections.

1.1.1 GAMMA RAYS AND X-RAYS

Gamma rays and x-rays are the most powerful electromagnetic radiation in the electromagnetic spectrum. Gamma rays cover a portion of the electromagnetic spectrum with wavelengths of less than 10 pm (1 pm = 10^{-12} m) or with corresponding frequencies greater than 3×10^{19} Hz, whereas x-rays cover wavelengths between 0.01 and 10 nm (1 nm = 10^{-9} m) or the frequency range of 3×10^{19}–3×10^{16} Hz. Both gamma rays and x-rays are ionizing radiation, which can liberate electrons from atoms or

Introduction to Light and Optical Theories

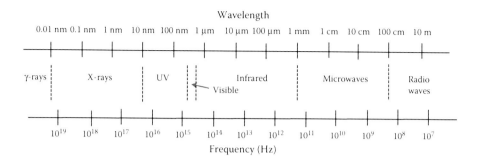

FIGURE 1.1 Spectrum of electromagnetic radiation from gamma rays to radio waves.

molecules, and can permanently damage or destroy living cells. They can travel in straight lines and their paths cannot be altered by electrical or magnetic fields. Gamma rays and x-rays can usually penetrate deep into matter, although the degree of penetration depends on their energy and the medium they are traveling through. Because of their low absorption and the ability of "seeing through" objects, x-ray-based technologies (i.e., x-ray imaging, computed tomography or CT, etc.) are widely used in medical diagnosis and also in food inspection. Gamma rays are useful in astronomy and physics for studying high-energy objects or regions, the irradiation of food, and diagnostic imaging in medicine.

1.1.2 UV Light

UV light refers to the section of an electromagnetic spectrum with the wavelengths approximately between 10 and 400 nm (or $\sim 3 \times 10^{16}$–8×10^{14} Hz), which is shorter than visible light but longer than x-rays. Sunlight contains UV light, but most of it is absorbed by the ozone layer and atmosphere. UV light can also be produced by electric arcs and light sources such as mercury-vapor lamps, tanning lamps, and black lights. Although not as powerful as x-rays, UV light can heat matter, cause chemical reactions, and make substances to glow or fluoresce, and it can damage living tissues. UV light can be further divided into UV-C (10–280 nm), UV-B (280–315 nm), and UV-A (315–400 nm). UV-C rays are the most harmful and are almost completely absorbed by the atmosphere. UV-B rays are harmful to living organisms and can damage deoxyribonucleic acid (DNA). UV-A rays can age skin cells and damage DNA, but they also have beneficial effects to living organisms by inducing the production of vitamin D in the skin.

UV light has found many applications, because it can cause chemical reactions and excite fluorescence in materials. UV-based fluorescence spectroscopy and imaging are widely used in environmental monitoring, DNA sequencing, genetic analysis, clinical diagnosis, food safety and quality inspection, plant health monitoring, and many other fields. UV light is used for disinfection and decontamination of surfaces and water, and it is also provided as a medical therapy, and so on. Generally, UV light has poor penetration in biological tissues, and it is thus not suitable for detecting internal structural and chemical characteristics of biological materials.

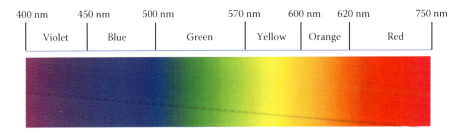

FIGURE 1.2 Spectrum of visible light from 400 to 750 nm.

1.1.3 VISIBLE AND INFRARED LIGHT

Visible light covers the wavelengths of 400–750 nm, corresponding to the frequency range of approximately 8×10^{14}–4×10^{14} Hz. Human eyes are the most sensitive to visible light, thus enabling humans to see objects in the surrounding environment. The variations of wavelength in the visible region give humans a sensation of color, ranging from violet (400–450 nm) to blue (450–500 nm), green (500–570 nm), yellow (570–600 nm), orange (600–620 nm), and red (620–750 nm) (Figure 1.2).

Infrared light refers to the region with wavelengths between 750 nm and 1 mm, corresponding to the frequency range of 4×10^{14}–3×10^{11} Hz. Infrared light can be further divided into three parts: near infrared (750–2500 nm), mid infrared (2.5–10 μm), and far infrared (10 μm – 1 mm).

The patterns of interaction of visible and near-infrared photons with materials are similar, and they primarily take place in the form of molecular vibration and oscillation. Many plant products absorb visible light at selected wavelengths, due to the presence of pigments (carotenoid, anthocyanin, and chlorophyll) in the plant tissue. Water absorption peaks occur at a number of wavelengths in the visible and infrared region (e.g., 750, 970, 1450, 1940 nm, etc.). Near-infrared light has strong interactions with hydrogen bonds such as N–H, C–H, O–H, F–H, etc., which are abundant in plant and animal products and many other biological materials. Over the past decades, near-infrared spectroscopy (NIRS) has been widely used for composition analysis of food and agricultural products, chemical or functional imaging, biomedical diagnosis, and process control and monitoring. Light in the visible and near-infrared region of 700–1400 nm has good penetration in plant and animal tissues (ranging from a few millimeters to several centimeters). In the biomedical field, this spectral region is known as "diagnosis window." The majority of light scattering techniques presented in this book utilize the visible and near-infrared light, because photons can travel a long distance and undergo multiple scattering events, before being absorbed or reemerging from the medium.

1.1.4 MICROWAVES

Microwaves refer to the electromagnetic radiation covering the wavelengths ranging from 1 mm to 1 m, with the frequency range of 300 GHz (1 GHz = 10^9 Hz) to 300 MHz (1 MHz = 10^6 Hz). Hence, the term "microwaves" technically is a

Introduction to Light and Optical Theories

misnomer in view of their wavelength. Water vapor and oxygen have strong absorption in this spectral region, which is especially prominent above 40 GHz. Microwave radiation is absorbed by molecules that have a dipole moment in liquids but does not cause ionization in matter. Hence, it is generally considered safe to humans and biological materials. Microwaves have found a wide range of applications in communications (satellites, radars, mobile phones, wireless local area network [LAN], Global Positioning System [GPS], etc.), food heating and drying (microwave ovens around 2.45 GHz), and property detection (i.e., dielectric properties in food and agricultural materials for detecting moisture content and other compositions). Microwave radiation can be generated from a variety of sources including magnetron (in microwave ovens for food heating), klystron, traveling-wave tube, gyrotron, etc.

1.1.5 RADIO WAVES

Radio waves cover the spectral region approximately from 1 m to 100 km, corresponding to the frequencies from 3 GHz to 3 kHz, which are the longest in the electromagnetic spectrum. Radio waves of different wavelengths have different propagation characteristics. Radio waves have low energy, but have excellent penetration in many materials because these materials do not absorb radio waves. Radio waves are widely used in radio communications and data transmission such as televisions (TVs), broadcasting, mobile phones, wireless network and radar, communication satellites, etc.

1.2 OPTICAL THEORIES

For studying light scattering and propagation in food and biological materials and developing instruments to measure optical scattering and absorption properties, it is necessary to understand and apply appropriate optical theories, from the simple ray theory to electromagnetic theory and to quantum theory. This section is, therefore, intended to provide a brief summary of key features and application scopes of these theories, and readers who are interested in gaining further knowledge of these optical theories are advised to read related reference books (Hecht, 2002; Saleh and Teich, 2007).

As described earlier, light is considered both as electromagnetic waves and packets of photons in classical and modern optical theories. This wave–photon duality characteristic provides the foundation for explaining different phenomena when light interacts with matter. On the basis of the historical timeline, the optical theories may be divided into *ray* or *geometrical theory*, *wave theory*, *electromagnetic theory*, and *quantum theory* (Figure 1.3). Ray theory is the earliest one developed for describing the interaction of light with objects whose dimensions are far greater than that of light beams. Ray theory provides an approximate description of various phenomena that are readily observed in our daily life, such as *reflection* and *refraction* when light propagates from one medium to another, image formation, etc. The theory provides an estimate of the location and direction of light. However, ray theory is limited in explaining many other phenomena that happen at much smaller dimensional scales. Consequently, wave theory was developed in the seventeenth

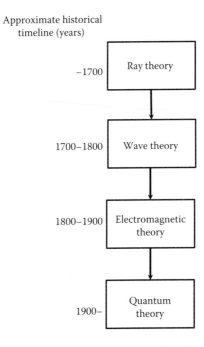

FIGURE 1.3 Approximate timeline for the development of optical theories.

and eighteenth century to explain phenomena such as *interference* and *diffraction*. Wave theory treats light, traveling in a medium, as a scalar function of position and time (or the wave function). However, it cannot provide an adequate explanation for the reflection and refraction of light at the boundaries between dielectric materials, nor does it explain phenomena such as *polarization* and *coherence*, which would require treating light as *vector waves*. In electromagnetic theory, light propagates in two mutually coupled vector waves, that is, electric and magnetic waves, each of them traveling at the speed of light. Wave theory is thus considered a special case or an approximation of electromagnetic theory. Electromagnetic theory enables us to establish the connection between the speed of light and the electromagnetic properties of materials in which light travels. It explains almost all the phenomena that are encountered in experimental observations. However, the theory cannot predict all outcomes of light interaction with matter, especially at the molecular or atomic level. Quantum theory, which emerged in the late nineteenth century and became mature in the early twentieth century, provides an ultimate theory that explains the outcome of all phenomena, in which light interacts with matter. It enables us to understand the fundamental interactions of light with biological materials at the molecular or even subatomic levels. It explains many phenomena involving the absorption and emission of light after it interacts with biological materials, such as fluorescence, phosphorescence, and inelastic scattering (e.g., Raman). Quantum theory also lays the foundation for many modern discoveries and technological breakthroughs, such as lasers, molecular spectroscopic techniques, nanotechnology, etc. In the following subsections, we briefly summarize the main features of each theory, which are

Introduction to Light and Optical Theories

needed in the understanding, development, and application of various light scattering techniques for food property, quality, and safety assessment.

1.2.1 Ray Theory

Ray theory is the simplest among the four optical theories; it explains some basic phenomena that we observe or experience in our daily life. The theory is adequate for explaining those situations in which the dimensions of objects are much greater than light wavelengths and visually detectable in our surrounding. Ray theory forms the basis for explaining light *reflection* and *refraction* at the interface or boundary between two media of different optical properties. It offers an adequate explanation for image formation and light transfer in optical fibers, etc.

Consider a light beam that travels from one medium (say air) with a refractive index designated as n_1, into the second medium of refractive index n_2. The refractive index of a medium describes the speed of light in the medium relative to that of light in vacuum, which is dimensionless and is given as follows:

$$n = \frac{c_0}{c}, \tag{1.2}$$

where c and c_0 are the velocity of light in the medium and vacuum, respectively. Since the speed of light traveling in a medium cannot be faster than that in vacuum, the refractive indexes for media are greater than 1. For food and biological materials, the refractive index values range between 1.30 and 1.50, and for water, $n = 1.33$. As the speed of light changes from one medium to another, its traveling direction also changes. This light-traveling direction change gives rise to refraction, reflection, and, to a lesser extent, scattering in food and biological materials, since they are heterogeneous in composition. Scattering is a complex phenomenon, which will be discussed in detail in Chapters 2 and 3. Hence, the refractive index is considered a fundamental property parameter in studying light propagation in food and biological materials.

As the light beam is incident with an oblique angle on the flat surface or interface between two optically homogeneous media (i.e., a constant, uniform refractive index), the light-traveling direction in the second medium would deviate from the original straight line in the first medium. The amount of deviation in the light-traveling direction, called *refraction*, will depend on the relative values of the refractive index for the two media. The refraction of light between the two homogeneous media is governed by *Snell's law* or the *law of refraction* (Figure 1.4)

$$\sin\theta_2 = \frac{n_1}{n_2}\sin\theta_1, \tag{1.3}$$

where θ_1 is the angle of light incidence from the normal of the surface for the first medium, and θ_2 is the angle of refraction for the second medium. A fraction of light is also reflected at the interface at the same angle as that of incidence, referred to as *surface* or *specular reflection* or *Fresnel reflection*, and the amount of surface

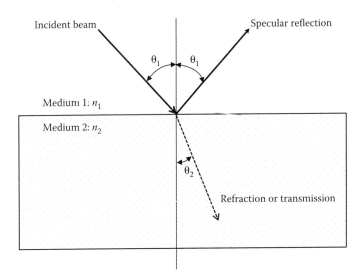

FIGURE 1.4 Snell's law of reflection and refraction at a flat interface between two optically homogeneous media with the refractive indexes of n_1 and n_2, respectively.

reflection depends on the angle of incidence as well as the relative values of the refractive index for the two media. The amount of light transmitted from medium 1 to medium 2 and the amount of light reflected at the interface can be calculated using a set of equations, by employing electromagnetic theory. We will delay the discussion on the equations of calculating Fresnel reflection till Chapter 2.

A special case that is of great significance in modern optic fiber technology is the so-called *total internal reflection*. It takes place when light passes from an optically denser medium (i.e., with a larger refractive index) to an optically thinner medium (i.e., with a smaller refractive index) and the incident angle is greater than a specific critical angle with respect to the normal to the surface (Figure 1.5). Under such a situation, the light cannot pass the second medium and is totally reflected. This critical angle, θ_c, can be determined from Snell's law in Equation 1.3:

$$\sin\theta_i = \frac{n_2}{n_1}\sin\theta_t. \tag{1.4}$$

When $\sin\theta_t = 1$ or $\theta_t = 90°$, the corresponding angle θ_i is called the critical angle θ_c, that is,

$$\theta_c = \sin^{-1}\left(\frac{n_2}{n_1}\right). \tag{1.5}$$

Hence, when the angle of light incidence is equal to or greater than θ_c, all light will be reflected back and no light passes through the second medium.

Introduction to Light and Optical Theories 9

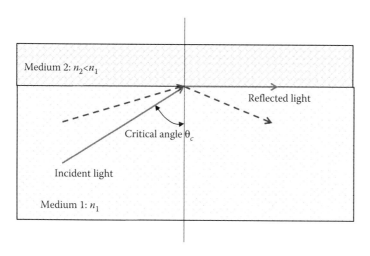

FIGURE 1.5 Total internal reflection occurs when the light incident angle is equal to, or greater than, the critical angle θ_c and the refractive index of the first medium is greater than that of the second medium (i.e., $n_1 > n_2$).

Modern optic fiber communications, fiber-optic-based detectors, and endoscopes are based on the principle of total internal reflection, which transmits optical signals for long distances with no or little loss.

1.2.2 Wave Theory

In wave theory, light is treated as scalar waves. The propagation of light waves is governed by the famous wave equation

$$\nabla^2 u(\vec{r},t) = \frac{1}{c^2}\frac{\partial^2 u(\vec{r},t)}{\partial t^2}, \quad (1.6)$$

where $\nabla^2 = (\partial^2/\partial x^2) + (\partial^2/\partial y^2) + (\partial^2/\partial z^2)$ is the Laplacian operator, $u(\vec{r}, t)$ is known as the wave function, which depends on spatial position $\vec{r}(x, y, z)$ and time t, and c is the velocity of light traveling in the medium, that is, $c = c_0/n$, in which c_0 is the speed of light in vacuum and n is the refractive index of the medium.

The optical or radiant intensity $I(\vec{r}, t)$, which is defined as the optical power per unit area (W/m²), is proportional to the squared wave function

$$I(\vec{r}, t) = |u(\vec{r}, t)|^2. \quad (1.7)$$

Equation 1.7 establishes the relationship between the optical intensity and the wave function, and it also explains the physical meaning of the wave function, which is not specified in Equation 1.6. The definition of radiometric quantities such as radiant intensity $I(\vec{r}, t)$ is further explained in Chapter 3.

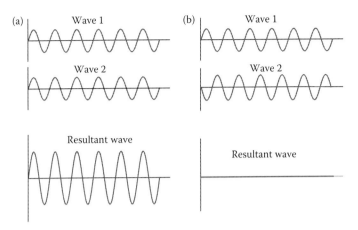

FIGURE 1.6 Concept of interference between two electromagnetic waves: (a) constructive interference and (b) destructive interference.

Since the wave equation in Equation 1.6 is linear, the principle of superposition applies. This means that when two or more optical waves are present at the same time in the same region, the total wave function is the sum of the individual wave functions. *Interference* occurs when two or more waves superpose to form a resulting wave of greater or lower amplitude. The interference is called *constructive* or *coherent* when two wave functions result in an increase in the magnitude of the light intensity, and it is *destructive* when the resulting wave is lower than that of single-wave functions. The phase difference between the two wave functions determines whether they are constructive or destructive (Figure 1.6). The principle of interference can be used to explain many natural phenomena, such as the bright and dark bands of light, called *fringe*, which is caused by the coexistence of multiple waves in a region. Optical interference has been widely applied in modern optical instrumentations, such as optical, radio, and acoustic interferometry.

Diffraction is a form of interference, which occurs when a light wave encounters an obstacle or a slit that is comparable in size to its wavelength or when light travels in a medium with a varying refractive index. For example, when a plane wave passes through a slit of width equal to the wavelength, it turns into a spherical wave (Figure 1.7). In many spectroscopic instruments, diffracting gratings are used to disperse the incoming broadband light into different wavelengths; the form of the light diffracted depends on the structure and number of the elements present on the grating. Speckle is another important phenomenon that occurs when a laser light illuminates a rough surface or the surface of biological materials in which molecules are in motion (see Chapter 14).

1.2.3 ELECTROMAGNETIC THEORY

While wave theory can explain most of the experimental observations about light, it has limitations in explaining phenomena such as polarization, in which light waves

Introduction to Light and Optical Theories

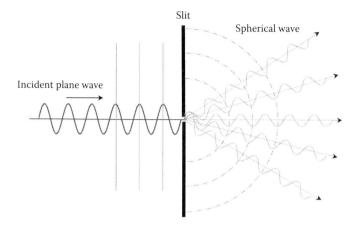

FIGURE 1.7 Diffraction of a plane wave into a circular wave when it passes a slit that is of the same scale as the wavelength of the traveling wave.

oscillate only in one or more than one orientation. James Clark Maxwell (1831–1879), a British scientist, first demonstrated that electromagnetic radiation consists of the electric and magnetic fields, and the two fields travel as wave vectors at the speed of light. The electric and magnetic waves are orthogonal to each other, perpendicular to the direction of light propagation in space (Figure 1.8).

Electromagnetic theory provides a comprehensive explanation of almost all experimental or natural phenomena. The governing equations for electromagnetic waves, also called *Maxwell's equations*, are a set of partial differential equations for electric and magnetic fields

$$\nabla \cdot E = \frac{\rho_c}{\epsilon} \quad \text{(Gauss's law)}, \tag{1.8}$$

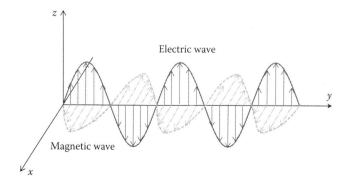

FIGURE 1.8 Propagation of electromagnetic radiation in the space in two perpendicular electric and magnetic waves at the same speed of light.

$$\nabla \cdot \boldsymbol{B} = 0 \quad \text{(Gauss's law for magnetism)}, \tag{1.9}$$

$$\nabla \times \boldsymbol{E} = -\frac{\partial \boldsymbol{B}}{\partial t} \quad \text{(Faraday's law of induction)}, \tag{1.10}$$

$$\nabla \times \boldsymbol{B} = \mu \left(\boldsymbol{J} + \epsilon \frac{\partial \boldsymbol{E}}{\partial t} \right) \quad \text{(Ampere's circuital law)}, \tag{1.11}$$

in which $\nabla \cdot$ is called the *divergence* given by

$$\nabla \cdot = \frac{\partial}{\partial x} + \frac{\partial}{\partial y} + \frac{\partial}{\partial z}, \tag{1.12}$$

and $\nabla \times$ is a vector field, called the *curl*

$$\nabla \times \boldsymbol{E} = \boldsymbol{i} \left(\frac{\partial H_z}{\partial y} - \frac{\partial E_y}{\partial z} \right) + \boldsymbol{j} \left(\frac{\partial E_x}{\partial z} - \frac{\partial E_z}{\partial x} \right) + \boldsymbol{k} \left(\frac{\partial E_y}{\partial x} - \frac{\partial E_x}{\partial y} \right), \tag{1.13}$$

where E and B represent the electric and magnetic fields, respectively, and are vector quantities, ρ_c is the current charge and J is the current density, and ε and μ are the permittivity and permeability of free space, respectively. A brief description on the concept of vectors and derivatives as well as various mathematical symbols is given in the Appendix of Chapter 3. Derivation and detailed explanation of Equations 1.8 through 1.11 can be found in many optics textbooks for undergraduate and graduate students (Hecht, 2002). It should be mentioned that the wave equation in Equation 1.6 is a special case for Maxwell's equations. Hence, electromagnetic theory encompasses wave theory and enables us to explain many other phenomena that otherwise cannot be explained by wave theory.

Maxwell's equations show the interdependence of electric and magnetic waves. The energy density of an electric field can be computed by

$$U_E = \frac{1}{2} \epsilon E^2. \tag{1.14}$$

And the energy density of a magnetic field is given by

$$U_B = \frac{1}{2\mu_0} B^2 = \frac{1}{2} \epsilon E^2. \tag{1.15}$$

The total energy density of the electric and magnetic fields is the sum of Equations 1.14 and 1.15

$$U = U_E + U_B = \epsilon E^2. \tag{1.16}$$

The radiant intensity of the electromagnetic field is the velocity of the light in the medium times the time average of the energy density $(1/2)\epsilon E_0^2$

$$I = \frac{1}{2} c\epsilon E_0^2. \tag{1.17}$$

As mentioned previously, *polarization* is an important phenomenon that is often encountered in our daily life and in scientific and industrial applications. Sunlight is unpolarized, that is, its waves oscillate in all directions randomly. When it is reflected at a surface, the reflected light (specular reflection) attains a degree of polarization. Hence wearing a pair of polarized sunglasses will enable us to filter out some of the polarized, surface reflectance, giving us a clearer view of the object. Some crystals absorb light in one state of polarization, but they transmit light in the other orthogonal state. This phenomenon is known as *dichroic*, and it is utilized in making polarized filters or polarizers. *Birefringence* refers to a phenomenon where the reflective index of a material depends on the polarization and propagation direction of light. When light enters a birefringent material, it is split by polarization into two mutually perpendicular rays. Birefringence has many applications in optical devices, such as liquid crystal displays, light modulators, polarizing filters, etc.

1.2.4 Quantum Theory

Electromagnetic theory provides the most complete description of light phenomena within the context of classical optics. However, it has failed to explain the interaction of light with matter at the nanoscale level. Quantum theory, also known as *quantum mechanism* or *quantum physics*, is able to deal with virtually all physical phenomena; it extends beyond the classical electromagnetic theory by explaining the interaction of light with matter at the atom and subatom levels. It provides a complete mathematical framework for describing light behavior and its interaction and exchange of energy with matter.

In quantum theory, light exhibits both wave-like and particle-like behavior. When traveling in space or a medium, light behaves like electromagnetic waves, consisting of electric and magnetic fields. Meanwhile, quantum theory treats light as photons. Photons have no mass at rest and carry electromagnetic energy and momentum (or spin) that governs its polarization properties. Photons interact with matter through the electric field of light by exerting forces on the electric charges and dipoles in atoms and molecules, causing them to vibrate or accelerate. Atoms and molecules can only take specific, discrete energy levels determined by quantum theory. The interaction of a photon with matter may take place in one of the two processes. In the first process, when the energy of the photon is equal to the difference between two allowed energy levels for an atom, the photon may be absorbed (or annihilated) by the atom, thus raising it to a higher energy level. In the second process, an atom can transition back to the original energy level, or to a lower energy level, resulting in the emission or creation of a new photon whose energy is equal to the difference between the energy levels. Atoms and molecules constantly undergo these upward and downward energy transitions, some of which are caused by thermal excitations and lead to emission and absorption. Hence, all objects with temperature above absolute zero emit or absorb electromagnetic radiation.

According to quantum mechanics, the allowed energy levels E of a physical system with mass m in the absence of time-varying interactions are governed by the Schrödinger equation

$$-\frac{\hbar^2}{2m}\nabla^2\psi(\vec{r})+V(\vec{r})\psi(\vec{r})=E\psi(\vec{r}), \quad (1.18)$$

where $\hbar = \upsilon/2\pi$, $\psi(\vec{r})$ is the complex wave function, $V(\vec{r})$ is the potential energy, and E is the energy that is quantized to discrete levels separated by the energy of a photon. For diatomic molecules (e.g., CO_2, N_2, and HCl), the discrete energy levels may be expressed as

$$E_\upsilon = \left(\upsilon + \frac{1}{2}\right)\hbar\omega,$$

$$(\upsilon = 0, 1, 2, \ldots) \quad (1.19)$$

where $\omega = (k/m_r)^{1/2}$ is the oscillation frequency and m_r is the reduced mass of the system. In the case of polyatomic molecules, there will be numerous allowed energy levels, which may be approximated as a series of diatomic, independent, and harmonic oscillators, as follows:

$$E(\nu_1, \nu_2, \nu_3, \ldots) = \sum_{i=1}^{3N-6}(\nu_i + 1)\hbar\omega.$$

$$(\nu_1, \nu_2, \nu_3, \ldots = 0, 1, 2, \ldots) \quad (1.20)$$

When the energy level is transitioned from 0 to 1 in any one of the vibrational states ($\nu_1, \nu_2, \nu_3, \ldots$), it is considered a fundamental-state transition and is allowed by selection rules. When the transition is from the ground state to $\nu_1 = 2, 3, \ldots$, and all others are zero, it is known as an overtone transition. Transitions from the ground state to a state where $\nu_i = 1$ and $\nu_j = 1$ simultaneously are called combination bands.

NIRS (750–2500 nm), a technique widely used for composition and quality analysis of food and agricultural products, is based on molecular overtone and combination vibrations. The vibrational bands in the near-infrared region are typically very broad and the absorptions are rather weak. On the other hand, mid-infrared spectroscopy (2.5–25 µm) is mainly based on fundamental vibrations and associated rotational–vibrational structures, while far-infrared spectroscopy (25–1000 µm) is based on rotational vibrations.

An object may emit light due to thermal effect (heating), chemical reaction, electric energy, absorption of photons, or ionized radiation. The light emitted by an object resulting from a nonthermal source is called *luminescent light* and the process is called *luminescence*. Luminescence can be divided into photoluminescence (due to photon absorption), electroluminescence (due to electric energy), chemiluminescence (due to chemical reaction), etc., depending on the type of nonthermal source.

Photoluminescence is related to light scattering and absorption, a radiation process during which photons are absorbed by a substance, and it then releases or emits light of the same or lower frequencies (or lower energy levels). A further description of photoluminescence processes is given in Chapter 2.

1.3 LIGHT SCATTERING AND ITS APPLICATIONS IN FOOD AND AGRICULTURE

After the introduction of the basic concepts of light and optical theories, it is now appropriate to give a brief overview of light scattering technology and its applications for food property, quality, and safety assessment.

Nature is full of light scattering phenomena. For instance, during a clear, sunny day, the sky looks particularly blue, because the scattering of blue light by fine particles in the atmosphere is predominant. Homogenized milk looks white because it has fat globules and casein micelles (protein) with a range of sizes that are comparable to the wavelengths of visible light, and these particles are good scatterers for the visible light from blue to red. On the other hand, skimmed milk looks bluish because the large-size fat globules that are better scatterers for the longer visible light have been removed. Well-milled wheat or rice flour looks much whiter than whole kernels or kernel fragments, because the fine particles in the flour are much-better scatterers of the visible light.

When a photon hits a particle or an interface between two media with different refractive indexes, it may be absorbed by the particle and subsequently converted into another form of energy (e.g., heat), which is called *absorption*, or it may be scattered to a different direction and then hits another particle, until it exits from the medium or is absorbed. Absorption is the annihilation of a photon by matter, which depends on the chemical composition of the medium. Water, protein, pigments, etc., absorb light at specific wavelengths. The scattering of photons by the particles or medium, however, is a physical process that largely depends on the wave characteristics of the photons and the structural properties of the medium they travel through, including density, particle size or shape, composition, etc. There are two types of light scattering: elastic and inelastic. *Elastic scattering* refers to the scattering events, in which the frequency of photons does not change or no net energy exchange occurs to the photons after they have interacted with particles, while in *inelastic scattering*, photons gain or lose energy and subsequently change the frequency after interaction with molecules or atoms in the material. *Raman scattering* is an inelastic-scattering phenomenon.

The study of light scattering and its applications has been an important field that crosses many disciplines from chemistry and biology to physics and engineering, to climatology and astronomy, and to food and agriculture. Researchers in different disciplines often use different approaches to study light scattering phenomena and subsequently develop different techniques to determine or measure light scattering properties or characteristics for different applications. Despite these differences, light scattering research is based on the same optical theories, as presented in the previous section. To study a light scattering phenomenon, we would need to invoke one or more of the four optical theories described above, although in most cases, it is sufficient to use electromagnetic theory. In this book, *light scattering techniques are broadly defined as those techniques that are developed according to the principle and theory of light transfer, or by utilizing light scattering phenomena, for achieving various application purposes.*

Research on light scattering has had a slow start, compared to other related technologies (i.e., NIRS). Prior to the advent of a powerful computer technology, it was rather difficult or inconvenient to measure light scattering properties of food and biological materials. Until 15 years ago, the measurement of light scattering had been mainly confined to physics, chemistry, and biomedical research laboratories equipped with sophisticated optical instrumentations. Biomedical researchers have been at the forefront in the development and adoption of many modern optical methods and techniques, including light scattering-based techniques. One noticeable exception is NIRS. The technique was first researched by a group of agricultural engineers in the late 1950s and the 1960s to meet the growing demand for fast, quantitative measurement of composition and quality of food and agricultural products. Karl Norris, an agricultural engineer with the U.S. Department of Agriculture Agricultural Research Service (USDA/ARS), is considered by many to be the father of the modern NIRS technique (Williams and Norris, 2001). By the 1970s, NIRS had become an important analytical instrument for laboratory analysis of food composition and quality attributes. During the 1980s and 1990s, rapid advances in the hardware and software of NIRS instrumentation had taken place after the advent of digital technology, and NIRS is now being routinely used for chemical composition and functional analysis.

While light scattering has long been recognized in NIRS measurement of opaque or diffuse biological and food materials, it has been largely ignored or treated as noise or useless signals that should be removed or avoided in the data acquisition and preprocessing. The lack of interest in measuring and utilizing light scattering properties in the early dates of research could be attributed to the fact that it is challenging to measure light scattering properties because scattering is often intertwined with absorption and the two are difficult to separate. Moreover, light scattering measurements also need complicated mathematical models and algorithms, which would be difficult to achieve without a powerful computer technology. Birth and colleagues at USDA/ARS are among the first who recognized the usefulness of light scattering properties for quality assessment of food and agricultural products. They published several papers in the late 1970s and early 1980s on the measurement of scattering properties for food and agricultural products, based on the Kubelka–Munk model (see Chapter 6) (Birth, 1978, 1982; Birth et al., 1978). During the 1990s, we have witnessed exponential growth in the research and applications of NIRS technique, but only sporadic research activities had been reported in the literature on light scattering for food property, quality, and safety assessment.

However, it has been a totally different situation in light scattering research in the biomedical field. A large number of scientific papers have been published since

Introduction to Light and Optical Theories

the late 1980s on utilizing or quantifying light propagation and scattering in biological tissues. Several novel optical techniques, along with appropriate theories, have been developed for *in vivo* measurements of the optical scattering and absorption properties of biological tissues over the past two decades. Significant progress has been made in the development of spatially-resolved, time-resolved, and frequency-domain spectroscopic techniques (Tuchin, 2007). Functional imaging with diffusing light has also gained significant attention in the biomedical research community.

Inspired by the research in the biomedical field, food and agricultural engineers began to renew their interest in light scattering research after the arrival of the twenty-first century. Several research groups from around the world have since been actively engaged in the development and application of light scattering-based techniques, including spatially-resolved, spatial-frequency domain, and time-resolved, to measure or characterize the properties and quality of food and agricultural products. A detailed description of these techniques and their applications for food quality and safety evaluation is given in Chapters 7 through 11.

In addition to the above-mentioned light scattering techniques, we have also seen significant research activities in the past decades on other light scattering-based techniques, notably dynamic light scattering (DLS) and Raman scattering, for food property, quality, and safety detection. DLS, also known as *photon correlation spectroscopy*, has been used for studying the size and structure of particles in colloidal solutions, because particles and macromolecules in colloidal solutions undergo the Brownian motion, which arises from collisions between the particles and the solvent molecules. The technique has received much attention after the advent of the laser in the 1960s. Early research in DLS was primarily done by researchers in chemistry and physics. Since the 1990s, much research has been reported on using DLS to characterize the structural and microrheological properties of colloidal food (see Chapter 13). DLS has also found its applications for plant materials, because they contain moving particles resulting from the Brownian motion. *Biospeckle*, a technique based on the DLS principle, has been used for quality assessment and monitoring of food products, especially fruits and vegetables (see Chapter 14).

Raman scattering is a form of inelastic scattering, and it is closely related to the structural and chemical properties of the scattering medium. Raman spectroscopy has become an important laboratory instrument for composition analysis and pathogen detection. Raman scattering imaging has been increasingly used in recent years for mapping chemical composition, and adulteration and pathogen detection in food and agricultural products (see Chapter 15).

As new and more effective light scattering-based techniques are being developed, we expect to see their increasing applications in food and agriculture. The following is a partial list of applications of light scattering-based techniques for food property, quality, and safety assessment:

- Measuring optical scattering and absorption properties of food products by integrating sphere technique and inverse adding-doubling algorithm
- Measuring optical properties, structure, and quality of food products by spatially-resolved and spatial-frequency domain techniques

- Measuring optical scattering and absorption properties of horticultural products and assessing maturity and postharvest quality using time-resolved reflectance spectroscopy
- Characterizing muscle structures and postmortem changes in meats and meat analogs
- Determining structural and microrheological properties of food by DLS technique
- Light scattering for structural and composition analysis of milk and milk products
- Quality assessment and monitoring of fruits and vegetables by biospeckle technique
- Food quality and safety assessment by Raman scattering technique
- Pathogen detection in food by elastic scattering and Raman scattering.

1.4 SUMMARY

Light scattering occurs as a result of the interaction of photons with matter, and it is dependent on the structure and chemical composition of the material. Over the past 15 years, significant progress and numerous applications in light scattering have been made for assessing properties, quality, and safety of food and agricultural materials. To understand light scattering phenomena and develop effective measurement techniques, it is important to have a basic knowledge of light and its interactions with biological materials. Hence, in this introductory chapter, we have given a brief description of the basic concepts of light and various optical theories. In Chapters 2 and 3, we will provide a further, more detailed description on the principles and theories of light scattering and propagation in turbid food and biological materials.

REFERENCES

Birth, G. S. The light scattering properties of foods. *Journal of Food Science* 16, 1978: 916–925.
Birth, G. S. Diffuse thickness as a measure of light scattering. *Applied Spectroscopy* 36, 1982: 675–682.
Birth, G. S., C. E. Davis and W. E. Townsend. The scatter coefficient as a measure of pork quality. *Journal of Animal Science* 46, 1978: 639–645.
Hecht, E. *Optics*, 4th edition. Reading, MA: Addison-Wesley, 2002.
Saleh, B. E. A. and M. C. Teich. *Fundamentals of Photonics*, 2nd edition. New York: Wiley-Interscience, 2007.
Tuchin, V. *Tissue Optics: Light Scattering Methods and Instruments for Medical Diagnosis*, Bellingham, WA: SPIE, 2007.
Williams, P. and K. Norris (eds.). *Near-Infrared Technology in the Agricultural and Food Industry*, 2nd edition. St. Paul, MN: AACC International Press, 2001.

2 Overview of Light Interaction with Food and Biological Materials

Renfu Lu

CONTENTS

2.1 Introduction 19
2.2 Reflection and Transmission... 21
    2.2.1 Fresnel Equations for Reflection and Transmission 21
    2.2.2 Diffuse Reflection and Transmission 24
2.3 Absorption 29
    2.3.1 Absorption and Emission of Photons 29
    2.3.2 Absorption Coefficient... 31
2.4 Scattering.. 33
    2.4.1 Rayleigh Scattering.. 34
    2.4.2 Mie Scattering 35
        2.4.2.1 Scattering Coefficient 35
        2.4.2.2 Scattering Phase Function and Anisotropy Factor 35
        2.4.2.3 Mie Theory Model 39
    2.4.3 Raman Scattering 39
2.5 Summary 40
Disclaimer... 40
References... 40

2.1 INTRODUCTION

Foods are produced and consumed in a wide variety of forms to provide the necessary nutrients and energy as well as sensory pleasure for humans. Food may be classified into various groups, according to their source of origin (i.e., plant based and animal based), structural or physical properties (solid, semisolid, and liquid), composition (fibrous, gelatinous, starchy, oleaginous, and crystalline), or postharvest-processing procedure (raw or whole, minimally processed, processed culinary ingredients, and ultraprocessed). In the U.S. Department of Agriculture (USDA) Dietary Guidelines for Americans (www.choosemyplate.gov), foods are classified into five major classes: fruits, vegetables, grains, protein foods, and dairy. On the basis of the optical properties, food may also be classified into *transparent* (e.g., cooking oils and beverages), *semitransparent* or *translucent* (jelly, fruit juices), and *turbid* or *opaque* (fruits, vegetables, nuts, grains, meats, dairy, etc.). In transparent foods, light passes through the material straightly without changing the direction or being scattered. On the other hand, in turbid or opaque foods, light would undergo multiple scattering and change direction at each scattering event, before it is being absorbed or reemerges from the product. The vast majority of foods are opaque, which means that light can undergo multiple scattering in them. The ability of light to penetrate a turbid food product mainly depends on its absorption and scattering properties. For instance, research has shown that light can penetrate the tissue of apple fruit up to several centimeters (Qin and Lu, 2009). Transmittance mode is now being used in some commercial fruit sorting and grading facilities to inspect the internal quality and condition of fruits such as apples.

Food and biological materials such as fruits, vegetables, and animal meats are composed of cells with similar functions, which are grouped together to form tissues. Animal tissues can be classified into four different types: muscle, connective, nervous, and epithelial. Plant tissues, on the other hand, are broadly categorized into epidermis, ground, and vascular systems. Cells are the basic structural, functional, and biological units or "building blocks" of biological materials, whose dimensions typically range between 1 and 100 μm. A cell consists of cytoplasm encompassed by the cell membrane. The cytoplasm comprises cytosols (the gel-like fluids) and the organelles, which form the internal structures of the cell and carry out various vital functions needed for all living things. The cell membrane, which contains biomolecules such as proteins and nucleic acids, functions to protect the cell from its surrounding. These cellular structures carry out different biological functions and differ greatly in chemical composition and size (ranging from nanometers to microns). Hence, they have great ramifications in the absorption and scattering of light in biological tissues.

When light hits the tissue, it interacts with the various cellular components of the tissue, and the pattern of interactions depends on the structural and chemical properties of individual cellular components. As photons travel through the tissue, they encounter the cellular structures, which serve as scattering or absorbing particles, resulting in the change in the propagation direction, or the absorption, of the photons by the molecules in the structures. The process repeats as the photons continue to move through the cellular structures until they are either being completely absorbed or reemerge from the tissue. Hence, the study of light transfer in biological and food materials is about quantification of the absorption and scattering of light in the tissue conglomerates. Each biological species or food product has its unique structural and chemical characteristics, which, in turn, determine its optical absorption and scattering properties.

This chapter discusses the basic phenomena occurring during the interaction of light with biological and food materials, which form the foundation for different light scattering techniques that have been developed for the property, quality, and safety assessment of food and agricultural products. We first discuss the reflection and transmission of light at the interface of two optically homogeneous media, followed by a discussion on diffuse reflection and transmission from the surface of food and biological materials. We then present the basic concepts about the interaction of light with food and biological materials in the form of absorption and scattering. Three types of scattering, that is, Rayleigh, Mie, and Raman, are discussed and their implications in determining the properties and quality of food and agricultural products are discussed.

2.2 REFLECTION AND TRANSMISSION

2.2.1 FRESNEL EQUATIONS FOR REFLECTION AND TRANSMISSION

In Chapter 1, we have briefly discussed the refraction and reflection of light at the interface between two optically homogeneous media with different reflective indexes. The refraction of light at the interface follows Snell's law of refraction. Since scattering is closely associated with the refraction, reflection, and transmission of light at an interface between two media or by particles in the media, we will spend more space here on this important phenomenon, also known as Fresnel reflection and transmission, and present the mathematical equations for calculating reflectance and transmittance.

Figure 2.1 shows the reflection and transmission of the light incident from an oblique angle at a flat surface of the medium with a refractive index of n_2. According to Snell's law of refraction (Equation 1.3 in Chapter 1), a fraction of the incident light is reflected at the surface in an angle equal to the angle of incidence θ_1, while the remaining light will enter the medium and propagate in the angle of θ_2.

While Snell's law gives the geometry of reflection and refraction, it does not provide quantitative information on the amount of light being reflected and transmitted.

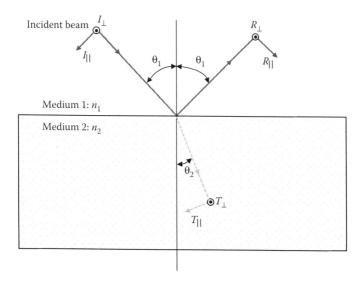

FIGURE 2.1 Fresnel reflection and transmission when a light beam is incident on the flat surface or interface between two optically homogeneous media with the refractive indexes of n_1 and n_2, respectively. In the figure, I, R, and T denote the incident light intensity, reflectance, and transmittance, respectively; the subscripts $\parallel$ and $\perp$ denote the parallel and perpendicular light components, and the symbol ⊙ represents the light component perpendicular to the plane of light incidence.

To determine the reflectance and transmittance, we must use Maxwell's equations and the boundary conditions associated with these equations. The incident light, with its intensity of I, is polarized with the two electric fields, which are perpendicular and parallel to the plane of incidence, respectively (Figure 2.1). Hence, the reflected and transmitted light will also have perpendicular and parallel components. Without going into the details, we simply give the following equations for calculating the light fraction that is reflected or transmitted at the interface between the two media, in terms of their parallel and perpendicular components. Detailed derivations of these equations can be found in many applied optics textbooks (Guenther, 1990; Hecht, 2002). The fraction of the parallel light component, $r_\parallel$, which is reflected at the interface is

$$r_\parallel = \frac{n_2 \cos\theta_1 - n_1 \cos\theta_2}{n_1 \cos\theta_2 + n_2 \cos\theta_1}, \tag{2.1}$$

and the fraction of the perpendicular light component, $r_\perp$, is

$$r_\perp = \frac{n_1 \cos\theta_1 - n_2 \cos\theta_2}{n_1 \cos\theta_1 + n_2 \cos\theta_2}. \tag{2.2}$$

The fraction of the parallel light component, $t_\parallel$, that is transmitted into the second medium, is

$$t_\parallel = \frac{n_1}{n_2}(1+r_\parallel) = \frac{2 n_1 \cos\theta_1}{n_1 \cos\theta_2 + n_2 \cos\theta_1}, \tag{2.3}$$

and the fraction of the perpendicular light component, $t_\perp$, is

$$t_\perp = 1 + r_\perp = \frac{2 n_1 \cos\theta_1}{n_1 \cos\theta_1 + n_2 \cos\theta_2}. \tag{2.4}$$

The overall reflectance R and transmittance T for the two polarized states of the incident light are therefore given by

$$R_\parallel = |r_\parallel|^2 \quad \text{and} \quad R_\perp = |r_\perp|^2, \tag{2.5}$$

$$T_\parallel = 1 - R_\parallel \quad \text{and} \quad T_\perp = 1 - R_\perp. \tag{2.6}$$

Equations 2.1 through 2.4 suggest that the proportion of the light that is reflected or transmitted largely depends on the incident angle as well as the ratio of the reflective index for the two media. The total specular reflectance for an oblique incident angle can be calculated from the following equation:

$$R = \frac{R_\parallel + R_\perp}{2} = \frac{1}{2}\left(\frac{\tan^2(\theta_1 - \theta_2)}{\tan^2(\theta_1 + \theta_2)} + \frac{\sin^2(\theta_1 - \theta_2)}{\sin^2(\theta_1 + \theta_2)}\right), \tag{2.7}$$

and

$$T = 1 - R = 1 - \frac{1}{2}\left(\frac{\tan^2(\theta_1 - \theta_2)}{\tan^2(\theta_1 + \theta_2)} + \frac{\sin^2(\theta_1 - \theta_2)}{\sin^2(\theta_1 + \theta_2)}\right). \tag{2.8}$$

When light is normally incident on the medium with n_2, the total reflectance is equal to

$$R = \frac{R_\parallel + R_\perp}{2} = \left(\frac{n_1 - n_2}{n_1 + n_2}\right)^2, \tag{2.9}$$

and the fraction of the light transmitted across the interface is given by

$$T = 1 - R = \frac{4 n_1 n_2}{(n_1 + n_2)^2}. \tag{2.10}$$

The derivations of Equations 2.9 and 2.10 have utilized Snell's law (Equation 1.3). Hence, when a light beam is normally incident from air ($n_1 = 1.0$) into water ($n_2 = 1.35$) or a biological tissue ($n_2 \approx 1.33$), approximately only 2% of the incident light is reflected, while 98% of the light is diffusely reflected from or transmitted into water or the tissue.

Fresnel Equations 2.7 through 2.10 for reflection and transmission only apply to the perfect flat surface of materials that are optically homogeneous or transparent (i.e., no scattering). When light is incident on such surface, all reflected light rays will leave at an angle equal to the incident angle. Such reflection is called *specular reflection* or *Fresnel reflection*. Specular reflection gives us the sensation of glossiness or shininess for the surface of an object. While all surfaces give specular reflection, only polished surfaces of non-light-absorbing metallic materials (e.g., aluminum or silver) can reflect light specularly with great efficiency. All other materials, even when perfectly polished, cannot specularly reflect light with great efficiency, because most of the light is either reflected into different angles at the surface or enters into the medium and then scatters into different directions before it is being absorbed by, or reemerges from, the medium.

In most applications, specular reflectance is not useful and should be avoided in measurement. There are, however, a few situations in which specular reflectance can provide important information on the condition and quality of food products. For instance, many raw or fresh food products (such as apple, orange, eggplant, tomato, etc.) produce a thin layer of natural wax at the surface when they reach the maturation stage (Figure 2.2). For these products, the level of glossiness or shininess can be

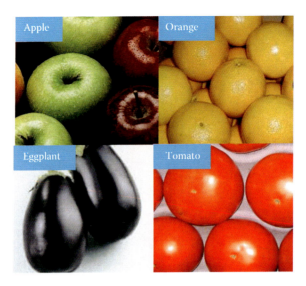

FIGURE 2.2 Fresh food products (apple, orange, eggplant, and tomato) showing various degrees of glossiness.

indicative of their condition or shelf life. Gloss can increase the aesthetic appeal of certain food products such as apple. Hence, it has been a common practice by the U.S. apple industry to apply artificial wax to apple fruit during postharvest packing to enhance the product's visual appeal.

Since gloss is an important parameter in assessing the quality and condition of many industrial and food products, a number of gloss-measuring techniques and commercial instruments are available for a variety of products, including automotive coating, furniture coating, plastics, metals, paper, and food products. Measurement of gloss requires the detection of specular reflectance from the surface of a product. Most commercial gloss-measuring instruments are only suitable for products with flat surfaces. For food and agricultural products, their surfaces are often irregular or curved, and an accurate measurement of gloss for these products is challenging. Special gloss measuring techniques have thus been developed for agricultural and food products, such as apple, banana, tomato, eggplant, onion, and bell green pepper (Jha and Matsuoka, 2002; Mizrach et al., 2009; Nussinovitch et al., 1996). A gloss-measuring prototype using a visible/near-infrared spectrometer coupled with a fiber-optic light/detector was assembled in the author's laboratory for measuring the gloss of apples (Mizrach et al., 2009) (Figure 2.3). Later, an improved laboratory gloss prototype based on an imaging technique was developed for measuring the gloss of apple and other food products (unpublished).

2.2.2 Diffuse Reflection and Transmission

In deriving Fresnel equations for reflection and transmission, we assume that the interface between the two media is perfectly flat (even at the microscopic level) and the media are optically homogeneous (i.e., the refractive index is constant within

Overview of Light Interaction with Food and Biological Materials 25

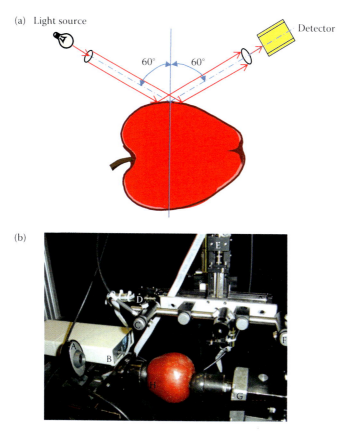

FIGURE 2.3 Measurement of gloss for apple fruit: (a) the concept of gloss measurement and (b) a gloss measurement prototype (Mizrach et al., 2009). (A: camera, B: visible and near-infrared spectrometer, C: collimating light source, D and E: horizontal and vertical motorized stages, F: light detecting lens, G: fruit holder/spinner, and H: sample).

each medium). However, the surface and internal structure of most food and biological materials do not meet these criteria. Hence, the actual interaction of an incident light at the interface between two media is much more complicated.

Consider two cases that are most likely to be encountered in the real world (Figure 2.4). For case one in Figure 2.4a, the surface is rough but the medium is optically homogeneous. When the light is incident on the rough surface, a small fraction of the light will be reflected according to Snell's law and a larger fraction of the incident light will be reflected back at different angles by the rough surface, which is called *diffuse reflection*. And the remaining light will be refracted and transmitted in the medium. In the second case, as shown in Figure 2.4b, the surface is rough and the medium is not optically homogeneous (most food and biological materials belong to this category of optical media). Under such a situation, both specular and diffuse reflection will occur at the rough surface. In the meantime, as the light enters the optically inhomogeneous medium, scattering and absorption will take place simultaneously in the medium. As a result, the light will be scattered into

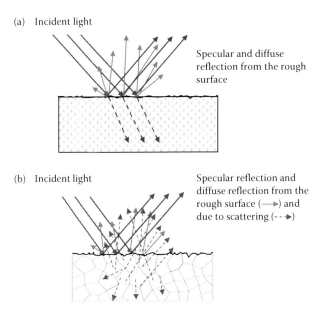

FIGURE 2.4 Diffuse reflection occurring on two types of medium: (a) specular and diffuse reflection at a rough surface of an optically homogeneous medium and (b) specular reflection and diffuse reflection from the rough surface and scattering in an optically inhomogeneous medium.

different directions; the surviving photons will undergo multiple scattering events and reemerge from the incident area or its neighboring area or emerge from the opposite side of the medium, which is often called *transmission*. The reemitted light will emerge from the surface in different directions and also occur as diffused light. Hence, strictly speaking, there are two types of diffuse reflection in case two: one is attributed to the roughness of the surface, which may be called *surface diffuse reflection*, while the other is caused by the scattering of light within the medium as well as the rough surface, which may be referred to as *body diffuse reflection*. In practical applications, it is difficult to separate the two types of diffuse reflection. It should, however, be pointed out that these two types of diffuse reflection carry different information about the medium. For instance, surface diffuse reflection is useful for assessing the roughness or texture of a surface, while body diffuse reflection can be used for assessing the properties of internal tissues.

While most surfaces give diffuse reflection, there is a special type of surface, called "*Lambertian surface*" or "*matte surface*." The radiant intensity or diffuse reflectance observed from such surface is proportional to the cosine of the angle between the observer's position and the surface normal (Figure 2.5a), that is,

$$I(\theta) = I_0 \cos\theta, \tag{2.11}$$

where I_0 is the radiant intensity from the surface element ds observed at the normal position and $I(\theta)$ is the radiant intensity observed from an off-normal angle

Overview of Light Interaction with Food and Biological Materials

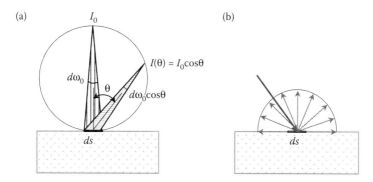

FIGURE 2.5 Diffuse reflection from a Lambertian surface: (a) the radiance from a Lambertian surface measured at an angle θ from the normal to the surface follows the cosine law and (b) the diffuse reflectance from a Lambertian surface is uniform in all directions.

of θ ($-90° \leq \theta \leq 90°$). From Equation 2.11, we can further draw a conclusion that the radiant intensity per unit angle from the normal position (P_o) is equal to that observed from a position at an angle of θ [$P(\theta)$]

$$P_o = \frac{I_0}{d\omega_0} = \frac{I}{d\omega_0 \cos\theta} = \frac{I_0 \cos\theta}{d\omega_0 \cos\theta} = \frac{I_0}{d\omega_0} = P(\theta). \quad (2.12)$$

Strictly speaking, the angle $d\omega_0$ in Equation 2.12 should be replaced with the solid angle. Since the concept of a solid angle will not be introduced until Chapter 3, here, we simply assume that the surface element has a unit width to simplify the discussion.

Equation 2.12 suggests that the radiation from a Lambertian surface is uniform in all directions, which is schematically shown in Figure 2.5b. Hence, when a Lambertian surface is viewed from any angle, it has the same apparent radiance. In other words, when the human eye observes the object from different angles, it has the same brightness. The sun is approximately a Lambertian radiator, because its brightness is almost the same when being viewed from different places. A black body is a perfect Lambertian radiator. Unfinished wood exhibits approximately Lambertian reflection, although not all rough surfaces show Lambertian characters. Standard reflection panels such as Spectralon panels (Labsphere Inc., North Sutton, NH, USA) that are commonly used in imaging or spectroscopic calibration have nearly perfect Lambertian reflection. The surface of some food and biological materials can be approximately treated as a Lambertian surface. In measuring the optical absorption and scattering properties of fruit with a curved surface using a hyperspectral imaging-based spatially-resolved technique (Chapter 7), Qin and Lu (2008) proposed a method to correct reflectance measurements from the curved surface of fruit, by assuming that reflection from the surface of apple fruit obeys Lambert's cosine law.

Diffuse reflectance, including both surface and body, is not only dependent on the surface characteristics of a product, but, more importantly, on its optical absorption and scattering properties. Measurement of diffuse reflectance by visible/near-infrared spectroscopy can thus provide an effective means for composition analysis and quality evaluation of food and agricultural products. In conventional visible/near-infrared spectroscopy, spectral measurements are mainly carried out in three sensing modes (i.e., *reflectance*, *transmittance*, and *interactance*), as schematically shown in Figure 2.6. Each sensing mode requires a different instrumentation setup and may have different implications in measuring food and agricultural products. For instance, in reflectance mode implementation, using the fiber-optic technique, the detecting fibers and the light delivery fibers are often integrated into the same probe, or the detecting probe is placed in such a way that it measures the light that is diffusely reflected from the same area illuminated by the incident light beam (Figure 2.6a). The measured signals thus contain both surface and body diffuse reflectance. This sensing mode is widely used for food and agricultural products because it is easy to implement. However, since the majority of the measured light is likely to come from the surface or a region that is very close to the surface, reflectance mode is not effective or suitable for detecting the internal tissue of food samples that are inhomogeneous, especially when the properties or condition of the surface and the sublayer are distinctly different. In transmittance mode (Figure 2.6b), the measured light has survived after passing the whole product and reemerged from the opposite side of the light incident area. Hence, transmittance measurements tend to provide more information about the internal condition of the product, compared with the reflectance mode. It is, however, more challenging to implement the transmittance mode, because it requires a high-intensity light source to obtain sufficient output signals. In addition, the light pathlength is directly influenced by the size and shape of the sample, which can complicate spectral data analysis. In interactance mode (Figure 2.6c), the positioning of the sensing probe is such that it measures the light reflected from an area that is separated by a specific distance from the light incident area. This sensing mode has the flexibility of controlling the light pathlength, providing a good compromise between reflectance and transmittance modes. Several

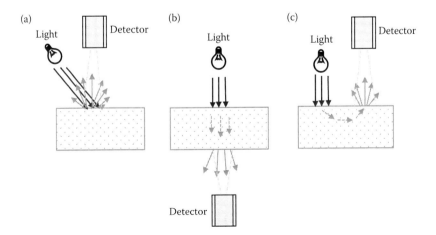

FIGURE 2.6 Three sensing modes commonly used in visible and near-infrared spectroscopic measurement: (a) reflectance, (b) transmittance, and (c) interactance.

studies have shown that the interactance mode offered better measurement results over reflectance and transmittance modes (Mendoza et al., 2012; Schaare and Fraser, 2000).

Despite the wide use of the three sensing modes with visible/near-infrared spectroscopy in the composition and property assessment of food and agricultural products, they only provide approximate measurements of the aggregate effect of light absorption and scattering in food and biological materials. Spectral measurements based on these sensing modes are of empirical nature because the data acquired depend on the setup of lighting and detecting probes, which often varies among the instrument manufacturers. Moreover, conventional visible/near-infrared spectroscopy does not provide separate information on the absorption and scattering properties, which would be desirable for many applications that require knowledge of both chemical and structural or physical characteristics.

Light scattering techniques presented in this book are primarily based on the principle and theory of light transfer or the utilization of light scattering phenomena for the assessment of food products. As shown in the following chapters, light scattering techniques normally require measuring light intensity distributions in the spatial, time, or frequency domain, and hence they are inevitably more complicated and demanding in instrumentation and experimental setup, compared with conventional visible/near-infrared spectroscopy. Consequently, light scattering techniques also offer some unique application opportunities, which, otherwise, could not be accomplished by a conventional spectroscopic technique.

2.3 ABSORPTION

2.3.1 ABSORPTION AND EMISSION OF PHOTONS

Absorption is the annihilation of photons by atoms, electrons, or molecules and the conversion into a different form of energy, such as heat, chemical energy, fluorescence, or phosphorescence. Absorption occurs only when the energy level of a photon matches a specific quantum state of an atom or molecule. Those photons whose frequencies do not match the energy gap are transmitted or scattered. After absorbing the photonic energy, the molecule will transition from its ground state to an excited state instantaneously (about 10^{-15} s). The transition is determined by the energy of the photon

$$E = h\upsilon = h\frac{c}{\lambda}, \qquad (2.13)$$

where E is the energy of the photon in J, h is Planck's constant ($h = 6.62618 \times 10^{-34}$ J), c is the speed of light in vacuum ($c = 3.0 \times 10^8$ m/s), υ is the frequency of the photon, and λ is the wavelength in m.

According to quantum theory, absorption may take place in three forms of energy transition: (1) electronic, (2) vibrational, and (3) rotational. Electronic transitions occur in atoms and molecules, while vibrational and rotational transitions only occur in molecules. Electronic transitions are generally more energetic, occurring in the ultraviolet, visible, and near-infrared regions. Vibrational and rotational transitions, on the other hand, typically occur at lower energy levels corresponding to the visible and infrared regions. The spectral regions where these transitions occur are known as absorption bands, and they are dependent on particular molecules or atoms in biological tissues. Near-infrared spectroscopy studies the absorption of photons by molecules in the near-infrared region (750–2500 nm) caused by vibrational or rotational transitions. In the near-infrared and mid-infrared regions, water is the dominant absorbing chromophore (Figure 2.7).

The excitation state can only be maintained for a very short time period (ranging between 10^{-8} and 10^{-12} s), and then the excited molecule starts to decay to a lower energy state. During the course of the energy decay, the system releases energy that is equal to the difference in energy between the two states. The release of energy may take place in the form of heat or new photons (called *photoluminescence*). The emission of photons or photoluminescence may occur in one or multiple processes, as shown in Figure 2.8. In one process, shown in Figure 2.8a, the excited molecule returns to the original, lower energy state by releasing photons of the same energy as the absorbed molecule. This process involves no net energy loss or gain by the system. *Rayleigh scattering*, which will be described in detail in a later section, is such a process and is known as elastic scattering because no net change in the energy occurs in the transition process. In the second process, as shown in Figure 2.8b, the molecule is instantaneously excited from the ground state to an excited state through electronic or vibrational transitions. The molecule at the excited state then goes through nonradiative relaxation, also called *internal conversion*, in which the excitation energy is dissipated as heat. The molecule can stay in the intermediate state for a long time, before it goes back to the original ground state by emitting new photons whose frequencies are normally lower than that of the original radiation. This process is called *fluorescence*. Plant-based materials are fluorescing, due to the presence of chlorophylls. When a plant product such as an apple is illuminated by shortwave

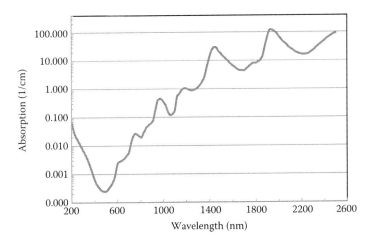

FIGURE 2.7 Spectrum of the absorption coefficient for water for the wavelengths of 200–2600 nm.

Overview of Light Interaction with Food and Biological Materials

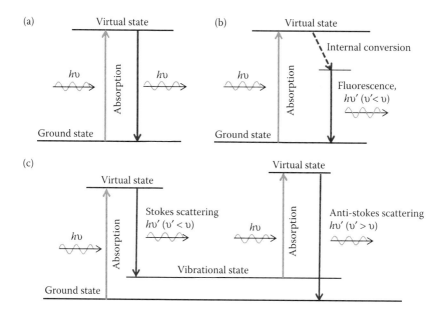

FIGURE 2.8 Three forms of photoluminescence, resulting from the absorption of the incident photons with energy $h\upsilon$: (a) Rayleigh scattering with no net energy change, (b) fluorescence generating lower-wavelength photons, and (c) Raman scattering resulting in longer-wavelength (Stokes) and/or shorter-wavelength (anti-Stokes) photons.

ultraviolet light, it emits longer-wavelength visible light (Noh and Lu, 2007). Since chlorophyll fluorescence is associated with the physicochemical activities of chlorophylls in the plant material, it is useful for assessing the growth, maturity and post-harvest quality, and condition of plant products (DeEll and Toivonen, 2003). In the third process of energy transition (Figure 2.8c), molecules are elevated to a higher virtual state after absorbing photons. As the excited molecules return to the original state, photons of different frequencies are produced and scattered. The scattered photons can have higher (called *anti-Stokes scattering*) or lower (*Stokes scattering*) energy than the absorbed photons. This process is termed *Raman scattering*. Raman scattering is inelastic scattering because there exist energy exchanges between the photon and the molecule. Raman scattering is further discussed in Section 2.4.3.

2.3.2 ABSORPTION COEFFICIENT

To characterize the absorption properties of a food or a biological material, we need to introduce a parameter, called *absorption coefficient* or μ_a. Consider a plane-wave light beam (i.e., having uniform intensity in any plane perpendicular to the direction of travel) that is normally incident on an absorbing medium, as shown in Figure 2.9. Assume that the medium is uniformly distributed with N identical light-absorbing particles. The total energy entering the medium, P_{in}, is given by

$$P_{in} = I_0 A, \tag{2.14}$$

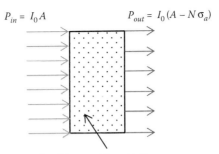

FIGURE 2.9 Concept of absorption expressed in terms of the effective cross-sectional area for absorbing particles in a homogenous medium, where N is the number of absorbing particles, σ_a is the effective absorption area for the average particle, and A is the cross-sectional area of the element.

where A is the incident area and I_0 is the incident energy in power per unit area. As the light beam encounters each particle in the medium, a small fraction of the light is absorbed. The net energy leaving the medium, P_{out}, is

$$P_{out} = I_0(A - N\sigma_a), \qquad (2.15)$$

where σ_a is the effective absorption cross-sectional area for each particle, in units of m² or cm². Hence, the energy absorbed by the medium is equal to

$$P_{abs} = P_{in} - P_{out} = I_0 N\sigma_a. \qquad (2.16)$$

Now, we may define the absorption coefficient, μ_a, by the following expression:

$$\mu_a = N\sigma_a = \frac{P_{abs}}{I_0}, \qquad (2.17)$$

where μ_a has the units of 1/m or, more commonly, 1/cm or 1/mm. Equation 2.17 shows that the absorption coefficient is, in essence, the total effective absorption cross-sectional area for all absorption particles in a unit volume of the medium. It should be mentioned that the effective absorption cross-sectional area is not equivalent to the actual geometrical cross-sectional area of the particles.

The reciprocal of Equation 2.17 gives the effective *mean free path for absorption* or *absorption length*

$$l_a = \frac{1}{\mu_a}. \qquad (2.18)$$

The mean free path l_a represents the average distance a photon needs to travel between the two absorbing particles before it is being absorbed.

While the above definition and expression of the absorption coefficient is helpful in studying light propagation in the medium containing absorbing particles, a more straightforward expression for the absorption coefficient can be derived when the medium is considered to be homogenous. Again, consider the case shown in Figure 2.9, where the medium is considered homogeneous, with a thickness of s. As the collimated light beam with a uniform intensity I_0 passes the medium, the *Beer–Lambert law* states that the change or attenuation of light intensity in the medium is directly proportional to the distance over which the light has traveled

$$dI(x) = -\mu_a I(x) dx, \tag{2.19}$$

where $dI(x)$ is the differential change in the intensity of a collimated light beam with the initial intensity of I_0 as it travels for an infinitesimal distance dx in the homogeneous absorbing medium. By rearranging Equation 2.19 and then taking integration, we have the following equation:

$$I = I_0 \text{Exp}(-\mu_a s). \tag{2.20}$$

Equation 2.20 indicates that as a collimated beam passes through a pure absorbing medium, its intensity decreases exponentially with the distance, over which the light has traveled, and the rate of decrease in light intensity is determined by the absorption coefficient μ_a.

In practical applications, transmittance measurements are often used for determining the absorption coefficient, which are expressed by the following equation:

$$T = \frac{I}{I_0} = \text{Exp}(-\mu_a s), \tag{2.21}$$

where T is the transmittance in a dimensionless unit. *Absorbance (Abs)*, or *optical density (OD)*, is a commonly used quantity in near-infrared spectroscopy, which is given by

$$Abs = OD = \log_{10}\left(\frac{I_0}{I}\right) = -\log_{10} T. \tag{2.22}$$

A further discussion on utilizing the Beer–Lambert law for food optical property measurements is presented in Chapter 6.

2.4 SCATTERING

Scattering occurs after light passes the medium that is optically inhomogeneous, due to the presence of discrete particles and/or variations in the refractive index. It results in the change of light propagation direction in the medium. Scattering is affected by the density and size of particles and cellular structures in the material. Hence, scattering is useful for detecting the structural characteristics and composition of food and biological materials.

There are two types of scattering in food and biological materials: *elastic* and *inelastic*. In elastic scattering, the material does not change its state after interaction with the photons and all photons scattered from the particles maintain the same wavelength. Rayleigh and Mie scattering are elastic. Raman scattering, on the other hand, is inelastic, because the photons scattered by the molecules or particles do not have the same wavelengths as the original ones.

2.4.1 Rayleigh Scattering

Rayleigh scattering, named after the British physicist Lord Rayleigh (John William Strutt, 1842–1919), occurs when photons interact with particles that are smaller than the wavelength of the radiation. Rayleigh scattering results from the electric polarization of the particles. The oscillating electric field of an electromagnetic wave acts on the charges within a particle, causing them to move at the same frequency. The particle therefore becomes a radiating dipole resulting in the scattering of light. The upper atmosphere of the earth contains many small gaseous particles, which result in the scattering of sunlight. Hence, we see a blue sky or a yellow tone of the sun.

When an unpolarized light with intensity I_0 is incident on a single spherical particle with diameter d (d is less than λ, the wavelength of the light), the intensity of the light, I, that is scattered into another angle θ is given by the following equation (Bohren and Huffman, 2004):

$$I = I_0 \frac{1+\cos^2\theta}{2s^2} \left(\frac{2\pi}{\lambda}\right)^4 \left(\frac{n^2-1}{n^2+1}\right)^2 \left(\frac{d}{2}\right)^6, \qquad (2.23)$$

where n is the refractive index of the particle and s is the distance from the particle to the detector. The Rayleigh scattering cross-sectional area is obtained by averaging all angles in Equation 2.23

$$\sigma_s = \frac{2\pi^5}{3} \frac{d^6}{\lambda^4} \left(\frac{n^2-1}{n^2+2}\right)^2. \qquad (2.24)$$

Equation 2.23 shows that the scattering of light is inversely proportional to λ^4. This means that scattering is much stronger in a short wavelength than in longer wavelengths. This explains why the sky looks blue, because ultraviolet and blue light is being strongly scattered by the gases in the upper atmosphere than other longer-wavelength visible light. Equation 2.24 also indicates that the effective scattering cross-sectional area is different from the actual diameter of the particle; it is also dependent on the wavelength of the incident light and the refractive index.

In biological materials, only striations in collagen fibrils and membranes are in the range of 100 nm or less. Hence, Rayleigh scattering only applies to these ultra-structure components.

2.4.2 MIE SCATTERING

Mie scattering is named after Gustav Mie, a German physicist (1869–1957), who developed a theoretical treatment for the scattering of light in homogeneous dielectric spheres, based on Maxwell's equations. Mie scattering theory applies to the particles whose size is comparable to the wavelength of light. Most cellular structures in biological materials are in the range between a few nanometers and a few micrometers. Hence, the scattering of light in biological and food materials can be adequately described by Mie scattering theory.

2.4.2.1 Scattering Coefficient

To describe the scattering of light by particles or a turbid medium with a varying refractive index, we need to introduce another parameter, that is, a *scattering coefficient* or μ_s. By following the same procedure outlined for defining the absorption coefficient in Section 2.3.2, we can define the scattering coefficient μ_s as the total number N of identical scattering particles for a unit volume of the medium multiplied by the effective scattering cross-sectional area σ_s of each particle, that is,

$$\mu_s = N\sigma_s = \frac{P_{sca}}{I_0}, \tag{2.25}$$

where P_{sca} is the light energy scattered by the aggregate of particles in the medium. Like the absorption coefficient, the scattering coefficient also has the units of m^{-1} or cm^{-1}. The *mean free path for scattering* is the reciprocal of μ_s

$$l_s = \frac{1}{\mu_s}. \tag{2.26}$$

The mean free path for scattering l_s is the average distance that light travels between two scattering particles before a scattering even occurs.

Likewise, the scattering of light in an optically homogeneous scattering medium can be described by the Beer–Lambert law

$$I = I_0 \text{Exp}(-\mu_s s), \tag{2.27}$$

where I_0 is the intensity of the collimated incident beam and s is the thickness of the scattering medium.

2.4.2.2 Scattering Phase Function and Anisotropy Factor

As a photon hits a scattering particle, it generally changes the propagation direction (there are, however, exceptions in which after hitting the particle, the photon continues to move forward without being deflected, which is termed as *ballistic scattering*). According to quantum theory, the actual direction into which a photon is scattered by the particle is a random process; the photon may go straightforward without the direction change (forward scattering, $\theta = 0°$), or it may go backward completely

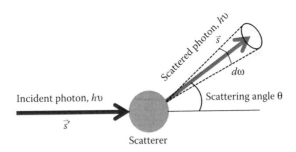

FIGURE 2.10 A scattering event occurs when an incident photon with energy $h\upsilon$ hits the scatterer along the direction $\vec{s}'$, and it is then scattered into the direction $\vec{s}$.

(backward scattering, $\theta = 180°$). In the vast majority of cases, the photon would be deflected by an angle θ from the incident direction, ranging between $0°$ and $180°$ (or between $-180°$ and $0°$) (Figure 2.10). In the spherical coordinate system, in addition to the deflection angle θ, a second azimuthal angle φ is needed for completely describing the orientation of the photon propagation direction. However, it is generally assumed that a scattering event is independent of azimuthal angle.

After a large quantity of photons hit the particle, a consistent scattering profile would eventually emerge. This scattering profile represents the cumulative effect of all scattering events on the particle, which is a function of the deflection angle θ. This scattering profile is commonly called the *scattering phase function*, designated as $p(\theta)$, which is a function of the angle θ. The scattering phase function $p(\theta)$ is unique for each particle, determined by the size, shape, and orientation of the particle. Since the scattering of individual photons at the particle is a random process, $p(\theta)$ is a probability distribution, which must satisfy the following condition:

$$\int_0^\pi p(\theta) 2\pi \sin\theta d\theta = 1. \tag{2.28}$$

When the scattering is isotropic (i.e., the particle scatters photons equally in all directions), the scattering phase function can be readily obtained from Equation 2.28

$$p(\theta) = \frac{1}{4\pi}. \tag{2.29}$$

Rayleigh scattering is close to isotropic scattering. In scattering-dominant biological materials (i.e., $\mu_s \gg \mu_a$), the scattering is often considered isotropic after photons have gone through sufficient multiple scattering events.

Most light scattering in food and biological materials is not isotropic. In fact, it is strongly forward scattering, which is called *anisotropic scattering*. The scattering phase function may be determined in an experiment by measuring angular distributions of light scattering from a thin sample of tissue, using an optical device such as a goniophotometer. To ensure accuracy, the thickness of the samples must be not more than $1/(\mu_s + \mu_a)$. The preparation of such samples is extremely difficult, if not possible. Hence, effort has been devoted to the development of various forms of the scattering phase function, such as Henyey–Greenstein and delta-Eddington, for describing the angle-dependent feature of light scattering in turbid media.

Before we discuss the scattering phase function, it is necessary to introduce another important parameter for the characterization of light scattering profiles, which is called the *anisotropy factor g*. The anisotropy factor measures the fraction of light that retains forward direction after a single scattering event; it provides a measure of the degree of anisotropy in scattering. Let us again consider the scattering event, as shown in Figure 2.10, in which a packet of photons hits the particle at zero-degree angle ($\theta = 0°$), and it is then scattered by the particle at a deflection angle θ. The fraction of the photon packet that is retained in the forward direction is related to $\cos\theta$. Since the scattering phase function considers the average scattering angles, it is logical to introduce the mean scattering angle g by the following equation:

$$g = \int_0^\pi p(\theta)\cos\theta 2\pi \sin\theta d\theta = <\cos\theta>, \qquad (2.30)$$

where

$$\int_0^\pi p(\theta) 2\pi \sin\theta d\theta = 1. \qquad (2.31)$$

The anisotropy factor g is commonly expressed as a function of $\cos\theta$, which can be derived from Equation 2.30

$$g = \int_{-1}^{1} p(\cos\theta)\cos\theta d(\cos\theta), \qquad (2.32)$$

where

$$\int_{-1}^{1} p(\cos\theta) d(\cos\theta) = 1. \qquad (2.33)$$

Values of the anisotropy factor g range between -1 and 1, where $g = -1$ is the total backward scattering; $g = 0$ is the isotropic scattering, while $g = 1$ represents the total forward scattering. $g = 0$ only means the integral over -1 to 1, since the inverse of $g = 0$ is not necessarily true (Welch et al., 2011). For most food and biological materials, g values are between 0.7 and 0.9, which means that forward scattering is dominant.

Henyey and Greenstein (1941) developed an expression that describes the angular dependence of scattering by small particles in interstellar dust clouds. The Henyey–Greenstein function has been proven suitable for describing the angular scattering of particles in biological tissues, and it has been widely used for studying light propagation in biological materials. The Henyey–Greenstein function is expressed as a function of deflection angle θ and anisotropy factor g:

$$p(\theta) = \frac{1}{4\pi} \frac{1-g^2}{(1+g^2-2g\cos\theta)^{3/2}}. \quad (2.34)$$

Equation 2.34 is also commonly expressed as a function of cosθ

$$p(\cos\theta) = \frac{1}{2} \frac{1-g^2}{(1+g^2-2g\cos\theta)^{3/2}}. \quad (2.35)$$

Figure 2.11 shows the Henyey–Greenstein scattering phase function (Equation 2.34) as a function of the scattering angle θ for various values of g. For isotropic scattering, the line in the figure is constant at 1/4π.

Modifications to the Henyey–Greenstein function and other forms of scattering phase functions have also been proposed. A detailed discussion on these functions can be found in Chapter 4.

At this point, it is also helpful to introduce another scattering parameter, called the *reduced scattering coefficient,* designated as μ'_s, which is the lumped property of the scattering coefficient μ_s and the anisotropy factor g

$$\mu'_s = \mu_s(1-g). \quad (2.36)$$

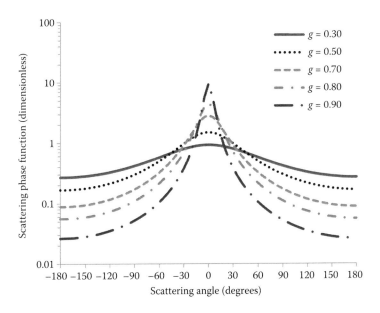

FIGURE 2.11 Relationship between the Henyey–Greenstein scattering phase function and the scattering angle for different values of the anisotropy factor g.

The introduction of μ'_s is necessary and also reduces the number of independent optical property parameters needed for dealing with food and biological materials, in which scattering is dominant, so that photon propagation can be approximated as a diffusion process. Under such a situation, scattering, after the initial step, becomes isotropic and the propagation of light in the medium can be adequately described by a simplified theory, that is, diffusion approximation theory. A detailed discussion of light transfer in scattering-dominant media by diffusion approximation theory is given in Chapter 3.

2.4.2.3 Mie Theory Model

Mie theory provides a framework and numerical solutions to the scattering problem for spheres of any size. The theory is derived from the well-known Maxwell's equations, and it allows computation of the electric and magnetic fields inside and outside the particle as well as the angular distribution of light after interacting with the particle. Mie theory calculates the scattering efficiency, the effective scattering cross-sectional area, and the anisotropy factor, after the information about the sphere (i.e., diameter, density, and refractive index) is provided as inputs. The derivation of Mie solution is quite involved mathematically and hence is skipped here. The mathematical derivation of Mie solution from Maxwell's equations can be found in several reference books (Bohren and Huffman, 2004; Wang and Wu, 2007). Computer codes for calculating Mie scattering are available for free downloading at a number of websites (e.g., http://omlc.org/software/mie/; http://www.scattport.org/index.php/light-scattering-software.).

2.4.3 RAMAN SCATTERING

Raman scattering or *Raman scattering effect* occurs, as the scattered photons shift the wavelength from the original exciting photons. Raman scattering is named after C. V. Raman (1888–1970), an Indian physicist, who discovered the phenomenon in the 1920s.

Raman scattering occurs when an intensive light beam impinges on a molecule. As a photon excites the molecule in either ground, vibrational, and rotational state or an excited state, it moves to a virtual, higher energy state for a short time period. As the excited molecule relaxes immediately, it results in an inelastic scattered photon. The inelastic scattered photon can have a lower (Stokes) or higher (anti-Stokes) frequency than the original photon. In Raman scattering, the resulting state of the scattered photon is different from the state of the original photon (Figure 2.8c). The energy or wavelength shift of the resulting photon is determined by the difference between the excited virtual energy state and the resulting state of the scattered photon. This is different from the emission of fluorescence where a molecule at an excited state goes through a series of internal conversions during relaxation and it then emits a photon as it returns to the original ground state. Raman scattering can be excited by a light source in any wavelength in the visible and infrared region, while fluorescence can only be excited by a shorter ultraviolet light. The emitted light from the fluorescence process generally has lower energy or longer wavelength than the exciting light, whereas the emitted light from Raman scattering can be lower or higher in energy or wavelength than the exciting light.

Raman spectroscopy employs the Raman scattering effect for analyzing materials. Raman spectroscopic measurement generates a spectrum of Raman scattered light and the spectrum depends on the molecules and their state in the sample. Thus, Raman spectroscopy is very useful for composition analysis and identification of liquid and solid food materials. Various Raman techniques have been developed and used for composition, quality, and safety inspection of food and agricultural products. A comprehensive review of Raman scattering techniques is given in Chapter 15.

2.5 SUMMARY

As light passes from one optically homogeneous medium to another homogeneous medium, reflection and refraction or transmission occur at the interface, which are described by a set of Fresnel equations and Snell's law. When a light beam is incident on the unpolished or rough surface of a homogeneous or nonhomogeneous material, both specular and diffuse reflection occur. Food and biological materials are composed of cells, filled with water and void air space. Cells are heterogeneous with different cellular structures, which function as scattering and/or absorbing particles when light interacts with them, whereas water and void space may be treated as optically homogeneous. Hence, when light hits the tissue of food and biological materials, scattering and absorption occur, resulting in both surface and body diffuse reflection. To adequately describe the absorption and scattering characteristics of food and biological materials, we need three optical parameters, that is, absorption coefficient, scattering coefficient, and the anisotropy factor, in addition to the refractive index. Forward scattering is dominant in most biological and food products, which enables the visible and near-infrared light in the spectral region of 700–1400 m to penetrate the tissue for large distances, before being absorbed or reemerging from the tissue. Scattering may occur in an elastic or inelastic form. Rayleigh and Mie scattering are elastic because they involve no net energy exchange between photons and molecules. Rayleigh scattering occurs when the size of particles is smaller than the wavelength of the radiation. Light scattering in food and biological materials can be adequately described by Mie scattering theory, which is appropriate for particles of any size. Raman scattering involves energy exchange between photons and molecules, and it is inelastic. Since light scattering and absorption are determined by the chemical and structural characteristics of food and biological materials, measurement of light scattering and absorption properties can thus provide an effective means for assessing their property and condition.

DISCLAIMER

The mention of commercial products is solely for providing the reader with factual information, and it does not imply endorsement or recommendation by the USDA.

REFERENCES

Bohren, C. F. and D. R. Huffman. *Absorption and Scattering of Light by Small Particles*. Morlenbach, Germany: Wiley-VCH Verlag GmbH & Co, 2004.

DeEll, J. R. and P. M. A. Toivonen (eds.). *Practical Applications of Chlorophyll Fluorescence in Plant Biology.* Boston, MA: Kluwer Academic Publishers, 2003.

Guenther, R. D. *Modern Optics.* New York: John Wiley & Sons, 1990.

Hecht, E. *Optics*, 4th edition. Reading, MA: Addison-Wesley, 2002.

Henyey, J. C. and J. L. Greenstein. Diffusion radiation in galaxy. *Astrophysics Journal* 93, 1941: 70–83.

Jha, S. N. and T. Matsuoka. Development of freshness index of eggplant. *Applied Engineering in Agriculture* 18, 2002: 555–558.

Mendoza, F., R. Lu and H. Cen. Comparison and fusion of four nondestructive sensors for predicting apple fruit firmness and soluble solids content. *Postharvest Biology and Technology* 73, 2012: 89–98.

Mizrach, A., R. Lu and M. Rubio. Gloss evaluation of curved-surface fruits and vegetables. *Food and Bioprocess Technology* 2, 2009: 300–307.

Noh, H. and R. Lu. Hyperspectral laser-induced fluorescence imaging for assessing apple quality. *Postharvest Biology and Technology* 43, 2007: 193–201.

Nussinovitch, A., G. Ward and S. Lurie. Nondestructive measurement of peel gloss and roughness to determine tomato fruit ripening and chilling injury. *Journal of Food Science* 61, 1996: 383–387.

Qin, J. and R. Lu. Measurement of the optical properties of fruits and vegetables using spatially resolved hyperspectral diffuse reflectance imaging technique. *Postharvest Biology and Technology* 49, 2008: 355–365.

Qin, J. and R. Lu. Monte Carlo simulation for quantification of light transport features in apples. *Computers and Electronics in Agriculture* 68, 2009: 44–51.

Schaare, P. N. and D. G. Fraser. Comparison of reflectance, interactance and transmission modes of visible–near infrared spectroscopy for measuring internal properties of kiwifruit (*Actinidia chinensis*). *Postharvest Biology and Technology* 20, 2000: 175–184.

Wang, L. V. and H. I. Wu. *Biomedical Optics: Principles and Imaging.* New Jersey: John Wiley & Sons, 2007.

Welch, J. A., M. J. C. van Gemert and W. M. Star. Chapter 3. Definitions and overview of tissue optics. *Optical–Thermal Response of Laser-Irradiated Tissue.* 2nd edition (eds. Welch, A. J. and M. J. C. van Gemert), pp. 145–201. New York, USA: Springer, 2011.

3 Theory of Light Transfer in Food and Biological Materials

Renfu Lu

CONTENTS

3.1 Introduction 44
3.2 Radiometric Quantities 44
    3.2.1 Radiant Energy and Power 44
    3.2.2 Radiant Intensity 45
    3.2.3 Radiance and Irradiance 46
    3.2.4 Fluence Rate and Flux 48
3.3 Radiative Transfer Theory 49
    3.3.1 Energy Loss Due to Scattering Out dE_{div} 50
    3.3.2 Energy Loss Due to Extinction dE_{ext} 50
    3.3.3 Energy Gain Due to Scattering in dE_{sca} 50
    3.3.4 Energy Gain Due to Internal Source dE_{sou} 51
    3.3.5 Standard Form of the RTE 51
3.4 Diffusion Approximation Theory 52
3.5 Boundary Conditions 54
    3.5.1 Zero Boundary Condition 55
    3.5.2 Partial Current Boundary Condition 55
    3.5.3 Extrapolated Boundary Condition 56
3.6 Analytical Solutions to Diffusion Equation for Semi-Infinite Scattering Media 57
    3.6.1 Illumination by cw Light 57
    3.6.2 Illumination by Short-Pulsed Light 59
    3.6.3 Illumination by Frequency-Modulated Light 61
    3.6.4 Illumination by Spatially Modulated Light 63
3.7 Finite Element Modeling of Light Propagation in Scattering Media 65
    3.7.1 Finite Element Formulation 66
    3.7.2 Application Examples 67
3.8 Summary and Conclusions 69
References 73
Appendix: Vectors and Derivatives 74

3.1 INTRODUCTION

In the previous two chapters, we have covered the basic concepts of light and optical theories as well as light interactions with the tissue in the form of absorption, reflection, and scattering. In this chapter, we first define the basic radiometric quantities that are needed for describing light propagation in food and biological materials. Radiative transfer theory is then derived, according to the principle of conservation of energy. As the radiative transfer theory equation is generally too complex to solve, diffusion approximation theory needs to be introduced to simplify the mathematical description of light propagation in food and biological materials. It is then followed with a discussion of the three boundary conditions commonly used for solving the diffusion equation. Moreover, analytical solutions to the diffusion equation under several special light illumination conditions are presented, which form the theoretical foundation for a number of modern, noninvasive, or *in vivo* optical property measurement techniques, including a spatially-resolved, time-resolved, frequency domain, and spatial-frequency domain. Numerical methods such as the finite element analysis offer flexibility in dealing with complex geometries and nonhomogeneous or layered scattering media, and hence are useful in studying light propagation in food and biological materials. Finally, application examples are given of using the finite element method (FEM) for modeling light propagation in semi-infinite scattering media and for predicting the reflectance at the surface.

3.2 RADIOMETRIC QUANTITIES

This section introduces the basic radiometric quantities that are used for describing light transfer in food and biological materials; they include radiant energy and power, radiant intensity, radiance and irradiance, and fluence rate and flux.

3.2.1 RADIANT ENERGY AND POWER

Radiant energy, denoted as Q, is the energy of electromagnetic radiation, expressed in joule or J. Radiant energy represents the cumulative output (or input) over a period of time, and hence it is insufficient for an accurate description of the energy output that varies with time.

Radiant power or *radiant flux*, denoted as P, is the radiant energy emitted, transmitted, reflected, or received per unit time. It has the unit of watt or W (J/s). Hence, a radiating source emitting a radiant power P (W) over a time period Δt (s) will produce a radiant energy Q

$$Q = P\Delta t. \tag{3.1}$$

The radiant power can thus be described as the change in radiant energy over a time interval Δt. In the differential form, it can be written as follows:

$$P = \left. \frac{Q(t+\Delta t) - Q(t)}{\Delta t} \right|_{\Delta t \to 0} = \frac{dQ}{dt}. \tag{3.2}$$

Theory of Light Transfer in Food and Biological Materials

The *spectral flux* or *spectral power* is the radiant flux per unit wavelength or frequency

$$P_\lambda = \frac{P(\lambda + \Delta\lambda) - P(\lambda)}{\Delta\lambda}\bigg|_{\Delta\lambda \to 0} = \frac{dP}{d\lambda}, \qquad (3.3)$$

where λ is the wavelength. The spectral flux has the unit of W/m or W/nm.

3.2.2 RADIANT INTENSITY

Radiant intensity (I) is the radiant power emitted, transmitted, reflected, or received by a source in a given direction per unit solid angle (Figure 3.1)

$$I = \frac{P}{\omega}, \qquad (3.4)$$

where ω is the solid angle. The radiant intensity is thus direction dependent, and it has the unit of W/sr, where sr stands for steradian, which is the unit for measuring a solid angle.

The *solid angle* is used for the three-dimensional space, like the angle for the two-dimensional space. In the two-dimensional space, an angle is formed by two rays that originate from the same point. Angles are measured in degrees or radians. One radian is the plane angle subtended by a circular arc whose length is equal to the radius of the circle, and a full circle is equivalent to 2π radians. In the three-dimensional space, the solid angle can be similarly defined as the angle in the two-dimensional space. A solid angle is formed by the locus of all straight lines that originate from the same point, called apex or vertex, and these lines are rotationally symmetric. Solid angles are measured in *steradians* or sr and, like the angle, they are dimensionless. A solid angle is a three-dimensional angle subtended by the surface segment of a sphere with radius r, intersected by a cone originating at the center of the sphere (Figure 3.2). A steradian is mathematically defined as

$$\omega = \frac{A}{r^2}, \qquad (3.5)$$

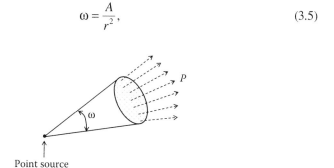

FIGURE 3.1 Radiant intensity emitted from a point source along a direction, passing a cone segment with the solid angle of ω.

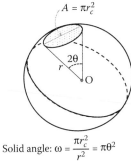

FIGURE 3.2 Concept of the solid angle measured from a sphere in units of steradians. A steradian is defined as the angle subtended by the surface area segment of a sphere with radius r intersected by a cone originating from the center of the sphere, where ω is the solid angle, r is the radius of the sphere, r_c is the radius of the surface segment of the sphere intersected by the cone, and θ is the polar angle.

where ω is the solid angle in steradians, A is the surface area of the spherical cap, and r is the radius of the sphere. Since the surface area of a sphere with the radius r is $4\pi r^2$, it has 4π steradians according to Equation 3.5. In the spherical coordinate system, the solid angle in the differential form for an arbitrarily oriented infinitesimal area of dA with the unit normal of $\vec{s}$ can be written as follows:

$$d\omega = \frac{dA}{r^2} = \sin\theta \, d\theta \, d\varphi, \qquad (3.6)$$

where θ is the polar angle (latitude) and φ is the azimuth angle (longitude). In deriving Equation 3.6, we have utilized the relationship of $dA = r^2 \sin\theta \, d\theta \, d\varphi$.

Hence, in the differential form, the radiant intensity can be expressed as

$$I = \frac{dP}{d\omega} = \frac{dP}{\sin\theta \, d\theta \, d\varphi}. \qquad (3.7)$$

In practical applications, the radiant intensity is also expressed as per unit wavelength or frequency, known as *spectral intensity*

$$I_\lambda = \frac{dI}{d\lambda} = \frac{d^2P}{\sin\theta \, d\theta \, d\varphi \, d\lambda}, \qquad (3.8)$$

which has the unit of W/(sr · m) or W/(sr · nm).

3.2.3 RADIANCE AND IRRADIANCE

Radiance, denoted as L, is the radiant power or flux emitted, reflected, transmitted, or received by or from a surface (Figure 3.3). It is defined as the power P per solid

Theory of Light Transfer in Food and Biological Materials

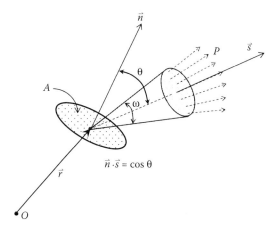

FIGURE 3.3 Radiance L from a surface with an area of A into a solid angle of ω, in the direction $\vec{s}$ with an angle of θ from the normal ($\vec{n}$) to the surface.

angle (ω) per unit projected area ($A \cos \theta$), where θ is the angle between the direction of radiation and the normal to the surface A. Hence, it is mathematically expressed as follows:

$$L = \frac{P}{\omega A \cos \theta}. \tag{3.9}$$

The radiance has the unit of W/(m².sr) and it is a directional quantity.

When radiance flows through an infinitesimal area dA at a location $\vec{r}$, in the direction of unit vector $\vec{s}$, within an infinitesimal solid angle $d\omega$, it is related to the power $dP(\vec{r},\vec{s},t)$

$$dP(\vec{r},\vec{s},t) = L(\vec{r},\vec{s},t)\, dA\, d\omega, \tag{3.10}$$

where the time variable t is included in $P(\vec{r},\vec{s},t)$ and $L(\vec{r},\vec{s},t)$ to indicate that these quantities are time dependent. When the direction $\vec{s}$ is not perpendicular to the area dA, the projected area is $dA \cos \theta$. Then, we have

$$dP(\vec{r},\vec{s},t) = L(\vec{r},\vec{s},t)\, dA \cos \theta\, d\omega. \tag{3.11}$$

Hence the radiance can be formally expressed as

$$dL(\vec{r},\vec{s},t) = \frac{d^2 P(\vec{r},\vec{s},t)}{dA \cos \theta\, d\omega}. \tag{3.12}$$

Likewise, the *spectral radiance* L_λ is given by

$$L_\lambda = \frac{dL}{d\lambda}, \quad (3.13)$$

which has the unit of W/(m²·sr·m) or W/(m²·sr·nm).

In most optical or imaging applications for food and biological materials, a light beam, either broadband (e.g., white light such as quartz tungsten) or monochromatic (e.g., laser), is used to illuminate a sample. When the incident radiation is constant over time (often termed *continuous-wave* or cw), the rate of energy per unit time per unit area of the sample surface is referred to as *irradiance*, denoted as E. Hence, the irradiance, also called *intensity*, is the power that irradiates a unit surface area

$$E = \frac{P}{A}, \quad (3.14)$$

which has the unit of W/m².

In the differential form, irradiance can be written as

$$E = \frac{dP}{dA}. \quad (3.15)$$

Likewise, the spectral irradiance E_λ is defined as

$$E_\lambda = \frac{dE}{d\lambda}, \quad (3.16)$$

where E_λ has the unit of W/(m².nm).

3.2.4 FLUENCE RATE AND FLUX

The *fluence rate* Φ is the total energy flowing into or out from a small area from all directions, which is given by the following equation:

$$\Phi(\vec{r},t) = \int_{4\pi} L(\vec{r},\vec{s}',t)\, d\omega, \quad (3.17)$$

where $\vec{s}'$ is the unit vector representing the direction of light incidence. The fluence rate has the unit of W/m²; it is independent of the direction of photon propagation.

In the spherical coordinate system, we have

$$\Phi(\vec{r},t) = \int_{\theta=0}^{\theta=\pi} \int_{\varphi=0}^{\varphi=2\pi} L(\vec{r},\vec{s}',t)\sin\theta\, d\varphi\, d\theta. \quad (3.18)$$

Theory of Light Transfer in Food and Biological Materials

The energy *flux J*, also known as the *current density*, represents the photon energy flowing in or out from a small area per unit time. It is given by

$$\mathbf{J}(\vec{r},t) = \int_{4\pi} \vec{s} L(\vec{r},\vec{s},t)\,d\omega, \tag{3.19}$$

which has the unit of W/m². Although having the same unit as the fluence rate, the flux is a vector quantity.

3.3 RADIATIVE TRANSFER THEORY

With the radiometric quantities defined in the previous section, we are now ready to derive the *radiative transfer equation*, also known as the *RTE*, based on the principle of conservation of energy. In deriving the RTE, the effects of polarization, coherence, and nonlinearity are ignored, and inelastic scattering (e.g., Raman scattering) is also not considered. In addition, the refractive index, the scattering and absorption coefficients, and anisotropy factor are assumed to be time independent; that is, these optical property parameters do not change with time. However, they can vary spatially.

Consider the changes of radiation energy flowing through an infinitesimal element volume $dV = dAds$ at position $\vec{r}$, within a differential solid angle element $d\omega$, where dA is the cross-sectional area of the infinitesimal element and dS is the length of the element. There are four forms of radiation contributing to the net change of energy within the infinitesimal element (Figure 3.4):

- The radiant energy scattering out from the system (dE_{div})
- The energy lost or consumed in the system (dE_{ext}) through extinction (attributed to both absorption and scattering away from the system)
- The energy gained through scattering into the system (dE_{sca})
- The energy generated by radiation source(s) within the system (dE_{sou})

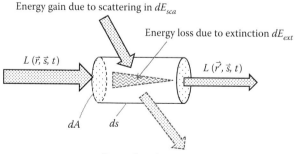

FIGURE 3.4 Radiant energy flowing in and out from a volume element with area dA and length dS.

Mathematically, the net energy change dE can be expressed as

$$dE = -dE_{div} - dE_{ext} + dE_{sca} + dE_{sou}. \tag{3.20}$$

3.3.1 ENERGY LOSS DUE TO SCATTERING OUT dE_{div}

As a light beam enters the element with the cross-sectional area dA and length ds (Figure 3.4), it will be scattered or diverged into different directions (no wavelength change is expected since we only consider elastic scattering in the derivation), resulting in an energy loss for the element. This loss of energy in a given time can be expressed as

$$dE_{div} = \frac{\partial L(\vec{r},\vec{s},t)}{\partial s} ds dA d\omega = \frac{\partial L(\vec{r},\vec{s},t)}{\partial s} dV d\omega, \tag{3.21}$$

where $dV = dsdA$ is the differential volume of the element. Equation 3.21 can be rewritten in divergence form as follows:

$$dE_{div} = \vec{s} \cdot \nabla L(\vec{r},\vec{s},t) dV d\omega = \nabla \cdot L(\vec{r},\vec{s},t) dV d\omega. \tag{3.22}$$

(See the appendix of the chapter for the definition of ∇, $\nabla \cdot$ and other derivative operators.)

3.3.2 ENERGY LOSS DUE TO EXTINCTION dE_{ext}

As shown in Chapter 2, the loss of light energy due to absorption or scattering over an infinitesimal distance is linearly proportional to the distance light has traveled. When the medium contains both scattering and absorbing particles, the same relationship by Snell's law still holds. However, under such a situation, the absorption or scattering coefficient needs to be replaced with the *extinction coefficient* μ_t, also called the *total attenuation coefficient*, which is given as

$$\mu_t = (\mu_a + \mu_s). \tag{3.23}$$

The energy loss in the volume element due to absorption and scattering per unit time is given by

$$dE_{ext} = (\mu_t ds) L(\vec{r},\vec{s},t) dA d\omega = \mu_t L(\vec{r},\vec{s},t) dV d\omega, \tag{3.24}$$

where $\mu_t ds$ is the probability of energy loss caused by absorption and scattering over the distance ds.

3.3.3 ENERGY GAIN DUE TO SCATTERING IN dE_{sca}

As radiation is incident on the infinitesimal element from direction $\vec{s}'$ and then scatters into the solid angle $d\omega$ in the direction $\vec{s}$, the energy gained by the system in a given time is

$$dE_{sca} = \mu_s dV \left[\int_{4\pi} L(\bar{r},\bar{s}',t) p(\bar{s}' \cdot \bar{s}) d\omega' \right] d\omega, \tag{3.25}$$

where $\mu_s dV$ represents the total scattering cross-sectional area in the infinitesimal volume dV, $(\bar{s}' \cdot \bar{s})$ is the cosine between the incident and scattering-out angles, and $p(\bar{s}' \cdot \bar{s})$ is the scattering phase function describing the probability of photons incident on the element from an angle of $\bar{s}'$ and being scattered into an angle of $\bar{s}$. The scattering phase function is described in detail in Chapter 2.

3.3.4 Energy Gain Due to Internal Source dE_{sou}

The energy produced from an internal radiating source in the volume element per unit time is given by

$$dE_{sou} = S(\bar{r},\bar{s},t) dV d\omega, \tag{3.26}$$

where $S(\bar{r},\bar{s},t)$ has the unit of W/(m³.sr).

3.3.5 Standard Form of the RTE

Within the solid angle in the volume element, the net change of energy per unit time is given by

$$dE = \frac{\partial L(\bar{r},\bar{s},t)}{c \partial t} dV d\omega. \tag{3.27}$$

By substituting all energy components given in Equations 3.21 through 3.27 into the energy balance equation in Equation 3.20, we obtain

$$\frac{1}{c}\frac{\partial L(\bar{r},\bar{s},t)}{\partial t} = -\nabla \cdot L(\bar{r},\bar{s},t) - \mu_t L(\bar{r},\bar{s},t) + \mu_s \int_{4\pi} L(\bar{r},\bar{s},t) p(\bar{s}' \cdot \bar{s}) d\omega' + S(\bar{r},\bar{s},t).$$

$$(3.28)$$

Equation 3.28 is the standard form of the RTE, also known as the *Boltzmann equation*. When the net energy change in the volume element is in steady state (i.e., independent of time), the term on the left hand of the equation becomes zero. Although the integro-differential equation in Equation 3.28 seems simple at the first glance, its simplicity, in fact, is an illusion. The complexity of the RTE arises from the fact that there are six variables in the equation, that is, $\bar{r},\bar{s},t$, or $(x, y, z, \theta, \varphi, t)$. It is therefore difficult, if not possible, to solve the RTE using either an analytical or numerical approach. For practical applications, we need to introduce simplifications

3.4 DIFFUSION APPROXIMATION THEORY

Diffusion approximation equation, or simply *diffusion equation*, is a simplified form of the RTE, which provides a practical means to model light propagation in scattering media. Diffusion equation can be derived using two different approaches: (1) the photon propagation in the medium is treated as a diffusion process, and Fick's law and the conservation of energy are invoked for obtaining the diffusion theory equation, and (2) the RTE is simplified to the diffusion theory equation based on several assumptions and simplifications (Star, 2011). In this section, we use the second approach to derive the diffusion theory equation.

Two different forms of diffusion theory equation have been reported in the literature, when the equation is derived from the RTE. One form of diffusion theory equation is derived by assuming that the radiance (Equation 3.28) can be split into a collimated component (nonscattering) and a scattering or diffusing component (Deulin and L'Huillier, 2006; Groenhuis et al., 1983; Star, 2011). The second form of diffusion theory equation is obtained without splitting the radiance into two components (Haskell et al., 1994; Wang and Wu, 2007). As a result, the two forms of diffusion equation yield slightly different expressions for modeling light propagation in a scattering medium, mainly in the region of the scattering medium covered by the collimated component. In the following derivation, we follow the approach of Haskell et al. (1994), that is, the radiance is not split into the nonscattering and scattering components.

To simplify the RTE in Equation 3.28, it is assumed that radiance $L(\vec{r},\vec{s},t)$ can be represented by a series of spherical harmonics (e.g., Legendre polynomials), which are a series of special functions used to represent functions defined on the surface of a sphere. To the first degree of approximation, the radiance $L(\vec{r},\vec{s},t)$ can be expressed by the first two terms of its Taylor series expansion (Wang and Wu, 2007)

$$L(\vec{r},\vec{s},t) = \frac{1}{4\pi}\Phi(\vec{r},t) + \frac{3}{4\pi}\bm{J}(\vec{r},t)\cdot\vec{s}, \qquad (3.29)$$

where $\Phi(\vec{r},t)$ is the fluence rate and $\bm{J}(\vec{r},t)$ is the photon current or the flux.

As discussed earlier, the fluence rate represents the total energy flowing into or out of a small volume, and hence it is independent of the photon-flowing direction. On the other hand, the photon current is direction dependent. Equation 3.29 shows that when scattering is greater than absorption (i.e., $\mu_s \gg \mu_a$), the radiance can be expressed as an isotropic fluence rate Φ plus a small directional flux $\bm{J}$. Consequently, the RTE can be simplified to the following equation by integrating over all solid angles

$$\frac{1}{c}\frac{\partial\Phi(\vec{r},t)}{\partial t} + \mu_a\Phi(\vec{r},t) = -\nabla\cdot\bm{J}(\vec{r},t) + S(\vec{r},t). \qquad (3.30)$$

In deriving Equation 3.30, we have used the following relationships:

$$\int_{4\pi} \frac{1}{c} \frac{\partial L(\vec{r},\vec{s},t)}{\partial t} d\omega = \frac{1}{c} \frac{\partial}{\partial t} \int_{4\pi} L(\vec{r},\vec{s},t) d\omega = \frac{1}{c} \frac{\partial \Phi(\vec{r},t)}{\partial t}, \qquad (3.31)$$

$$\int_{4\pi} \vec{s} \cdot \nabla L(\vec{r},\vec{s},t) d\omega = \nabla \cdot \int_{4\pi} \vec{s} L(\vec{r},\vec{s},t) d\omega = \nabla \cdot \boldsymbol{J}(\vec{r},t), \qquad (3.32)$$

$$\mu_s \int_{4\pi} \left(\int_{4\pi} L(\vec{r},\vec{s},t) p(\vec{s}' \cdot \vec{s}) d\omega' \right) d\omega = \mu_s \Phi(\vec{r},t), \qquad (3.33)$$

and

$$\int_{4\pi} S(\vec{r},\vec{s},t) d\omega = S(\vec{r},t), \qquad (3.34)$$

in which we have assumed that the source is isotropic, that is,

$$S(\vec{r},\vec{s},t) = \frac{S(\vec{r},t)}{4\pi}. \qquad (3.35)$$

Equation 3.30 no longer contains the directional vector $\vec{s}$, but it still has two radiometric quantities $\Phi(\vec{r},t)$ and $\boldsymbol{J}(\vec{r},t)$. To further simplify it, we introduce another assumption that

$$\frac{1}{|\boldsymbol{J}(\vec{r},t)|} \frac{\partial |\boldsymbol{J}(\vec{r},t)|}{\partial t} \ll c(\mu_a + \mu_s). \qquad (3.36)$$

This assumption means that the fractional change in the flux $\boldsymbol{J}(\vec{r},\vec{s},t)$ within an infinitesimal time interval is negligibly small. This leads to Fick's law for the flux $\boldsymbol{J}$

$$\boldsymbol{J}(\vec{r},t) = -D\nabla \Phi(\vec{r},t), \qquad (3.37)$$

in which D is called the *diffusion coefficient*, and is given by

$$D = \frac{1}{3[\mu_a + (1-g)\mu_s]} = \frac{1}{3(\mu_a + \mu_s')}, \qquad (3.38)$$

where μ_s' is referred to as the *reduced scattering coefficient*

$$\mu_s' = (1-g)\mu_s. \qquad (3.39)$$

A negative sign appears in Equation 3.37 because the flux decreases along the negative gradient. Substituting Equation 3.37 into Equation 3.30 yields

$$\frac{\partial \Phi(\vec{r},t)}{\partial t} = \nabla \cdot [D\nabla \Phi(\vec{r},t)] - \mu_a \Phi(\vec{r},t) + S(\vec{r},t). \quad (3.40)$$

Equation 3.40 is referred to as the *diffusion approximation equation* or *diffusion equation*. If the diffusion coefficient is spatially invariant, Equation 3.40 has a simpler form of expression

$$\frac{1}{c}\frac{\partial \Phi(\vec{r},t)}{\partial t} = D\nabla^2 \Phi(\vec{r},t) - \mu_a \Phi(\vec{r},t) + S(\vec{r},t). \quad (3.41)$$

The diffusion equation simplifies the RTE that is expressed in the form of radiance $L(\vec{r},\vec{s},t)$ into the form of fluence rate $\Phi(\vec{r},t)$, which is direction invariant. In addition, in the diffusion equation, the anisotropy factor g is no longer treated as a separate optical parameter. Instead, it is combined with the scattering coefficient μ_s to form a new optical parameter, called the reduced scattering coefficient (μ'_s) (Equation 3.39). Hence, there are only two independent optical parameters, that is, the absorption and reduced scattering coefficients, in the diffusion equation. In deriving the diffusion Equation 3.41, it is assumed that the fractional change in the flux $\boldsymbol{J}(\vec{r},t)$ is far less than one (hence Equation 3.37 is valid) and the radiance can be expressed in the first order of Taylor series (Equation 3.29). These two assumptions imply that scattering is dominant over absorption in the medium, that is, $\mu'_s \gg \mu_a$, which suggests that photons undergo multiple scattering events before being absorbed.

3.5 BOUNDARY CONDITIONS

The diffusion equation in Equation 3.41 is much easier to solve, compared with the RTE in Equation 3.28, when an appropriate boundary condition is imposed for a scattering medium. There are two types of boundary or interface for modeling the propagation of light in the scattering medium: one is the interface between a nonscattering medium and a scattering or diffusive medium with the same refractive index, which is referred to as *the refractive index-matched boundary*, and the other is the interface between two scattering media with different refractive indexes, referred to as the *refractive index-mismatched boundary*. For the refractive index-matched boundary, no light scatters into the scattering medium. Hence, we have the following mathematical expression:

$$L(\vec{r},\vec{s},t) = 0 \quad \text{for } (\vec{s}\cdot\vec{n}) > 0, \quad (3.42)$$

where $\vec{n}$ is the unit normal vector pointing into the scattering medium. Since the radiance is direction dependent, the integral of the radiance pointing to the scattering medium should also be zero

$$\int_{(\bar{s}\cdot\bar{n})>0} L(\bar{r},\bar{s},t)(\bar{s}\cdot\bar{n})d\omega = 0. \tag{3.43}$$

For a refractive index-mismatched boundary, the radiance entering from the ambient medium into the scattering medium should be equal to the radiance leaving the scattering medium

$$\int_{(\bar{s}\cdot\bar{n})>0} L(\bar{r},\bar{s},t)(\bar{s}\cdot\bar{n})d\omega = \int_{(\bar{s}\cdot\bar{n})<0} R_F L(\bar{r},\bar{s},t)(\bar{s}\cdot\bar{n})d\omega, \tag{3.44}$$

where R_F is the Fresnel reflection at the interface, which is given by Equation 2.7 in Chapter 2.

As shown earlier in the derivation of the diffusion equation, it is difficult to exactly meet the boundary conditions given by Equation 3.43 or 3.44. Hence, various boundary conditions that are approximate to those given in Equations 3.43 and 3.44 have been proposed for solving the diffusion equation (Aronson, 1995; Deulin and L'Huillier, 2006; Haskell et al., 1994; Schweiger et al., 1995; Wang and Wu, 2007). In the following subsections, we discuss three types of boundary conditions, that is, *zero boundary condition* (ZBC), *partial current boundary condition* (PCBC), and *extrapolated boundary condition* (EBC).

3.5.1 ZERO BOUNDARY CONDITION

The ZBC requires that the fluence rate $\Phi(\bar{r},t)$ is zero at the boundary. Mathematically, it is written as

$$\Phi(\bar{r},t) = 0 \quad \text{at the boundary } \partial\Sigma. \tag{3.45}$$

While Equation 3.45 is simple mathematically, it may not be realistic in real applications because zero fluence rate cannot be perfectly satisfied when the scattering medium is illuminated by a light beam. Nevertheless, this boundary condition may be useful in providing approximations for biological materials under certain applications (Haskell et al., 1994).

3.5.2 PARTIAL CURRENT BOUNDARY CONDITION

To derive the PCBC, we first consider the refractive index-matched boundary and then extend to the refractive index-mismatched boundary. From Equations 3.29 and 3.43, we have

$$\mathbf{J}(\bar{r},t) = \frac{1}{4}\Phi(\bar{r},t) + \frac{1}{2}\mathbf{J}(\bar{r},t)\cdot\bar{n} = 0. \tag{3.46}$$

Substituting the radiant flux $\mathbf{J}(\vec{r},t)$ in Equation 3.37 into Equation 3.46 gives

$$\Phi(\vec{r},t) - 2D\nabla\Phi(\vec{r},t)\cdot\vec{n} = 0, \qquad (3.47)$$

or

$$\Phi(\vec{r},t) - 2D\frac{\partial\Phi(\vec{r},t)}{\partial z} = 0. \qquad (3.48)$$

Equation 3.48 is the PCBC for the refractive index-matched interface.

For the index-mismatched boundary, Equation 3.48 can be modified into the following expression:

$$\Phi(\vec{r},t) - 2DC_F\frac{\partial\Phi(\vec{r},t)}{\partial z} = 0, \qquad (3.49)$$

where

$$C_F = \frac{1+R_f}{1-R_f}, \qquad (3.50)$$

in which R_f is a parameter describing the internal reflection at the boundary and is given by (Schweiger et al., 1995)

$$R_f \approx -1.4399n^{-2} + 0.7099n^{-1} + 0.6681 + 0.636n, \qquad (3.51)$$

in which n is the ratio of the refractive index between the scattering medium and the ambient medium.

3.5.3 Extrapolated Boundary Condition

The PCBC is still difficult to satisfy in obtaining analytical solutions to the diffusion equation. An alternative, simpler boundary condition, called the *EBC*, has been used in obtaining analytical solutions to the diffusion equation. This boundary condition is based on the fact that Equation 3.49 is the Taylor series expansion of the first order for $\Phi(z = -2D,t)$, that is,

$$\Phi(z = -2D,t) = \Phi(z = 0) - 2C_F D\frac{\partial\Phi(\vec{r},t)}{\partial z}\bigg|_{z=0} = 0. \qquad (3.52)$$

This means that the fluence rate is approximately zero at the extrapolated boundary of $z = -2D$, which belongs to the category of homogeneous Dirichlet boundary conditions. This EBC is useful for solving the diffusion equation under the illumination of an infinitely small cw (or steady state), or pulsed light beam, which is presented in the following section.

3.6 ANALYTICAL SOLUTIONS TO DIFFUSION EQUATION FOR SEMI-INFINITE SCATTERING MEDIA

In this section, we present solutions of the diffusion equation under four special illumination conditions (i.e., cw or steady-state, pulsed, frequency modulated, and spatially modulated), and these solutions have formed the theoretical foundation for a number of noninvasive or *in vivo* optical property measurement techniques, including spatially-resolved, time-resolved, frequency domain, and spatial-frequency domain.

3.6.1 ILLUMINATION BY CW LIGHT

A problem that is of special importance in determining the optical properties of scattering media is the normal illumination of a homogeneous semi-infinite medium by an infinitesimal cw or steady-state light beam, as shown in Figure 3.5. We present an analytical solution of the problem, by following the approach proposed by Farrell et al. (1992).

Under the steady-state illumination of a light beam for a scattering medium, the diffusion equation in Equation 3.41 becomes

$$D\nabla^2 \Phi(\bar{r}) = \mu_a \Phi(\bar{r}) - S(\bar{r}). \tag{3.53}$$

As discussed in Section 3.5.3, under the PCBC, the fluence rate is zero at the extrapolated boundary of $z = -2D$ (Equation 3.52). To meet this requirement, Farrell et al. (1992) proposed the concept of using a negative "image source" so that the fluence rate at the extrapolated boundary of $z = -2D$ is equal to zero. As shown in Figure 3.5, the problem is now represented by two point sources for an infinite scattering medium; one positive source is located at the depth of $z = z_0 = 1/\mu'_s$, while the second, negative image source is located at the distance $z = 2z_b - z_0$ above the real boundary of the medium, $z = 0$. The parameter z_b is given by

$$z_b = 2C_F D = \frac{2}{3}\frac{(1+R_f)}{(1-R_f)}\frac{1}{(\mu_a + \mu'_s)}. \tag{3.54}$$

Thus, the solution to the diffusion equation becomes to find the sum of the fluence rate generated by both positive and negative image sources in the infinite scattering medium. The fluence rate generated by an isotropic point source at the location of $(0, z_0)$ in an infinite medium can be described by Green's function (Farrell et al., 1992)

$$\Phi(r, z_0) = \frac{1}{4\pi D} \frac{\exp(-\mu'_{eff} r_1)}{r_1}, \tag{3.55}$$

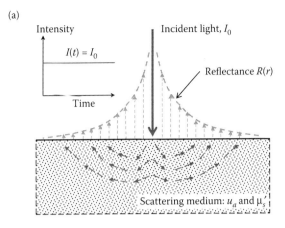

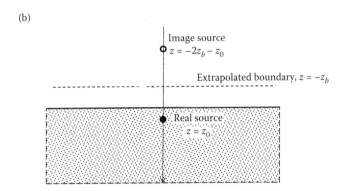

FIGURE 3.5 (a) A semi-infinite scattering medium under the normal incidence of a steady-state (or cw) infinitesimal light beam with the intensity of I_0. (b) Application of an EBC with a real source at $z = z_0$ and an image source at the distance of $z = 2z_b - z_0$ above the real boundary for obtaining an analytical solution to the diffusion equation, where $z_0 = 1/\mu_t'$ and z_b is given in Equation 3.54.

where

$$r_1 = [(z-z_0)^2 + r^2]^{1/2}, \qquad (3.56)$$

and

$$\mu_{eff}' = [3\mu_a(u_a + \mu_s')]^{1/2}, \qquad (3.57)$$

where μ_{eff}' is the *effective attenuation coefficient*. Therefore, the solution for a point light source in a semi-infinite medium is the sum of the solutions for both positive and negative sources, which is given as follows:

$$\Phi(r, z_0) = \frac{1}{4\pi D}\left(\frac{\exp(-\mu_{eff}' r_1)}{r_1} - \frac{\exp(-\mu_{eff}' r_2)}{r_2}\right), \qquad (3.58)$$

in which

$$r_2 = [(z+z_0-2z_b)^2 + r^2]^{1/2}. \quad (3.59)$$

Since the reflectance R or the flux (Equation 3.37) at the surface ($z = 0$) is what we actually measure, it can be calculated by the following equation:

$$R(r)|_{z=0} = -D\nabla\Phi(r,z)|_{z=0}. \quad (3.60)$$

Substituting Equation 3.58 into Equation 3.60 yields

$$R(r)|_{z=0} = \frac{1}{4\pi}\left[z_0\left(\mu'_{eff} + \frac{1}{r_1}\right)\frac{\exp(-\mu'_{eff}r_1)}{r_1^2} + (z_0+2z_b)\left(\mu'_{eff} + \frac{1}{r_2}\right)\frac{\exp(-\mu'_{eff}r_2)}{r_2^2}\right]. \quad (3.61)$$

Equation 3.61 has been widely used for determining the optical properties of the scattering media (Farrell et al., 1992; Lin et al., 1997; Mourant et al., 1997; Nichols et al., 1997; Qin and Lu, 2006). The predicted reflectance profiles obtained from Equation 3.61 generally match well with that of Monte Carlo (MC) simulations, which are considered to provide accurate results for predicting light propagation in scattering media (Qin and Lu, 2009). The discrepancies between the results predicted by Equation 3.61 and MC simulation mainly occur within one or two mean free paths from the light beam.

It should be noted that the reflectance predicted by Equation 3.61 only represents the flux leaving the medium surface in its normal direction. Haskell et al. (1994) pointed out that it was better to express the reflectance as the integral of the radiance over the backward hemisphere. In addition, as we have shown earlier, in deriving the diffusion Equation 3.41, the radiance is expressed as two terms, that is, the fluence rate and the photon current or the flux (Equation 3.29). Taking into consideration these two factors, Kienle and Patterson (1997) proposed an improved equation to predict the diffuse reflectance at the surface of a semi-infinite scattering medium

$$R_m(r)|_{z=0} = 0.118\Phi(r)|_{z=0} + 0.306R(r)|_{z=0}, \quad (3.62)$$

for a relative refractive index of $n = 1.4$ between the scattering medium and the ambient medium, where $R(r)|_{z=0}$ is given in Equation 3.61 and $\Phi(r)|_{z=0}$ is given by Equation 3.58. Several studies (Cen and Lu, 2010; Kienle and Patterson, 1997) have shown that the modified equation in Equation 3.62 gave a more accurate prediction of the reflectance profiles than Equation 3.61.

In Chapter 7, a review is presented of several optical property measurement techniques based on the steady-state spatially-resolved principle presented in this section.

3.6.2 Illumination by Short-Pulsed Light

Now, let us consider the illumination of a semi-infinite scattering by a short-pulsed light beam, as shown in Figure 3.6. The solution to the diffusion equation in

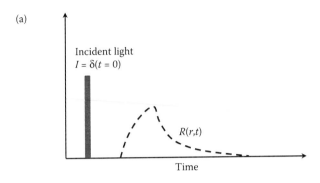
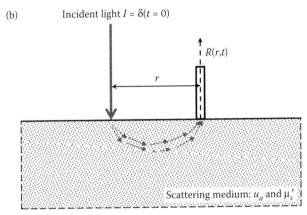

FIGURE 3.6 (a) A semi-infinite scattering medium under the normal incidence of an infinitesimal pulse light beam with the intensity of $I = \delta(t = 0)$, where δ is the Dirac delta function that is infinite at time $t = 0$ and zero when $t \neq 0$, with the resultant attenuated, time-delayed reflectance [$R(r,t)$] at a distance of r from the light incident point. (b) The concept of time-resolved reflectance measurement for estimating the optical properties of a scattering medium.

Equation 3.41 for an infinite medium illuminated by an infinitely short-pulsed point light, that is, $S(\vec{r},t) = \delta(r = 0, t = 0)$, is given by Patterson et al. (1989)

$$\Phi(r,z,t) = \frac{z_0}{(4\pi Dct)^{3/2}} \exp\left(-\frac{r^2}{4Dct} - \mu_a ct\right). \tag{3.63}$$

Equation 3.63 is a form of Green's function. By extending the EBC, as shown in Section 3.6.1, to the current case, the fluence rate at the surface of the semi-infinite medium is then the sum of the contributions from the two isotropic sources (one is located at $z = z_0$ and the other at $z = 2z_b - z_0$)

$$\Phi(r,z,t) = \frac{c}{(4\pi Dct)^{3/2}} \exp(-\mu_a ct)\left[\exp\left(-\frac{(z-z_0)^2 + r^2}{4Dct}\right) - \exp\left(-\frac{(z+z_0+2z_b)^2 + r^2}{4Dct}\right)\right].$$

$$\tag{3.64}$$

Using Fick's law in Equation 3.37, we obtain the following expression for reflectance $R(r,t)$:

$$R(r,t)|_{z=0} = \frac{1}{(4\pi Dc)^{3/2} t^{5/2}} \exp(-\mu_a ct) \left[z_0 \exp\left(-\frac{r_1^2}{4Dct}\right) + (z_0 + 2z_b) \exp\left(-\frac{r_2^2}{4Dct}\right) \right], \quad (3.65)$$

where the parameters or variables are explained in Section 3.6.1. The so-called *time-resolved reflectance technique* is based on the principle presented here. The technique has been extensively used for determining the optical absorption and scattering coefficients of biological and food materials (Cubeddu et al., 2001; Patterson et al., 1991). The absorption coefficient can be obtained by rearranging Equation 3.65

$$-\frac{d \ln R(r,t)}{dt} = \mu_a c + \frac{5}{2t} - \frac{r^2 + z_0^2}{4Dct^2}, \quad (3.66)$$

where

$$z_0 = \frac{1}{(\mu_a + \mu_s')}. \quad (3.67)$$

In Chapter 8, time-resolved technique based on the mathematical equations derived in this section is presented for measuring the quality and maturity of horticultural products.

3.6.3 ILLUMINATION BY FREQUENCY-MODULATED LIGHT

Another case that is of great interest in measuring the optical properties of scattering media is the illumination of a semi-infinite medium by a frequency-modulated light beam, as shown in Figure 3.7. In this case, the same equation for fluence rate in Equation 3.64 for pulsed light illumination can be rewritten in terms of modulation and phase. This can be achieved by taking the Fourier transform of Equation 3.64, which leads to the expression for modulation $M(r,f)$ and phase $\phi(r,f)$ (Patterson et al., 1991)

$$M(r,f) = \frac{(1 + \psi_0^2 + 2\psi_t)^{1/2}}{(1 + \psi_\infty)} \exp(\psi_\infty - \psi_t), \quad (3.68)$$

and

$$\phi(r,f) = \psi_t - \tan^{-1}\left(\frac{\psi_r}{1 + \psi_t}\right), \quad (3.69)$$

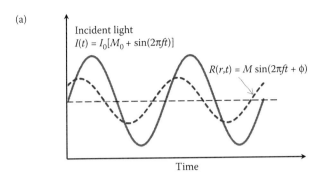

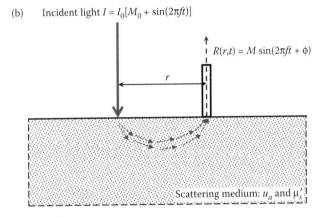

FIGURE 3.7 (a) A semi-infinite scattering medium under the normal incidence of an infinitesimal light beam varying with time as a function of $I(t) = I_0[M_0 + \sin(2\pi ft)]$, where I_0 is the amplitude of the sinusoidal light, M_0 is a constant, f is the frequency, and t is the time, with the resultant reflectance (R) at a distance r from the light incident point, which has an amplitude of M and a phase delay of ϕ. (b) The concept of frequency-domain reflectance measurement for estimating the optical properties of a scattering medium.

where

$$\psi_0 = \left\{ \frac{1}{D}(r^2 + z_0^2) \frac{\left[(c\mu_a)^2 + (2\pi f)^2\right]^{1/2}}{c} \right\}^{1/2}, \tag{3.70}$$

$$\psi_r = -\psi_0 \sin\left(\frac{\theta}{2}\right), \tag{3.71}$$

$$\psi_i = \psi_0 \cos\left(\frac{\theta}{2}\right), \tag{3.72}$$

$$\theta = \tan^{-1}\left(\frac{2\pi f}{c\mu_a}\right), \quad (3.73)$$

$$\psi_\infty = \psi_0(f=0) = \psi_t(f=0) = \left[\mu_{eff}'^2\left(r^2 + z_0^2\right)\right]^{1/2}, \quad (3.74)$$

in which f is the modulation frequency. For the known values of r and f, we can estimate the optical absorption and scattering coefficients from the above set of equations by using the method suggested by Patterson et al. (1991).

Frequency-domain technique that is based on the principle outlined above has been developed for measuring the optical properties of biological tissues (Cletus et al., 2010; Pogue and Patterson, 1994). However, the technique so far has not been reported for measuring food and agricultural products.

3.6.4 Illumination by Spatially Modulated Light

Consider a semi-infinite scattering medium illuminated at its surface by a steady-state, planar sinusoidal lighting pattern (Figure 3.8), which is described by the following equation:

$$I(x,z)|_{z=0} = I_0[a_0 + \sin(2\pi f x + \alpha)], \quad (3.75)$$

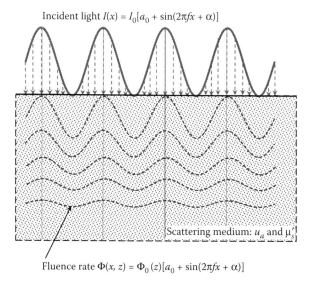

FIGURE 3.8 Semi-infinite scattering medium under normal incidence of a sinusoidal light wave with its intensity varying spatially as a function of $I(x) = I_0[a_0 + \sin(2\pi f x + \alpha)]$, where I_0 is the amplitude of the sinusoidal light beam, a_0 is a constant, f is the spatial frequency, x is the distance in the horizontal direction, and α is the phase angle. The fluence rate $\Phi(x,z)$ in the medium also follows a sinusoidal wave function and decays exponentially with depth in the vertical direction z and Φ_0 is the fluence rate at location $x = 0$ and is given in Equation 3.82.

where I_0 is the amplitude of the sinusoidal light source, f is the spatial frequency in 1/m, α is the phase angle, and a_0 is a constant to reflect the fact that the light incident intensity cannot be negative. Under such an illumination condition, the diffusion equation in Equation 3.41 is independent of the spatial dimension y (Cuccia et al., 2009), and it can be rewritten as

$$\nabla^2 \Phi(x,z) - \mu_{eff}'^2 \Phi(x,z) = -3\mu_t' S. \qquad (3.76)$$

Svaasand et al. (1999) reported that a planar illumination at the surface of a semi-infinite scattering medium can be treated as an extended source. This means that the source term in Equation 3.76 can be expressed as

$$S = I(x,z)|_{z=0} = I_0(z)[a_0 + \sin(2\pi f x + \alpha)], \qquad (3.77)$$

where $I_0(z)$ can be described by the following equation:

$$I_0(z) = P_0 \mu_s' \exp(-\mu_t' z), \qquad (3.78)$$

in which P_0 is the incident optical intensity.

To satisfy Equation 3.76 with the source term given by Equation 3.77, the fluence rate must also be in the form of a sinusoidal function with the same frequency and phase. Hence we have

$$\Phi(x,z) = \Phi_0(z)[a_0 + \sin(2\pi f x + \alpha)]. \qquad (3.79)$$

Substituting Equations 3.77 and 3.79 into Equation 3.76 yields

$$\frac{d^2 \Phi_0(z)}{dz^2} - \mu_{eff}'^{*2} \Phi_0(z) = -3\mu_s' \mu_t' P_0 \exp(-\mu_t' z), \qquad (3.80)$$

where

$$\mu_{eff}'^{*2} = \mu_{eff}'^2 + (2\pi f)^2. \qquad (3.81)$$

Equation 3.80 is an ordinary differential equation, which only depends on z, and it can be readily solved, which yields

$$\Phi_0(z) = \frac{3 P_0 a'}{((\mu_{eff}'^{*2}/\mu_t'^2) - 1)} \exp(-\mu_t' z) + C \exp(-\mu_{eff}'^* z), \qquad (3.82)$$

where C is a constant determined by the boundary condition, and a', referred to as the *reduced albedo*, is given by

Theory of Light Transfer in Food and Biological Materials

$$a' = \frac{\mu'_s}{\mu'_t}. \tag{3.83}$$

By applying the PCBC in Equation 3.48, the constant C is determined as

$$C = \frac{3P_0 a'_s(3+2C_F)}{(1-(\mu'^{*2}_{\mathit{eff}}/\mu'^2_t))(3+2C_F(\mu'^*_{\mathit{eff}}/\mu'_t))}, \tag{3.84}$$

where C_F is defined in Equation 3.50.

The diffuse reflectance R at the surface of the scattering medium can be obtained by using the PCBC (Haskell et al., 1994):

$$R(f)\big|_{z=0} = \frac{-J|_{z=0}}{P_0} = -\frac{1}{2C_F}\Phi\big|_{z=0} = \frac{3a'/(2C_F)}{((\mu'^*_{\mathit{eff}}/\mu'_t)+1)((\mu'^*_{\mathit{eff}}/\mu'_t)+3/(2C_F))}. \tag{3.85}$$

Equation 3.85 can be used to measure the optical absorption and scattering coefficients of tissues. This is accomplished by acquiring images from a tissue sample illuminated by the sinusoidal lighting of different frequencies. Since the acquired images contain both dc (or planar) and ac components, a demodulation method is needed to separate the dc and ac components, which allows to estimate the optical property parameters from Equation 3.85 (Cuccia et al., 2009).

Spatial-frequency domain technique, based on the principle presented above, was recently developed for determining the optical absorption and scattering properties and internal imaging of biological tissues and food products (Anderson et al., 2007; Cuccia et al., 2009; Lin et al., 2013; Saager et al., 2010; Weber et al., 2009). Compared with other techniques, spatial-frequency domain technique is relatively simple in instrumentation, and it allows to map the optical properties over a large field of view, as opposed to point measurements by other techniques.

3.7 FINITE ELEMENT MODELING OF LIGHT PROPAGATION IN SCATTERING MEDIA

Analytical solutions for the diffusion equation presented above provide a simple, elegant form for describing the reflectance of light at the surface of, and the fluence rate in, semi-infinite scattering media under the illumination of an ideal light beam. These analytical solutions, if used appropriately, offer a fast, convenient means for predicting light propagation in scattering media and for noninvasive or nondestructive measurement of the optical properties through inverse algorithms. However, care needs to be taken when using these analytical equations to ensure that the boundary conditions, under which the solutions are derived, are fully met. Moreover, analytical solutions are only available for the scattering media that can be approximated to be semi-infinite. Such restrictions can be problematic for the food and biological materials that are of a finite size and/or an irregular shape. In deriving the analytical solutions for the special cases presented in the previous section, we assume that light beams are infinitely small (except spatially modulated lighting) and are incident on the medium in the normal direction. Actual light beams are of a finite size and are often obliquely incident on the medium. These deviations can affect the prediction of light propagation, and thus the accuracy of estimating the optical absorption and scattering properties of food and biological tissues.

Numerical methods, such as finite element, finite difference, and boundary element, are very useful for solving boundary value problems for partial differential equations in engineering, physics, and other fields of science. FEM, in particular, is a powerful numerical method for accurate modeling of light propagation in scattering media with complex geometries under normal or oblique illumination of a light beam of finite size. In this section, we give an overview of finite element formulation procedures and present application examples on finite element modeling of light propagation in food and biological materials. Results from different boundary conditions are compared against MC simulations, which are considered to be the "gold" standard in modeling light propagation in scattering media. MC simulation methods are presented in Chapter 4.

3.7.1 Finite Element Formulation

Finite element formulation is a numerical technique for finding approximate solutions to boundary value problems for partial differential equations. Generally, the finite element formulation involves the following steps: first, it subdivides or discretizes the whole region of interest into simple, small elements, known as finite elements (such as rectangles, triangles, etc.), which creates a mesh of elements. Each element is defined or represented by its nodes, which are the vertices of the element. Second, simple (often linear) equations are used to locally approximate the original partial differential equation for individual nodes of each element. A variational method, such as the Galerkin method, is used to minimize the residual of approximation by fitting trial functions to the original partial differential equation. Third, these simple equations are then systematically assembled into a global system of linear equations. Finally, solutions are obtained by solving the global system of equations, in terms of quantities of interest for each element node.

To develop a finite element formulation for the diffusion equation in Equation 3.41, we first divide the region of interest into small finite elements of a specific shape (e.g., triangle, rectangle, etc.). The fluence rate Φ for the nodes of each element is approximated as a linear combination of shape functions $\psi_i(\vec{r})$ such that

$$\Phi^a(\vec{r},t) = \sum_{i=1}^{N} \Phi^a_i(t)\psi_i(\vec{r}_j), \qquad (3.86)$$

where $i = 1, 2, \ldots, N$, in which N is the total number of nodes for the element; $\Phi_i(t)^a$ is the value of Φ at the node of i; and $\psi_i(\vec{r}_j) = \delta_{ij}$ is the nodal shape function for each element. δ_{ij} is the Kronecker delta function; $\delta_{ij} = 1$ when $i = j$ and $\delta_{ij} = 0$ when $i \neq j$. Replacing Φ by $\Phi^a(\vec{r},t)$ in Equation 3.41 yields

Theory of Light Transfer in Food and Biological Materials

$$\frac{\partial \Phi^a(\vec{r},t)}{\partial t} - D\nabla^2 \Phi^a(\vec{r},t) + \mu_a \Phi^a(\vec{r},t) - S(\vec{r},t) = Res(\vec{r},t), \quad (3.87)$$

where $Res(\vec{r},t)$ is the residual of approximation.

A variational method, such as the weak Galerkin method, is then applied to Equation 3.87, to force the weighted average of the residual in Equation 3.87 to be zero over the domain Ω for the function $\psi_i(\vec{r})$. Hence, we have

$$\int_\Omega \psi_i \left(\frac{\partial \Phi^a(\vec{r},t)}{c\partial t} - D\nabla^2 \Phi^a(\vec{r},t) + \mu_a \Phi^a(\vec{r},t) \right) d\Omega - \int_\Omega \psi_i S(\vec{r},t) d\Omega = 0, \quad (3.88)$$

for each node. Integrating Equation 3.88 by parts and then substituting $\Phi^a(\vec{r},t)$ by the expression in Equation 3.86 yields

$$\frac{1}{c} \int_\Omega \psi_i \sum_{j=1}^N \psi_j \left(\frac{\partial \Phi_j^a(t)}{\partial t} \right) d\Omega - \int_\Omega D\psi_i \sum_{j=1}^n \Phi_j^a(t) \nabla^2 \psi_j \, d\Omega + \int_\Omega \psi_i \mu_a \sum_{j=1}^N \Phi_j^a \psi_j \, d\Omega$$
$$= \int_\Omega \psi_i S(\vec{r},t) d\Omega. \quad (3.89)$$

Equation 3.89 is a system of linear equations, which can be rewritten as

$$B \frac{\partial \Phi}{\partial t} + [M+C]\Phi^a = Q, \quad (3.90)$$

where B, M, C, and Q are the coefficient matrices for the system of linear equations in Equation 3.90. For time-independent problems, Equation 3.90 becomes

$$[M+C]\Phi^a = Q. \quad (3.91)$$

By solving the system of linear equations in Equation 3.90 or 3.91, we obtain values of the fluence rate at each node for the entire domain Ω.

3.7.2 Application Examples

In this section, we present an application example to show how finite element analysis can be used to model light propagation in a semi-infinite scattering medium normally illuminated by either an infinitesimal light beam or a finite-size beam. We compare the three boundary conditions (Section 3.5) for prediction of the fluence rate and reflectance at the surface, as well as evaluate the effect of light beam size on modeling results. A detailed description of the finite element modeling approach and results can be found in Wang et al. (2015).

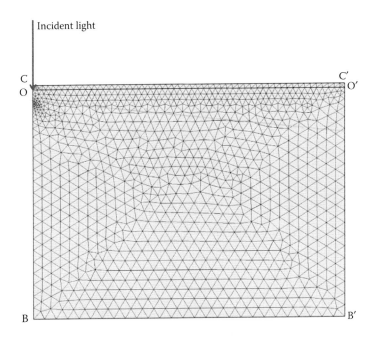

FIGURE 3.9 Finite element mesh for a semi-infinite scattering medium under normal illumination of a point light beam at the position $z = 0$ (the vertical position) and $r = 0$ (the horizontal direction), where the line OO' represents the real boundary ($z = 0$) and CC' is the extrapolated boundary ($z = z_0 = 1/\mu'_t$). The problem is treated as an axisymmetric problem with the axis of symmetry along the line of CB, while the lines BB' and O'B' are the truncated boundaries for the semi-infinite medium (Adapted from Wang, A., R. Lu and L. Xie. Finite element modeling of light propagation in fruit under illumination of continuous-wave beam. ASABE Paper #152189254, 17pp. St. Joseph, MI: ASABE, 2015.).

Consider a semi-infinite scattering medium, which is represented by a mesh of triangular elements, as shown in Figure 3.9. First, we consider the normal incidence of an infinitesimal light beam on the scattering medium. An analytical solution for the problem was given by Farrell et al. (1992), as discussed in Section 3.6.1. In this example, we use the solution from Kienle and Patterson (1997), as shown in Equation 3.62, since it has been proved to give more accurate results. It should be mentioned that the solution by Kienle and Patterson (1997) is derived using the EBC. To avoid unnecessary computational time, the semi-infinite medium is reduced to a rectangular medium with the dimension of 50 mm in depth and 70 mm in the horizontal direction. The selection of such dimensions would ensure that the fluence rate at the boundaries (other than the surface) would be zero or negligible. The problem is, in fact, an axisymmetric problem.

The three boundary conditions, including the ZBC, PCBC, and EBC, are imposed on the surface of the semi-infinite medium, as follows:

- For ZBC, the fluence rate is set to zero at the boundary OO' (Figure 3.9)
- For PCBC, the condition given in Equation 3.49 is applied to the boundary OO'

- For EBC, the fluence rate is set to zero at the extrapolated boundary CC', which is separated from the real boundary OO' by the distance of $z = -2D$

One important consideration in the finite element modeling is the source term. Since the light beam is collimated, it cannot be accurately described in the diffusion equation, in which the source term is assumed to be isotropic. To circumvent this problem, it is common to assume that the collimated beam is applied at one mean free path (i.e., 1 mfp) below the scattering surface. To take the finite size beam into consideration, we also assume that the beam uniformly illuminates the scattering medium at 1 mfp below the surface.

Figure 3.10 compares the results for the fluence rate along the vertical direction at a horizontal distance of 2 mm from the axis of symmetry, predicted by the MC simulation (considered to be the exact solution) and the FEM, for the two scattering media with different combinations of absorption and reduced scattering coefficients for the three boundary conditions described. Compared with the MC simulations, ZBC yields the poorest results, which are especially prominent at smaller depths. PCBC and EBC produce similar results for both scattering media and compare well with MC simulation results.

Figure 3.11 further compares the reflectance at the surface of the two scattering media predicted by the FEM, the analytical solution (Equation 3.62), and MC simulation. Finite element simulations for PCBC and EBC produced similar results for reflectance, as that by the analytical solution; they all compare very well with the MC simulation results. On the other hand, the results for ZBC deviate significantly from MC simulations. The diffusion model in all four cases does not accurately predict the reflectance at the surface when the distance is less than 1–2 mfps.

Similar results are also found when a finite-size beam is normally incident on the surface of a semi-infinite scattering medium (Figure 3.12). Within the incident area, the FEM produces a relatively large error, compared with MC simulations. Beyond the light incident area, the finite element predictions for EBC or PCBC match MC simulation very well. The discrepancy between the MC and finite element results becomes more prominent as the beam size increases from 1 to 4 mm. It should be mentioned that MC simulations for the finite beam were carried out by using the convolution of the response of an infinitely small beam.

3.8 SUMMARY AND CONCLUSIONS

The RTE provides an accurate description of light propagation in scattering media, but it is generally difficult to solve for practical applications. For food and biological materials, in which scattering is dominant, the RTE can be simplified to diffusion approximation theory or diffusion equation. Analytical solutions to the diffusion equation for several special situations (i.e., cw, pulsed, frequency modulated, and spatial-frequency modulated) have been derived. These solutions provide the theoretical foundation for noninvasive optical property measurement techniques such as spatially-resolved, time-resolved, frequency domain, and spatial-frequency domain. For many applications where complex boundaries are involved and light illumination patterns cannot be considered to be a point source, FEM should be considered.

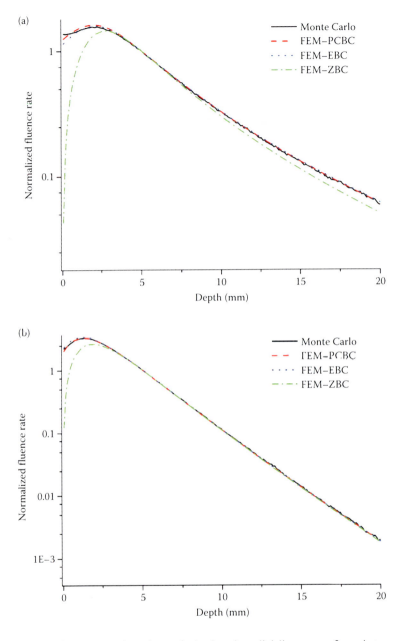

FIGURE 3.10 Fluence rate along the vertical axis at the radial distance $r = 2$ mm in two semi-infinite scattering media with (a) $\mu_a = 0.005$ mm^{-1} and $\mu'_s = 0.5$ mm^{-1} and (b) $\mu_a = 0.02$ mm^{-1} and $\mu'_s = 2$ mm^{-1} under the normal illumination of an infinitesimal light beam, as shown in Figure 3.9, simulated by MC and FEM for the PCBC, EBC, and ZBCs (Adapted from Wang, A., R. Lu and L. Xie. Finite element modeling of light propagation in fruit under illumination of continuous-wave beam. ASABE Paper #152189254, 17pp. St. Joseph, MI: ASABE, 2015.).

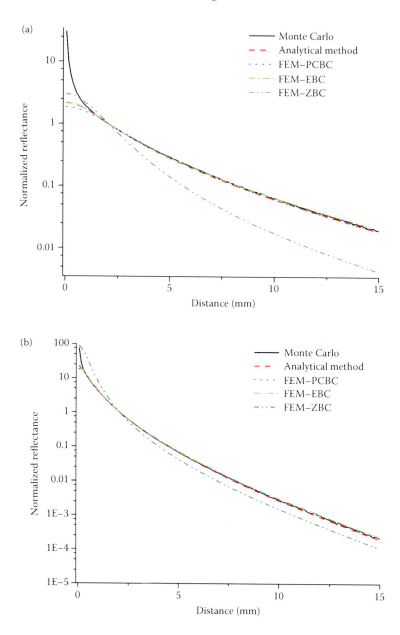

FIGURE 3.11 Spatially-resolved reflectance for two semi-infinite scattering media with (a) $\mu_a = 0.005$ mm^{-1} and $\mu'_s = 0.5$ mm^{-1} and (b) $\mu_a = 0.02$ mm^{-1} and $\mu'_s = 2$ mm^{-1} under the normal illumination of an infinitesimal light beam, as shown in Figure 3.9, simulated by MC, analytical method, and FEM for the PCBC, EBC, and ZBCs (Adapted from Wang, A., R. Lu and L. Xie. Finite element modeling of light propagation in fruit under illumination of continuous-wave beam. ASABE Paper #152189254, 17pp. St. Joseph, MI: ASABE, 2015.).

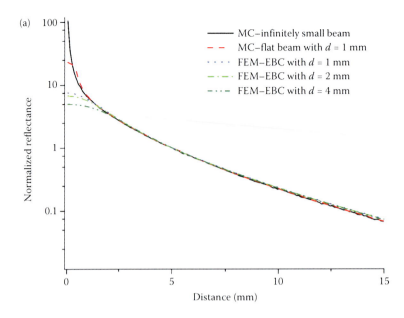

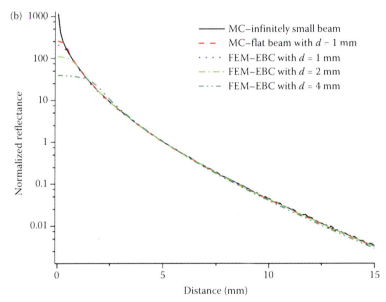

FIGURE 3.12 Spatially-resolved reflectance obtained by MC and FEM for two semi-infinite scattering media with (a) $\mu_a = 0.005$ mm^{-1} and $\mu_s' = 0.5$ mm^{-1} and (b) $\mu_a = 0.02$ mm^{-1} and $\mu_s' = 2$ mm^{-1} under the normal illumination of finite-size beams of size 0, 1, 2, or 4 mm (Adapted from Wang, A., R. Lu and L. Xie. Finite element modeling of light propagation in fruit under illumination of continuous-wave beam. ASABE Paper #152189254, 17pp. St. Joseph, MI: ASABE, 2015.).

The method is flexible and also fast, and thus can be very useful in modeling light propagation in the scattering media. However, the selection of appropriate boundary conditions is critical to obtaining accurate finite element results.

REFERENCES

Anderson, E. R., D. J. Cuccia and A. J. Durkin. Detection of bruises on Golden Delicious apples using spatial-frequency-domain imaging. *Proceedings of SPIE* 6430, 2007: 64300. Bellington, WA: SPIE

Aronson, R. Boundary conditions for diffusion of light. *Journal of Optical Society of America A* 12, 1995: 2532–2539.

Cen, H. and R. Lu. Optimization of the hyperspectral imaging-based spatially-resolved system for measuring the optical properties of biological materials. *Optics Express* 18, 2010: 17412–17432.

Cletus, B., R. Künnemeyer, P. Martinsen, A. McGlone and R. Jordan. Characterizing liquid turbid media by frequency-domain photon migration spectroscopy. *Journal of Biomedical Optics* 14, 2010: 024041–024047.

Cubeddu, R., C. D'Andrea, A. Pifferi, P. Taroni, A. Torricelli, G. Valentini, M. Ruiz-Altisent et al. Time-resolved reflectance spectroscopy applied to the nondestructive monitoring of the internal optical properties in apples. *Applied Spectroscopy* 55, 2001: 1368–1374.

Cuccia, D. J., F. Bevilacqua, A. J. Durkin, F. R. Ayers and B. J. Tromberg. Quantitation and mapping of tissue optical properties using modulated imaging. *Journal of Biomedical Optics* 14, 2009: 1354–1356.

Deulin, X. and J. P. L'Huillier. Finite element approach to photon propagation modeling in semi-infinite homogeneous and multilayered tissue structures. *European Physical Journal: Applied Physics* 33, 2006: 133–146.

Farrell, T. J., M. S. Patterson and B. Wilson. A diffusion theory model of spatially resolved, steady-state diffuse reflectance for the noninvasive determination of tissue optical properties *in vivo*. *Medical Physics* 19, 1992: 879–888.

Groenhuis, R. A. J., H. A. Ferwerda and J. J. Ten Bosch. Scattering and absorption of turbid materials determined from reflection measurements 1: Theory. *Applied Optics* 22, 1983: 2456–2462.

Haskell, R. C., L. O. Svaasand, T. Tsay, T. Feng, M. S. McAdams and B. J. Tromberg. Boundary conditions for the diffusion equation in radiative transfer. *Journal of Optical Society of America* 11, 1994: 2727–2741.

Kienle, A. and M. S. Patterson. Improved solutions of the steady-state and the time-resolved diffusion equations for reflectance from a semi-infinite turbid medium. *Journal of Optical Society of America A* 14, 1997: 246–254.

Lin, A. J., A. Ponticorvo, S. D. Konecky, H. Cui, T. B. Rice, B. Choi, A. J. Durkin and B. J. Tromberg. Visible spatial frequency domain imaging with a digital light microprojector. *Journal of Biomedical Optics* 18, 2013: 096007.

Lin, S., L. Wang, S. L. Jacques and F. K. Tittel. Measurement of tissue optical properties by the use of oblique-incidence optical fiber reflectometry. *Applied Optics* 36, 1997: 136–143.

Mourant, J. R., T. Fuselier, J. Boyer, T. M. Johnson and I. J. Bigio. Predictions and measurements of scattering and absorption over broad wavelength ranges in tissue phantoms. *Applied Optics* 36, 1997: 949–957.

Nichols, M. G., E. L. Hull and T. H. Foster. Design and testing of a white-light, steady-state diffuse reflectance spectrometer for determination of optical properties of highly scattering systems. *Applied Optics* 36, 1997: 93–104.

Patterson, M. S., B. Chance and B. C. Wilson. Time resolved reflectance and transmittance for the non-invasive measurement of tissue optical properties. *Applied Optics* 28, 1989: 2331–2336.

Patterson, M. S., J. D. Moulton, B. C. Wilson, K. W. Berndt and J. R. Lakowicz. Frequency-domain reflectance for the determination of the scattering and absorption properties of tissue. *Applied Optics* 30, 1991: 4474–4476.

Pogue, B. W. and M. S. Patterson. Frequency-domain optical absorption spectroscopy of finite tissue volumes using diffusion theory. *Physics in Medicine and Biology* 39, 1994: 1157–1180.

Qin, J. and R. Lu. Hyperspectral diffuse reflectance imaging for rapid, noncontact measurement of the optical properties of turbid materials. *Applied Optics* 45, 2006: 8366–8373.

Qin, J. and R. Lu. Monte Carlo simulation for quantification of light transport features in apples. *Computers and Electronics in Agriculture* 68, 2009: 44–51.

Saager, R. B., D. J. Cuccia and A. J. Durkin. Determination of optical properties of turbid media spanning visible and near-infrared regimes via spatially modulated quantitative spectroscopy. *Journal of Biomedical Optics* 15, 2010: 017012.

Schweiger, M., S. R. Arridge, M. Hiraoka and D. T. Deply. The finite element method for the propagation of light in scattering media: Boundary and source conditions. *Medical Physics* 22, 1995: 1779–1792.

Star, W. M. Diffusion theory of light transport. In Welch, A. J. and M. J. C. van Gemert (eds.) *The Book Optical–Thermal Response of Laser-Irradiated Tissue*, Chapter 6, 2nd edition, pp. 145–201. New York, NY: Springer, 2011.

Svaasand, L. O., T. Spott, J. B. Fishkin, T. Pham, B. J. Tromberg and M. W. Berns. Reflectance measurements of layered media with diffuse photon-density waves: A potential tool for evaluating deep burns and subcutaneous lesions. *Physics in Medicine and Biology* 44, 1999: 801–813.

Wang, A., R. Lu and L. Xie. Finite element modeling of light propagation in fruit under illumination of continuous-wave beam. ASABE Paper #152189254, 17pp. St. Joseph, MI: ASABE, 2015.

Wang, L. V. and H. I. Wu. *Biomedical Optics: Principles and Imaging*. Hoboken, N J: John Wiley & Sons, 2007.

Weber, J. R., D. J. Cuccia, A. J. Durkin and B. J. Tromberg. Noncontact imaging of absorption and scattering in layered tissue using spatially modulated structured light. *Journal of Applied Physics* 105, 2009: 102028.

APPENDIX: VECTORS AND DERIVATIVES

A3.1 VECTORS

A *vector* or *vector field*, denoted as $\vec{A}$, can be expressed as

$$\vec{A} = A_x \vec{i} + A_y \vec{j} + A_z \vec{k}, \tag{A3.1}$$

where $\vec{i}, \vec{j}$, and $\vec{k}$ are the unit vectors in the directions of x, y, and z axes for a three-dimensional Cartesian coordinate system, and A_x, A_y, and A_z are the vector components in the direction of the three coordinates, which are scalar (Figure A3.1). In the spherical coordinates, a vector can be expressed as

$$\vec{A} = A_r \vec{r} + A_\theta \vec{\theta} + A_\varphi \vec{\varphi}, \tag{A3.2}$$

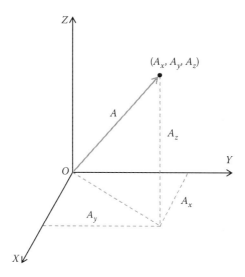

FIGURE A3.1 Vector $\bar{A}$ in the Cartesian coordinate system.

where $\bar{r}, \bar{\theta}$, and $\bar{\varphi}$ are the unit vectors in the direction of r, θ, and φ in the spherical coordinate system, in which r is the radius or radius coordinate, θ is the polar angle, and φ is the azimuth angle (Figure A3.2). In the standard matrix notation, a row vector $\bar{A}$ can be written as

$$\bar{A} = [A_x, A_y, A_z]. \tag{A3.3}$$

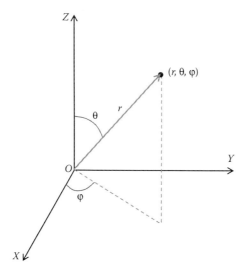

FIGURE A3.2 Vector $\bar{A}$ in the spherical coordinate system.

The dot product of two vectors $\vec{A} = [A_x, A_y, A_z]$ and $\vec{B} = [B_x, B_y, B_z]$ is defined by

$$\vec{A} \cdot \vec{B} = A_x B_x + A_y B_y + A_z B_z. \tag{A3.4}$$

The dot product of two vectors may also be defined by

$$\vec{A} \cdot \vec{B} = \|\vec{A}\|\|\vec{B}\|\vec{s} \cdot \vec{s}' = \|\vec{A}\|\|\vec{B}\|\cos\theta, \tag{A3.5}$$

where $\|\vec{A}\|$ and $\|\vec{B}\|$ are the magnitude of the two vectors $\vec{A}$ and $\vec{B}$, respectively, which are given by

$$\|\vec{A}\| = \sqrt{\vec{A} \cdot \vec{A}} \quad \text{and} \quad \|\vec{B}\| = \sqrt{\vec{B} \cdot \vec{B}}, \tag{A3.6}$$

and $\vec{s}$ and $\vec{s}'$ are the unit vectors of $\vec{A}$ and $\vec{B}$, respectively (i.e., $\|\vec{s}\| = 1$ and $\|\vec{s}'\| = 1$), with $\vec{s} \cdot \vec{s}' = \cos\theta$ being the angle between the two vectors.

A3.2 DERIVATIVES

The *gradient*, denoted as grad or ∇, is a vector differential operator of a scalar function $f(x, y, z)$. In the Cartesian coordinate system, the gradient is the vector field whose components are the partial derivatives of $f(x, y, z)$

$$\text{grad}(f) = \nabla f = \frac{\partial f}{\partial x}\vec{i} + \frac{\partial f}{\partial y}\vec{j} + \frac{\partial f}{\partial z}\vec{k}, \tag{A3.7}$$

and in the spherical coordinate system, the gradient of $f(r, \theta, \varphi)$ is

$$\text{grad}(f) = \frac{\partial f}{\partial r}\vec{r} + \frac{1}{r}\frac{\partial f}{\partial \theta}\vec{\theta} + \frac{1}{r\sin\theta}\frac{\partial f}{\partial \varphi}\vec{\varphi}. \tag{A3.8}$$

The *divergence*, denoted as div or $\nabla\cdot$, is a vector operator that measures the magnitude of a vector field's source or sink at a given point. In a three-dimensional Cartesian coordinate system, the divergence of a vector field, $F(x, y, z)$, is given by

$$\text{div}(F) = \nabla \cdot F = \frac{\partial F}{\partial x} + \frac{\partial F}{\partial y} + \frac{\partial F}{\partial z}, \tag{A3.9}$$

and for the spherical coordinates, the divergence of a vector field $F(r, \theta, \varphi)$ is given by

$$\nabla \cdot F = \frac{1}{r^2}\frac{\partial}{\partial r}(r^2 F_r) + \frac{1}{r\sin\theta}\frac{\partial}{\partial \theta}(\sin\theta F_\theta) + \frac{1}{r\sin\theta}\frac{\partial F_\varphi}{\partial \varphi}. \tag{A3.10}$$

The divergence is a linear operator with the following additive characteristics:

$$\text{div}(aF + bG) = a\ \text{div}(F) + b\ \text{div}(G), \tag{A3.11}$$

where a and b are real numbers and F and G are the vector fields. The divergence follows the product rule:

$$\text{div}(\rho F) = \text{grad}(\rho) \cdot F + \rho\, \text{div}(F), \tag{A3.12}$$

where ρ is a scalar function and F is a vector field or function.

The *Laplace operator* or *Laplacian*, denoted as ∇^2 or $\nabla \cdot \nabla$, is a second-order differential operator, which is defined as the divergence of the gradient (∇f). It has the following expression:

$$\nabla^2 f = \nabla \cdot \nabla f = \frac{\partial^2 f}{\partial x^2} + \frac{\partial^2 f}{\partial y^2} + \frac{\partial^2 f}{\partial z^2}, \tag{A3.13}$$

for the Cartesian coordinates, and

$$\nabla^2 f = \nabla \cdot \nabla f = \frac{1}{r^2}\frac{\partial^2}{\partial r^2}\left(r^2 \frac{\partial^2 f}{\partial r^2}\right) + \frac{1}{r^2 \sin\theta}\frac{\partial}{\partial \theta}\left(\sin\theta \frac{\partial^2 f}{\partial \theta}\right) + \frac{1}{r^2 \sin^2\theta}\frac{\partial^2 f}{\partial \varphi^2}, \tag{A3.14}$$

for the spherical coordinate system.

4 Monte Carlo Modeling of Light Transfer in Food

Rodrigo Watté, Ben Aernouts, and Wouter Saeys

CONTENTS

4.1 Introduction 79
4.2 Principles of MC Simulations 82
4.3 Tracing Photons through a Tissue 83
    4.3.1 Localizing Photons in a Tissue 83
    4.3.2 Launching a Photon 86
    4.3.3 Photon Step Size 86
    4.3.4 Photon Absorption 87
    4.3.5 Photon Scattering 88
    4.3.6 Reflection and Refraction at Layer Boundaries 92
4.4 Improving Computational Speed 93
    4.4.1 Photon Packets 93
    4.4.2 Convolution 94
    4.4.3 Scaling MC 95
    4.4.4 Perturbation MC 96
    4.4.5 Hybrid MC Methods 96
    4.4.6 Parallel-Computed MC Algorithms 96
    4.4.7 Variance Reduction Techniques 97
4.5 MC Simulations on Food Products 97
    4.5.1 More Accurate Parametric Phase Functions 99
    4.5.2 Nonparametric Phase Functions 100
    4.5.3 From Plane Parallel to Realistic Geometries 103
4.6 Conclusions 103
Acknowledgments 105
References 105

4.1 INTRODUCTION

The wave properties of light propagating through a turbid medium tend to be averaged out after several scattering interactions. Consequently, the modeling of light propagation in turbid media can be simplified by ignoring the wave-like behavior. Therefore, it is acceptable to model the propagation of light through turbid media as a stream of particles, each with a quantum of energy (Ishimaru, 1987; Patterson et al., 1991; Wang et al., 1995). Under these assumptions, the radiative transport theory (RTT) accurately describes the propagation (i.e., scattering and absorption) of light in turbid tissues. Diffraction and interference effects are typically ignored in the RTT.

Light can interact with the biological tissue in several ways: (1) at the interface of two media with different refractive indices, photons will be refracted and reflected; (2) within a medium individual photons can be absorbed by the material and the electromagnetic energy is transformed into other forms of energy (e.g., heat, fluorescence, etc.); and (3) the photons that have not been absorbed can be scattered, because microscopic variations of the dielectric properties in the media cause a change in their direction of propagation. The optical properties of the medium are therefore expressed in terms of the following parameters: (1) refractive index n, (2) absorption coefficient μ_a, (3) scattering coefficient μ_s, and (4) scattering phase function $p(\theta)$. The Henyey–Greenstein (HG) phase function is a commonly used approximation, which simplifies the scattering phase function based on a single-anisotropy factor g (Ishimaru, 1987; Welch and van Gemert, 1995).

The radiative transport equation (RTE) provides a mathematical expression of photon transport through turbid media by considering the energy balance of incoming, outgoing, internal source, absorbed, and emitted photons for an infinitesimal volume in the medium (see Chapter 3 for a further detailed description). It models the flow of photons inside a tissue in terms of an integro-differential equation describing the change in radiance L as a function of the previously mentioned optical properties. Since no analytical solution is available, the RTE must be simplified based on some assumptions or solved numerically (Ishimaru, 1987; Patterson et al., 1991). Approximate analytical solutions are popular (Farrell et al., 1992; Ishimaru, 1987; Patterson et al., 1991; Skipetrov and Chesnokov, 1998; Welch and van Gemert, 1995), because they are computationally fast. By making assumptions concerning the photons and the medium in which they are flowing, several variables in the RTE can be replaced by a smaller number of independent variables. In the first category of approximation methods, the angular variation of the radiation flux (caused by scattering) is reduced to an upward and downward diffuse flux (e.g., two-stream approximations (Kocifaj, 2012), Kubelka–Munk (Yang and Miklavcic, 2005), and Delta–Eddington approximations (Cong et al., 2007; Lehtikangas and Tarvainen, 2013)). This simplification makes it possible to solve the RTE analytically. Another category of approximation methods starts from the assumption that the scattering phase function can be expanded in a series of Legendre polynomials, from which only a limited number of polynomials are taken into account (Bevilacqua and Depeursinge, 1999; Toublanc, 1996). A common practice is to expand the scattering phase function to the first order. This approximation reduces the RTE to a diffusion equation where the propagation of photons only depends on the absorption coefficient μ_a, and the diffusion coefficient D (Farrell et al., 1992; Ishimaru, 1987; Welch and van Gemert, 1995):

$$D = \frac{1}{3(\mu_a + \mu'_s)} \quad (4.1)$$

$$\mu'_s = (1-g)\mu_s$$

As the scattering information in D is described by the reduced scattering coefficient μ'_s (Farrell et al., 1992; Ishimaru, 1987; Welch and van Gemert, 1995), the effect of the anisotropy factor g is not considered independently from the effect of the scattering coefficient μ_s. Owing to the simplification of the angular variation of the radiation flux, these approximations are rather inaccurate in describing highly anisotropic scattering phenomena that are common in food emulsions, suspensions, and tissues. As a result, these approximations should only be used in turbid tissues, where the chance for a photon to be absorbed is much lower than the chance of it being scattered. Consequently, after several scattering interactions with almost no absorption, the anisotropic character of the photon propagation tends to be averaged out and the light distribution in these highly scattering media tends to become isotropic. This important assumption is valid for light distributions in highly scattering media, as long as these are observed at large distances from the light source so the original anisotropic character of photons is averaged out. A commonly used rule of thumb to decide on the validity of the diffusion approximation is that the distance from the source should be $>10\times(\mu_a+\mu'_s)^{-1}$ (Bevilacqua and Depeursinge, 1999). The diffusion approximation is not valid for optically thin samples or short source–detector distances ($<10\times(\mu_a+\mu'_s)^{-1}$).

The adding-doubling method, which was first proposed by Stokes in 1862, handles the angular distribution of the radiation flux more accurately (Kubelka and Munk, 1931). First, it calculates the reflectance and transmittance for a single, thin homogeneous layer. Once they are known for a single slab, the reflectance and transmittance are calculated for a slab twice as thick, by juxtaposing two identical slabs and summing the contributions from each slab, until eventually a layer with a thickness equal to the sample thickness is created. Layers with different optical properties can be added to each other, so that the total reflectance and transmittance can be calculated even for multilayered tissues. The main advantage of the adding-doubling method, compared to other approximations of the RTE, is that it no longer limits the light propagation to highly scattering media. As a result, this method is useful in the near-infrared (NIR) and mid-infrared region (≥ 1000 nm) (Fernandez-Oliveras et al., 2013; Honda et al., 2009), where water absorption dominates the light propagation. Furthermore, there is no longer a restriction on the scattering anisotropy. Since biological tissues tend to be very anisotropic (g ranges from 0.7 to 0.95), this is a significant improvement in accuracy compared to the previously mentioned approximations. However, the main disadvantages are that the methodology is restricted to uniform layers and that only the angular profiles of the reflectance and transmittance are modeled. If one is interested in the spatial light distribution that is not only dependent on the depth, and also on the radial distance, this method is not satisfactory. A detailed description of the adding-doubling method, coupled with the integrating sphere technique, for measuring the optical properties of food is given in Chapter 6.

The discrete ordinates method is based on the discretization of the directional variation of the radiative intensity. A solution to the transport problem is found by solving the transport equation for a set of discrete directions spanning the total range of scattering angles (Fiveland and Jessee, 1994; Modest, 2003). Calculation of the radiative intensity for a certain direction is achieved gradually along the optical path, using the results obtained from the neighboring points. Reducing the number of discrete elements can reduce the complexity of the methodology, but this will come at the expense of a loss in precision (Modest, 2003). Another major drawback is that the methodology is only valid for weakly anisotropic tissues (rendering it imprecise for modeling light propagation in foods).

Finally, probabilistic methods such as Monte Carlo (MC) simulation are popularly used to model light propagation in turbid tissues. MC methods are a category of computational methods that involve random sampling of physical quantities (Bevilacqua and Depeursinge, 1999; Flock et al., 1989; Hielscher et al., 1996; Ishimaru, 1987; Patterson et al., 1991; Wang et al., 1995; Watté et al., 2012; Welch and van Gemert, 1995). This is a very interesting concept for solving problems involving multiple scattering. With multiple scattering, the randomness of the interaction tends to be averaged out by a large number of scattering events, such that the final path of the radiation appears to be a deterministic distribution of intensity. Consequently, a large number of photons has to be simulated to converge toward a stable solution. MC methods have become popular in many different fields, including tissue optics. In MC simulations, the propagation of photons is discretized in small steps and the processes are stochastically described using probability density functions (Ishimaru, 1987; Patterson et al., 1991; Wang et al., 1995; Welch and van Gemert, 1995). For each step of photon movement in the homogeneous tissue, a pathlength and deflection angle are calculated from the probability density functions that characterize the optical properties of the tissue. These steps represent scattering events that occur due to microscopic differences in the homogeneous tissue. The MC methodology is based on the assumption that the macroscopic optical properties are uniform over the turbid tissue. This assumption is valid as long as the tissue is homogeneous on a macroscale. The procedure of random sampling is repeated until results are obtained with an acceptably low variance. This approach is very flexible since boundary conditions can easily be defined. However, to obtain acceptable results, a large number of photons should be simulated, thus making it computationally intensive. Nevertheless, as MC methods provide a flexible and accurate solution for modeling light propagation in turbid media, they are considered to be the golden standard (Bevilacqua and Depeursinge, 1999; Wang et al., 1995).

This chapter, therefore, covers the principles and implementation of MC methods for modeling light propagation in turbid tissues, with an emphasis on food products.

4.2 PRINCIPLES OF MC SIMULATIONS

An MC methodology is a stochastic simulation technique in which a physical process is repeated in different, independent computations. The collection of the different simulation outcomes is averaged out, resulting in an expected value. The method describes the propagation of photons as individual particles that solely interact with the tissue in which they are located. The MC methodology typically does not treat the photons as a wave, and ignores features such as polarization and phase (although variants exist where this is taken into account). Each layer of a tissue is characterized by the following optical properties: (1) refractive index n, (2) absorption coefficient μ_a, (3) scattering coefficient μ_s, and (4) scattering phase function $p(\theta)$. These are macroscopic optical properties that are assumed to extend uniformly over the biological tissue. In a classic algorithm, no details are computed concerning the distribution of light inside individual cells.

The photon movement—the displacement from one photon–tissue interaction to the next one—is described by probability functions (using the optical properties). The step size between two photon–tissue interactions or the deflection angle at each scattering event, are two examples of optical phenomena that are described with probability functions.

When the photons propagate through the biological tissue, an absorption and fluence profile is generated. After the photon exits the tissue, that contribution is added to the spatially- and angularly-resolved reflectance or transmittance profile. All these physical quantities are stored in a grid with a predefined spatial resolution. The number of photons required in an MC simulation depends on the sought-after information. For instance, the total reflectance can be accurately simulated with as low as 10,000 photons. With the same number of photons, the simulated spatial distribution of the absorption profile will still be very noisy. To reduce this noise to the same level as observed for the total reflectance with 10,000 photons, the number of photons has to be increased with one or several orders of magnitude. Obviously, this factor depends on the spatial resolution: a grid with a finer, higher resolution will require more photons to obtain the same level of noise. This trade-off between accuracy and computation time is considered to be one of the most important drawbacks of the MC methodology.

4.3 TRACING PHOTONS THROUGH A TISSUE

An overview of the steps involved in computing the path of individual photons in a MC simulation is illustrated in Figure 4.1.

4.3.1 Localizing Photons in a Tissue

One of the crucial aspects of the MC algorithm is the ray tracer. This is the part of the code responsible for localizing the photons at each point of the simulation (Watté et al., 2012). It is important to trace the positions of the photons as one of the main goals of using the MC technique is to model the spatial and/or temporal character of light distribution. The most important physical quantities that are recorded, are reflectance and transmittance (both total and spatially/angularly-resolved), and the photon absorption/fluence (Badal et al., 2009; Flock et al., 1989; Hielscher et al., 1996; Kienle and Patterson, 1996; Wang et al., 1995; Watté et al., 2012). The simulation traces the three-dimensional (3D) position of the photons. The photon deposition is recorded in the grid elements of a spatial array ($A(x, y, z)$ [$1/cm^3$, or more specifically J/cm^3 per J of delivered energy]), from which the fluence can be computed by dividing the local absorbed quantity of photon energy by the local absorption coefficient μ_a (Badal et al., 2009; Flock et al., 1989; Hielscher et al., 1996; Kienle and Patterson, 1996; Wang et al., 1995; Watté et al., 2012).

Standard MC algorithms use different coordinate systems simultaneously, as illustrated in Figure 4.2. The actual ray tracer uses a 3D Cartesian coordinate system (Wang et al., 1995). Typically, the origin of the coordinate system is either the incident

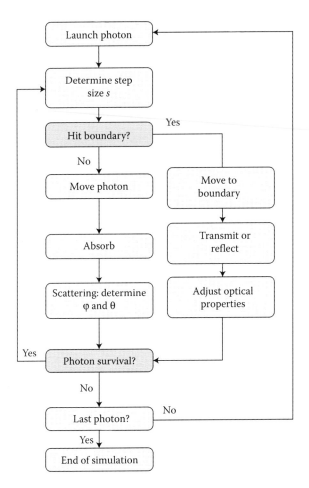

FIGURE 4.1 Generic flowchart of an MC algorithm describing how photon movement is simulated. Typically, 10^5–10^7 photons are launched. The step size, absorbed fraction, and deflection angle θ are dependent on the optical properties, which, in turn, determine the photon position for each step and at each time.

point on the tissue surface, or the center point of the light source. The absorption profile could be stored in a similar 3D grid, but the tissue structure is often simplified with a cylindrical symmetry (Wang et al., 1995). In many applications, the tissue can be considered homogeneous or has a plane-parallel layered structure of homogeneous layers. Therefore, symmetry with respect to the center point (the light source) can be assumed. As a result, the photon deposition is stored in a two-dimensional array, $A(r, z)$, where r and z are the radial and longitudinal (z-axis) coordinates of the cylindrical coordinate system, respectively. The Cartesian coordinate system and the cylindrical coordinate system share the origin and the z-axis. The reflectance and transmittance are stored in yet another coordinate system as the depth is an irrelevant parameter in this case. Instead, these outputs are recorded as a function of the r-coordinate and α, which is the angle between the photon-exiting direction and the

Monte Carlo Modeling of Light Transfer in Food

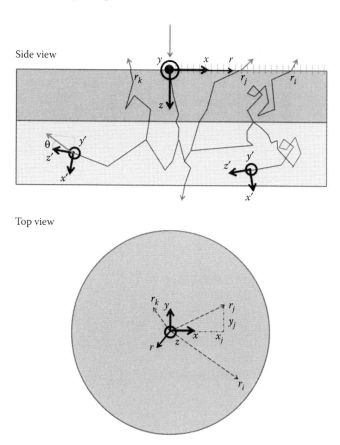

FIGURE 4.2 Schematic representation of a two-layered tissue with the Cartesian coordinate system (x, y, z), cylindrical coordinate system (r, z), and moving Cartesian coordinate system (x', y', z'). The deflection angle θ is illustrated, with respect to the Cartesian coordinate system. The azimuthal φ is omitted here, but illustrated in Figure 4.5. The deflection angle θ and the azimuthal φ compose the moving spherical coordinate system. The projections of θ and φ on (x', y', z') will determine the direction cosines. These direction cosines provide a straightforward methodology to update the new position of the photon in the general (x, y, z) coordinate system. The absorption output of each photon is stored in a (r, z) coordinate system, since a cylindrical symmetry is assumed.

normal. A moving spherical coordinate system, whose z-axis is dynamically aligned with the photon propagation direction, is used for sampling the propagation direction change of a photon packet. In this spherical coordinate system, the deflection angle θ and the azimuthal angle φ due to scattering are first sampled. Then, the photon direction is updated in terms of the directional cosines in the Cartesian coordinate system (Wang et al., 1995). This process is schematically illustrated in Figure 4.2. A more detailed explanation will be given in Section 4.2. The reason for storing the different outputs in different coordinate systems comes from the stochastic nature of the MC algorithm. Reducing the number of dimensions means that fewer photons are required to achieve stable simulation results with a higher signal-to-noise ratio.

4.3.2 LAUNCHING A PHOTON

The initial position of the photon depends on the type of light source. Typically, MC simulations model a point light source, in which case the Cartesian position is defined as the center point of the coordinate system (0, 0, 0). The initial direction of the photon depends on the simulation type. In the simplest case, the photon is injected orthogonally into the tissue. The photon direction is specified by the directional cosines (μ_x, μ_y, μ_z), which are the projections of the vector on the three coordinate axes (the projection on the (x', y', z') in Figure 4.2). At the launching of a photon, the directional cosines are therefore (0, 0, 1), since the photons are initially propagating along the z-axis.

4.3.3 PHOTON STEP SIZE

While propagating through a turbid food material (e.g., suspension, emulsion, foam, or tissue), photons will encounter (micro)structures with different refractive indices. When the photon approaches the interface between two materials with a different refractive index, a scattering event takes place. In an MC model, the photon travels to a predefined step with size s before being deflected. The photon step size is sampled from the probability distribution for the photon's free path s (Wang et al., 1995). This is determined by the total attenuation coefficient μ_t, which is defined as the sum of the absorption coefficient μ_a and the scattering coefficient μ_s. It is the probability of photon interaction per unit infinitesimal pathlength. Its inverse is known as the mean free path, or the average distance a photon travels inside the tissue without interacting with it (Wang et al., 1995)

$$\mu_t = \frac{-dP\{s \geq s'\}}{P\{s \geq s'\}ds'} \quad (4.2)$$

or

$$d[\ln(P\{s \geq s'\})] = -\mu_t ds' \quad (4.3)$$

This equation can be integrated over s' in the range [0, s_1], resulting in an exponential distribution that can be rearranged to obtain the cumulative distribution function of the free path s (Wang et al., 1995)

$$P\{s \geq s_1\} = \exp(-\mu_t s_1)$$
$$P\{s < s_1\} = 1 - \exp(-\mu_t s_1) \quad (4.4)$$

which can be rearranged to provide a value for the step size s. To sample a step size, a pseudorandom number ξ is introduced. With this pseudorandom number distributed between 0 and 1, it becomes possible to sample the step size with the desired distribution (Wang et al., 1995):

$$s_1 = \frac{-\ln(1-\xi)}{\mu_t} \qquad (4.5)$$

Since ξ is uniformly distributed, the distribution of ξ is identical to the distribution of $(1-\xi)$. Therefore, Equation 4.5 can be rewritten as (Wang et al., 1995)

$$s_1 = \frac{-\ln(\xi)}{\mu_t} \qquad (4.6)$$

4.3.4 Photon Absorption

In an MC simulation, the propagation of individual photons can be traced through the turbid medium. Owing to the scattering events, a photon travels a long path in the tissue. As the absorption probability is constant for a given step size, the number of absorbed photons is large. Therefore, a very large number of photons has to be traced, such that a sufficiently large number of photons would reach the point of detection, allowing accurate estimates of the diffuse reflectance at larger distances from the source. To make this process more efficient, the negative effects of the discrete nature of photon absorption can be avoided by tracing "photon packets" (Flock et al., 1989; Hielscher et al., 1996; Kienle and Patterson, 1996; Watté et al., 2012). Each photon packet is then assigned with a weight w, which decreases each time an absorption event takes place. Instead of sampling from a probability distribution to decide whether a photon is absorbed or not, the weight w of the photon packet is reduced each time the photon packet has propagated a step in the turbid medium. This can be calculated as (Flock et al., 1989; Hielscher et al., 1996; Kienle and Patterson, 1996; Wang et al., 1995; Watté et al., 2012)

$$\Delta w = w \cdot \frac{\mu_a}{\mu_t} \qquad (4.7)$$

Eventually, when a critical threshold has been reached, the photon packet is considered to have become negligible and the simulation for that packet stops. This not only allows to calculate the remaining photon weight at each step, but also to calculate the photon weight at each specific point (x, y, z) on the photon path. In a 3D grid, one can determine the remaining photon weight at the boundaries of each section of the grid. If the difference in photon weight is calculated, one obtains the fraction of the photon weight that has been lost because of absorption in that specific section of the grid. After applying this principle for many photons, a map can be created to visualize the photon absorption in the tissue.

Cylindrical symmetry can be exploited to define an absorption map where the photon absorption is plotted as a function of the radial distance r and the depth z, as illustrated in Figure 4.3. It should be noted that, in Figure 4.3, the absorbed fraction of the photon weights has been represented on a logarithmic scale, as otherwise the absorption close to the point of illumination would dominate the figure. Each

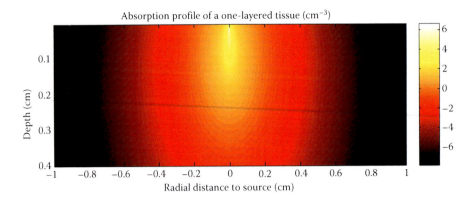

FIGURE 4.3 Visualization of the photon absorption on a log10-scale. The sample tissue consists of a single layer. The absorption profile is stored in the (r, z) cylindrical coordinate system, which has been mirrored to provide this figure. The stochastic noise is clearly visible, especially at a larger distance from the source, because of the low spatial resolution of the absorption grid ($\Delta z = 1$ μm and $\Delta r = 1$ μm).

pixel represents the fraction of energy stored in the pixel (dimensionless), afterwards corrected for the volume of the pixel. This means that in this representation, the absorption profile depends on the resolution of the grid. This grid is not depicted, since the grid lines would perturb the representation of the absorption profile. It is also possible to visualize the fluence rate for each section of the grid, as illustrated in Figure 4.4. This fluence rate is defined as the remaining photon weight that passes through each pixel. This is analogous to the visualization of the absorption in a tissue, but provides an even better understanding of the photon propagation inside a biological tissue.

4.3.5 PHOTON SCATTERING

The phase function describes the chance that a photon is deflected under a certain angle after a scattering interaction. For scatterers of a relatively simple form and known dielectric properties, this function can be calculated from electromagnetic theory by solving Maxwell's equations (Liu and Ramanujam, 2006; van de Hulst et al., 1957). For spherical scatterers with a diameter of about the same size or larger than the photon wavelength, the phase function can be obtained from Mie theory for spherical particles (van de Hulst, 1957) or from T-matrix codes (Mishchenko and Travis, 1994; Mishchenko et al., 2004) for scatterers of an arbitrary shape.

To implement this scattering phase function more easily into MC algorithms, an important distinction is made between the deflection angle θ and the azimuthal angle φ (Wang et al., 1995). In Figure 4.5, a photon originally propagates along the z-axis. At the origin of the axes, a scattering event occurs. The angle θ describes the

Monte Carlo Modeling of Light Transfer in Food

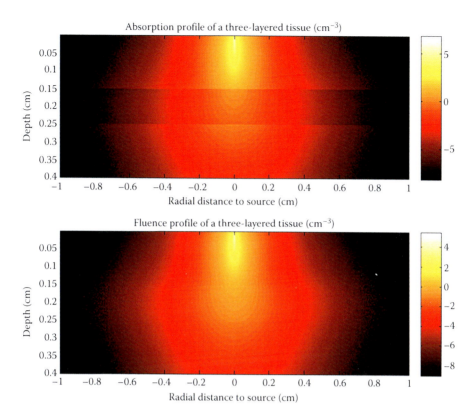

FIGURE 4.4 Visualization of the absorption profile (top) and fluence profile for a three-layered tissue on log10-scale. The sample tissue consists of three different layers, with different optical properties, hence the different fluence and absorption profiles. The absorption profile is stored in the (r, z) cylindrical coordinate system, which has been mirrored to provide this figure. The stochastic noise is clearly visible, especially at a larger distance from the source, because of the small spatial resolution of the absorption grid ($\Delta z = 1$ µm and $\Delta r = 1$ µm).

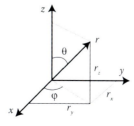

FIGURE 4.5 Scattering of a photon as a function of the deflection angle θ and the azimuthal φ.

deflection with respect to the z-axis. However, an entire cone of scattering events can be defined by this deflection angle. To uniquely describe a scattering event, one has to define the azimuthal angle φ as well. This azimuthal angle is selected from a random distribution, since photons are equally distributed on the scattering cone in the case of spherical scatterers. This would not be the case for more complex scatterers (e.g., cylindrical). As most MC algorithms assume symmetrical scatterers (i.e., the scattering is independent of the angle at which the photon particle hits the scatterer), the deflection angle $\theta \in [0, \pi]$ is typically derived from the scattering phase function, while the azimuthal angle $\varphi \in [0, 2\pi]$ is sampled from a uniform distribution.

Although it was originally developed to describe the scattering of light by interstellar dust clouds, the HG scattering phase function has also been widely used for biological tissues (Bevilacqua and Depeursinge, 1999; Farrell et al., 1992; Jacques, 1989; Liu and Ramanujam, 2006; Patterson et al., 1989; Prahl et al., 1993; Wang and Jacques, 1993; Wang et al., 1995). The HG phase function is described by the following equation (Henyey and Greenstein, 1941):

$$p_{HG}(\Omega) = \frac{1}{4\pi} \frac{1-g^2}{(1+g^2 - 2g\cos(\Omega))^2} \quad (4.8)$$

In this equation, the parameter g represents the anisotropy factor and the parameter Ω is the deflection angle, expressed in steradians. The function is normalized such that the integral over 4π steradians is unity. However, this HG phase function, like all other anisotropic scattering phase functions, is inconvenient to utilize in the RTE. As previously mentioned, the RTE is typically solved by simplifying the scattering problem. Any phase function can be theoretically expanded into a basic set of spherical harmonics (χ_n) and Legendre polynomials $P_n(\cos \theta)$ (Liu and Ramanujam, 2006; van de Hulst, 1957). The main motivation for using these expansions is that they allow the description of the phase function using fewer parameters, such that it can be handled easily in the RTE. An interesting feature of the HG phase function—and the reason for its popularity—is the fact that it can be easily expanded into a series of Legendre terms, which result in a series with an elegant solution when using approximations

$$p_{HG}(\Omega) = \sum_{n=1}^{\infty} (2n+1)g^n P_n(\Omega) \quad (4.9)$$

where $P_n(\Omega)$ is the Legendre polynomial of the nth order. The diffusion theory is a result of simplifying the HG phase function, with its two-term Legendre expansion. A significant advantage of the MC methodology is that it does not require any simplification of the phase function. However, the original Equation 4.8 has to be rewritten to define the phase function as a function of the deflection angle θ, expressed in radians. The probability distribution for the cosine of the deflection angle, $\cos \theta$, is described by

$$p_{HG}(\cos(\theta)) = \frac{1}{2}\frac{1-g^2}{(1+g^2-2g\cos(\theta))^{1.5}} \quad (4.10)$$

where the anisotropy factor g is defined as the average cosine of the deflection angle, characterizing the asymmetry of the scattering function. Theoretically, it can have values ranging from −1 to 1. However, only positive values have been observed in biological tissues. Isotropic scattering is defined by an equal scattering in all directions, which corresponds to an anisotropy value of 0. If g approaches 1/−1, the scattering is more forward/backward oriented. The major advantage of this approach is that it provides an analytical solution that can be expressed as (Wang et al., 1995)

$$\cos(\theta) = \begin{cases} \frac{1}{2g}\left\{1+g^2 - \left[\frac{1-g^2}{1-g+2g\xi}\right]^2\right\} & \text{if } g > 0 \\ 2\xi - 1 & \text{if } g = 0 \end{cases} \quad (4.11)$$

A more detailed description for obtaining this analytical solution can be found in Wang et al. (1995). The solution can be obtained by computing the accumulated distribution and integrating Equation 4.10 from −1 to $\cos(\theta)$

$$P_{HG}(\cos(\theta)) = \frac{1}{2}\int \frac{1-g^2}{(1+g^2-2g\cos(\theta))^{1.5}} d(\cos(\theta)) \quad (4.12)$$

where $P_{HG}(\cos(\pi)) = 0$ and $P_{HG}(\cos(0)) = 1$. This procedure guarantees that the phase function can be linked to a pseudorandom number ξ, ranging from 0 to 1. The parameter ξ is provided by a pseudorandom number generator in the algorithm. The goal is to translate this random number, distributed between 0 and 1, into a deflection angle, following the distribution of Equation 4.10. The analytical solution in Equation 4.11 guarantees that the distribution of the deflection angle θ follows that of the HG phase function, instead of the random distribution of ξ. The simple analytical form of the HG phase function is the main reason why it has been used by many researchers. Wang and Jacques (1993) experimentally determined that the HG function accurately described scattering in the biological tissue. For biological tissues, the HG phase function with g values around 0.8–0.9 has been reported to provide the best approximation of the actual phase functions (Bashkatov et al., 2011; Beauvoit et al., 1995; Kim et al., 2010; Mourant et al., 1995; Tuchin, 2007). However, there is no physical basis for the HG phase function parameterization, except that it provides a good fit to many naturally occurring phase functions. Therefore, it could be interesting to replace it with a phase function derived from the underlying physical phenomena.

Once the deflection angle and the azimuthal angle have been defined, a new direction of the photon packet can be determined. In the first step, the relative directional cosines can be determined (Equation 4.13). These represent the change in direction described as a function of the (x', y', z') coordinate system of Figure 4.2. It is clear from these equations that the direction along the z'-axis strictly depends on the scattering angle θ. The equations can be derived from simple geometry (Wang et al., 1995)

$$\mu'_x = \sin\theta \cdot \cos\varphi$$
$$\mu'_y = \sin\theta \cdot \sin\varphi \quad (4.13)$$
$$\mu'_z = \sin(\mu_z) \cdot \cos\varphi$$

When translated to the general (x, y, z) coordinate system—the general coordinate system necessary for tracking the positions of the photons—the new direction of the photon packet will become (Wang et al., 1995)

$$\mu'_x = \frac{\sin\theta}{\sqrt{1-\mu_z^2}} \cdot (\mu_x\mu_z\cos\varphi - \mu_y\sin\varphi) + \mu_x\cos\theta$$

$$\mu'_y = \frac{\sin\theta}{\sqrt{1-\mu_z^2}} \cdot (\mu_y\mu_z\cos\varphi + \mu_x\sin\varphi) + \mu_y\cos\theta \quad (4.14)$$

$$\mu'_z = -\sin\theta\cos\varphi\sqrt{1-\mu_z^2} + \mu_z\cos\theta$$

Once the new directional cosines have been calculated and the step size s has been established, the new photon position (x^*, y^*, z^*) can be calculated as follows:

$$x^* = x + \mu_x \cdot s$$
$$y^* = y + \mu_y \cdot s \quad (4.15)$$
$$z^* = z + \mu_z \cdot s$$

4.3.6 Reflection and Refraction at Layer Boundaries

When the scattering medium has discrete dimensions, a photon packet traveling through it may hit a boundary of the medium, either between the scattering medium and the ambient medium (e.g., air and water), or between two different scattering media. At this interface, the photon packet can either be reflected back into the original medium or refracted into the next one. In the latter case, the photon packets either continue their propagation into the next layer, or are observed as reflectance or transmittance. In MC simulations, this is handled as follows: if the photon step size is large enough for a boundary interaction to occur, the photon packet is moved to the boundary. Once at the boundary, the probability is computed that the photon packet is internally reflected, depending on the angle of incidence. This angle of incidence α_i evidently depends on the directional cosine.

$$\alpha_i = \cos^{-1}(|\mu_z|) \tag{4.16}$$

The relationship between the angle of incidence α_i and the angle of transmission α_t is determined by Snell's law. This relation is defined by the refractive indices of both media (n_i and n_t):

$$n_i \cdot \sin \alpha_i = n_t \cdot \sin \alpha_t \tag{4.17}$$

The fraction R of the incident power that is reflected from the interface can be calculated with Fresnel's equations. This reflectance depends on the polarization of the incident ray. While some MC algorithms take this polarization into account (Gangnus et al., 2004; Ramella-Roman et al., 2005a, b; Wang and Jacques, 2003), most algorithms just assume an average of s-polarized—the light is polarized with the electric field of the light perpendicular to the plane of incidence—and p-polarized—the incident light is polarized parallel to the plane of incidence—light. Under this assumption, the fraction of light that is internally reflected can be computed as

$$R = \frac{1}{2} \left[\frac{\sin^2(a_i - a_t)}{\sin^2(a_i + a_t)} + \frac{\tan^2(a_i - a_t)}{\tan^2(a_i + a_t)} \right] \tag{4.18}$$

While the fraction R is reflected back into the original layer, the remaining fraction $(1 - R)$ is transmitted to the next layer. In MC simulations, the individual photon packets are not split in a reflected and a refracted fraction; instead, this phenomenon is also handled in a stochastic way. This is done by generating a random number ξ and comparing this to the threshold value R

$$\begin{aligned} &\text{if } \xi \leq R \Rightarrow \text{internal reflection} \\ &\text{if } \xi > R \Rightarrow \text{transmission} \end{aligned} \tag{4.19}$$

The logic behind this is the same as for the other steps in the MC simulations. Namely, the average effect will be correct if a sufficiently large number of photon packets are simulated.

4.4 IMPROVING COMPUTATIONAL SPEED

4.4.1 PHOTON PACKETS

In tracing individual photons, the photon tracking ends when the photon is absorbed. This makes it statistically unlikely for simulated photons to propagate a large distance away from the source, resulting in an inefficient methodology to obtain the light propagation information at further distances. To overcome this problem, most MC algorithms for light propagation trace photon packets instead of single photons

(Farrell et al., 1992; Jacques, 1989; Liu and Ramanujam, 2006; Patterson et al., 1989; Prahl et al., 1993; Wang and Jacques, 1993; Wang et al., 1995). In this way, the simulation time does not suffer from the all-or-nothing nature of single photon absorption events (Wang et al., 1995). Each photon packet is initially assigned a weight, w, equal to unity. With each absorption event, a fraction of the weight will be lost. This process continues until a certain threshold value is reached (e.g., 0.001). Typically, the decision to continue the photon propagation depends on a Russian Roulette mechanism. This technique gives the photon packet a certain chance to survive (e.g., 1 in 15). The surviving photon packet continues its propagation with a new weight. As the photon packets travel a longer distance in the medium, a larger fraction of this weight is dissipated into the medium and recorded into the absorption grid. If the photon packet escapes the tissue, the photon packet still has a remaining weight w, which is recorded in the reflectance or transmittance matrix depending on the side where it exits the medium.

4.4.2 Convolution

To accurately simulate the light propagation for a given light beam, the spatial and angular distribution of this beam should be known. The resulting light transfer can then be simulated by injecting photons (packets) at different positions and orientations. However, since a sufficiently large number of photons should be simulated for every position and orientation, this would lead to very long simulation times. Therefore, an alternative approach has been proposed (Ahnesjö, 1986; Boyer and Mok, 1985; Jones, 1977; Mackie et al., 1985). First, the light propagation is simulated for an infinitesimally small beam where each photon starts at the point $(x, y) = (0, 0)$. The simulation for this infinitely small beam is then considered as an impulse response. In the case of foods and other biological products that are turbid tissues, it is reasonable to assume that the photons propagate independently through the medium. Since the photons undergo many scattering interactions in turbid tissues, the phase relationships between the photons vary in a rapid stochastic manner and the wave characteristics will not have a significant effect on the light propagation. Under this assumption, the following properties hold: (1) if the intensity of the light source is multiplied with a certain factor, the measured quantities will increase with the same factor. (2) The response of two photon beams is the sum of the responses of the individual photon beams. (3) If a photon beam is shifted horizontally over a distance r, the response will shift over the same distance in the same direction. This allows the calculation of the light propagation for any illumination beam by convoluting the intensity distribution of the beam with the impulse response of the turbid medium (Wang et al., 1995).

Therefore, if it is assumed that the photon beam is collimated, the response of an infinitely narrow photon beam will be Green's function (a function approximating the Dirac–Delta impulse response) of the tissue system. The response of the finite-size photon beam can then be computed from the convolution of Green's function according to the profile of the finite-size photon beam. If the source has an intensity profile $S(x, y)$, the response can be obtained through convolution as follows (Wang et al., 1995):

$$C(x; y; z) = \int_{-\infty}^{+\infty} \int_{-\infty}^{+\infty} G(x-x', y-y', z) S(x', y', 0) dx' dy' \quad (4.20)$$

where $G(x - x', y - y', z)$ is Green's function, (x, y) represents the observation point, and (x', y') represents the source point.

4.4.3 SCALING MC

Another mechanism to reduce the computation time for MC simulations, is to employ a scaling method. This scaling method can be used to calculate the diffuse reflectance profile for several sets of optical properties from a single simulation. Battistelli et al. (1985) proposed two scaling formulas, a simpler one that can be applied when the scattering function of the medium has a high forward peak and a second form of the scaling formula when the first case is not valid. A more powerful approach was proposed by Graaff et al. (1993), which took advantage of the mechanism of MC method. Since each step of the photon trajectory is linked to $(\mu_a + \mu_s)^{-1}$, scaling the optical properties of the tissue can result in a scaling of the total pathlength s of each photon. As a result, the radial displacement r of each photon t can be scaled as well. In equation form, this can be written as

$$\begin{aligned} s_i &= \frac{\mu_{t0}}{\mu_{t1}} \cdot s_0 \\ r_i &= \frac{\mu_{t0}}{\mu_{t1}} \cdot r_0 \\ t_i &= \frac{\mu_{t0}}{\mu_{t1}} \cdot t_0 \\ W_i &= \frac{\mu_{s1}}{\mu_{s0}} \frac{\mu_{t0}}{\mu_{t1}} \cdot W_0 \end{aligned} \quad (4.21)$$

where W_0 is the photon weight after the scaling process. As long as the tissue can be considered to be homogeneous, the scaling method is well suited for simulating diffuse reflectance profiles. The scaling MC method records these different trajectory parameters for each individual, simulated photon. After performing the simulation for one set of properties, the stored results can be adjusted for a change in optical properties. Kienle and Patterson (1996) used the same principles to create an MC database for the inverse estimation of optical properties from reflectance spectra. Liu and Ramanujam (2007) used the same principles for the inverse estimation of optical properties from spatially-resolved reflectance spectra and developed a multi-layered scaling method to calculate the total diffuse reflectance for a wide range of optical properties.

4.4.4 Perturbation MC

Similar to the scaling method, the perturbation MC algorithm computes a baseline simulation and then calculates the result for a new turbid medium with similar optical properties. An important difference is that the positions of all the different photons are no longer scaled. Instead, only the weight is changed, based on the perturbation theory (Hayakawa et al., 2001; Kumar and Vasu, 2004; Sassaroli, 2011; Seo et al., 2007)

$$w_{new} = \left(\frac{\mu_{sn}}{\mu_{so}}\right)^{j} \cdot \exp[-(\mu_{tn} - \mu_{to}) \cdot S] \cdot w_{old} \quad (4.22)$$

where S is the photon path length and j is the number of collisions that a photon experienced. This expression poses an important advantage as the reflectance profile can be adjusted more easily. As a result, it can be efficiently extended to a multilayered case. However, contrary to the scaling method of Graaff et al. (1993), the perturbation method is an approximation of which the accuracy depends on the difference between the optical properties of the new medium and those used in the original baseline simulation.

4.4.5 Hybrid MC Methods

Hybrid MC methods combine the accuracy of MC models with the fast computation speed of analytical or approximate solutions such as the diffusion approximation. One approach uses a series of MC simulations for a set of optical properties to correct the results from the diffusion theory (Flock et al., 1988). In another approach, the accuracy of MC simulations at small source–detector distances is combined with the computation speed of the diffusion theory at larger distances. The former provides higher accuracy for short source–detector distances at a reasonable computational cost, while the latter converges with the results of MC simulations for longer source–detector distances, but at a significantly lower computational cost (Wang and Jacques, 1993). The third approach was proposed by Hayashi et al. (2003), for simulating light propagation in a two-layered human skin model. The tissue contained both high-scattering regions—simulated with diffusion approximation—and low-scattering regions—simulated with the MC algorithm. Finally, Tinet et al. (1996) elaborated a fast hybrid algorithm that works in two stages: the first stage consists of an information generator, which establishes the contribution of each scattering event to the total reflectance and transmittance. Specific series of scattering events, leading to a contribution to an optically interesting area (e.g., photons being captured by a detector), are thereby mimicked. With this procedure, the number of photons required to obtain a specified accuracy is significantly reduced. In the second stage, the generated information is used to analytically compute any desired result.

4.4.6 Parallel-Computed MC Algorithms

MC algorithms simulate the propagation of (packets of) photons by approximating these as particles that do not interact with each other. As a result, each photon or packet of photons can be traced independently from the others. This makes MC algorithms well suited for parallelization. Parallel computing has recently received increasing attention thanks to the emergence of powerful graphics cards or graphics-processing units (GPUs) and the platforms that allow coding for GPUs. Martinsen et al. (2009) implemented the MC algorithm in C using the CUDA toolkit to allow simulations on an NVIDIA graphics card. This GPU-based MC algorithm was reported to be 70 times faster than a classic central-processing unit (CPU)-based MC algorithm. Fang and Boas (2009) used the GPU computation power to simulate time-resolved light propagation in a meshed medium. An additional advantage of these MC algorithms for GPU is that they can be combined with different performance-enhancing techniques. Cai and He (2012) developed a fast perturbation MC algorithm on GPU to improve the computational power with a factor of 1000. More recently, Internet-based parallel computing has become another alternative to exploit the parallelization possibilities of the MC algorithm (Pratx and Xing, 2011).

4.4.7 Variance Reduction Techniques

Typically, the variance of the optical outputs in MC simulations is reduced by increasing the number of photons in a simulation. The aim of a variance reduction technique is to reduce this variance without increasing the number of photons by increasing the number of rare events of interest without adding systematic errors. Geometry-splitting MC algorithms reduce the required number of photons by reducing the variance in the optically relevant areas of the simulation. The geometry-splitting technique can increase the fraction of relevant photons (e.g., for computing the reflectance at large source–detector distances, or computing the photons that have passed through a specific layer at a certain depth, or photons captured by a detector). In this way, the total number of photons needed for the simulation can be reduced. A geometry-splitting algorithm will largely reduce the variance in the simulated signals, especially when the goal is to simulate rarely occurring events. The particle-splitting technique is another commonly used variance reduction technique (Lima et al., 2011; Liu and Ramanujam, 2006) where the number of photons increased, contrary to the Russian Roulette mechanism. The medium is divided into several volumes, where the variance can be reduced in certain areas by increasing the chance of sampling in these important volumes. Specific photons can be split into N photons of different energies (e.g., the original photon weight divided by N) and directions. This mechanism is then activated only for photons moving through a region of interest.

4.5 MC SIMULATIONS ON FOOD PRODUCTS

Optical measurement of food quality is challenging due to the presence of multiple layers (e.g., skin around fruit flesh, skin over the subcutaneous tissue) and the effects of multiple scattering in the structured tissues. To gain insight into the light–tissue interaction, light propagation models, such as the MC methodology, are a powerful tool. MC simulations have been widely used for modeling light propagation in biomedical research, but their use in the food and agricultural domain has so far been limited. One of the reasons for the slower adoption of MC simulations in agricultural and food research was the lack of knowledge on the exact optical properties of those products. However, this lack of insight into the optical properties has not prevented wide application of optical measurement techniques in the food and agricultural sector. Optical sensors in the visible (Vis) and NIR range are now widely used in quality control systems, allowing noninvasive and nondestructive analysis. However, only a few studies have aimed at understanding how light propagates in food and agricultural tissues.

Inspired by the progress in the biomedical domain, researchers in the food and agricultural domain have recently turned to MC algorithms for modeling the propagation of light inside an intact tissue. The main driver for the research comes from the fact that absorption and scattering properties of the different tissue layers affect light propagation through the entire tissue, each of them in a distinct manner, while other measurement techniques are irrevocably destructive. For example, *in vitro* experiments have been conducted to determine the light penetration depth by measuring sliced tissue samples with successive thickness (Lammertyn et al., 2000). The usefulness of the results obtained with this approach has, however, been criticized, because of the destructive sampling of the tissue slices and the limited dynamic range of the used detectors (Fraser et al., 2001). MC simulations were thus proposed as a better alternative to determine the light penetration depth in fruit tissue (Fraser et al., 2001). MC simulations can also be used to study the impact of the enveloping tissue on the optical characterization of the deeper layers. This concept was applied to study the effect of the optical properties of mandarin skin on light penetration into the fruit with the aim to select the most effective sample presentation mode (Fraser et al., 2003). Zamora-Rojas et al. (2014) used MC simulations to study how pork skin influences the optical characterization of the more relevant subcutaneous tissues with the aim to design a spatially-resolved reflectance probe for transcutaneous measurement of the fatty acid composition of the subcutaneous fat in Iberian pigs.

Qin and Lu (2009) determined the patterns of diffuse reflectance, internal absorption, and penetration depth for a range of optical properties typical of "Golden Delicious" apples by means of MC simulations. Baranyai and Zude (2009) analyzed kiwifruit with backscattering imaging and modeled it with the MC method. The measured and calculated profiles were compared to estimate the anisotropy factor, to differentiate between premium quality and overripe kiwifruits. MC simulations have also been used in the optimization of a hyperspectral imaging-based spatially-resolved system for determination of the optical properties of biological materials (Cen and Lu, 2010). Finally, MC simulations also allow the selection of specific wavelengths in the Vis/NIR region that penetrate the irrelevant layers more easily, thereby providing more information on the layers of interest (Fraser et al., 2001, 2003; Zamora-Rojas et al., 2014).

Although MC simulations can be very valuable to obtain insight into light propagation in food products, popular codes such as MCML might not be ideal for all purposes. In the following sections, some limitations of these codes will be discussed and potential improvements will be presented.

4.5.1 MORE ACCURATE PARAMETRIC PHASE FUNCTIONS

Food suspensions and emulsions contain a large variety of scattering particles (polydisperse), while biological tissues have a hierarchical structure that consists of cells composed of many cell organelles (e.g., mitochondria, cell walls, nuclei, etc.). Each microscopic particle/organelle has its own scattering phase function, which is determined by the physical properties of the scatterer. The contribution of each type of scatterer to the average scattering properties of a tissue is thus determined by its individual scattering properties and its concentration (Banerjee and Sharma, 2010; Beauvoit et al., 1995; Cheong et al., 1990; Gélébart et al., 1996; Mourant et al., 1995). An average phase function, such as the HG phase function, is typically used as a simpler and computationally faster alternative for simulating the Mie phase function for all different types and sizes of scatterers inside a biological tissue (Banerjee and Sharma, 2010; Cheong et al., 1990; Gélébart et al., 1996; Lima et al., 2011; Liu and Ramanujam, 2006; Pratx and Xing, 2011). This alternative has the advantage of being computationally faster and simpler to interpret, but this simplicity comes at the expense of a reduction in accuracy (Banerjee and Sharma, 2010; Gélébart et al., 1996; Passos et al., 2005; Schmitt and Kumar, 1998; Sharma and Banerjee, 2003, 2005; Sharma et al., 2007; Wang, 2000).

The main advantage of the HG phase function is that it provides an elegant solution when expressed in a set of spherical harmonics and Legendre polynomials. However, after the value of the first moment ($\chi_1 = g$) has been chosen, all other moments ($\chi_i = g_n$) are determined. Bevilacqua and Depeursinge (1999) demonstrated that for reflectance simulations, the number of moments that has to be taken into account is related to the source–detector distance. The closer the source is to the detector, the more anisotropic the radiance will be. So, a larger number of moments should be taken into account. For this purpose, Bevilacqua and Depeursinge (1999) modified the HG phase function into a weighted sum of the HG phase function p_{HG} and a term describing the contribution of the second moment g_2 only, independent from g_1. This contribution is typical for scattering by smaller particles, known as Rayleigh scattering ($\sim[1 + \cos^2 \theta]$) (van de Hulst, 1957)

$$p_{MHG}(\theta) = \alpha \cdot p_{HG}(\theta) + (1-\alpha) \cdot \frac{3}{4\pi} \cos^2 \theta \qquad (4.23)$$

where α is a weighting factor that guarantees the normalization of the phase function. The modified HG phase function can thus be seen as a weighted sum of the effects of strong forward scattering by larger particles (relative to the wavelength of light, e.g., cells) and Rayleigh scattering, originating from smaller structures (e.g., smaller cell organelles, such as mitochondria) (Bevilacqua and Depeursinge, 1999). In the original HG phase function, choosing g automatically determines every moment of the anisotropy factor ($g = g_{HG} = g_1$). In this modified version, the relationship has been changed to (Bevilacqua and Depeursinge, 1999)

$$\begin{aligned} g_1 &= \alpha \cdot g_{MHG} \\ g_2 &= \alpha \cdot g_{MHG}^2 + \frac{2}{5}(1-\alpha) \end{aligned} \qquad (4.24)$$

Other parametric phase functions have also been proposed. For example, the two-term HG phase function combines two independent HG phase functions with anisotropy factors g_1 and g_2 (Kattawar, 1975). This makes it possible to attribute a small or negative anisotropy factor to one of both terms to describe the backward scattering by smaller structures. The modified HG phase function can be seen as a special case of this two-term HG phase function

$$p_{TTHG}(\theta) = \alpha \cdot p_{HG1}(\theta) + (1-\alpha) \cdot p_{HG2}(\theta) \qquad (4.25)$$

Another phase function, typically used for describing the scattering by red blood cells, is the Gegenbauer kernel phase function (Reynolds and McCormick, 1980)

$$p_{GK}(\cos\theta) = \frac{\alpha g}{\pi \dfrac{\mu_s}{\mu_a + \mu_s}} \cdot \frac{(1-g^2)^{2\alpha}}{[(1+g)^{2\alpha} - (1-g^{2\alpha})][1 + g^2 - 2g\cos\theta]^{(1+\alpha)}} \qquad (4.26)$$

where α is a fitting parameter, g is the anisotropy factor, θ represents the deflection angle, and μ_s and μ_a are the scattering and absorption coefficients, respectively. The parameters are varied until the simulated angular distribution of scattered light fits the measured one. Typical α-values are found around 0.49–0.50 and typical g-values are between 0.97 and 0.99 (Reynolds and McCormick, 1980).

4.5.2 Nonparametric Phase Functions

Although many parametric phase functions can be proposed, they will always be an approximation of the real phase function. Therefore, several researchers have proposed to replace the analytical, parametric phase function by a nonparametric one (Banerjee and Sharma, 2010; Gélébart et al., 1996; Passos et al., 2005; Schmitt and Kumar, 1998; Sharma and Banerjee, 2003, 2005; Sharma et al., 2007; Wang, 2000; Watté et al., 2012). This nonparametric phase function is typically a discretized version of a phase function. Similar to MC methods with parametric phase functions, the cumulative distribution of the phase function is calculated. The main difference is that the integration of the phase function is now a numerical process, instead of an analytical operation, since the goal is to obtain a methodology applicable to a large variety of phase functions. A pseudorandom number ξ is linked to the phase function with the following formula:

$$\xi = F(\theta_1) = \frac{\int_0^{\theta_1} p(\theta)d\theta}{\int_0^{\pi} p(\theta)d\theta} \qquad (4.27)$$

Similarly to the analytical process, a random number ranging from 0 to 1 is linked to the distribution. However, instead of searching for an analytical solution, a look-up table is typically used for this process. This nonparametric methodology can be used for (1) a discretized version of a parametric phase function (e.g., the HG phase function), (2) a phase function calculated from the Mie theory, or (3) a phase function that has been measured with a goniometer.

Mie theory allows the reconstruction of the scattering phase function for scatterers of a certain size (monodisperse). However, most particulate systems contain particles of different sizes. The scattering phase function for such polydisperse systems can be reconstructed by simulating the contributions of the different scatterer sizes and summing these. This can be extended to polydisperse systems characterized by a particle size distribution by discretizing this distribution (Aernouts et al., 2014) as illustrated in Figure 4.6. One can clearly see the effect of the number of fractions used on the scattering phase function. It should be noted that for a small number of fractions, the scattering phase function is extremely lobed as is typical for the phase function of a monodisperse system simulated with Mie theory. However, by refining the particle size distribution, the phase function gets smoothed out due to the interactions between the phase functions of different sizes. The fact that the phase function for polydisperse systems is fairly smooth explains why parametric phase functions can provide a reasonable approximation.

The scattering inside food products is caused by air bubbles, suspended particles, emulsified droplets, cell organelles, and cell components. Emulsions, such as homogenized milk, ice cream, sauces, etc., are a collection of scatterers (typically spherical) in a medium such as water or oil. Fruits and vegetables are a collection of cells, cell organelles, air pores, etc. All these can be described as a complex collection of scatterers in a homogeneous medium. Therefore, the previously mentioned approach for simulating the scattering phase function of polydisperse particulate systems could provide a more realistic scattering phase function that takes into account the size distribution of the scatterers. However, this will also only be an approximation as many of these scatterers are not perfectly spherical. The scattering phase function for most types and sizes of scatterers in food products can be calculated with Mie theory (spherical and cylindrical scatterers) (van de Hulst, 1957), T-matrix method (arbitrary geometries) (Mishchenko and Travis, 1994; Mishchenko et al., 2004) or Rayleigh–Gans–Debye approximation (ellipsoids).

Some tissues have a microstructure that consists of ordered elongated subunits, such as the myofibrils in muscles or the collagen fibers in skin, tendons, or ligaments. These tissues show an anisotropic light propagation that cannot be described by the normally used isotropic random media models (see further details in Chapter 11 on light propagation in meat muscle). For example, Kienle et al. (2003) simulated the light propagation through human dentin, where the main scatterers are the tubules, which can be approximated as (infinitely) long cylinders. They calculated the scattering by an infinitely long cylinder, applied the resulting phase function in an MC code, and compared their simulations with goniometric measurements. The scattering pattern was found, both experimentally and theoretically, to have an anisotropic character. While it is possible to simulate the scattering phase function for such complex-scattering microstructures, it might be more efficient to determine the real scattering function of such tissues experimentally. This empirical scattering phase function could then also be converted into a look-up table to be used in the MC simulations (Watté et al., 2012).

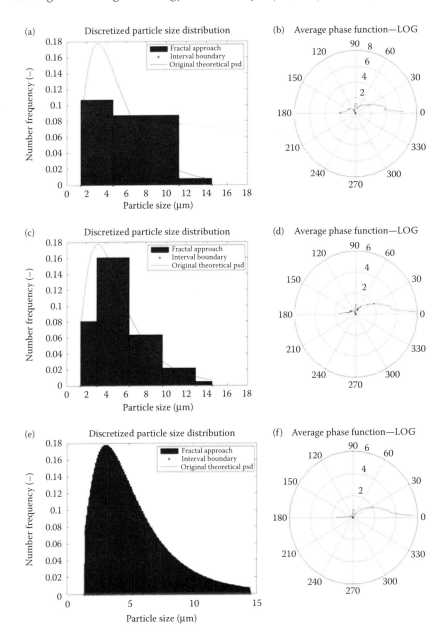

FIGURE 4.6 Particle size distribution of a polydisperse food system and corresponding phase function on logarithmic scale (hence the LOG nomenclature in the figure). The phase function describes the relative fraction of photons that are scattered in a certain deflection angle. The scattering is mainly forward (0°) for the different compartments; (a) and (b) discretized particle size distribution (three intervals) and resulting phase function; (c) and (d) discretized particle size distribution (five intervals) and resulting phase function; (e) and (f) final discretized particle size distribution (127 intervals) and resulting phase function. (Adapted from Aernouts, B., R. Watté, R. Van Beers et al. *Optics Express* 22(17), 2014: 20223–20238.)

4.5.3 From Plane Parallel to Realistic Geometries

Modeling the light propagation in plant products such as fruits and vegetables is a complex matter, since they are anisotropic, heterogeneous, noncontinuous, and complex biological systems. Simplistic MC models, assuming that the foods are multilayered, do not fully grasp the complexity and the noncontinuity of these food tissues.

A typical MC algorithm consists of two distinct parts: physics modeling and geometry tracking. The geometry tracker, also known as the ray tracer, keeps track of the position and properties of the photon packets that are transported through the medium (Badal et al., 2009; Sassaroli, 2011). A general-purpose code, such as the MCML code (Wang et al., 1995), is employed to track photon movement in simple geometries (planes and cylinders). These MC algorithms, however, have the potential for simulating the light propagation through more complex geometries. Researchers have extended these algorithms to voxelized geometries (Badal and Badano, 2009; Badal et al., 2009). Since voxelized images are not well suited for modeling targets with curved boundaries or locally refined structures, MC methods have also been improved to simulate the light propagation in a mesh with complex structures (Badal et al., 2009; Fang and Boas, 2009). Prior to applying the MC methodology, an MC mesh is derived from the voxel-based images, acquired by tomographic techniques (e.g., microscopy and x-ray tomography). These mesh-based MC or PenMesh methods would better simulate the light propagation within the complex structures. Watté et al. (2015) further improved the accuracy of the mesh-based MC methods by incorporating the flexible phase function choice discussed in Section 4.4.2. This meshed MC with the flexible phase function choice (MMC-fpf) allows the incorporation of the contribution of the different types of scatterers in the different tissues, which together form the biological sample under study (e.g., a plant leaf, an apple, a mouse head, etc.). The actual 3D microstructure images, derived from high-resolution x-ray tomography, are used to provide the structural information for the meshes. This strongly reduces the need for further simplifications and assumptions, and thereby improves the validity of the light propagation model and the experimental validation. This MMC-fpf concept is illustrated for a cereal sphere in Figure 4.7. This simple example, in which the air structures are large and nonuniformly distributed over the cereal, is very challenging for the classic MC algorithms that ignore the nonuniform distribution of the air pores. Since the cereal is being illuminated by a single light point source (point of illumination is at the back of the cereal, and not visualized in Figure 4.7), its absorption profile can vary locally because of the presence of the air pores. As a result, certain zones of the cereal are much more difficult for photons to access. This explains the differences in absorption in Figure 4.7.

4.6 CONCLUSIONS

Different methods have been used for modeling light transport in turbid media, all of which are based on the RTE. The MC method is the most flexible and accurate technique, allowing a stochastic computation of the photon distribution in a tissue with specified optical properties. The MC methodology is based on the assumption

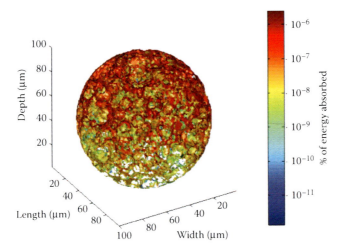

FIGURE 4.7 Percentage of photon energy stored in each voxel in a cereal sphere, expressed on a log10-scale. The sample tissue consists of a uniform dough and air bubbles. The air bubbles are relatively large, and therefore influence the light propagation. Because of their size, they are no longer homogeneously distributed in the cereal, and classic models would fail to take these into account.

that the macroscopic optical properties are uniform over the turbid tissue. This assumption is valid, as long as the tissue is homogeneous on a microscale. Conventionally, food samples or emulsions are described as semi-infinite multilayered tissues whose optical properties include (1) refractive index n, (2) the absorption coefficient μ_a, (3) the scattering coefficient μ_s, and (4) anisotropy factor g. This approach is well suited for modeling the light propagation in turbid foods, in which light undergoes multiple scattering. The MC methodology simulates a large number of photons that move as packets of particles, until this stochastic process converges to a stable solution. The accuracy can be achieved by increasing the number of photons in the simulation and hence a larger computation time, which is the main drawback of the technique.

Several solutions have been proposed to improve the computation speed of MC algorithms (scaling MC, perturbation MC, hybrid MC, variance reduction, and parallel computing). Furthermore, the MC methods have been improved by replacing the simplistic HG phase function, which depends on a single anisotropy factor g, with more complex parametric phase functions (i.e., two-term HG, modified HG, and Gegenbauer kernel). Alternative nonparametric phase functions have also been proposed based on the Mie theory for monodisperse solutions of a spherical scatterer and for polydisperse systems with a discretized particle size distribution, and the T-matrix method for arbitrary geometries. This is an important improvement for modeling the light propagation in emulsions, such as homogenized milk, ice cream, sauces, etc. Mesh-based MC methods, which take into account macrostructural information, have recently emerged and could be very valuable to accurately simulate light propagation in realistic, complex food geometries, such as fruit and vegetables, cereals and meat products, which are anisotropic, heterogeneous, and noncontinuous.

Despite the limitations (e.g., ignoring the wave properties of photons, failure to take the complexity of macro- and microstructures into consideration, etc.), MC models provide the best solution for modeling light propagation in food tissues and are the most powerful tool for addressing the challenges posed by the hierarchical structure of multiple, arbitrarily shaped layers, in which multiple scattering takes place. They provide insight into the light–tissue interaction, and can, therefore, be useful in the estimation of optical properties using advanced sensors or in the optimization of the design of such sensors.

ACKNOWLEDGMENTS

The authors gratefully acknowledge the Institute for the Promotion of Innovation through Science and Technology in Flanders (IWT-Flanders) for financial support through the GlucoSens (SB-090053) and Chameleon (SB-100021) projects and the PhD grant of Rodrigo Watté (SB grant 101552). Ben Aernouts has been funded as PhD fellow of the Research Foundation—Flanders (FWO, grant 11A4813N). The provision of the breakfast cereal mesh by Dr. Els Herremans, Dr. Pieter Verboven, and Professor Bart Nicolaï from KU Leuven Department of Biosystems, MeBioS in Belgium is highly acknowledged.

REFERENCES

Aernouts, B., R. Watté, R. Van Beers et al. Flexible tool for simulating the bulk optical properties of polydisperse spherical particles in an absorbing host: Experimental validation. *Optics Express* 22(17), 2014: 20223–20238.

Ahnesjö, A. Collapsed cone convolution of radiant energy for photon dose calculation in heterogeneous media. *Medical Physics* 16(4), 1986: 577–592.

Badal, A. and A. Badano. Accelerating Monte Carlo simulations of photon transport in a voxelized geometry using a massively parallel graphics processing unit. *Medical Physics* 36(11), 2009: 4878–4880.

Badal, A., I. Kyprianou, D. P. Banh et al. PenMesh—Monte Carlo radiation transport simulation in a triangle mesh geometry. *IEEE Transactions on Medical Imaging* 28(12), 2009: 1894–1901.

Banerjee, S. and S. K. Sharma. Use of Monte Carlo simulations for propagation of light in biomedical tissues. *Applied Optics* 49(22), 2010: 4152–4159.

Baranyai, L. and M. Zude. Analysis of laser light propagation in kiwifruit using backscattering imaging and Monte Carlo simulation. *Computers and Electronics in Agriculture* 69(1), 2009: 33–39.

Bashkatov, A. N., E. A. Genina, and V. V. Tuchin. Optical properties of skin, subcutaneous, and muscle tissues: A review. *Journal of Innovative Optical Health Sciences* 4(1), 2011: 9–38.

Battistelli, E., P. Bruscaglioni, A. Ismaelli et al. Use of two scaling relations in the study of multiple-scattering effects on the transmittance of light beams through a turbid atmosphere. *Journal of the Optical Society of America A* 2(6), 1985: 903–912.

Beauvoit, B., S. M. Evans, T. W. Jenkins et al. Correlation between the light-scattering and the mitochondrial content of normal-tissues and transplantable rodent tumors. *Analytical Biochemistry* 226(1), 1995: 167–174.

Bevilacqua, F. and C. Depeursinge. Monte Carlo study of diffuse reflectance at source–detector separation close to one transport mean free path. *Journal of the Optical Society of America A* 16(12), 1999: 2935–2945.

Boyer, A. L. and E. C. Mok. Calculation of photon dose distributions in an inhomogeneous medium using convolution calculations. *Medical Physics* 13(4), 1985: 503–509.

Cai, F. and S. He. Using graphics processing units to accelerate perturbation Monte Carlo simulation in a turbid medium. *Journal of Biomedical Optics* 17(4), 2012: 040502.

Cen, H. and R. Lu. Optimization of the hyperspectral imaging-based spatially-resolved system for measuring the optical properties of biological materials. *Optics Express* 18(16), 2010: 17412–17432.

Cheong, W.-F., S. A. Prahl, and A. J. Welch. A review of the optical properties of biological tissues. *IEEE Journal of Quantum Electronics* 26(12), 1990: 2166–2185.

Cong, W., H. Shen, A. Cong et al. Modeling photon propagation in biological tissues using a generalized Delta–Eddington phase function. *Physical Review E, Statistical, Nonlinear, and Soft Matter Physics* 76(5), 2007: 051913.

Fang, Q. and D. A. Boas. Monte Carlo simulation of photon migration in 3D turbid media accelerated by graphics processing units. *Optics Express* 17(22), 2009: 20178–20190.

Farrell, T. J., M. S. Patterson, and B. Wilson. A diffusion theory model of spatially resolved, steady-state diffuse reflectance for the noninvasive determination of tissue optical properties *in vivo*. *Medical Physics* 19(4), 1992: 879–888.

Fernandez-Oliveras, A., M. Rubiño, and M. M. Pérez. Scattering and absorption properties of biomaterials for dental restorative applications. *Journal of the European Optical Society: Rapid Publications* 8(1), 2013: 13056.

Fiveland, W. A. and P. J. Jessee. Finite element formulation of the discrete-ordinates method for multidimensional geometries. *Journal of Thermophysics and Heat Transfer* 8(3), 1994: 426–433.

Flock, S. T., B. C. Wilson and M. S. Patterson. Hybrid Monte Carlo-diffusion theory modeling of light distributions in tissue. *Laser Interaction with Tissue*. SPIE. Los Angeles, CA, June 3, 1988.

Flock, S. T., M. S. Patterson, B. Wilson et al. Monte Carlo modelling of light propagation in highly scattering tissues model prediction and comparison with diffusion theory. *IEEE Transactions on Biomedical Engineering* 36(12), 1989: 1162–1167.

Fraser, D. G., R. B. Jordan, R. Künnemeyer et al. Light distribution inside mandarin fruit during internal quality assessment by NIR spectroscopy. *Postharverst Biology and Technology* 27(2), 2003: 185–196.

Fraser, D. G., R. Künnemeyer, R. B. Jordan et al. NIR (near infra-red) light penetration into an apple. *Postharverst Biology and Technology* 22(1), 2001: 191–195.

Gangnus, S. V., S. J. Matcher, and I. V. Meglinski. Monte Carlo modeling of polarized light propagation in biological tissues. *Laser Physics* 14(6), 2004: 886–891.

Gélébart, B., E. Tinet, J. M. Tualle et al. Phase function simulation in tissue phantoms: A fractal approach. *Journal of the European Optical Society Part A* 5(4), 1996: 377–388.

Graaff, R., M. H. Koelink, F. F. M. de Mul et al. Condensed Monte-Carlo simulations for the description of light transport. *Applied Optics* 32(4), 1993: 426–434.

Hayakawa, C. K., J. Spanier, F. Bevilacqua et al. Perturbation Monte Carlo methods to solve inverse photon migration problems in heterogeneous tissues. *Optics Letters* 26(17), 2001: 1335–1337.

Hayashi, T., Y. Kashio, and E. Okada. Hybrid Monte Carlo-diffusion method for light propagation in tissue with a low-scattering region. *Applied Optics* 42(16), 2003: 2888–2896.

Henyey, L. G. and J. L. Greenstein. Diffuse radiation of the galaxy. *Astrophysics Journal* 93(1), 1941: 70–83.

Hielscher, A. H., H. Liu, B. Chance et al. Time-resolved photon emission from layered turbid media. *Applied Optics* 35(4), 1996: 719–728.

Honda, N., K. Ishii, A. Kimura et al. Determination of optical property changes by laser treatments using inverse adding-doubling method. *Optical Interactions with Tissue and Cells XX. SPIE.* San Jose, CA, February 12, 2009.

Ishimaru, A. *Wave Propagation and Scattering in Random Media.* London: Academic Press, 1987.

Jacques, S. L. Time-resolved reflectance spectroscopy in turbid tissues. *IEEE Transactions on Medical Imaging* 36(12), 1989: 1155–1161.

Jones, T. A. A computer method to calculate the convolution of statistical distributions. *Mathematical Geology* 9(6), 1977: 635–647.

Kattawar, G. W. A three-parameter analytic phase function for multiple scattering calculations. *Journal of Quantitative Spectroscopy and Radiative Transfer* 15(9), 1975: 839–849.

Kienle, A. and M. S. Patterson. Determination of the optical properties of turbid media from a single Monte Carlo simulation. *Physics in Medicine and Biology* 41(10), 1996: 2221–2227.

Kienle, A., F. K. Forster, R. Diebolder et al. Light propagation in dentin: Influence of microstructure on anisotropy. *Physics in Medicine and Biology* 48(2), 2003: 7–14.

Kim, A., M. Roy, F. Dadani et al. A fiberoptic reflectance probe with multiple source–collector separations to increase the dynamic range of derived tissue optical absorption and scattering coefficients. *Optics Express* 18(6), 2010: 5580–5594.

Kocifaj, M. Two-stream approximation for rapid modeling the light pollution in local atmosphere. *Astrophysics and Space Science* 341(2), 2012: 301–307.

Kubelka, P. and F. Munk. Ein beitrag zur optik der farbanstriche. [An article on optics of paint layers.] *Zeitschrift fur technische Physik* 12, 1931: 593–601.

Kumar, P. Y. and R. M. Vasu. Reconstruction of optical properties of low-scattering tissue using derivative estimated through perturbation Monte-Carlo method. *Journal of Biomedical Optics* 9(5), 2004: 1002–1012.

Lammertyn, J., A. Peirs, J. De Baerdemaecker et al. Light penetration properties of NIR radiation in fruit with respect to non-destructive quality assessment. *Postharvest Biology and Technology* 18(2), 2000: 121–132.

Lehtikangas, O. and T. Tarvainen. Utilizing Fokker–Planck–Eddington approximation in modeling light transport in tissues-like media. *Diffuse Optical Imaging IV. SPIE.* Munich, Germany, May 12, 2013.

Lima, I. T. Jr., A. Kalra, and S. S. Sherif. Improved importance sampling for Monte Carlo simulation of time-domain optical coherence tomography. *Biomedical Optics Express* 2(5), 2011: 1069–1081.

Liu, Q. and N. Ramanujam. Sequential estimation of optical properties of a two-layered epithelial tissue model from depth-resolved ultraviolet–visible diffuse reflectance spectra. *Applied Optics* 45(19), 2006: 4776–4790.

Liu, Q. and N. Ramanujam. Scaling method for fast Monte Carlo simulation of diffuse reflectance spectra from multilayered turbid media. *Journal of the Optical Society of America A* 24(4), 2007: 1011–1025.

Liu, Q. and F. Weng. Combined Henyey–Greenstein and Rayleigh phase function. *Applied Optics* 45(28), 2006: 7475–7479.

Mackie, T. R., J. W. Scrimger, and J. J. Battista. A convolution method of calculating dose for 15-MV x rays. *Medical Physics* 12(2), 1985: 188–196.

Martinsen, P., J. Blaschke, R. Künnemeyer et al. Accelerating Monte Carlo simulations with an NVIDIA (R) graphics processor. *Computer Physics Communications* 180(10), 2009: 1983–1989.

Mishchenko, M. I. and L. D. Travis. *T*-matrix computations of light scattering by large spheroidal particles. *Optics Communications* 109(1), 1994: 16–21.

Mishchenko, M. I., G. Videen, V. A. Babenko et al. *T*-matrix theory of electromagnetic scattering by particles and its applications: A comprehensive reference database. *Journal of Quantitative Spectroscopy and Radiative Transfer* 88(1), 2004: 357–406.

Modest, M. F. *Radiative Heat Transfer.* San Diego, CA: Academic Press, 2003.

Mourant, J. R., I. J. Bigio, J. Boyer et al. Spectroscopic diagnosis of bladder cancer with elastic light scattering. *Lasers in Surgery and Medicine* 17(4), 1995: 350–357.

Palmer, G. M. and N. Ramanujam. Monte Carlo-based inverse model for calculating tissue optical properties. Part I: Theory and validation on synthetic phantoms. *Applied Optics* 45(5), 2006: 1062–1071.

Passos, D., J. C. Hebden, R. Guerra et al. Tissue phantom for optical diagnostics based on suspension of microspheres with a fractal distribution. *Journal of Biomedical Optics* 10(6), 2005: 064036.

Patterson, M. S., B. Chance, and B. C. Wilson. Time resolved reflectance and transmittance for non-invasive measurement of tissue optical properties. *Applied Optics* 28(12), 1989: 2331–2336.

Patterson, M. S., B. C. Wilson, and D. R. Wyman. The propagation of optical radiation in tissue I. Models of radiation transport and their application. *Lasers in Medical Science* 6(2), 1991: 155–168.

Prahl, S. A., M. J. C. van Gemert, and A. J. Welch. Determining the optical properties of turbid media by using the adding-doubling method. *Applied Optics* 32(4), 1993: 559–568.

Pratx, G. and L. Xing. Monte Carlo simulation of photon migration in a cloud computing environment with MapReduce. *Journal of Biomedical Optics* 16(12), 2011: 125003.

Qin, J. and R. Lu. Monte Carlo simulation for quantification of light transport features in apples. *Computers and Electronics in Agriculture* 68(1), 2009: 44–51.

Ramella-Roman, J. C., S. A. Prahl, and S. L. Jacques. Three Monte Carlo programs of polarized light transport into scattering media: Part I. *Optics Express* 13(12), 2005a: 4420–4438.

Ramella-Roman, J. C., S. A. Prahl, and S. L. Jacques. Three Monte Carlo programs of polarized light transport into scattering media: Part II. *Optics Express* 13(25), 2005b: 10392–10405.

Reynolds, L. O. and N. J. McCormick. Approximate two-parameter phase function for light scattering. *Journal of the Optical Society of America* 70(10), 1980: 1206–1212.

Sassaroli, A. Fast perturbation Monte Carlo method for photon migration in heterogeneous turbid media. *Optics Letters* 36(11), 2011: 2095–2097.

Schmitt, J. M. and G. Kumar. Optical scattering properties of soft tissue: A discrete particle model. *Applied Optics* 37(13), 1998: 2788–2798.

Seo, I., J. S. You, C. K. Hayakawa et al. Perturbation and differential Monte Carlo methods for measurement of optical properties in a layered epithelial tissue model. *Journal of Biomedical Optics* 12(1), 2007: 014030-15.

Sharma, S. K. and S. Banerjee. Role of approximate phase functions in Monte Carlo simulation of light propagation in tissues. *Journal of Optics A: Pure Applied Optics* 5(3), 2003: 294–302.

Sharma, S. K. and S. Banerjee. Volume concentration and size dependence of diffuse reflectance in a fractal soft tissue model. *Medical Physics* 32(6), 2005: 1767–1774.

Sharma, S. K., S. Banerjee, and M. K. Yadav. Light propagation in a fractal tissue model: A critical study of the phase function. *Journal of Optics A: Pure Applied Optics* 8(1), 2007: 1–7.

Skipetrov, S. E. and S. S. Chesnokov. Analysis, by the Monte Carlo method, of the validity of the diffusion approximation in a study of dynamic multiple scattering of light in randomly inhomogeneous media. *Quantum Electronics* 28(8), 1998: 733–737.

Tinet, E., S. Avrillier, and J. M. Tualle. Fast semianalytical Monte Carlo simulation for time-resolved light propagation in turbid media. *Journal of the Optical Society of America A* 13(9), 1996: 1903–1915.

Toublanc, D. Henyey–Greenstein and Mie phase functions in Monte Carlo radiative transfer computations. *Applied Optics* 18(18), 1996: 3270–3274.

Tuchin, V. *Tissue Optics—Light Scattering Methods and Instruments for Medical Diagnosis* (2nd ed.), Bellingham, WA: SPIE Press, 2007.

van de Hulst, H. C. *Light Scattering by Small Particles*. New York, NY: Dover Publications Inc., 1957.

Wang, L. and S. L. Jacques. Hybrid model of Monte Carlo simulation and diffusion theory for light reflectance by turbid media. *Journal of the Optical Society of America A* 10(8), 1993: 1746–1752.

Wang, L., S. L. Jacques, and L. Zheng. MCML—Monte Carlo modeling of photon transport in multi-layered tissues. *Computer Methods and Programs in Biomedicine* 47(2), 1995: 131–146.

Wang, R. K. Modelling optical properties of soft tissue by fractal distribution of scatterers. *Journal of Modern Optics* 47(1), 2000: 103–120.

Wang, X., L. Wang, C.-W. Sun et al. Polarized light propagation through scattering media: Time-resolved Monte Carlo simulations and experiments. *Journal of Biomedical Optics* 8(4), 2003: 608–617.

Watté, R., B. Aernouts, and W. Saeys. A multilayer Monte Carlo method with free phase function choice. *Optical Modelling and Design II. SPIE*. Brussels, Belgium, May 10, 2012.

Watté, R., B. Aernouts, R. Van Beers et al. Modeling the propagation of light in realistic tissue structures with MMC-fpf: A meshed Monte Carlo method with free phase function. *Optics Express* 23(13), 2015: 17467–17486.

Welch, A. J. and M. J. C. van Gemert. *Optical–Thermal Response of Laser Irradiated Tissue*. New York, NY: Plenum Press, 1995.

Yang, L. and S. J. Miklavcic. Revised Kubelka–Munk theory. III. A general theory of light propagation in scattering and absorptive media. *Journal of the Optical Society of America A* 22(9), 2005: 1866–1873.

Zamora-Rojas, E., A. Garrido-Varo, B. Aernouts et al. Understanding near infrared radiation propagation in pig skin reflectance measurements. *Innovative Food Sciences and Emerging Technologies* 22(1), 2014: 137–146.

5 Parameter Estimation Methods for Determining Optical Properties of Foods

Kirk David Dolan and Haiyan Cen

CONTENTS

5.1 Introduction 111
    5.1.1 Forward Problem versus Inverse Problem 112
        5.1.1.1 Forward Problem 112
        5.1.1.2 Inverse Problem 113
    5.1.2 Importance of the Inverse Problem and Parameter Estimation 114
5.2 Theory of Parameter Estimation 115
    5.2.1 Standard Statistical Assumptions 115
    5.2.2 Scaled Sensitivity Coefficients 116
    5.2.3 Ordinary Least Squares 117
    5.2.4 Confidence Intervals of the Parameters 118
    5.2.5 Residual Analysis 118
    5.2.6 Sequential Estimation 119
5.3 Case Study 120
    5.3.1 Choice of the Model 121
    5.3.2 Data Transformation and Weighting Method 122
    5.3.3 Scaled Sensitivity Coefficients 123
    5.3.4 Estimation of the Parameters 125
    5.3.5 Statistical Results for the Optical Parameters
        and the Dependent Variable 126
5.4 Summary 129
References 130

5.1 INTRODUCTION

One of the goals of engineering research is to invent, design, or develop new technology for practical applications (Petroski, 2010). To achieve this goal, there are two distinct activities in engineering research: the first is simulation and the second is data collection. Simulation requires some type of model, typically a mathematical one. Data collection requires instrumentation and the building of prototypes,

112 Light Scattering Technology for Food Property, Quality and Safety Assessment

whether they are software or hardware. In today's competitive world, neither area alone is sufficient. Simulation alone lacks experimental verification that could show deficiencies in the model. Data collection alone cannot determine the results of other experimental conditions that are not tested. Combining the two areas gives the most insight into how to efficiently advance the technology.

Because the training for data collection and simulation is specialized, researchers are typically expert in one area or the other. The "bridge" that brings these two activities together is parameter estimation, which can be used to determine the parameter values in the model, and the error structure. The use of proper parameter estimation techniques allows the researcher to gain the most benefit from both domains simultaneously.

Solving inverse light transport problems, which deal with estimating the optical parameters, including the absorption and reduced scattering coefficients of food and biological materials, is of great interest to researchers because of its practical importance in developing effective optical-sensing techniques for the nondestructive evaluation of food property, quality, and safety. However, there exist challenges in the accurate estimation of optical parameters due to the complexity of mathematical models and potential experimental errors in measuring the dependent variables (i.e., diffuse reflectance) in the model.

Therefore, the purpose of this chapter is to describe, by theory and example, how to estimate optical property parameters in a given mathematical model. The chapter first defines the forward and inverse problem, then presents a brief description of the theory of parameter estimation, and finally gives a step-by-step example for estimating the optical absorption and reduced scattering coefficients for food from the diffuse reflectance profile generated by Monte Carlo (MC) simulation under the steady-state illumination of an infinitesimal light beam on semi-infinite scattering media. These parameters not only provide quantitative information on light propagation in food, but can also be used in predictive models for assessing food property, quality, and safety.

5.1.1 Forward Problem versus Inverse Problem

The words modeling and simulation have become popular in the engineering literature, with many books recently using the words in the title. There have been stunning successes in the past century because of the ability to model and simulate physical events. Within the world of modeling, there are two distinct categories of problems: the forward problem and the inverse problem. Many, if not most, books on modeling and simulation exclusively deal with the forward problem. This chapter deals with parameter estimation, which is an inverse problem.

5.1.1.1 Forward Problem

In this chapter, the *forward problem* is defined as "calculation of the dependent variable, given known parameter values." It has also been described in a generalized way as, "determining effects from known causes." An example is the modified Gompertz model of the dependent variable reflectance, R, as a function of one independent variable, r, the scattering distance:

$$R(r) = \alpha + \beta[1 - \exp(-\exp(\gamma - \delta r))]. \tag{5.1}$$

The modified Gompertz model in Equation 5.1 has been used to describe the spatially-resolved reflectance profiles of food products under the illumination of a steady-state light beam (see Chapter 10 for further details). If the parameters α, β, γ, and δ are known or assumed to be known, then the forward problem is the computation of $R(r)$ for any combination of values for α, β, γ, and δ. The interval spacing for the independent variables can be as small as desired, and an unlimited number of $R(r)$ can be computed. In most cases, the forward problem requires no iteration and has no convergence problems, even if the model is a system of ordinary or partial differential equations (PDEs). In summary, a forward problem requires only a mathematical model, a range for each independent variable, and the values of all parameters. It is important to recognize that *no data are needed* for the forward problem; so, there is no need for experimental work to perform a forward problem. The forward problem can be done entirely on a computer. Of course, experimental work would be needed to confirm the results of the forward problem.

The forward problem is the most common problem presented in engineering and science textbooks. Many commercial software programs are available, using numerical methods such as finite element, to simulate very complex geometries to solve multiphysics problems with systems of PDEs. Regardless of how complex these problems are, they are still forward-problem solvers, and require all parameters to be known beforehand.

5.1.1.2 Inverse Problem

The definition of the inverse problem is "the determination of constants in mathematical formulae, given a set of data (x_i, y_i)," where x_i and y_i are the pairs of observed data, and represent the independent and dependent data, respectively. There may be multiple independent variables. A more generic way of describing the inverse problem is "the determination of unknown causes from known effects." Back to the example of Equation 5.1, the inverse problem would be to determine the parameters α, β, γ, and δ, given a set of reflectance data $R(r)$. If the model is nonlinear with respect to the parameters, an initial guess is required for the parameters, and there must be iteration to find the best-fit parameters based on a criterion, such as minimization of the sum of least squares of residuals, where residual = observed value−predicted value. Predicted values are the $R(r)$ computed by the model (Equation 5.1). For nonlinear models, the possibility of nonconvergence exists, and will be discussed later in this chapter.

A model is linear with respect to all parameters if the first derivative of the model with respect to each parameter does not contain that parameter. Another way of stating this is that each parameter appears in the model raised to the power of one. Otherwise, the model is nonlinear with respect to its parameters. A linear model has an explicit solution to all its parameters, and does not need initial guesses of the parameters. The model in Equation 5.1 is nonlinear, because although α and β appear to the power of one, yet γ and δ are within the exponential function. The first derivative of R with respect to γ and δ contains γ and δ. This model, as well as most light scattering models, is nonlinear, requiring initial guesses of parameters and an iterative solution to the inverse problem.

Because the minimization criterion for an inverse problem involves the residual vector, then *inverse problems, by definition, require data*. The data can be experimental or simulated. This point will be discussed in the next section. In summary, an inverse problem requires a mathematical model, ranges of independent variables, a set of data, and for nonlinear models, initial guesses of the parameters.

5.1.2 Importance of the Inverse Problem and Parameter Estimation

Why is the inverse problem important? Who needs to know the inverse problem? Consider the following two questions:

1. Do I have a mathematical model with parameters in it?
2. Do I have data?

If the answers to both questions are "yes," then it is an inverse problem. Since many, if not most, researchers in engineering experimental laboratories answer both questions "yes," it would seem that the inverse problem is ubiquitous among engineers and among many scientists.

How does a researcher know whether the problem is a forward problem or an inverse problem? If the problem has data with error, it is an inverse problem. If the problem has no data, it must be a forward problem, because there is no residual vector. Another case is when parameters are given to the researcher so that the forward problem can be done. In that case, someone else did the inverse problem to estimate the parameters. A trivial example would be the determination of the gravitational constant, $g = 9.81$ m/s^2, which has already been done by experiment. So, even in these cases, an inverse problem existed before the forward problem was performed.

One more source of confusion is thinking that *optimization* and the *inverse problem* are identical. They are not. The confusion stems from the fact that similar nonlinear regression algorithms may be used for minimization in optimization and parameter estimation. However, the difference is that optimization does not require data and has no errors. The result of each optimization run is considered errorless. In contrast, parameter estimation always considers the errors in the data. Furthermore, the coefficients in optimization may or may not have physical meaning. Conversely, the parameters in parameter estimation usually have physical meaning, and their errors and confidence intervals are indispensable for interpretation of the estimated parameter values.

Despite the universal need for inverse problems for engineers and scientists, there are very few university courses on parameter estimation nationwide. Entire books are written on modeling without any mention of the inverse problem, as if modeling consisted entirely of the forward problem. Yet, experimental data must eventually be used to confirm forward-problem results. In summary, the inverse problem is indispensable for virtually all engineering and scientific research; yet, the teaching of it is neglected and far overshadowed by teaching of the forward problem. Yet, both the forward problem and the inverse problem are important. The purpose of this chapter

Parameter Estimation Methods for Determining Optical Properties of Foods 115

is to give an inverse problem tutorial for researchers and practitioners who are interested in measuring the optical properties of food and agricultural products.

5.2 THEORY OF PARAMETER ESTIMATION

The reasons for knowing the theory of parameter estimation include

1. To know beforehand which parameters can be estimated
2. To know beforehand which parameters will be estimated most easily and most accurately
3. To perform an optimal experimental design such that parameter correlation and errors are minimized
4. To troubleshoot when parameter estimation is not working

Assume that the true model (unknown) is η. Then the observed values Y are described by

$$Y(x,\beta) = \eta(x,\beta) + \varepsilon(x,\beta), \tag{5.2}$$

where x is the independent variable, β is the parameter, and ε is the error vector. The error vector is the difference between the observed values and the true model values:

$$\varepsilon(x,\beta) = Y(x,\beta) - \eta(x,\beta). \tag{5.3}$$

Statistically speaking, we cannot know the true model. Even if the model is based on theory that is well understood, there may be assumptions that are not exact for the experimental setup, such as assuming a two-dimensional model, and there are small errors due to not taking into account the third dimension. Therefore, we define the model we are using as $f(x,b)$, where b is the estimate of the true parameters. The residual vector, e

$$e(x,b) = Y(x,b) - f(x,b), \tag{5.4}$$

is a good estimate of the error vector, ε.

5.2.1 STANDARD STATISTICAL ASSUMPTIONS

The results of the parameter estimation will be most accurate if the following assumptions about the measurement errors are true: (1) additive errors, (2) zero mean, (3) constant variance, (4) errors are not correlated, and (5) the error shows a Gaussian distribution (Beck and Arnold, 2007). When little or no prior information is available about the errors, the ordinary least squares (OLS) is recommended. These assumptions will be tested by residual analysis after parameter estimation.

If the standard statistical assumptions are not met, data transformations can be applied to meet the assumptions. For example, if the absolute value of the variance increases with Y, then assumption 3 is violated. To account for this, the sum of squares of the relative errors = $(Y_{obs}-Y_{pred})/Y_{obs}$ can be used instead (Motulsky and Christopoulos, 2004). The Y_{obs} in the denominator will hopefully "scale" the relative errors to be nearly constant along the x-axis. Another common method to handle nonconstant variance or multiplicative errors (violation of assumption 1) is to take the logarithm of the Y values, which also tends to produce a more constant variance along the x-axis.

5.2.2 Scaled Sensitivity Coefficients

Parameter estimation involves inferring the parameter values from changes in the dependent variable caused by small changes in the parameters. In the extreme worst case, if a small parameter change causes no change in the dependent variable, obviously that parameter cannot be estimated. That parameter has no effect on the model, and can be removed or set to any value. In the extreme best case, a very small change in the parameter causes a very large change in the model. In that case, the parameter can be easily estimated with a small error. The exception to this last statement is if that parameter is completely correlated ($\rho = 1.0$) with another parameter. In that case, the model change caused by each of the two correlated parameters is exactly the same; so, the separate effect of the two parameters cannot be distinguished. A combination of the two parameters, such as a ratio or product, may still be able to be estimated, but the parameters cannot be simultaneously estimated separately.

The previous paragraph can also be mathematically expressed as follows. The sensitivity coefficient X_i for β_i is (Beck and Arnold, 1977)

$$X_i = \frac{\partial \eta}{\partial \beta_i}. \tag{5.5}$$

If X_i is independent of all the parameters β, then the model is linear with respect to the parameters. Otherwise, the model is nonlinear.

The sensitivity matrix, $\mathbf{X}$, is an $n \times p$ matrix of sensitivity coefficients (n is the number of data, and p is the number of parameters), where each column corresponds to one of the parameters in the model (Beck and Arnold, 2007):

$$\mathbf{X} = \begin{pmatrix} \left(\frac{\partial \eta_1}{\partial \beta_1}\right) & \cdots & \left(\frac{\partial \eta_1}{\partial \beta_p}\right) \\ \vdots & \ddots & \vdots \\ \left(\frac{\partial \eta_n}{\partial \beta_1}\right) & \cdots & \left(\frac{\partial \eta_n}{\partial \beta_p}\right) \end{pmatrix}. \tag{5.6}$$

Correlation between any two parameters can also be identified by comparing the shapes of the plotted sensitivity coefficients. Any two parameters with similar shapes are correlated to each other, and will be more difficult to estimate. If X_i can

Parameter Estimation Methods for Determining Optical Properties of Foods

be summed together with different constant coefficients to equal zero, then not all the parameters can be estimated separately

$$D_1 \frac{\partial \eta}{\partial \beta_1} + D_2 \frac{\partial \eta}{\partial \beta_2} + \ldots D_p \frac{\partial \eta}{\partial \beta_p} = 0 \tag{5.7}$$

where D_i are constants. This condition should be checked to make sure that there is no linear dependence of the parameters.

To determine whether the sensitivity coefficient is "large," X must be scaled to have the same units as the dependent variable. Multiply Equation 5.5 by β_i to obtain the scaled sensitivity coefficient S_i

$$S_i = \beta_i \frac{\partial \eta}{\partial \beta_i}. \tag{5.8}$$

Each S_i is drawn on the same plot, together with the predicted value η. Negative values are treated as positive values when determining "size." The parameter with the largest S and the least correlation will be the easiest parameter to estimate, with the lowest error.

In most cases, it is wise to plot S using a numerical-difference approximation, such as forward difference, unless the derivative is simple to derive analytically. Using a finite difference avoids errors in attempting to do analytical derivatives.

It is the authors' experience that parameters with maximum S values less than ~5% of the total η span will be difficult to estimate, and should be either removed from the model or fixed at some value and not estimated.

5.2.3 Ordinary Least Squares

The most common criterion to estimate parameters is to minimize the sum of squares of the errors. There are various nonlinear regression methods that will perform OLS for the standard situation where all assumptions are true. Although it is not necessary to know all the inner workings of the routine, knowing some key indicators will go a long way in diagnosing error messages when OLS does not work.

One common error that occurs during parameter estimation is that the sensitivity matrix (Equation 5.6) is reported to be "ill-conditioned" and the "parameter estimates may be unreliable." In this case, first examine all the scaled sensitivity coefficients X with reasonable estimates of the parameters. If any X_i is too small, that is, less than 5%, consider setting that parameter to a constant value and removing it from the estimates. Doing this will decrease the number of columns in the Jacobian by one for each parameter set constant. Next, examine whether any two of the sensitivity coefficients are highly correlated, by seeing if the shapes of the two X are similar. If the ratio of two X is nearly a constant, you may have to fix one of the corresponding parameters to a constant.

Next, check that the condition number is less than approximately 1 million (Bonate, 2011). If the condition number is much higher than 1 million, the estimation is not reliable (Beck and Arnold, 1977). One or more of the parameters may be held constant or removed to mitigate this problem. Also, if the parameters are of different orders of magnitude, consider normalizing them so that all of them are on a scale of 0–1 when they go into the Jacobian. The normalization can be done inside the equation for the dependent variable by dividing by the estimated value of the parameter.

As long as all the remaining parameters have X that are large and uncorrelated, then the OLS should run without trouble. If trouble still remains, it is a good practice to choose only one parameter, and make sure the code works for that parameter. Then successively add parameters based on the size of X, testing the code for each additional parameter. When the troublesome parameter is found, it should be set to a constant value or removed from the model.

5.2.4 CONFIDENCE INTERVALS OF THE PARAMETERS

Knowing the value of a parameter without its confidence interval or error estimate makes the parameter estimate uninterpretable (van Boekel, 1996). All estimates should be reported with their confidence intervals. The most common method is to report asymptotic confidence intervals for nonlinear models (Dolan, 2003)

$$b_i \pm \sigma_i t_{(1-0.5\alpha),v}, \tag{5.9}$$

where b_i is the parameter estimate for the ith parameter; t is the value of the t-statistic at confidence level $1-0.5\alpha$, where α is typically 0.05; and σ_i is the standard error of the ith parameter, and is computed from the diagonal of the symmetric parameter variance–covariance matrix, which looks like this for two parameters:

$$\begin{pmatrix} \sigma_1^2 & \sigma_{12} \\ \sigma_{21} & \sigma_2^2 \end{pmatrix}. \tag{5.10}$$

Although the asymptotic intervals are approximate, they are the most computationally efficient, and give an easily interpretable symmetric interval. A more accurate method is to use bootstrapping and MC methods (Mishra et al., 2011). Briefly, the bootstrap is a random resampling with replacement of either data or the residuals. Each resampling of n data or residuals, results in a new synthetic data set. Nonlinear regression is applied to the synthetic set, to obtain a new set of parameters. Doing this 500–1000 times gives a distribution of parameter values. By sorting 1000 parameter values from the lowest to the highest, and identifying the bottom 2.5% value and top 2.5% value, one can obtain the asymmetric bootstrap confidence interval at those two levels.

5.2.5 RESIDUAL ANALYSIS

After conduction of the parameter estimation, and viewing of the predicted Y and the observed Y, the residuals should be plotted both in a scatter plot and in a histogram. Each of the standard statistical assumption should be evaluated based on these residual plots and results. Additive errors may be confirmed by viewing the residuals versus the dependent variable, to make sure that the errors do not increase with increasing Y. Zero mean can be confirmed by computing the mean to see how close it is to zero. Constant variance can be confirmed by viewing residuals versus the independent variable, to make sure that there is nearly a constant bandwidth. Correlation can be seen by a pattern of residuals rather than a random scatter. Define runs = number of times residuals cross the zero line. Then a useful criterion is that the residuals are not correlated when the number of runs $\geq (n + 1)/2$. Gaussian distribution can be seen by the shape of the histogram, or examined quantitatively through other statistical tests.

5.2.6 SEQUENTIAL ESTIMATION

Another method that has certain advantages over OLS is sequential estimation (Beck and Arnold, 2007). The advantages of this method include being able to estimate all the parameters as each data point is added, thereby being able to track the development of the parameters during the experiment. To gain these advantages, the following prior information must be supplied: (1) an estimate of the covariance matrix, P, of the parameters; (2) an estimate of the covariance matrix of the dependent variable errors, ψ; and (3) an estimate of the parameter values. The method is sufficiently forgiving that only reasonable estimates are needed, usually based on the researcher's experience or previous experimental results. Once the values are approximated, maximum *a posteriori* estimation (MAP) is used.

For linear problems, the sequential procedure is developed using a matrix inversion lemma. The index m is the number of observations taken at any one time, such as multiresponse of the two dependent variables, where m = 2, and p is the number of parameters. The procedure is given as the following, with the dimension of each matrix given on the left side of the equation (Beck and Arnold, 2007) (all terms in Equations 5.11 through 5.16 are matrices):

$$[p \times m] \quad A_{i+1} = P_i X_{i+1}^T, \tag{5.11}$$

$$[m \times m] \quad \Delta_{i+1} = \psi_{i+1} + X_{i+1} A_{i+1}, \tag{5.12}$$

$$[p \times m] \quad K_{i+1} = A_{i+1} \Delta_{i+1}^{-1}, \tag{5.13}$$

$$[m \times 1] \quad e_{i+1} = Y_{i+1} - \hat{Y}_{i+1} = Y_{i+1} - X_{i+1} b_i, \tag{5.14}$$

$$[p \times 1] \quad b_{i+1} = b_i + K_{i+1} e_{i+1}, \tag{5.15}$$

$$[p \times p] \quad P_{i+1} = P_i - K_{i+1} X_{i+1} P_i. \tag{5.16}$$

The b_{i+1} is the updated parameter value at the new data point $i + 1$, where e is the residual vector, X is the sensitivity matrix, and Y and $\hat{Y}$ are the observed and predicted dependent variables, respectively.

For nonlinear problems, an analogous routine is given in Beck and Arnold (1977). Their routine uses the above lemma inside an "iteration" loop that runs the above lemma multiple times until a tolerance is attained for the final parameter values. The parameter estimates are normalized and plotted together with the independent variable. An appropriate model is one where all the parameter estimates nearly reach a constant before the end of the experiment, and preferably after half the experiment has been completed. If any normalized parameter has not reached a constant by the end of the experiment, the model then needs revising for those data.

5.3 CASE STUDY

We now present a case study to show how the parameter estimation procedures presented above can be used to estimate the optical absorption and scattering coefficients of food samples. The spatially-resolved reflectance data for the steady-state illumination of an infinitesimal light beam on a semi-infinite scattering medium were generated from MC simulations.

The propagation of light in a medium can be described by the diffusion approximation equation when scattering is dominant (see Chapter 3). Direct solutions to the diffusion equation coupled with appropriate boundary conditions, referred to as forward problems, provide a quantitative description of light transport in the medium. The corresponding inverse problems, also called inverse radiation transport problems, deal with estimating the optical properties of scattering media. In this section, the simulation study reported by Cen et al. (2010) is presented as an example of the parameter estimation for determining the optical properties of food products. Much of the content presented herein is reproduced, with the permission of the publisher, with a few changes from the original paper of Cen et al. (2010).

Thirty-six different combinations of the absorption coefficient (μ_a) and the reduced scattering coefficient μ_s' and their transport mean free path [$1\,\text{mfp}' = (\mu_a + \mu_s')^{-1}$] were selected (Table 5.1), which span a large range of values: $0.004 \leq \mu_a \leq 0.800$ mm^{-1}, $0.40 \leq \mu_s' \leq 4.00$ mm^{-1} and $5 \leq \mu_s'/\mu_a \leq 100$. These values were chosen based on the published data for the optical properties of fruit and other food products (Budiastra et al., 1998; Cubeddu et al., 2001; Qin and Lu, 2008). The scattering medium was assumed to be semi-infinite, with the refractive index of 1.35, which is similar to that of fruit (Mourant et al., 1997), and the refractive index for the ambient medium was 1.0, equal to that for air. MC simulation (see Chapter 4) was used to validate the diffusion model and inverse algorithm, which offers the most flexible and accurate approach for quantifying the optical features of light transport that are difficult to measure directly. The low-noise diffuse reflectance profiles were first generated by MC simulation with the optical properties given in Table 5.1, under the normal illumination of an infinitesimal light beam onto the semi-infinite scattering medium (see Chapters 3 and 7 for the theory and technique of spatially-resolved measurement). The spatially-resolved reflectance profiles were then fitted by the inverse algorithm for the diffusion model to deduce the optical properties of food products.

TABLE 5.1
Thirty-Six Combinations of the Absorption (μ_a) and Reduced Scattering Coefficients (μ_s') and Their Corresponding Transport Mean Free Path (mfp') Used in MC Simulations (Unit: mm^{-1} for μ_a and μ_s', and mm for mfp')

Group No.	μ_a	μ_s'	mfp'	Group No.	μ_a	μ_s'	mfp'
1	0.080	0.40	2.08	19	0.008	0.40	2.45
2	0.140	0.70	1.19	20	0.014	0.70	1.40
3	0.200	1.00	0.83	21	0.020	1.00	0.98
4	0.400	2.00	0.42	22	0.040	2.00	0.49
a5	0.600	3.00	0.28	23	0.060	3.00	0.33
a6	0.800	4.00	0.21	24	0.080	4.00	0.25
7	0.040	0.40	2.27	25	0.006	0.40	2.46
8	0.070	0.70	1.30	26	0.010	0.70	1.41
9	0.100	1.00	0.91	27	0.014	1.00	0.99
10	0.200	2.00	0.45	28	0.029	2.00	0.49
11	0.300	3.00	0.30	29	0.043	3.00	0.33
a12	0.400	4.00	0.23	30	0.057	4.00	0.25
13	0.020	0.40	2.38	31	0.004	0.40	2.48
14	0.035	0.70	1.36	32	0.007	0.70	1.41
15	0.050	1.00	0.95	33	0.010	1.00	0.99
16	0.100	2.00	0.48	34	0.020	2.00	0.50
17	0.150	3.00	0.32	35	0.030	3.00	0.33
18	0.200	4.00	0.24	36	0.040	4.00	0.25

Source: Adapted from Cen, H., R. Lu and K. Dolan. *Inverse Problems in Science and Engineering* 18(6), 2010: 853–872.

[a] These groups were excluded in the data analysis.

5.3.1 CHOICE OF THE MODEL

Diffusion approximation to the radiation transport equation is the most widely used model to describe light propagation in biological materials. A steady-state diffusion model presented by Haskell et al. (1994) is given below

$$D\nabla^2\Phi(r) - \mu_a\Phi(\mathbf{r}) = -S(\mathbf{r}), \quad (5.17)$$

where $\mathbf{r} = (x,y,z)$, $D = 1/[3(\mu_a + \mu_s')]$ is the diffusion constant, and $S(\mathbf{r})$ is the isotropic scattering source. This PDE can be solved either by numerical or analytical methods. Numerical methods suffer the main drawback of computational complexity with a long computational time, especially for solving inverse problems. Several analytical solutions derived from Equation 5.17 for the homogenous media are available. Farrell et al. (1992) derived an analytical solution from the diffusion equation using the extrapolated boundary condition, and the diffuse reflectance from the medium was calculated as the current across the boundary that originated from a single isotropic point source located at a depth of one transport mean free path in the medium. Later, Kiele and Patterson (1997) proposed an improved analytical solution by calculating the radiance as the sum of an isotropic fluence rate $\Phi(r, z = 0)$ and the flux $J(r)$ as presented in Section 3.6.1 of Chapter 3. Therefore, the reflectance $R(r)$ at the surface of the semi-infinite medium can be considered as a function of the source–detector distance (r) as well as two unknown optical parameters of the medium as shown in Equation 5.18

$$R(r) = C_1\Phi(r,\mu_a,\mu'_s,z=0) + C_2 J(r,\mu_a,\mu'_s), \qquad (5.18)$$

where the fluence rate $\Phi(r,\mu_a,\mu'_s,z=0)$ and the flux $J(r,\mu_a,\mu'_s)$ are two functions related to r, μ_a, and μ'_s, and more details can be found in Section 7.2.1 of Chapter 7. The two unknown optical parameters can then be estimated by solving an inverse light transport problem with the measured diffuse reflectance and source–detector distance.

5.3.2 Data Transformation and Weighting Method

Data transformation is usually applied when the raw data do not meet the statistical assumptions given in Section 5.2.1 or when the data range covers several orders of magnitude (Motulsky and Christopoulos, 2004). In this chapter, the reflectance data (dependent variable Y) decreased dramatically along the source–detector distance (independent variable X), and the statistical assumptions that the Gaussian distribution of errors of the Y-data and constant variance of errors along the X-axis were violated when nonlinear regression was applied. Therefore, we applied the logarithm and integral transformation to the MC simulation data and the diffusion model (Equation 5.19) so that the simulation data would satisfy the statistical assumptions. In the logarithm transformation, the natural logarithm of Equation 5.19 called logarithm-transformed diffusion model (LTDM) is given as

$$R_{log}(r) = \log\left[C_1\Phi(r,z=0) + C_2 J(r,z=0)\right]. \qquad (5.19)$$

The integral of Equation 5.19, defined as the integral-transformed diffusion model (ITDM), is calculated by (Gobin et al., 1999)

$$\begin{aligned}R_{int}(r) &= \int_0^r R(\rho)\rho d\rho \\ &= C_1\left[f(r,z_0) - f(r,z')\right] + C_2[g(r,z_0) + g(r,z')],\end{aligned} \qquad (5.20)$$

where

$$z' = z_0 + 2z_b, \qquad (5.21)$$

$$f(r,z) = \frac{1}{4\pi D\mu_{eff}} \left\{ \exp(-\mu_{eff}z) - \exp\left[-\mu_{eff}(r^2+z^2)^{1/2}\right] \right\}, \quad (5.22)$$

$$g(r,z) = \frac{1}{4\pi} \left\{ \exp(-\mu_{eff}z) - z(r^2+z^2)^{-1/2} \exp\left[-\mu_{eff}(r^2+z^2)^{1/2}\right] \right\}. \quad (5.23)$$

If the scatter of Y-data is Gaussian and the variance of the scatter is the same at all values of X, then the correct parameters can be found by the least-squares estimates without any data transformation. However, in some cases, the variance of the scatter often increases as Y increases. With this type of data, the least-squares method is inappropriate because it tends to give undue weight to points with large Y values on the sum-of-squares value and ignores points with small Y values. To overcome this problem, a proper weighting method can also be applied in the nonlinear regression. One common method is the relative weighting that minimizes the sum of squares of the relative distances of the data from the curve ($\Sigma[(R_{obs} - R_{pred})/R_{obs})]^2$, where R_{obs} is the experimental data and R_{pred} is the predicted reflectance from the diffusion model) (Motulsky and Christopoulos, 2004). The relative-weighting diffusion model (RWDM) was used for the nonlinear least-squares curve fitting to extract the optical properties, and the result was also compared with those obtained from the data transformation methods with the absolute weighting $\left[\Sigma(R_{obs} - R_{pred})^2\right]$.

5.3.3 Scaled Sensitivity Coefficients

In performing sensitivity analysis, sensitivity coefficients for the absorption and reduced scattering coefficients for the original diffusion model (ODM), LTDM, ITDM, and RWDM were calculated as functions of the source–detector distance (ranging between 1 mfp′ and 10 mm) for the corresponding transformed diffuse reflectance profiles. Figure 5.1 shows the sensitivity coefficients as well as the diffuse reflectance for two sets of optical properties with $\mu_a = 0.006$ mm^{-1}, $\mu_s' = 0.40$ mm^{-1} and $\mu_a = 0.057$ mm^{-1}, and $\mu_s' = 4.00$ mm^{-1}. In all four models, the magnitudes of the reduced scattering sensitivity coefficients are generally closer to that of the corresponding values of R. Moreover, the shapes of the two sensitivity coefficients are quite different. These observations show that values for the sensitivity coefficient of μ_s' are "large" (i.e., on the order of R) and uncorrelated to those for the sensitivity coefficient of μ_a (different shapes), which are desirable conditions for estimating μ_s' in all four models with the exception of the sensitivity coefficients of μ_s' at the distance larger than 3 mm in Figure 5.1(a2). Furthermore, the sensitivity coefficients for μ_s' are, in general, closer to the values of reflectance R at small source–detector distances than those at far source–detector distances, while the situation is different for the sensitivity coefficient of μ_a. This is because under the scattering-dominant condition ($\mu_s' \gg \mu_a$), diffuse reflectance close to and far from the incident point of the light source does not have equal sensitivity to the optical properties of the medium. Overall, signals close to the source strongly depend on the reduced scattering property, whereas those far from the source exhibit a large dependence on the absorption

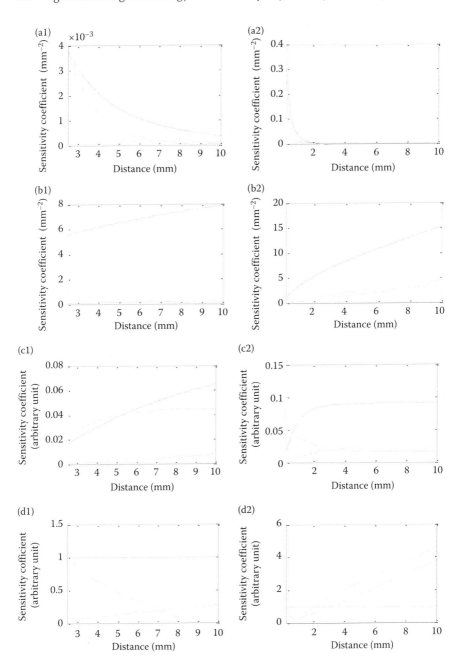

FIGURE 5.1 Sensitivity coefficients of the optical parameters ($\mu_a = 0.006$ mm^{-1} and $\mu_s' = 0.40$ mm^{-1} for the left pane of plots [a1, b1, c1, and d1], and $\mu_a = 0.057$ mm^{-1} and $\mu_s' = 4.00$ mm^{-1} for the right pane of plots [a2, b2, c2, and d2]) as functions of the source–detector distance for ODM (a1, a2), LTDM (b1, b2), ITDM (c1, c2), and RWDM (d1, d2). Solid curves are for reflectance R, dashed curves are for μ_a and dot curves are for μ_s'. (Adapted from Cen, H., R. Lu and K. Dolan. *Inverse Problems in Science and Engineering* 18(6), 2010: 853–872.)

effect. This information is useful for determination of the minimum and maximum source–detector distances for the curve fitting, and also provides a guide for the selection of appropriate data transformation and weighting methods.

Values of the sensitivity coefficient for μ_a are relatively smaller compared with those for μ'_s due to much smaller values of μ_a than those of μ'_s, especially for the ODM (Figure 5.1a). However, with the use of the data transformation or relative-weighting methods, values for the sensitivity coefficient of absorption increase, as shown in Figure 5.1b–d. This means that better estimations of μ_a can be achieved with the data transformation or relative-weighting methods. Since in most cases, values of the reduced scattering sensitivity coefficient are still larger in magnitude than those of the absorption sensitivity coefficient, the reduced scattering coefficient, μ'_s, on average, would be estimated more accurately than μ_a for the diffusion model.

5.3.4 Estimation of the Parameters

Owing to the large difference between reflectance profiles calculated from the diffusion model and MC simulations for groups 1–4 with $\mu'_s/\mu_a = 5$, only the optical properties with $10 \leq \mu'_s/\mu_a \leq 100$ (a total of 29 groups in Table 5.1) were used in the parameter estimation. Figure 5.2 shows the relative errors of estimating the absorption and reduced scattering coefficients of the scattering media by fitting ODM, LTDM, ITDM, and RWDM to the MC simulation data with the true values of optical properties. The average relative errors for the 29 sets of optical properties extracted from ODM, LTDM, ITDM, and RWDM are 16.8%, 10.4%, 10.7%, and 11.4% for μ_a, respectively, and 8.1%, 6.6%, 7.0%, and 7.1% for μ'_s, respectively. Better estimations of μ'_s are obtained, which are in agreement with the sensitivity analysis. The patterns of the relative errors in estimating μ_a and μ'_s obtained from these models are similar for 29 simulation groups. Larger errors of μ_a for ODM appear in the groups with the largest and smallest mfp' for the same ratios of μ'_s/μ_a, with the exception of $\mu'_s/\mu_a = 10$. This is probably because the preselected 1.5 mm minimum source–detector distance is too small for those groups with large mfp' compared with the optimal position of 1–2 mfp' and because the diffusion theory is unable to account for the nondiffusing component of the reflectance that is encountered for these small source–detector separations. Relatively large errors of estimating μ'_s with ODM are also obtained for those groups with a small value of mfp' (Figure 5.2b), due to the removal of the reflectance data close to the light source. Therefore, it is important to select an optimal source–detector distance range in the experimental design and curve-fitting process. In Figure 5.2, it is also observed that the error curves of μ_a and μ'_s are strikingly symmetric, which indicates that they are highly correlated. Hence, when one of the parameters is overestimated, the other is likely to be underestimated in the nonlinear least squares.

The errors of estimating μ_a and μ'_s dramatically decrease when the data transformation and weighting methods are applied to most of the data groups. The logarithm transformation of the raw data gives the best results with the smallest average errors of estimating the two optical parameters. However, the data transformation and weighting methods do not yield smaller errors for group 11 ($\mu_a = 0.300$ mm^{-1},

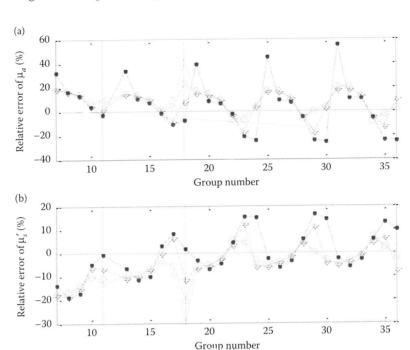

FIGURE 5.2 Relative errors of estimating 29 groups of μ_a (a) and μ'_s (b) shown in Table 5.1 by the original model, and the data transformation and relative-weighting methods: ODM (●), LTDM (o), ITDM (*), and RWDM (Δ). (Adapted from Cen, H., R. Lu and K. Dolan. *Inverse Problems in Science and Engineering* 18(6), 2010: 853–872.)

$\mu'_s = 3.00\,\text{mm}^{-1}$) and group 18 ($\mu_a = 0.200\,\text{mm}^{-1}$, $\mu'_s = 4.00\,\text{mm}^{-1}$), when compared with those obtained using the original model.

5.3.5 Statistical Results for the Optical Parameters and the Dependent Variable

Statistical analysis was conducted to interpret the results of parameter estimations. The relative errors, residual, sum-of-squares, variance–covariance matrix, and confidence intervals were calculated. Nonlinear least-squares estimates are based on a set of statistical assumptions. The results of parameter estimations were further tested by examining the validity of these assumptions. In this chapter, the independent variable (source–detector distance *r*) is known precisely, and all the "error" thus comes from the dependent variable (reflectance *R*). For different data transformation/weighting methods, the runs test (change of sign plus 1) of groups 25 and 30 was conducted based on the plots of the residual versus source–detector distance shown in Figure 5.3. The average numbers of runs of these two groups for the ODM, LTDM, ITDM, and RWDM are 4.5, 42, 3, and 41, respectively. The expected number of runs for the independent disturbances along the source–detector distance is about

Parameter Estimation Methods for Determining Optical Properties of Foods

half of the measurements. There are 86 observations for each group, and the numbers of runs from ODM and ITDM are considerably smaller than 86/2 = 43, while the numbers of runs from LTDM and RWDM are close to the expected numbers of runs. Similar numbers of runs are obtained for other cases. It indicates that the residuals in ODM and ITDM are highly correlated, which violates the assumption of uncorrelated errors. Hence, it is not appropriate to select the original model or integral transformation for the parameters estimation. The frequencies of residuals for groups 25 and 30 from ODM, LTDM, ITDM, and RWDM are shown in Figure 5.4. The distributions of the residuals approximately follow a normal distribution except for the one shown in Figure 5.4(c2). Further examination of the distributions of residuals was conducted for other groups of optical parameters; they follow similar patterns as shown in Figure 5.4: normal distributions are always observed for LTDM and RWDM, while this normal distribution assumption is violated by using ODM and ITDM in several cases. Hence, it can be concluded that the logarithm transformation and relative-weighting methods are more suitable and reliable for accurate estimation of the absorption and reduced scattering coefficients of turbid media.

As examples, the estimated optical parameters for groups 25 and 30 (Table 5.1) extracted using the nonlinear least squares for the LTDM method are shown in Table 5.2.

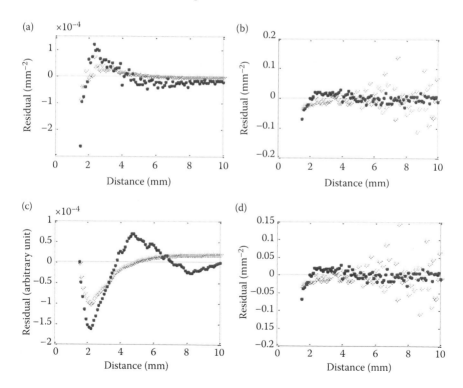

FIGURE 5.3 Residual plots for the reflectance data versus source–detector distances from (a) ODM, (b) LTDM, (c) ITDM, and (d) RWDM with $\mu_a = 0.006$ mm^{-1} and $\mu'_s = 0.40$ mm^{-1} (dot curve) and $\mu_a = 0.057$ mm^{-1} and $\mu'_s = 4.00$ mm^{-1} (asterisk curve). (Adapted from Cen, H., R. Lu and K. Dolan. *Inverse Problems in Science and Engineering* 18(6), 2010: 853–872.)

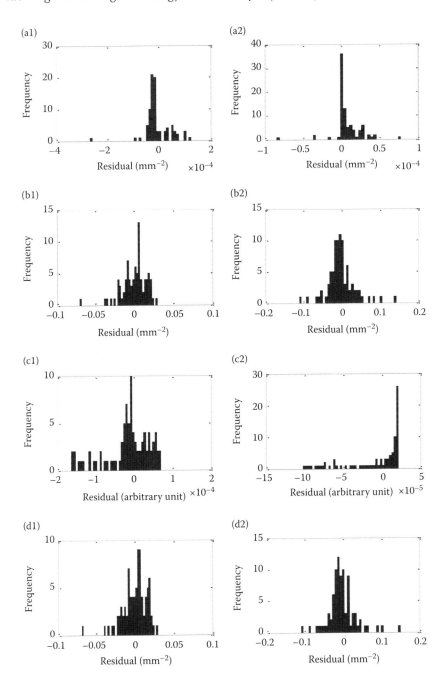

FIGURE 5.4 Residual histograms for the reflectance data from ODM (a), LTDM (b), ITDM (c), and RWDM (d) with (1) $\mu_a = 0.006$ mm^{-1} and $\mu'_s = 0.40$ mm^{-1} for the left pane of plots (a1, b1, c1, and d1), and (2) $\mu_a = 0.057$ mm^{-1} and $\mu'_s = 4.00$ mm^{-1} for the right pane of plots (a2, b2, c2, and d2). (Adapted from Cen, H., R. Lu and K. Dolan. *Inverse Problems in Science and Engineering* 18(6), 2010: 853–872.)

TABLE 5.2
Statistical Results for Estimating the Optical Parameters Using the LTDM Method

Group No.	Parameter	True Value (mm^{-1})	Estimated Value (mm^{-1})	Standard Error (mm^{-1})	Relative Error (%)	95% Asymptotic Confidence Interval
25	μ_a	0.006	0.007	0.0015	16.7	[0.0069, 0.0072]
	μ'_s	0.40	0.38	0.023	−5.0	[0.377, 0.381]
30	μ_a	0.057	0.059	0.0129	3.5	[0.0560, 0.0616]
	μ'_s	4.00	3.88	0.660	−3.0	[3.734, 4.002]

Source: Adapted from Cen, H., R. Lu and K. Dolan. *Inverse Problems in Science and Engineering* 18(6), 2010: 853–872.

Figure 5.5 shows the corresponding three-dimensional (3-D) plots of the sum of squares, which provide information about the characteristics of the standard error, confidence intervals, and relative ease of convergence of the nonlinear regression routine (Mishra et al., 2008). The sum-of-squares surface in Figure 5.5a is much shallower along both μ_a-axis and μ'_s-axis than in Figure 5.5b, causing the difficult convergence of group 25 to find the lowest point on that surface, while it is easy to visualize the optimization of μ_a and μ'_s in Figure 5.5b for group 30. For the two groups, the surfaces of both plots are shallower along the μ_a-axis than along the μ'_s-axis, resulting in the standard errors and confidence intervals for μ_a to be larger than those for μ'_s, which are consistent with the results shown in Table 5.2. Better convergence can be obtained if the curve shows steeper changes along both the μ_a-axis and μ'_s-axis.

5.4 SUMMARY

This chapter provided a solid "bridge" connecting simulation and data collection. The two research areas were brought together by showing how to estimate optical properties in nonlinear models from experimental data. The case study has demonstrated that the parameter estimation technique is important for accurately estimating the optical absorption and reduced scattering coefficients of foods.

The importance of parameter estimation includes that it can generalize data taken from only a few specific conditions, thereby allowing others to predict what will occur at conditions that were not used. The theory of parameter estimation includes plotting scaled sensitivity coefficients to determine which parameters can be estimated, which parameters will be most accurate, and where during the experiment, the optimal region is for estimating each parameter. This plotting does not require data, but does require reasonable estimates of the parameters.

Only after the above preparation of these forward problems, can parameter estimation be conducted efficiently and reliably. A residual analysis can reveal how well the model performed, as well as show the relative errors of the parameters and the correlation matrix. Sequential estimation is an option to give further insight into how

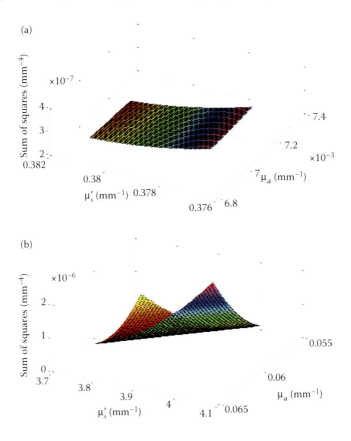

FIGURE 5.5 3-D plot of sum of squares for (a) group 25 and (b) group 30 (see Table 5.1) using the LTDM method. (Adapted from Cen, H., R. Lu and K. Dolan. *Inverse Problems in Science and Engineering* 18(6), 2010: 853–872.)

the parameters change with each additional datum. The case study showed how each of the parts of this chapter was implemented.

The authors hope that other researchers will test the theory and findings of this chapter. It is our desire that the methods described will assist other researchers in the optical modeling area.

REFERENCES

Beck, J. V. and K. J. Arnold. *Parameter Estimation in Engineering and Science*. New York: John Wiley & Sons, 1977.

Beck, J. V. and K. J. Arnold. *Parameter Estimation in Engineering and Science Revised Chapter 6*. Okemos, MI: Beck Engineering Consultants Company, 2007.

Bonate, P. L. *Pharmacokinetic–Pharmacodynamic Modeling*. New York: Springer, 2011.

Budiastra, I. W., Y. Ikeda and T. Nishizu. Optical methods for quality evaluation of fruits (Part 1)—Optical properties of selected fruits using the Kubelka–Munk theory and their relationships with fruit maturity and sugar content. *Journal of JSAM* 60(20), 1998: 117–128.

Cen, H., R. Lu and K. Dolan. Optimization of inverse algorithm for estimating optical properties of biological materials using spatially-resolved diffuse reflectance. *Inverse Problems in Science and Engineering* 18(6), 2010: 853–872.

Cubeddu, R., C. D'Andrea, A. Pifferi, P. Taroni, A. Torricelli, G. Valentini, M. Ruiz-Altisent et al. Time-resolved reflectance spectroscopy applied to the nondestructive monitoring of the internal optical properties in apples. *Applied Spectroscopy* 55(10), 2001: 1368–1374.

Dolan, K. D. Estimation of kinetic parameters for nonisothermal food process. *Journal of Food Science* 68, 2003: 728–741.

Farrell, T. J., M. S. Patterson and B. Wilson. A diffusion-theory model of spatially-resolved, steady-state diffuse reflectance for the noninvasive determination of tissue optical properties *in vivo*. *Medical Physics* 19, 1992: 879–888.

Gobin, L., L. Blanchot and H. Saint-Jalmes. Integrating the digitized backscattered image to measure absorption and reduced-scattering coefficients *in vivo*. *Applied Optics* 38(19), 1999: 4217–4227.

Haskell, R. C., L. O. Svaasand, T. T. Tsay, T. C. Feng and M. S. McAdams. Boundary-conditions for the diffusion equation in radiative-transfer. *Journal of the Optical Society of America A—Optics Image Science and Vision* 11(10), 1994: 2727–2741.

Kiele, A. and M. S. Patterson. Improved solutions of the steady-state and the time-resolved diffusion equations for reflectance from a semi-infinite turbid medium. *Journal of the Optical Society of America A* 14, 1997: 246–254.

Mishra, D. K., K. D. Dolan and L. Yang. Confidence intervals for modeling anthocyanin retention in grape pomace during nonisothermal heating. *Journal of Food Science* 73(1), 2008: E9–E15.

Mishra, D. K., K. D. Dolan and L. Yang. Bootstrap confidence intervals for the kinetic parameters of degradation of anthocyanins in grape pomace. *Journal of Food Process Engineering* 34, 2011: 1220–1233.

Motulsky, H. and A. Christopoulos. *Fitting Models to Biological Data Using Linear and Nonlinear Regression*. New York: Oxford University Press, 2004.

Mourant, J. R., T. Fuselier, J. Boyer, T. M. Johnson and I. J. Bigio. Predictions and measurements of scattering and absorption over broad wavelength ranges in tissue phantoms. *Applied Optics* 36(4), 1997: 949–957.

Petroski, H. Engineering is not science. http://spectrum.ieee.org/at-work/tech-careers/engineering-is-not-science. 2010. Accessed September 6, 2015.

Qin, J. and R. Lu. Measurement of the optical properties of fruits and vegetables using spatially resolved hyperspectral diffuse reflectance imaging technique. *Postharvest Biology and Technology* 49(3), 2008: 355–365.

van Boekel, M. A. J. S. Statistical aspects of kinetic modeling for food science problems. *Journal of Food Science* 61, 1996: 477–486.

6 Basic Techniques for Measuring Optical Absorption and Scattering Properties of Food

Changying Li and Weilin Wang

CONTENTS

6.1   Introduction 133
6.2   Beer's Law for Absorption Measurement... 134
6.3   Integrating Sphere Technique.. 135
    6.3.1   Theory.. 136
    6.3.2   Optical Measurements Using Integrating Spheres 138
6.4   Direct Measurement of Optically Thin (Single Scattering) Tissue 141
6.5   K–M Method... 144
    6.5.1   K–M Model.. 144
    6.5.2   K–M Measurement Techniques 146
    6.5.3   K–M Applications in Food and Agriculture............................. 147
6.6   IAD Method.. 149
    6.6.1   Theory of IAD 149
    6.6.2   Example Applications of Applying IAD Technique
             to Measure the Optical Properties of Food 152
6.7   Conclusions.. 156
References.. 156

6.1   INTRODUCTION

Absorption and scattering are two fundamental interactions of light with biological tissues. The basic optical properties related to absorption and scattering are the absorption coefficient (μ_a), scattering coefficient (μ_s), total attenuation coefficient ($\mu_t = \mu_a + \mu_s$), scattering phase function ($p(\cos\theta)$), anisotropy factor (g), and reduced scattering coefficient ($\mu'_s = \mu_s(1-g)$) (Welch and Van Gemert 1995). In addition, the refractive index, n, is also a basic optical parameter that describes the change of direction when light passes from one medium to another. In this chapter, we will introduce various basic techniques that have been developed to measure optical scattering and absorption properties of biological tissues, as well as the underlying theory of these techniques.

The photometric techniques that have been used to measure optical properties of biological tissues can be classified as *in vivo* and *ex vivo* methods. *In vivo* methods do not require sectioning of the tissue and are usually used as important diagnostic procedures in biomedical research. *Ex vivo* methods study sectioned tissues in glass slides and have the advantage of studying the effect of individual tissue structures. In this chapter, we will focus on *ex vivo* methods.

Ex vivo methods can be further separated into direct and indirect methods. In the direct method, optical properties can be measured experimentally without using a light propagation model. For instance, Beer's law can be used to directly measure optically thin (single scattering) samples. Beer's law and the direct method are introduced in Sections 6.2 and 6.4, respectively.

The integrating sphere technique is important for all the *ex vivo* methods to accurately measure reflection and transmission and therefore is introduced in Section 6.3.

Indirect methods use light propagation models to measure optical properties either in a noniterative or iterative manner. For instance, the noniterative method calculates the optical properties explicitly given by equations with measured optical responses. One example of such a method is the Kubelka–Munk (K–M) method that is introduced in Section 6.5. In indirect iterative methods, such as the inverse adding-doubling (IAD) method, the optical properties are obtained by repeatedly solving the radiative transport equation (RTE) until the solution matches the measured reflectance and transmittance values. The IAD method is introduced in Section 6.6.

6.2 BEER'S LAW FOR ABSORPTION MEASUREMENT

The Beer–Lambert law (also known as Beer's law) states that the unscattered light is attenuated exponentially when it travels through a material. As it relates the attenuation of light to the properties of the material through which the light travels, it can be used to estimate the concentration of a liquid chemical sample (absorbing only) and has been the foundation of spectrophotometric analysis in chemistry. This law is also used to understand and measure the absorption coefficient of biological tissues. To make Beer's law applicable to analyzing transmittance measurements (i.e., estimation of the absorption coefficient by measuring transmittance), it is typically required that there is no scattered light, or if the sample is optically thin (the thickness of the sample is less than the scattering mean free path, i.e., $d \ll 1/\mu_s$), to avoid multiple scattering (Wang and Wu 2012).

Assuming that light passes through a slab of a biological sample with a thickness of d and there are no reflections at the surface, Beer's law can be expressed as

$$I = I_o e^{-\mu_t d} \tag{6.1}$$

where I_o is the incident light intensity, I is the transmitted light intensity, d is the thickness of the sample, and μ_t is the total attenuation coefficient that is a combination of the absorption coefficient and scattering coefficient ($\mu_t = \mu_a + \mu_s$) (Figure 6.1).

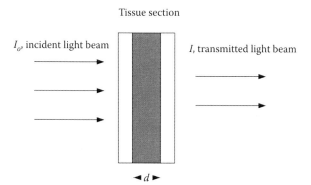

FIGURE 6.1 Illustration of Beer–Lambert law.

Given no scattered light, $\mu_t = \mu_a$. The negative sign indicates the reduction of the transmitted light intensity.

I/I_o is defined as the collimated transmittance T_c (also known as unscattered transmittance $T_c = I/I_o$), which represents the percentage of light that survives after the beam is transmitted through the sample without scattering. On the basis of Equation 6.1, the total attenuation coefficient μ_t (or the absorption coefficient μ_a, given no scattering) can be obtained from a tissue sample using the following equation:

$$\mu_t = -\frac{1}{d} \ln T_c \qquad (6.2)$$

This is basically Beer's law that can be used to calculate the absorption coefficient of the biological tissue.

When the transmission is measured at the interfaces where a mismatched refractive index exists, T_c needs to be corrected using the following equations:

$$r = \frac{R_g + R_t - 2R_g R_t}{1 - R_g R_t} \qquad (6.3)$$

where R_g and R_t are the Fresnel reflections at the air–glass and glass–tissue interfaces, respectively. The collimated transmittance T_c can be corrected using the parameter r and the measured transmission T before calculating μ_t:

$$T = \frac{(1-r)^2}{1 - r^2 T_c^2} T_c \qquad (6.4)$$

6.3 INTEGRATING SPHERE TECHNIQUE

The integrating sphere technique is an established technique for measuring optical radiation (Jacquez and Kuppenheim 1955). Integrating spheres, whose inner

136 Light Scattering Technology for Food Property, Quality and Safety Assessment

surfaces are coated with highly reflective diffuse materials, are designed to achieve homogenous distribution of light radiation on the sphere's inner wall. Integrating spheres have been used as an optical calibration and measurement tool for more than a century and the use of integrating spheres is very versatile in measuring the optical properties of tissues (Tuchin 2007). This section mainly focuses on introducing the basic theory and methods of using the integrating sphere for spectral measurements.

6.3.1 Theory

For an ideal integrating sphere, a light beam falling on the inner surface should be evenly scattered in all directions, which is called Lambertian reflection (Figure 6.2a) and the surface is called Lambertian surface. When the scattered light flux arrives at the other part of the inner surface, it is scattered in all directions again. After multiple Lambertian reflections, the light flux will be evenly distributed (spatially integrated) on the homogenous inner surface of the sphere.

Within an integrating sphere, radiation is exchanged between differential elements on diffuse surfaces. The flux leaving from one point (P1) and arriving at another point (P2) can be calculated by the exchange factor

$$dFlux_{p2\_p1} = \frac{\cos\theta_1 \cos\theta_2}{\pi S^2} dP_2 \qquad (6.5)$$

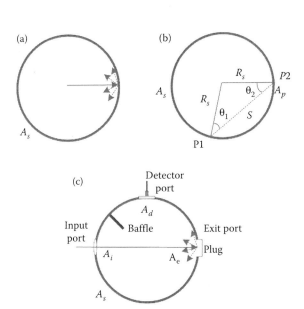

FIGURE 6.2 Schematics of (a) an ideal integrating sphere with the Lambertian surface, (b) the radiation exchange within the integrating sphere, and (c) a typical three-ports integrating sphere.

where S is the linear distance between the points $P1$ and $P2$ (Figure 6.2b). Since $S = 2R_s \cos\theta$ (R_s is the radius of the sphere), then we have

$$dFlux_{p2\_p1} = \frac{1}{4\pi R_s^2} dP_2 \qquad (6.6)$$

In an ideal integrating sphere, the fraction of flux received by area A_2 at a point is an integration of radiations from all directions, which is the fraction of the whole surface area of the sphere (A_s):

$$Flux_{p2} = \frac{A_2}{A_s} \qquad (6.7)$$

A standard integrating sphere often has three ports (Figure 6.2c): input port, exit port, and the port for the detector. Plugs with high reflective materials are also needed to cover the port that should be closed. For real integrating spheres, the surfaces do not have perfect Lambertian reflection. Thus, specular reflection still occurs, especially at the surface where the incident light beam first hits. To prevent measurement errors caused by specular reflection, baffle(s) coated with a high reflective material are often placed at specific location(s) inside the sphere to further diffuse the specular reflection and avoid direct illumination entering into the detector. In certain applications, the fourth port is used so that the specular reflection beam can exit the sphere.

A well-designed sphere should be coated with highly reflective materials for the desired wavelength region. The coating of the inner surface of the sphere has to be very uniform to approach a perfect Lambertian surface. The reflectivity (ρ_r) of the coating material is the most important factor to determine how close the scattering is to being a perfect Lambertian reflection. Taking the reflectivity into consideration, for a continuous-wave light entering into the integrating sphere, its radiance (L) on a diffusive surface of the sphere can be described by

$$L = \frac{\phi \rho_r}{\pi A_s} \qquad (6.8)$$

where ϕ is the input flux and A_s is the sphere area.

The geometry of the sphere is also critical to ensure that every element within a sphere receives the same intensity as the other elements (with the same size) of the sphere. Generally, to keep the spherical geometry, the total area of all ports (i.e., the input port (A_i), the detector port (A_d), and the exit port (A_e)) should not exceed 5% of the whole area of the sphere's internal surface. The fraction of ports (f) to the whole surface area of the sphere is defined as

$$f = \frac{A_i + A_e + A_d}{A_s} \qquad (6.9)$$

For a real integrating sphere with multiple ports, the radiation level on the inner wall, L_r, is dependent on the input flux, the sphere area A_s, the inner wall reflectivity ρ_r, and the port-to-sphere fraction (f)

$$L_r = \frac{\phi}{\pi A_s} \frac{\rho_r}{1 - \rho_r(1-f)} \quad (6.10)$$

in which $\rho_r / 1 - \rho_r(1-f)$ is often defined as the sphere multiplier (M). For each integrating sphere, its sphere multiplier should be a constant determined by ρ_r and f. The sphere multiplier describes the enhancement of the light flux after multiple reflections within the integrating sphere. In other words, assuming the sphere is coated by a nonreflective material and the incident light is evenly spread out to the whole inner surface, the irradiation on that nonreflective sphere's inner wall will be much weaker. The sphere multiplier describes how many times the light radiation can be enhanced in an integrating sphere compared to a nonreflective sphere. For most commonly used integrating spheres, their M values are in the range of 10–30. For instance, an integrating sphere with $\rho_r = 0.98$ and $f = 0.02$, its M equals 24.75. If f increases to 0.05, its M drops to 14.2. Thus, in real integrating sphere applications, it is generally preferred to use small-size ports.

Another important property of the coating material for integrating spheres is how evenly the light will be scattered at all angles. This property can be described by the bidirectional reflectance distribution function (BRDF), which is defined as the ratio of the scattered radiance at a given direction to the ideal scattered radiance in that direction by perfect Lambertian reflection. The unit of radiance coefficient and BRDF is 1/steradian. The detailed definition and measurements of BRDF can be found in the American Society for Testing and Materials (ASTM) Standard E1392-90 "Standard Practice for Angle-Resolved Optical Scatter Measurements on Specular or Diffuse Surfaces."

6.3.2 Optical Measurements Using Integrating Spheres

The most common use of integrating spheres in the food industry is to measure the spectral reflectance and transmittance of food materials. There are many practical advantages of using the integrating sphere technique than directly measuring the sample by a spectrometer. For example, in a regular spectrometer measurement where the incident light directly impinges on the sample surface, the detected reflectance often has a dependency on the angle and distance between the incident beam and the detector. When an integrating sphere is used, the flux reflected on the sample is all captured and normalized by the sphere. Thus, the angular dependency no longer becomes an issue. Also, the detector–object distance is often fixed in the integrating sphere measurement. Even if there is a small change between the sample–sphere distance, it will not affect the results of the measurements as long as all reflected light bounces back into the sphere. Using integrating spheres, the spectral measurements can also have high tolerance on the shape of the light beam and the inhomogeneity of the sample, since both incident light beam and the reflected/

Measuring Optical Absorption and Scattering Properties of Food

scattered light will be normalized on the inner surface of the sphere before being captured by the detector.

The total reflectance (R) of the incident light on a sample can be measured by integrating spheres in different configurations (Figure 6.3). In reality, R is often a relative value to describe the ratio of the reflected flux to the incident flux under a given condition

$$R = \frac{I - I_{dk}}{I_{ref} - I_{dk}} \rho_r \tag{6.11}$$

in which I is the reflected radiance collected by the detector, I_{dk} is the dark current signal of the detector, and I_{ref} is the reference radiance.

I_{ref} can be measured by projecting the incident light beam on the wall of the integrating sphere (Figure 6.3a), or using a white reference target with known reflectivity as the sample (Figure 6.3b). The dark signal I_{dk} can be measured by covering all ports of the integrating sphere and not providing any light source inside the sphere. To measure the total reflected flux I, the key is to adjust the incident angle of the light beam, so that the specular reflectance can fall on the wall of the sphere. In some applications, this is guaranteed by using the second baffle in the middle of the sphere (Figure 6.3c).

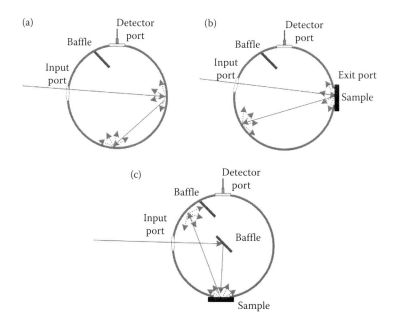

FIGURE 6.3 Typical integrating sphere configurations to measure the total reflectance: (a) the incident beam projecting on the internal wall of the sphere (the reference radiance), (b) the incident beam traveling straight and falling on the sample at the exit port, and (c) the incident beam is redirected to the sample port by the baffle.

If the specular reflectance escapes from the integrating sphere, then the detector only measures the diffuse flux (I_d), which can be calculated as the diffuse reflectance (R_d):

$$R_d = \frac{I_d - I_{dk}}{I_{ref} - I_{dk}} \rho_r \qquad (6.12)$$

The diffuse reflectance can be measured with various configurations using a single sphere in real applications (Figure 6.4). The challenge is to keep the diffused (scattered) light within the sphere as much as possible while allowing the specular reflectance to go out. Thus, a narrow collimated incident beam is often preferred. In some circumstances, an additional port may be needed as the exit port for the specular reflectance (Figure 6.4b).

The total transmittance (T) of the sample can be conveniently measured with the integrating sphere (Figure 6.5) by calculating the ratio of the transmitted flux (I_t) to the incident flux (I_{ref}):

$$T = \frac{I_t - I_{dk}}{I_{ref} - I_{dk}} \qquad (6.13)$$

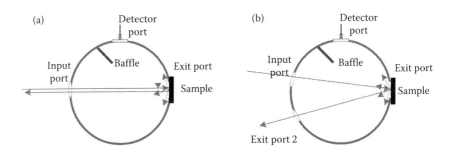

FIGURE 6.4 Integrating sphere configurations to measure the diffuse reflectance: (a) the same port is used for both incident and specularly reflected beams, and (b) two different ports are used for the incident and specularly reflected beams.

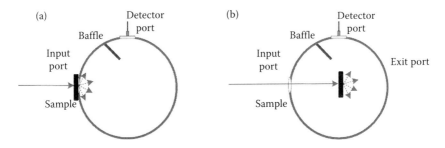

FIGURE 6.5 Typical integrating sphere configurations to measure the total transmittance: (a) the sample is placed at the entrance port and (b) the sample is placed within the sphere.

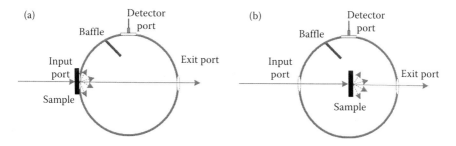

FIGURE 6.6 Examples of integrating sphere configurations for measuring diffuse transmittance.

Similar to reflectance, if the direct transmitted light beam goes out of the sphere (Figure 6.6), the ratio of the diffusely transmitted part of the transmitted flux (I_{df}) to the incident flux can be measured as the diffuse transmittance (T_d):

$$T_d = \frac{I_{df} - I_{dk}}{I_{ref} - I_{dk}} \tag{6.14}$$

6.4 DIRECT MEASUREMENT OF OPTICALLY THIN (SINGLE SCATTERING) TISSUE

Theoretically, optically thin (single scattering) tissues can be directly measured to obtain their optical properties (absorption coefficient, scattering coefficient, total attenuation coefficients, and scattering phase function). On the basis of Beer's law, the total attenuation coefficient (μ_t) can be derived by the following equation:

$$\mu_t = -\frac{1}{d} \ln T_c \tag{6.15}$$

where T_c is the collimated transmittance that can be measured based on the set up shown in Figure 6.7a. The transmitted light from the normally incident pencil beam is detected by the highly collimated detector to avoid scattered photons. d is the thickness of the thin sample. To be eligible for an optically thin (single scattering) tissue, d must be less than the mean-scattering path ($d \ll 1/\mu_s$).

Using an integrating sphere, the absorption coefficient (μ_a) can be directly obtained by measuring the photons transmitted or scattered by the sample, as illustrated in Figure 6.7b (Welch and Van Gemert 1995). In this setup, the sample is placed inside an integrating sphere with the specular reflectance rejected through the input port. The purpose of the baffle is to prevent light directly entering the detector without being rescattered from the sphere. Because the only loss of photons is due to absorption, based on Beer's law, the following equation can be used to derive the absorption coefficient:

$$\mu_a = -\frac{1}{d}\ln\frac{N_a}{\gamma N_0} \tag{6.16}$$

where N_a is the total number of photons detected, γ is the fraction of all photons that the detector detects in an integrating sphere, N_0 is the number of incident photons, and d is the thickness of the sample.

Using an integrating sphere, the scattering coefficient (μ_s) can be measured in a similar manner as the absorption coefficient measurement. As illustrated in Figure 6.7c, the detector measures the total scattered light while letting the unscattered photons (collimated transmittance) exit via a small coaxial port. Assuming that the tissue is a high albedo material (i.e., $\mu_s \gg \mu_a$), the following equation can be used to obtain the scattering coefficient:

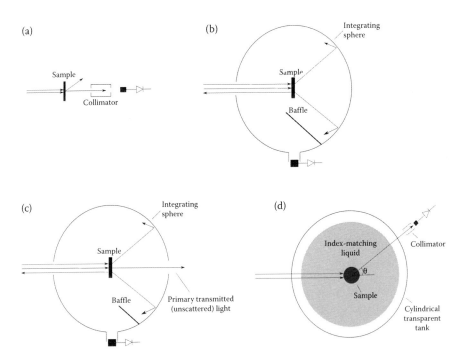

FIGURE 6.7 Illustration of methods used for measuring optical properties directly in single-scattering tissue sections: (a) μ_t, measuring the pencil beam transmission using a collimated detector; (b) μ_a, measuring the transmitted and scattered light using an integrating sphere; (c) μ_s, measuring the total scattered light; and (d) measuring the scattering phase function $p(\cos\theta)$. (With kind permission from Springer Science + Business Media. *Optical–Thermal Response of Laser-Irradiated Tissue*, Vol. 1, 1995, Welch, A. J. and M. J. C. Van Gemert.)

$$\mu_s = -\frac{1}{d}\ln\left(1 - \frac{N_s}{\gamma N_0}\right) \qquad (6.17)$$

where N_s is the number of scattered photons detected, γ is the fraction of all photons that the detector detects in an integrating sphere, N_0 is the number of incident photons, and d is the thickness of the sample.

The scattering phase function $p(\cos\theta)$ defines the amount of light scattered at an angle θ from the incident direction. On the basis of the setup shown in Figure 6.7d, the sample is placed in a cylindrical tank containing liquid that has the same refractive index as the sample to reduce the error from the refraction of the scattered light. A strongly collimated detector is rotated around the sample to measure the scattered light at different angles.

The anisotropy factor g can be derived by calculating the average cosine of the phase function over the entire sphere using the following equation:

$$g = \int_{4\pi} p(s,s')(s \cdot s')dw' \qquad (6.18)$$

where $p(s,s')$ is the phase function, s is the unit vector in the direction of the incident light, s' is the unit vector in the direction from the scattered light, and dw' is the differential solid angle in the direction s'.

Another important optical property is the refractive index (n) that describes the change of direction when light passes from one medium to another. There are different ways to measure the refractive index, but the following three methods are generally used for biological samples. The most common method is refractometry that measures the light refraction angles by using a prism. For instance, in the Abbe refractometer, a thin layer of a sample is placed between two prisms (an illuminating prism and a refracting prism). An illumination source is projected on the illuminating prism that has a ground surface to provide uniformly scattered light. The refracting prism is designed to have a higher refractive index than that of the sample to be measured. A detector is placed behind the refracting prism to show light and dark areas and allow the user to read the refractive index of the sample. Another approach of measuring the refractive index is ellipsometry that uses the polarized incident light beam and detects the reflected beam by spanning the plane of incidence (McCrackin et al. 1963; Pristinski et al. 2006). Optical coherence tomography (OCT) is also used to measure the refractive index by comparing its thickness z and its optical path delay (z') on an OCT image ($n = (z + z')/z$) (Tearney et al. 1995). A detailed description of these methods is beyond the scope of this chapter and interested readers are suggested to refer to these references.

It is important to note that although conceptually these methods of measuring optical properties of optically thin samples seem to be straightforward, realistically it is extremely difficult to measure these optical responses accurately so as to make these methods practical. Major challenges arise primarily due to the handling of the thin tissue and uncertainty in measurements. To meet the requirement of the optically thin (or single scattering) sample, the thickness of the sample is generally in the order of magnitude of 10 µm. Preparing and handling of such thin samples may inevitably change the optical properties of the sample. It could be extremely difficult to support the sample using the glass slide and to place it in the integrating sphere as shown in Figure 6.7, while keeping the sample surface smooth. In addition, the detected signals are expected to be very low due to the low-scattering capability of the sample, and the signal could be easily masked by the fluctuation of the incident light beam or the ambient light. Therefore, it is very important to monitor the stability of the light source. Furthermore, the interior response of the integrating sphere may not be perfectly uniform and thus, the signal could be affected by the spatial distribution of the scattered light within the integrating sphere. A small error in measured signals could lead to a relatively large error in the calculated optical properties. Another challenge is to accurately align the sample, the detector, and the integrating sphere to detect the collimated transmittance or to let the unscattered transmitted photons exit via a coaxial port. Owing to these various uncertainties and challenges, direct measurement of optically thin tissue sections is not commonly used. Practically, the μ_s and μ_a of food materials are often calculated by applying the inverse algorithm derived from more complicated light propagation models (such as Monte Carlo and adding-doubling [AD] model), using the reflectance and transmittance measured by the integrating sphere. The AD method using the integrating sphere technique is described in Section 6.6.

6.5 K–M METHOD

6.5.1 K–M Model

The K–M model describes the propagation of diffuse irradiance through a one-dimensional isotropic slab of a sample (Tuchin 2007). It provides a simple method to estimate optical properties of a sample by measuring diffuse reflectance (R) and transmittance (T). Owing to its simplicity, the model has been used to measure and estimate the optical properties of biological samples that have dominant scattering over absorption (high albedo materials).

The K–M model can separate absorption and scattering for light beam attenuation by using a one-dimensional two-flux theory. The theory assumes that the interaction of the incident beam and the tissue can be modeled by two fluxes: the incident diffuse flux (I_d) traveling forward and the counter-propagating diffuse flux (J_d) traveling backward (Figure 6.8). The incident diffuse flux is reduced due to absorption and scattering but is strengthened by backscattering of the second flux (counter-propagating flux) in the forward propagation direction. In a similar manner, changes in the counter-propagating flux can be explained. In this model, the variable K is used to describe the fraction of each flux lost due to absorption per unit path length, whereas variable S is used to describe the fraction lost due to scattering. This model assumes that K and S are uniform throughout the sample and all light fluxes are diffuse irradiance. In addition, the model neglects the amount of light lost from the edges of the sample during reflectance measurements.

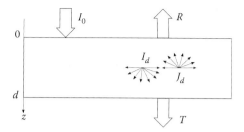

FIGURE 6.8 Two-flux K–M model (I_0: incident light; R: reflectance; T: transmittance; I_d: diffuse light flux in the forward direction; and J_d: diffuse light flux in the backward direction).

Following the K–M model, the K (absorption coefficient) and S (scattering coefficient) parameters of a sample can be determined by the following equations:

$$S = \frac{1}{bd} \ln\left[\frac{1 - R_d(a-b)}{T_d}\right] \tag{6.19}$$

$$K = S(a-1) \tag{6.20}$$

$$a = \frac{1 - T_d^2 + R_d^2}{2R_d} \tag{6.21}$$

$$b = \sqrt{a^2 - 1} \tag{6.22}$$

where the parameters a and b can be determined by measured optical responses from the tissue, such as diffuse transmittance T_d and diffuse reflectance R_d; d is the thickness of the sample; K and S are K–M absorption and scattering coefficients (units of inverse length m⁻¹), respectively; and can be determined by intermediate parameters a and b.

Although the K–M model provides a simple solution to characterize absorption and scattering using the two parameters K and S, these parameters are different from the commonly used and previously discussed transport coefficients (μ_a, μ_s, g). Several researchers (Klier 1972; Van Gemert and Star 1987) have attempted to relate the K–M coefficients to transport coefficients using the following equations:

$$K = 2\mu_a \tag{6.23}$$

$$S = \frac{3}{4}\mu_s(1-g) - \frac{1}{4}\mu_a \tag{6.24}$$

$$\mu_s' = \mu_s(1-g) > \mu_a \tag{6.25}$$

On the basis of the above equations, if K and S are given, the transport coefficients μ_a and μ'_s can be determined. To obtain anisotropy factor g, however, an additional collimated transmittance T_c has to be measured. On the basis of Beer's law, the total attenuation coefficient μ_t ($\mu_t = \mu_a + \mu_s$) can be derived from T_c and therefore μ_s and g can be separately calculated.

6.5.2 K–M Measurement Techniques

From the K–M theory (Equations 6.19 through 6.22), the absorption coefficient K and scattering coefficient S can be determined by three independent variables: the sample thickness d, as well as diffuse reflectance R_d and diffuse transmittance T_d, both of which can be measured using an integrating sphere.

Figure 6.9 illustrates the measurement geometry and the experiment setup that consists of an incident light source, a detector, and an integrating sphere. The sample is placed in position B whereas the standard barium sulfate plate would be placed in position A for transmittance measurements. For reflectance measurements, the sample would be placed in position A while the incident light beam is directly shed on the sample without the sample in position B. It should be noted that the sample is placed in glass containers. To absorb transmitted light through

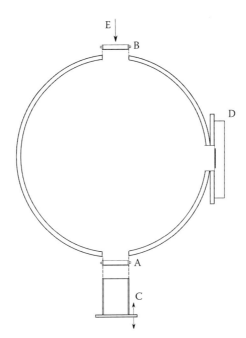

FIGURE 6.9 Reflectance and transmittance measurement using an integrating sphere setup for the K–M method. A: sample cup for reflectance, B: for transmittance, C: a cavity beneath the sample cup to absorb any transmitted radiation during reflectance measurements, D: a detector, and E: illuminating radiation. (Adapted from Birth, G.S., *Int. Agrophys.*, 2, 59–67, 1986.)

the sample during the reflectance measurements, a special tube C with a black interior is designed to create a black cavity (Birth 1986). To improve the accuracy of measurements, it was suggested that the sample surface be tangent to the integrating sphere surface wall. In addition, the sample thickness should be greater than the diffuse thickness δ ($\delta = 1/S$) to ensure that the radiation of the sample is totally diffuse.

Alternatively, if the sample is of infinite thickness, that is, the reflectance does not change by increasing the thickness of the sample, the diffuse transmittance would be zero and Equation 6.23 can be rearranged as

$$\frac{K}{S} = \frac{(1-R_\infty)^2}{2R_\infty} \tag{6.26}$$

where K/S is defined as the K–M function and R_∞ is the reflectance of a sample that is infinitely thick.

If the infinitely thick sample is not available, direct measurement of R_∞ is not possible. Instead, the three reflectance quantities can be measured using the integrating sphere to derive R_∞, based on the following three equations (Judd and Wyszecki 1975):

$$R_\infty = a - b \tag{6.27}$$

$$a = \frac{1}{2}\left(R + \frac{R_0 - R + R_g}{R_0 R_g}\right) \tag{6.28}$$

$$b = \sqrt{a^2 - 1} \tag{6.29}$$

where R is the reflectance of the sample layer backed by a white background, R_0 is the reflectance of the sample layer backed by an ideal black background, and R_g is the reflectance of the white background. a and b are two intermediate variables.

In many cases, the K–M function (the ratio of the absorption and scattering coefficients) is adequate to characterize the color of food samples, although the separate absorption coefficient (K) and scattering coefficient (S) cannot be readily obtained using this technique.

6.5.3 K–M Applications in Food and Agriculture

The K–M model was originally developed in 1931 to predict the optical properties and color of any materials with primary applications in the industry of paints and inks (Judd and Wyszecki 1975). The application of the K–M model in food and agriculture started in the 1970s and has been continually used until today.

Gerald Birth with the U.S. Department of Agriculture Agricultural Research Service was one of the earliest researchers applying the K–M method to measure the optical properties of foods. He not only applied the K–M method to measure optical properties of several foods, but also conducted in-depth studies to improve the method and enhance the accuracy of the measurement technique. For instance, the light scattering and absorption properties of white potato tubers were measured and the effect of geometrical dimension on scattering coefficient measurement was also studied (Birth 1978). The researcher also developed a concept of diffuse thickness (the minimum thickness to ensure total diffuse radiation) of a sample in measuring light scattering. As a case study, the diffuse thickness of the potato tissue was the reciprocal of the K–M scattering coefficient (Birth 1982). Light scattering properties over a broad spectrum of four types of small grains (oats, barley, rye, and wheat) were measured using the K–M method. Both intact and ground grains were measured and the study revealed that the ground grain was better than the intact grain data in meeting the criteria of using the K–M method (Birth 1986).

Another early study of the K–M method in agriculture was to investigate light scattering coefficients of model particulate systems. The K–M multiple-scattering coefficients for glass-bead models were experimentally determined by considering variations from the particulate diameter and the difference between the refractive indices of the host material and the particulate phase (Law and Norris 1973).

In the past three decades, the optical properties measured by the K–M method have been used in various fruits and vegetables for quality control and evaluation. For instance, the K/S ratio (or K–M function) was used to assess translucency in fresh-cut tomato during refrigerated storage. The results indicated that the K/S ratio increased during storage, but remained constant for intact fruits (Lana et al. 2006). In another related study, the K–M function was used to distinguish three distinct maturity groups of tomatoes during maturation (Hetherington and Macdougall 1992). The data showed that the scattering decreased as the tomatoes became mature.

The optical properties of apple and Japanese pear across the spectra of 240–2600 nm were measured using the K–M theory (Budiastra 1998). The study revealed that the presence of the skin led to an increase in the absorption coefficient (K) in the visible region but a decrease in the near-infrared (NIR) region. In comparison, the scattering coefficient (S) increased in the entire spectral region due to the skin. In another study, the optical properties measured by the K–M method as well as the CIE-Lab color coefficients were also used to evaluate the optical color and translucency of sliced kiwifruit with treatments of osmotic dehydration and freezing–thawing (Talens et al. 2002). The K–M method was also used to estimate chlorophyll concentrations in orange fruits due to the effect of ethylene (Knee et al. 1988).

In addition to fruits and vegetables, the K–M method was also applied to dairy products. The K–M function (K/S) was used to evaluate color development during heating in milk proteins. It was found that the K/S ratio was an effective indicator of browning of milk (Pauletti et al. 1999).

It is important to note that although the K–M model offers a simple method to measure optical properties, it has several limitations. For instance, the internal reflection in the sample is neglected and the model does not account for reflections at boundaries where refraction index mismatches could exist. The model also assumes isotropic scattering (therefore phase function is not considered) and diffuse irradiance, which may not be met in real samples. When these assumptions are unmet, the results obtained from the model may not give a good approximation of what occurs for light propagation in biological tissues. Nevertheless, the optical properties obtained from such a simple method are usually used as the first rough estimation in some more accurate methods such as the IAD method that will be introduced in the next section.

6.6 IAD METHOD

6.6.1 Theory of IAD

In the past three decades, many models and algorithms have been proposed to separate and calculate light scattering and absorption coefficients of samples based on reflectance and transmittance (Tuchin 2007). Among various proposed methods, IAD method (Pickering et al. 1993; Prahl et al. 1993) is one of the most common and accurate methods for measuring the optical properties of biological tissues, which has accuracy comparable to Monte Carlo method but requires much less computational time.

As indicated by its name, IAD is the inverse algorithm of the AD method. The AD method for tissue optics was introduced by van de Hulst (1980), which solves the RTE in a slab geometry. In the AD method (Figure 6.10a), the calculation starts from two simulated identical layers, which are thin enough to apply a single-scattering approximation within each of them. The transmitted light from the first

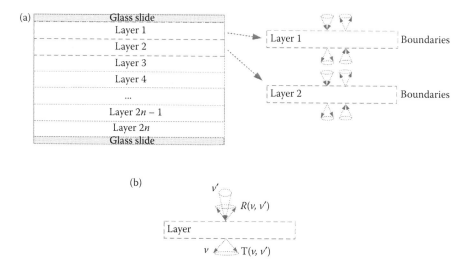

FIGURE 6.10 (a) Schematic of the parallel slab geometry of the AD method and (b) a simplified schematic of reflectance and transmittance on the slab geometry, in which v and v' denote the incident and transmitted angle of the light.

layer serves as the incident light on the second layer, while the light reflected by the second layer is considered as the incident light of the first layer in the opposite direction.

For each layer in the slab geometry, if d is the physical thickness of the material, its optical distance τ can be defined as $\tau = (\mu_s + \mu_a)d$. Let v' and v denote the cosine of the incident and transmitted angle of the light, respectively (Figure 6.10b), the total reflection R_d, the total transmission T_d, and the collimated transmission T_c are (Prahl 1995):

$$R_d = \int_0^1 \int_0^1 R(v',v) 2v' dv' 2v dv \qquad (6.30)$$

$$T_d = \int_0^1 \int_0^1 T(v',v) 2v' dv' 2v dv \qquad (6.31)$$

$$T_c = \int_0^1 T(1,v) 2v dv \qquad (6.32)$$

With the above definitions and assumptions, the reflection and transmission for light incident upon a boundary and moving toward a boundary can be calculated for certain quadrature angles using multiple equations (Prahl et al. 1993; Prahl 1995). Then the two layers are combined and considered as one element (i.e., layers 1 and 2 become the layer L-1; layers 3 and 4 become the layer L-2). After that, two new sequential elements can be doubled to apply the calculations again (L-1 and L-2 are considered as a pair and to be further combined as one layer). The process is repeated until the desired optical thickness is reached (doubling). Then, additional layers are added to simulate the boundary conditions such as glass slides where the optical properties of the layers are different from those of the main tissue samples (adding).

The AD method has been proven to be a powerful tool to calculate multiple scattering within the slab-shaped sample. The detailed equations and algorithm of the AD method has been intensively described in the literature, such as Prahl (1995). For a slab-shaped sample that is too thick to apply single-scattering approximation, if its optical properties (i.e., μ_s, μ_a, and anisotropy coefficient g) are given, the AD method can be used to calculate its light reflection and transmission.

The AD method was extended by Prahl et al. (1993) in an inverse manner to calculate the optical properties of biological samples based on their reflectance and transmittance, which is called the IAD method. Depending on what spectral measurements are known, the IAD method can be applied to calculate various optical properties of the sample. For instance, if the total reflectance and the total transmittance spectra are measured and anisotropy factor g is known, μ_s and μ_a can be calculated. If the total reflectance, the total transmittance, and the collimated transmittance are measured, μ_s, μ_a, and g can be calculated.

A general procedure of the IAD algorithm includes the following steps:

1. Measure the reflectance and transmittance of the sample by integrating sphere(s)
2. Make an initial guess of the optical properties (μ_s, μ_a, and g)
3. With the optical properties in step 2, calculate the reflectance and transmittance of the sample by using the AD method
4. Compare the reflectance and transmittance with the real measurements in step 1. If the error is larger than the preset threshold, adjust the optical properties (μ_s, μ_a, and g) by following an appropriate search method
5. Repeat steps 3 and 4 until a desired match is achieved

To implement step 1, double-integrating sphere setups, as shown in Figure 6.11, can be used to obtain both reflectance and transmittance measurements simultaneously (Pickering et al. 1993). The double-integrating sphere setup with a single beam is simple to construct and use. In real applications, the reflectance and transmittance of the sample can also be measured by using a single-integrating sphere at corresponding configurations.

In general, the IAD method has been proven to be fast and accurate for measuring the optical properties of thin and slab-shaped biological samples. The

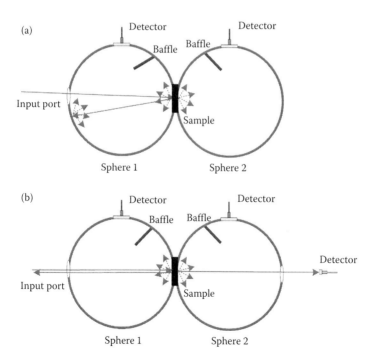

FIGURE 6.11 Examples of using double-integrating spheres: (a) the configuration for measuring the total reflectance and the total transmittance and (b) the configuration for measuring the diffuse reflectance and the diffuse transmittance.

appealing fact of the IAD method is that it considers the boundary conditions and can be used under any albedo and scattering anisotropy conditions (Pickering et al. 1993). A known limitation of the IAD method is that its calculations could be affected by the loss of the light scattered through the boundary of a sample and glass slides on the sample holder (Pickering et al. 1993; Prahl 2011). The light radiation losses are related to the geometry (i.e., size, thickness) of the sample and slides. To reduce the IAD calculation error caused by the light loss, the sample has to be a flat slab and completely covers the port of the integrating sphere. In addition, the diameter of the incident beam should be small and the diameter of the exit and the entrance ports of the integrating sphere should be much larger than that of the incident beam.

6.6.2 Example Applications of Applying IAD Technique to Measure the Optical Properties of Food

Figure 6.12 illustrates a single-integrating-sphere-based spectroscopic system used for measuring the optical properties of vegetable tissues, which mainly consisted of integrating sphere, fiber guide, and spectrometer. The system was used to measure the reflectance and transmittance spectra of healthy and diseased onion

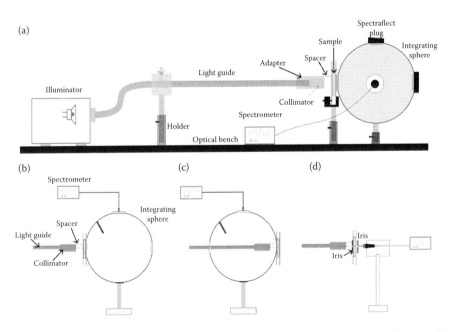

FIGURE 6.12 A single-integrating-sphere-based system used for measuring the optical properties of onion tissues: (a) the schematic of the whole system, and the configurations for (b) the total transmittance, (c) the total reflectance, and (d) the collimated transmittance. (From Wang, W., and C. Li. 2014. Optical properties of healthy and diseased onion tissues in the visible and near-infrared spectral region. *Transactions of the ASABE* 57 (6):1771–1782. Reprinted with permission.)

tissues in the wavelength range of 550–1650 nm and to calculate the μ'_s, μ_a, and g properties of the samples using the IAD algorithm (Wang and Li 2014).

The integrating sphere shown in Figure 6.12 (model 4P-GPS-060-SF, Labsphere, North Sutton, NH, USA) had an internal diameter of 152 mm, with four 25.4-mm-diameter ports at 0°, 90°, 180°, and the north pole. The internal wall of the sphere was coated with Spectraflect® (Labsphere, North Sutton, NH, USA), a highly reflective material with reflectivity greater than 98%. The incident light (550–1650 nm) was a white light beam provided by a 150-W DC-regulated fiber-optic illuminator (model DC-950, Dolan-Jenner Industries, Boxborough, MA, USA) with a goose neck light guide. As discussed in the previous section, the IAD method requires a narrow light beam much smaller than the size of the sample. Therefore, collimators (models F240SMA-B and F240SMA-C, Thorlabs, Newton, NJ, USA) were used in the study to collimate the divergent light of the illuminator to a light beam with 1.5 mm diameter. Because of the limitation of the spectral working range of the off-the-shelf collimators, the system used two collimators (focus length 7.93 mm and numeric aperture 0.5) to cover the whole testing spectral range.

An aluminum spacer with 8-mm-long needle-shaped pin was used to accurately control the beam size and the distance between the collimator and the sample in all measurements (Figure 6.12b–d). There were two spectrometers used as the detector: the visible and NIR spectrometer (model USB4000, Ocean Optics, Dunedin, FL, USA) and the NIR spectrometer (model CD024252, Control Development, Inc., South Bend, IN, USA). The visible–NIR spectrometer was used together with the collimator F240SMA-B and the NIR spectrometer was used with the collimator F240SMA-C. Owing to the limitations of the devices, the spectra in the spectral range of 880–950 nm had low signal-to-noise ratios (S/N), which were too noisy for IAD calculations. Therefore, the spectral data in this range were removed in the study.

In the study of Wang and Li (2014), this system was used to measure the optical properties of the three types of flesh tissues of Spanish yellow onions: (1) healthy, (2) sour skin (bacterial disease), and (3) neck rot (fungal disease), respectively. For each type of samples, 20 jumbo-size and disease-free onions were first selected. The healthy onions were stored in a refrigerator (6 ± 1°C) for 6–8 days. Onions in the sour skin group were inoculated with *Burkholderia cepacia* (*B. cepacia*) and samples in the neck rot group were inoculated with *Botrytis aclada* (*B. aclada*). Both *B. cepacia* and *B. aclada* used in the study were isolated from naturally diseased plants and cultured on potato dextrose agar medium. The suspension of inoculums (1.5 mL per onion) was infiltrated into the second fleshy scale of each onion. Sour skin samples were incubated at 30 ± 1°C for 4–6 days before testing and the neck rot samples were incubated in a humidity chamber (20 ± 1°C, relative humidity > 80%) for 9–15 days until disease symptoms were visible on the onion.

For each onion, four 30 × 30 mm flesh tissues were extracted and shaved to be slabs using a blade. The thickness of each onion tissue sample was measured with an electronic micrometer (model 35–025, iGaging, San Clemente, CA, USA). Then, the total transmittance T, the total reflectance R, and the collimated transmittance T_c were measured. On the basis of the measured R, T, and T_c spectra of the onion tissue, the IAD method was applied to estimate the optical properties (μ'_s, μ_a, and g) of the tissue sample using the open-source IAD program provided by Prahl (2011). In all IAD calculations for onion tissues, the refractive indices of onion dry skin and flesh were set to be 1.334 and 1.352, respectively, based on our experimental measurements using the Abbe refractometer (Abbe Mark II, Reichert, Inc., New York, NY, USA). The reflectivity of the integrating sphere was set as 98% based on the data sheet provided by the manufacturer.

The estimated μ_a of the onion flesh is illustrated in Figure 6.13a and c. In the visible region, the mean μ_a values of the sour flesh were significantly greater than those of the flesh of healthy onions. This is reasonable since flesh scales of sour skin onions have pale yellow to light-brown symptoms. The difference between the mean μ_a of sour skin-infected and healthy onion flesh was not significant in the NIR wavelength region of 950–1300 nm. For the mean μ_a values of the neck rot flesh were about 10 times greater than those of healthy onion flesh (Figure 6.13c) in the visible region, which was probably caused by sporulation and production of sclerotia by the fungus *B. aclada*. In the NIR wavelengths shorter than 1200 nm, the mean μ_a of the neck rot flesh was still significantly greater than those of healthy ones, which might be caused by the chemical constituents of *B. aclada*. In the NIR range longer than 1300 nm, the μ_a of all three types of onion flesh were greater than those of the visible region due to the strong light absorption around 1450 nm caused by water, and were not significantly different from each other.

As shown in Figure 6.13c, the μ'_s of the flesh of sour skin onions was significantly smaller than those of healthy onions, which indicated the change of the physical properties of the sour skin onion flesh: the bacterium breaks down the key structural components of onion cells and cell walls. Similarly, the mean μ'_s of the neck rot flesh (Figure 6.13d) was generally smaller than those of the healthy flesh, indicating a lower light scattering in the onion flesh with neck rot than in the healthy flesh. On the basis of optical properties calculated by IAD methods, bacterial (sour skin) and fungal (neck rot) infections significantly changed the light absorption and scattering properties of onion flesh. These results provided a foundation to further develop an optical method to distinguish pathogen-infected onion flesh from healthy onion flesh.

In summary, the above example illustrates the application of the IAD method to investigate the optical properties of vegetable tissues in a recent study. Other applications have also been reported in the literature. For instance, Saeys et al. (2008) measured the μ_a and μ_s of apple skin and flesh in 350–2200 nm using the IAD method. Wang and Li (2013) measured the optical properties of the skin and flesh of four common types of onions at 633 nm and concluded that the optical properties of onion tissues were significantly different between onion cultivars. In addition to being used as an accurate lab-analytical tool for measuring the optical properties of biological tissues, the IAD method also provides good reference measurements in the development of other techniques for measuring the optical properties of food items, such as the spatially-resolved method and Monte Carlo method (Lu 2008; Qin and Lu 2009; Cen et al. 2013).

Measuring Optical Absorption and Scattering Properties of Food

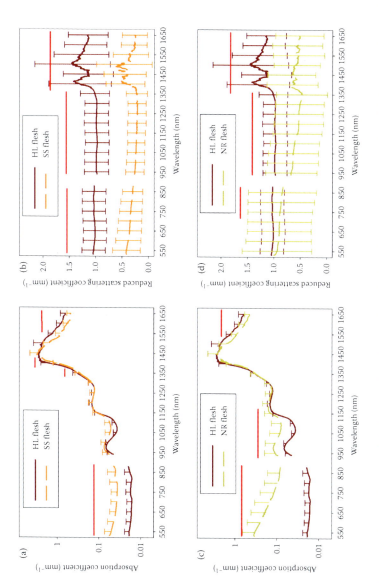

FIGURE 6.13 Average values and standard deviations of (a) the μ_a of the flesh of the onions with sour skin (SS flesh); (b) the μ'_s of the SS flesh; (c) the μ_a of the flesh of onions with neck rot (NR flesh); and (d) the μ'_s of the NR flesh. The red lines indicate the spectral regions where the μ_a or μ'_s of diseased flesh were significantly different from those of the healthy ones (HL) (using the significance level of 0.05). (From Wang, W. and C. Li. 2014. Optical properties of healthy and diseased onion tissues in the visible and near-infrared spectral region. *Transactions of the ASABE* 57 (6):1771–1782. Reprinted with permission.)

6.7 CONCLUSIONS

This chapter presents basic approaches and related theories of measuring optical properties of foods focusing on *ex vivo* methods. Beer's law is fundamental to directly measuring optical properties of optically thin (single scattering) samples and is also the basis to derive the total attenuation coefficient (μ_t) using collimated transmittance. The integrating sphere technique is fundamental to all *ex vivo* techniques introduced in this chapter including both direct and indirect methods (K–M method and IAD method). Although the direct method measuring the optically thin samples seems straightforward in concept, experimentally it is more challenging than the indirect methods mainly due to the difficulty in handling thin samples.

As a noniterative indirect method, the K–M method offers a simple approach to acquire optical properties (absorption coefficient K and scattering coefficient S) of food products and has been used in the past four decades in agriculture and food. However, the main drawback of this method is low accuracy when the assumptions of the model are unmet. As an iterative indirect method, the IAD is generally regarded as a fast and accurate method to measure the optical properties of thin and slab-shaped biological samples. The IAD method has been used to acquire optical properties of fruits and vegetables, as well as to provide reference measurements in the development of other *in vivo* techniques for measuring the optical properties of food items, such as the spatially-resolved method that will be introduced in the following chapters.

REFERENCES

Birth, G. S. 1978. The light scattering properties of foods. *Journal of Food Science* 43 (3):916–925.

Birth, G. S. 1982. Diffuse thickness as a measure of light scattering. *Applied Spectroscopy* 36 (6):675–682.

Birth, G. S. 1986. The light scattering characteristics of ground grains. *International Agrophysics* 2 (1):59–67.

Budiastra, I. W. 1998. Optical methods for quality evaluation of fruits. *Journal of Japanese Society of Agricultural Machinery* 60 (2):117–128.

Cen, H., R. Lu, F. Mendoza, and R. M. Beaudry. 2013. Relationship of the optical absorption and scattering properties with mechanical and structural properties of apple tissue. *Postharvest Biology and Technology* 85:30–38.

Hetherington, M. J. and D. B. Macdougall. 1992. Optical properties and appearance characteristics of tomato fruit (*Lycopersicon esculentum*). *Journal of the Science of Food and Agriculture* 59 (4):537–543.

Jacquez, J. A. and H. F. Kuppenheim. 1955. Theory of the integrating sphere. *Journal of the Optical Society of America* 55:460–470.

Judd, D. B. and G. Wyszecki. 1975. *Color in business, science, and industry*, 3rd ed. Wiley, New York, USA.

Klier, K. 1972. Absorption and scattering in plane parallel turbid media. *Journal of the Optical Society of America* 62 (7):882–885.

Knee, M., E. Tsantili, and S. G. S. Hatfield. 1988. Promotion and inhibition by ethylene of chlorophyll degradation in orange fruits. *Annals of Applied Biology* 113 (1):129–135.

Lana, M. M., M. Hogenkamp, and R. B. M. Koehorst. 2006. Application of Kubelka–Munk analysis to the study of translucency in fresh-cut tomato. *Innovative Food Science and Emerging Technologies* 7 (4):302–308.

Law, S. E. and K. H. Norris. 1973. Kubelka–Munk light-scattering coefficients of model particulate systems. *Transactions of the ASAE* 16 (5):914–921.

Lu, R. 2008. Quality evaluation of fruit by hyperspectral imaging. In *Computer Vision Technology for Food Quality Evaluation*, S. Da-Wen (ed.), 319–348. Amsterdam, Netherlands: Academic Press.

McCrackin, F. L., E. Passaglia, R. R. Stromberg, and H. L. Steinberg. 1963. Measurement of the thickness and refractive index of very thin films and the optical properties of surfaces by ellipsometry. *Journal of Research of the National Bureau of Standards, Section A* 67: 363–377.

Pauletti, M. S., E. J. Matta, E. Castelao, and D. S. Rozycki. 1999. Color in concentrated milk proteins with high sucrose as affected by glucose replacement. *Journal of Food Science* 64 (1):90–92.

Pickering, J. W., S. A. Prahl, N. van Wieringen, J. F. Beek, H. J. C. M. Sterenborg, and M. J. C. van Gemert. 1993. Double-integrating-sphere system for measuring the optical properties of tissue. *Applied Optics* 32 (4):399.

Prahl, S. A., M. J. C. van Gemert, and A. J. Welch. 1993. Determining the optical properties of turbid media by using the adding–doubling method. *Applied Optics* 32 (4):559–568.

Prahl, S. A. 1995. Chapter 5. The adding–doubling method. In *Optical–Thermal Response of Laser-Irradiated Tissue*, A. J. Welch and M. J. C. van Gemert (ed.), 101–129. New York: Plenum Press.

Prahl, S. 2011. Everything I think you should know about inverse adding-doubling. Wilsonville, OR, USA. http://omlc.org/software/iad/

Pristinski, D., V. Kozlovskaya, and S. A. Sukhishvili. 2006. Determination of film thickness and refractive index in one measurement of phase-modulated ellipsometry. *Journal of the Optical Society of America A* 23 (10):2639–2644.

Qin, J. and R. Lu. 2009. Monte Carlo simulation for quantification of light transport features in apples. *Computers and Electronics in Agriculture* 68:44–51.

Saeys, W., M. A. Velazco-Roa, S. N. Thennadil, H. Ramon, and B. M. Nicolai. 2008. Optical properties of apple skin and flesh in the wavelength range from 350 to 2200 nm. *Applied Optics* 47:908–919.

Talens, P., N. Martınez-Navarrete, P. Fito, and A. Chiralt. 2002. Changes in optical and mechanical properties during osmodehydrofreezing of kiwi fruit. *Innovative Food Science and Emerging Technologies* 3 (2):191–199.

Tearney, G. J., M. E. Brezinski, B. E. Bouma, M. R. Hee, J. F. Southern, and J. G. Fujimoto. 1995. Determination of the refractive index of highly scattering human tissue by optical coherence tomography. *Optics Letters* 20 (21):2258–2260.

Tuchin, V. V. 2007. *Tissue Optics: Light Scattering Methods and Instruments for Medical Diagnosis*. Bellingham, WA: SPIE/International Society for Optical Engineering.

van de Hulst, H. C. 1980. *Multiple Light Scattering: Tables, Formulas, and Applications*. Vol. 1. New York: Academic Press.

Van Gemert, M. J. C. and W. M. Star. 1987. Relations between the Kubelka–Munk and the transport equation models for anisotropic scattering. *Lasers in the Life Sciences* 1 (98):287–298.

Wang, L. V. and H.-I. Wu. 2012. *Biomedical Optics: Principles and Imaging*. Hoboken, NJ: John Wiley & Sons.

Wang, W. and C. Li. 2013. Measurement of the light absorption and scattering properties of onion skin and flesh at 633 nm. *Postharvest Biology and Technology* 86:494–501.

Wang, W. and C. Li. 2014. Optical properties of healthy and diseased onion tissues in the visible and near-infrared spectral region. *Transactions of the ASABE* 57 (6):1771–1782.

Welch, A. J. and M. J. C. Van Gemert. 1995. *Optical–Thermal Response of Laser-Irradiated Tissue*. Vol. 1. New York: Springer.

7 Spatially Resolved Spectroscopic Technique for Measuring Optical Properties of Food

Haiyan Cen, Renfu Lu, Nghia Nguyen-Do-Trong, and Wouter Saeys

CONTENTS

7.1 Introduction 159
7.2 Theory and Modeling for SRS 160
    7.2.1 Steady-State Light Transfer in Homogenous Media 161
    7.2.2 Steady-State Light Transfer in Layered Media 163
7.3 Instrumentation for SRS 165
    7.3.1 Optical Fiber Probe 165
        7.3.1.1 Translation Stage SRS 165
        7.3.1.2 Fiber-Array Probe SRS 166
        7.3.1.3 Illustration of Fiber-Array Probe SRS Measurement 168
    7.3.2 Monochromatic Imaging 169
    7.3.3 Hyperspectral Imaging 170
    7.3.4 Spatial Frequency-Domain Imaging 173
7.4 Measurement of Optical Properties of Food Products 175
    7.4.1 Optical Properties of Food Products 175
    7.4.2 Relationship between Optical Properties and Structural Properties of Apple Fruit 179
7.5 Conclusions and Future Trends 180
References 181

7.1 INTRODUCTION

Optical properties are important for understanding the behaviors and responses of food products when they are exposed to electromagnetic radiation. Light absorption and scattering are two primary phenomena during the interaction of photons with biological materials, which are characterized by the absorption coefficient (μ_a) and reduced scattering coefficient (μ_s'). Light absorption is mainly related to the chemical composition of the material, while scattering is influenced by the structural and physical properties. Hence, quantification of the optical properties of food products is key to understanding light propagation, designing effective optical devices, and developing new prototypes for quality and property evaluation of food products.

Over the years, many techniques and the associated instrumentation have been developed to measure the optical properties of food products. On the basis of measurement principles, these techniques can be classified into *direct* and *indirect*, either using empirical models or fundamental radiative transfer theory. Direct methods are normally required to measure reflectance and transmittance from samples with simple geometries (e.g., thin slabs). They are relatively easy to implement but are destructive. In contrast, indirect methods can be performed on intact samples nondestructively, but need sophisticated instrumentation and complex mathematical models derived from the fundamental radiative transfer theory. Recent studies have been mainly focused on indirect methods such as spatially-resolved, time-resolved, and frequency domain spectroscopy, because they are applicable to a wide range of biological materials without the need for sample preparation (Cen and Lu, 2010; Cletus et al., 2009; Nicolai et al., 2008). Spatially-resolved spectroscopy (SRS) is more viable for food applications because of low cost in instrumentation and easy implementation with the reflectance mode in measurement. SRS measurement can be implemented in different sensing configurations, including a fiber-optic probe, monochromatic imaging, hyperspectral imaging, and spatial frequency-domain imaging (SFDI). In recent years, we have seen an increasing interest from researchers in SRS technique for measuring the optical properties of food products ranging from fruit to meat, and to liquid or colloidal food products (e.g., milk, fruit juice, etc.), as well as for assessing the composition and quality of these food products (Cen et al., 2012b; Falconet et al., 2008; Nguyen Do Trong et al., 2014a; Qin and Lu, 2007; Xia et al., 2007).

In this chapter, we provide an overview of different measurement configurations based on the SRS principle, including a fiber-optic probe, monochromatic imaging, hyperspectral imaging, and SFDI. The chapter begins with an introduction of the theory and modeling of SRS technique, followed by a detailed description of instrumentation and measurement for the optical absorption and scattering properties of food. An application example from a recent study is then given as to how the optical properties are related to the fundamental structural properties of fruit tissues. Finally, we discuss the challenges and future trends in research for measuring the optical properties of food products using a spatially-resolved spectroscopic technique.

7.2 THEORY AND MODELING FOR SRS

A spatially-resolved spectroscopic technique was first proposed by Reynolds et al. (1976) for understanding light absorption and scattering in a finite blood medium. Later, Langerholc (1982) and Marquet et al. (1995) reported that spatially-resolved measurement could be used to solve both two-dimensional (2-D) and three-dimensional (3-D) multiple-scattering problems. In principle, the technique is achieved by measuring the diffusely reflected light from the sample surface at different source–detector distances irradiated with a point light source or narrow collimated beam, as shown in Figure 7.1. The two main optical parameters, including the absorption coefficient

SRS Technique for Measuring Optical Properties of Food

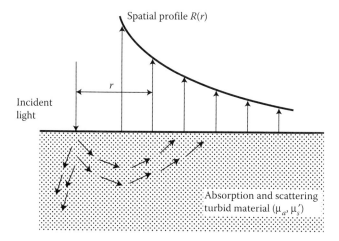

FIGURE 7.1 Measurement principle for a spatially-resolved technique. (From Cen, H. and R. Lu. *Optics Express* 18(16), 2010: 17412–17432.)

(μ_a) and reduced scattering coefficient (μ'_s), can then be extracted from the measured spatially-resolved diffuse reflectance profile based on radiative transfer theory through the use of diffusion approximation or Monte Carlo simulation, coupled with appropriate inverse parameter estimation algorithms, which are detailed in Chapters 3 through 5. Numerical methods are generally required for solving the radiation transfer equation, or Monte Carlo simulation is needed for tracing the transport of a large quantity of photons, to estimate the absorption and scattering parameters of the material. The latter needs no or fewer physical approximations on photon transport in the material, but they could be subject to statistical uncertainties during the estimation of reflectance. In addition, they also require substantial computational time. An alternative, popular approach is to use an analytical solution derived from the diffusion approximation equation for the spatially-resolved reflectance. While this requires that assumptions are made, the estimates of μ_a and μ'_s can be obtained quickly by using a proper inverse algorithm. In this section, we give a detailed description on the analytical solutions of the diffusion model, which can be used to estimate the absorption and reduced scattering coefficients from spatially-resolved diffuse reflectance, for homogenous media and two-layered homogenous media.

7.2.1 Steady-State Light Transfer in Homogenous Media

When scattering is dominant over absorption (i.e., $\mu'_s \gg \mu_a$), the radiative transport may be simplified as a diffusion process. This diffusion approximation has been widely used for modeling light propagation in homogenous or layered media. For the case of steady-state light transfer in semi-infinite homogenous media, Farrell et al. (1992) developed a model to describe the radial dependence of diffuse reflectance at the surface of the media under normal illumination by an infinitely small light beam (Figure 7.1). In this model, the diffuse reflectance from the medium is calculated as the flow across the boundary, and it originated from a single isotropic point source located at a depth of one transport mean free path in the medium. The model is suitable for both refractive index matched or mismatched surfaces. An analytical solution from the diffusion model is given as follows (Farrell et al., 1992):

$$R_{f,ho}(r) = \frac{a'}{4\pi}\left[\frac{1}{\mu'_t}\left(\mu_{eff} + \frac{1}{r_1}\right)\frac{\exp(-\mu_{eff}r_1)}{r_1^2} + \left(\frac{1}{\mu'_t} + \frac{4A}{3\mu'_t}\right)\left(\mu_{eff} + \frac{1}{r_2}\right)\frac{\exp(-\mu_{eff}r_2)}{r_2^2}\right] \quad (7.1)$$

where f and ho in $R_{f,ho}$ represent Farrell model and homogenous media, respectively; r is the source–detector distance; $a' = \mu'_s/(\mu_a + \mu'_s)$ is the transport albedo; $\mu_{eff} = [3\mu_a(\mu_a + \mu'_s)]^{1/2}$ is the effective attenuation coefficient; $\mu'_t = \mu_a + \mu'_s$ is the total attenuation coefficient; $r_1 = (z_0^2 + r^2)^{1/2}$ and $r_2 = [(z_0 + 2z_b)^2 + r^2]^{1/2}$ are the distances from the observation point at the interface to the isotropic source and the image source; $z_0 = (\mu_a + \mu'_s)^{-1}$ is the mean free path, $z_b = 2A/3\mu'_t$; and $A = 0.2190$ for $n = 1.35$ is the internal reflection coefficient related to the relative index of the tissue–air interface n, which can be calculated from an empirical equation developed by Groenhuis et al. (1983). In Equation 7.1, the diffuse reflectance $R_{f,ho}(r)$ at the surface of the semi-infinite turbid medium is a function of the source–detector distance r as well as the optical parameters of the investigated medium (i.e., the absorption (μ_a) and reduced scattering coefficients (μ'_s), and the relative refractive index (n)). Although it is known that the refractive index is wavelength dependent, in most reported studies, the relative refractive index n has been assumed to be constant for a given biological tissue or food product (Mourant et al., 1997). It should be mentioned that the potential inaccuracy caused by this assumption could exist. For many fruits and food products, the value of 1.35 has been chosen (Qin and Lu, 2008).

Later, Kienle and Patterson (1997) proposed an improved analytical solution by expressing the reflectance as the integral of the radiance over the backward hemisphere based on the study of Haskell et al. (1994). In this case, the radiance can be expressed as the sum of the isotropic fluence rate and the flux. The isotropic fluence rate at the surface ($z = 0$) using the extrapolated boundary, in which the fluence rate is forced to be zero, is expressed by

$$\Phi(r, z = 0) = \frac{1}{4\pi D}\left[\frac{\exp(-\mu_{eff}r_1)}{r_1} - \frac{\exp(-\mu_{eff}r_2)}{r_2}\right] \quad (7.2)$$

The diffuse reflectance calculated as the flux across the boundary is given by

$$R_{flux}(r) = \frac{1}{4\pi}\left[z_0\left(\mu_{eff} + \frac{1}{r_1}\right)\frac{\exp(-\mu_{eff}r_1)}{r_1^2} + (z_0 + 2z_b)\left(\mu_{eff} + \frac{1}{r_2}\right)\frac{\exp(-\mu_{eff}r_2)}{r_2^2}\right] \quad (7.3)$$

The final solution for the steady-state spatially-resolved diffuse reflectance for a homogenous medium derived by Kienle and Patterson (1997) is calculated as follows:

$$R_{k,ho}(r) = C_1\Phi(r, z = 0) + C_2 R_{flux}(r) \quad (7.4)$$

where k and ho in $R_{k,ho}$ represent Kienle model and homogenous media, respectively $C_1 = (1/4\pi)\int_{2\pi}[1 - R_{fres}(\theta)]\cos\theta d\Omega$ and $C_2 = (3/4\pi)\int_{2\pi}[1 - R_{fres}(\theta)]\cos^2\theta d\Omega$ are constants determined by the relative refractive index mismatch at the tissue–air interface, in which $R_{fres}(\theta)$ is the Fresnel reflection coefficient for a photon with an incident angle θ relative to the normal to the boundary, and Ω is the solid angle. Details on calculating these parameters can be found in Haskell et al. (1994). For a relative refractive index $n = 1.35$, a typical value for many biological materials (Mourant et al., 1997), C_1 and C_2 are equal to 0.1277 and 0.3269, respectively.

With the derived analytical solutions, the absorption and reduced scattering coefficients can be extracted from the acquired spatially-resolved diffuse reflectance profiles using an inverse algorithm. These analytical solutions of the homogenous diffusion model have been used in studies on determining the optical properties of liquid foods, fruits, and vegetables (Cen et al., 2012a; Erkinbaev et al., 2014; Herremans et al., 2013; Nguyen Do Trong et al., 2012; Qin and Lu, 2007; Qin et al., 2009).

7.2.2 STEADY-STATE LIGHT TRANSFER IN LAYERED MEDIA

Biological tissues, by nature, are heterogenous in structure and thus, their optical properties vary spatially within the tissues. However, for many biological materials such as fruit, their tissues may be approximated to be layered media (e.g., the skin and flesh of fruit); within each layer, its tissue structure and optical properties are approximately homogenous. It is therefore desirable or even necessary to study light propagation in the layered media and measure the optical properties of each layer.

Light transfer in multilayered media is inherently more complex than in homogenous media, which results in much greater challenges both mathematically and experimentally for the measurement of optical properties. Numerical methods (e.g., Monte Carlo, asymptotic approximation, and finite element) are commonly used to extract the optical properties of two-layered or three-layered media (González-Rodríguez and Kim, 2008; Liao and Tseng, 2014; Raulot et al., 2010). Although an accurate estimation of optical properties may be achieved by numerical methods, it is computationally time consuming and requires more complex inverse algorithms with the increased number of free parameters in a layered model. Studies were reported on obtaining analytical solutions to the diffusion model for heterogenous or layered media. Kienle et al. (1998) derived an analytical form of the two-layered diffusion model, which allows fast-forward computation of spatially-resolved diffuse reflectance and inverse algorithm implementation to estimate the optical properties of each layer.

Consider a two-layered turbid medium that is impinged upon perpendicularly by an infinitely small light beam (Figure 7.2). The thickness of the first layer (d) is assumed to be larger than one transport mean free path (i.e., $z_0 = 1/(\mu_{a1} + \mu_{s1}')$, in which μ_{a1} and μ_{s1}' are absorption and reduced scattering coefficients of the first layer). Under a steady-state condition, the diffusion equation for a two-layered turbid medium is given by (Kienle et al., 1998)

$$D_1\nabla^2\Phi_1(\mathbf{r}) - \mu_{a1}\Phi_1(\mathbf{r}) = -\delta(x,y,z-z_0) \quad 0 \leq z < d \quad (7.5)$$

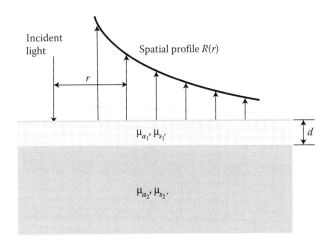

FIGURE 7.2 Scheme of light propagation in a two-layered turbid medium.

$$D_2 \nabla^2 \Phi_2(\mathbf{r}) - \mu_{a2}\Phi_2(\mathbf{r}) = 0 \quad d \le z \tag{7.6}$$

where Φ_i is the fluence rate of layer i, $D_i = 1/[3(\mu_{ai} + \mu_{si'})]$ is the diffusion constant, and δ is a generalized function. Equations 7.5 and 7.6 can be converted into ordinary differential equations using a 2-D Fourier transform. The following boundary conditions are applied for obtaining solutions for the fluence rate of each layer in the frequency domain $[\phi_i(z,s)]$ (Haskell et al., 1994; Kienle et al., 1998): the same refractive index for the first layer and second layer, the zero fluence rate of the first layer at the extrapolated boundary, the zero fluence rate of the second layer as $z \to \infty$, and the continuity of fluence rate at the boundary between the first and second layer.

After the differential equations in the frequency domain have been solved, the 2-D inverse Fourier transform is applied to obtain the following fluence solution in Equations 7.5 and 7.6 on the Cartesian coordinate system

$$\Phi_i(r,z) = \frac{1}{(2\pi)^2} \int_{-\infty}^{\infty}\int_{-\infty}^{\infty} \phi_i(z,s)\exp[-i(s_1 x + s_2 y)]ds_1 ds_2 = \frac{1}{2\pi}\int_0^{\infty} \phi_i(z,s)sJ_0(sr)ds \tag{7.7}$$

where $r = (x^2 + y^2)^{1/2}$ and J_0 is the zeroth-order Bessel function. This inverse transform is achieved by numerically evaluating the integral using an adaptive Gauss–Kronrod quadrature (Shampine, 2008). Spatially-resolved diffuse reflectance is obtained from the integration of radiance over the solid angle accepted by the fiber. Then, the final solution of the two-layered diffusion model is given as (Kienle et al., 1998)

$$R_{k,la}(r) = C_1 \Phi_1(r,z=0) + C_2 D_1 \frac{\partial}{\partial z}\Phi_1(r,z)\big|z=0 \tag{7.8}$$

where k and la in $R_{k,la}$ represent Kienle model and layered media, respectively; the parameters C_1 and C_2 are given in Equation 7.4 in Section 7.2.1 and Haskell et al. (1994). On the basis of Equation 7.8, five unknown parameters, including absorption and reduced scattering coefficients of the two layers and the thickness of the first layer (μ_{a1}, $\mu_{s1'}$, μ_{a2}, $\mu_{s2'}$, and d, where μ_{a2} and $\mu_{s2'}$ are absorption and reduced scattering coefficients of the second layer), need to be determined from the measured diffuse reflectance $R_{k,la}(r)$ as a function of the source–detector distance r. The two-layered model in Equations 7.7 and 7.8 was also used to model photon propagation in turbid media under steady-state conditions (Hollmann and Wang, 2007; Kienle et al., 1998; Tseng et al., 2008).

7.3 INSTRUMENTATION FOR SRS

In practice, spatially-resolved measurement employs a point light source or narrow collimated beam of constant intensity and single or multiple detectors at different source–detector distances. Optical fiber arrays and noncontact reflectance imagery are the two common types of sensing configurations in spatially-resolved measurement systems (Doornbos et al., 1999; Fabbri et al., 2003; Malsan et al., 2014; Pilz et al., 2008). With the optical fiber array configurations, a single spectrometer, multiple spectrometers, or a spectrograph–camera combination are needed to measure diffuse reflectance at different distances from the light incident point. Optical properties at multiple wavelengths or over a specific spectral region can be obtained using this method. Yet, the measurements need good contact between the detecting probe and the sample, which may not be practical or convenient for some solid food products. With the reflectance imagery configurations, a charge-coupled device (CCD) camera is typically used to acquire diffuse reflectance from the scattering medium generated by a point light beam. The measurement can be achieved without contacting the investigated medium, which is particularly advantageous for food products, because of the safety and sanitation requirements. In this section, instrumentation for four spatially-resolved measurement configurations, including a fiber-optic probe, monochromatic imaging, hyperspectral imaging, and SFDI, is reviewed.

7.3.1 OPTICAL FIBER PROBE

7.3.1.1 Translation Stage SRS

Spatially-resolved spectroscopic measurements in contact mode can be implemented by carrying out multiple conventional point spectroscopy measurements at several distances from the illumination point. In the translation stage SRS measurement modality, a spectrophotometer is used in combination with two optical fibers: one optical fiber illuminates the sample at a fixed location and another optical fiber collects the diffusely reflected light. One of the two fibers is connected to a translation stage that allows careful control of the source–detector distance. By performing diffuse reflectance measurements at different source–detector distances, spatially-resolved profiles are acquired over the full-wavelength range. The concept of translation stage SRS is schematically illustrated in Figure 7.3.

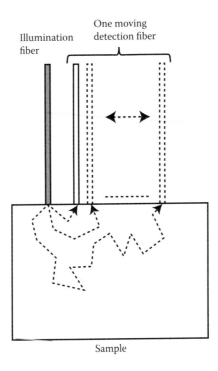

FIGURE 7.3 Schematic representation of translation stage SRS measurements. The zigzag lines represent the traveling paths of the photons in the scattering medium.

The translation stage SRS modality provides some major advantages: the number of spectroscopic measurements, the measured positions or source–detector distances of interest, and the maximum source–detector distance can be flexibly changed with respect to the optical properties of the studied sample to assure high signal-to-noise ratios (SNRs) in the acquired data. Thanks to the sequential acquisition of the diffusely reflected light, the measurements can be performed with a standard fiber-optic spectrophotometer. These typically have a higher SNR than the hyperspectral cameras used in other configurations for SRS. Moreover, thanks to the sequential scanning, the integration time can be adjusted depending on the source–detector distance to avoid saturation at close source–detector distances and noisy data at further measurement locations. However, this sequential spectroscopic data acquisition leads to a large measurement time that is only acceptable if the sample does not change during this time. In addition, fluctuations in the light source output during the measurement time period can introduce errors in the measured reflectance profiles. As a result, the sequential scanning and careful positioning of the illumination and/or detection fiber limit the applicability of translation stage SRS to scientific research.

7.3.1.2 Fiber-Array Probe SRS

Fiber-array probe SRS resolves the issue of careful fiber positioning by integrating an illumination fiber and multiple detection fibers at several fixed source–detector distances into a probe. Such a configuration for fiber-array probe SRS measurements is schematically illustrated in Figure 7.4. The other end of the detection fibers can then be coupled into a multiplexer that allows sequential coupling of the light captured by the different detection fibers into a spectrophotometer. While this configuration solves the fiber-positioning issue, the issue of sequential scanning remains. To overcome this issue, the spectrophotometer and multiplexer can be replaced by a hyperspectral camera. In this configuration, the detection fibers are coupled into the slit of a spectrograph that splits the light of the different fibers into its wavelength components and projects them onto the camera chip. The sequential scanning now turned into simultaneous acquisition of these signals, which significantly reduces the required measurement time. This makes this fiber-array probe SRS configuration more suitable for industrial and *in vivo* applications than the translation stage SRS configuration.

The following factors should be carefully considered and optimized in the design of a fiber-array probe SRS setup: specifications of the illumination fiber and the detection fibers (fiber diameter, numerical aperture…), number of detection fibers, and the arrangement of these fibers in the probe at predefined source–detector distances. Therefore, information on the targeted range of optical properties is essential to provide a good design of the SRS fiber-array probe for a given application. Good fiber-array probe SRS measurements require accurate calibration of the setup to compensate for wavelength-dependent variations in intensities of the illumination light and differences in the efficiencies of detection fibers.

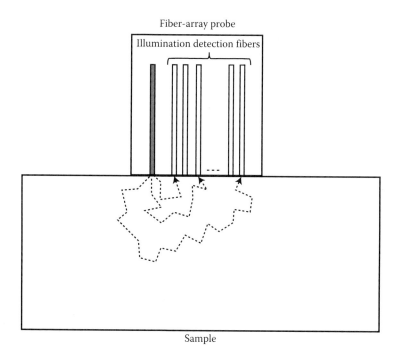

FIGURE 7.4 Schematic illustration of fiber-array probe SRS measurements. The zigzag lines represent the traveling paths of the photons in the scattering medium.

The fiber-array probe SRS configuration is suitable for measuring liquid or semi-liquid foods, thanks to the good contact between the probe tip and the food surfaces regardless of the surface motion of the liquid and semiliquid foods. The small fiber-array probe with short source–detector distances is also suitable for contact SRS measurement on thin slices of solid foods (e.g., dried fruit slices, leafy vegetables) in the reflectance mode thanks to the contact measurement configuration of the fiber-array probe SRS that provides higher SNRs in the acquired diffuse reflectance spectra than contactless measurements. Fiber-array probe SRS also requires a lower illumination power and has more flexibility to avoid capturing light that has been reflected by the material behind the sample. However, the contact between the fiber-array probe and the product poses a risk of cross-contamination between different products and requires regular and careful cleaning.

7.3.1.3 Illustration of Fiber-Array Probe SRS Measurement

The spatially-resolved diffuse reflectance data in the range of 650–970 nm acquired with a fiber-array probe integrating one illumination fiber and five detection fibers (all have the diameter of 200 μm) with the source–detector distances ranging from 0.3 to 1.2 mm for the case of a fresh apple containing both skin and flesh layers is illustrated in Figure 7.5.

The z-axis in Figure 7.5 shows the diffuse reflectance values computed by using an integrating sphere as a reference. It should be noted that relative reflectance values larger than 1 are observed, because the reflected intensities acquired in the integrating sphere (reference) are lower than those acquired on the apple sample. Fiber 1 is the closest detection fiber and fiber 5 is the farthest one from the illumination fiber. The diffuse reflectance spectrum acquired by each detection fiber has two absorption peaks indicating the absorption by chlorophyll (at 670 nm) and by water (at 970 nm). In the 700–900 nm range, the reflectance values are higher and quite flat, showing a lower absorption by the chemical constituents of the apple tissue in this

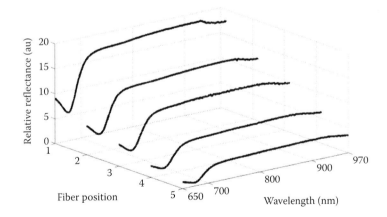

FIGURE 7.5 Spatially-resolved diffuse reflectance spectra of a fresh apple containing both skin and flesh layers. (From Nguyen Do Trong, N. et al. *Postharvest Biology and Technology* 91, 2014a: 39–48.)

SRS Technique for Measuring Optical Properties of Food

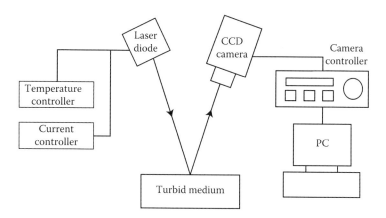

FIGURE 7.6 Diagram of a monochromatic imaging-based spatially-resolved spectroscopic (MISRS) system. (From Fabbri, F., M. A. Franceschini and S. Fantini. *Applied Optics* 42(16), 2003: 3063–3072.)

range. At each wavelength, the diffuse reflectance decreases with increasing the fiber position number or source–detector distance, because the photons that reach this fiber have traveled a longer path through the sample and thus have had more chance to be scattered or absorbed, as illustrated by the zigzag lines in Figure 7.4.

Recently, the potential of fiber-array probe SRS has been demonstrated for a quick estimation of the optical properties of turbid media such as food matrices (Watté et al., 2013), nondestructive characterization of the microstructure and texture of dried apple slices (Nguyen Do Trong et al., 2014b) and aerated sugar-containing foods (Herremans et al., 2013), and the quality of fresh apples during storage (Nguyen Do Trong et al., 2014a).

7.3.2 Monochromatic Imaging

Monochromatic imaging-based spatially-resolved spectroscopic (MISRS) technique as a noncontact method is suitable for measuring the optical properties of solid food products at one wavelength. Figure 7.6 shows a schematic diagram for an MISRS system. A laser diode is usually used as a light source to emit a beam at a specific wavelength to the sample surface. The detection of spatially-resolved diffuse reflectance is performed with a CCD camera that allows one to obtain an array of reflectance values simultaneously from the surface of an imaged medium.

Although the configuration of the MISRS method is relatively simple, issues related to the optics of detector, electronics, and adjacent pixels effects on the signal of each pixel need to be solved in the system development. Moreover, the true reflectance value of each pixel should be considered as a function of the signals of an area surrounding it, instead of just one pixel, which is also called the point-spread function (PSF) (Pilz et al., 2008). The PSF of a camera can have a significant effect on the accuracy of image intensity calibration and measurement (Du and Voss, 2004), which would finally affect the measurement accuracies of optical properties with the MISRS method. Many methods have been used to obtain the PSF of a camera system (Huang et al., 2002; Zandhuis et al., 1997). The study conducted by Pilz et al. (2008) indicated that errors in determining absorption and reduced scattering coefficients could be reduced by considering the PSF of a camera in the diffusion model.

To fulfill the requirement of the diffusion theory for steady-state spatially-resolved measurements, a laser beam with a selected wavelength is usually employed as a point light source. This would restrict MISRS to measure the optical properties only at single or several wavelengths. The MISRS method was used in the biomedical field for measuring the optical properties of Intralipid-ink phantoms, tissue phantoms, and of the human skin (Foschum et al., 2011; Pilz and Kienle, 2007; Zhang et al., 2010). Some studies were also reported on employing MISRS for describing scattering profiles in the applications of fruit-quality characterization (Baranyai and Zude, 2009; Lorente et al., 2013).

7.3.3 Hyperspectral Imaging

Most research on noncontact reflectance imagery mode was only able to provide optical property information at single or several wavelengths. Hyperspectral imaging-based spatially-resolved spectroscopic (HISRS) technique offers an effective solution for obtaining the optical properties of food for a broad spectral range simultaneously with a relatively fast measurement. A typical hyperspectral imaging system in the line scan mode is shown in Figure 7.7. It mainly consists of a high performance CCD camera that has a large dynamic range and low noise levels (achieved via deep cooling of the CCD detectors), an imaging spectrograph, a zoom or prime lens, a

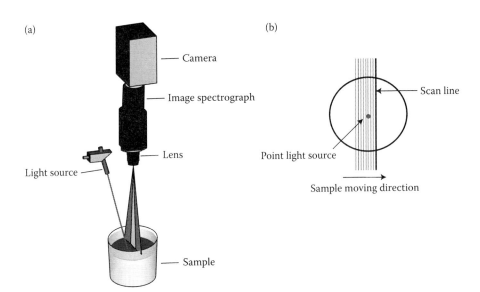

FIGURE 7.7 HISRS method: (a) schematic showing the major components of the system and (b) top view of a multi-line-scanning mode for acquiring spatially-resolved reflectance profiles. (From Cen, H. and R. Lu. *Optics Express* 18(16), 2010: 17412–17432.)

light source, and an optical fiber coupled with a focusing lens for delivering a small beam to the sample. This technique, combining imaging and spectroscopy methods, acquires both spectral and spatial information simultaneously, and it is, therefore, ideally suitable for measuring spatially-resolved diffuse reflectance profiles for a broad spectral range. To improve the repeatability of measurements, multiple line scans as shown in Figure 7.7b are usually taken from the investigated sample as the sample is moving at a predetermined velocity during image acquisition.

The procedure of extracting optical properties from the acquired hyperspectral image is described in the following example. Figure 7.8a shows a typical hyperspectral reflectance image for a liquid model sample made up of Intralipid (Sigma-Aldrich Inc., St. Louis, MO, USA), a standardized soybean oil-in-water emulsion, as scatterer, and blue dye (Direct Blue 71 and Naphthol Green B, Sigma-Aldrich Inc., St. Louis, MO, USA) as absorber. Each horizontal line taken from the image represents one spatially-resolved reflectance profile at a specific wavelength, as shown in Figure 7.8b. Hence, the reflectance image with a spectral resolution of 4.55 nm, in effect, consists of 111 spatially-resolved reflectance profiles in the 500–1000 nm wavelength range. Since the reflectance profiles are symmetric to the light incident point (or the peak point in Figure 7.8b), the two sides can be averaged to extract the optical absorption and reduced scattering coefficients. Each averaged spatially-resolved reflectance profile is then fitted by the diffusion model given in Equation 7.1 or 7.4, using an inverse algorithm, from which the spectra of absorption and reduced scattering coefficients for 500–1000 nm are obtained.

While the HISRS technique is useful for optical characterization of food with a relatively simple design and easy implementation, two key factors (i.e., light beam and source–detector distance) need to be carefully considered in the design and optimization of the HISRS optical system to meet the requirements of diffusion theory. The HISRS system uses a continuous-wave point light beam to illuminate

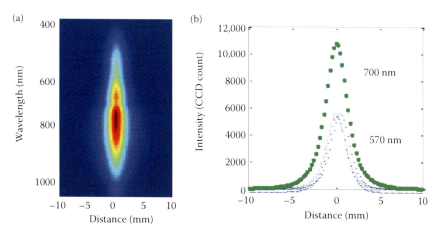

FIGURE 7.8 Hyperspectral reflectance image of a liquid model sample: (a) 2-D display with intensities being indicated by pseudo colors and (b) extracted spatially-resolved reflectance profiles at 570 and 700 nm. (From Cen, H. and R. Lu. *Optics Express* 18(16), 2010: 17412–17432.)

the sample. The shape and size of the beam can directly affect measurement accuracies. It is therefore important to examine and optimize the light beam. Extensive studies have been carried out to investigate the light beam characteristics through Monte Carlo simulation and/or experimental measurement of actual beam profiles (Cen and Lu, 2010). It was found that the beam in the system should be less than 1 mm in size to keep the error on the estimated bulk optical properties below 5%. Although a smaller beam is preferred, other factors such as light intensity or throughput and measurement repeatability also need to be considered in the optical design. In addition, accurate information on the source–detector distance, including minimum source–detector distance (r_{min}), maximum source–detector distance (r_{max}), and spatial resolution, are also required for determining the range of the spatially-resolved reflectance profile. Cen and Lu (2010) suggested that the optimal minimum source–detector distance should be around 1–4 mfp', which is consistent with the suggestion of Farrell et al. (1992) to choose r_{min} larger than 1 mfp', but different from the suggestion of Nichols et al. (1997) (0.75–1 mfp'). However, in practice, there is no *a priori* information about the exact values of μ_a and μ_s' over a specified wavelength range, and thus, it is difficult to determine the optimal minimum source–detector distance based on these loose criteria. Cen and Lu (2010) also recommended that the optimal maximum source–detector distance, which varies with the values of μ_a and μ_s', should be approximately between 10 and 20 mfp', and it is in general agreement with the studies of Nichols et al. (1997) and Pham et al. (2000). However, it should be noted that, in practice, the optimal maximum source–detector distance may be largely determined by the SNR with the minimum value of 20 since the SNR decreases with the increasing source–detector distance. Moreover, spatial resolutions between 0.07 and 0.25 mm per pixel were found to be appropriate for measuring the optical properties, as a higher resolution (e.g., 0.01 mm/pixel) did not result in improved measurement accuracy (Cen and Lu, 2010).

By incorporating optimization of the optical design and the algorithm, an optical property-measuring instrument, named "Optical Property Analyzer" (OPA), was developed based on the HISRS technique. This general-purpose optical instrument (Figure 7.9) consists of a line-scanning hyperspectral imaging system with an electron-multiplying CCD sensor and two separate light sources (Cen et al., 2012a). A point light source is used to generate spatially-resolved diffuse reflectance images for the optical property measurement, whereas a line light source is for general hyperspectral imaging applications. The performance of the OPA has been evaluated thoroughly for accuracy, stability, precision/reproducibility, and sensitivity by using liquid model samples. The average estimated error for all model samples was 24% for μ_a and 7% for μ_s'. It should be mentioned that the absolute values of μ_a for the test samples were very small (at least one order smaller than μ_s' for most of the test samples), thus causing the relatively large error of estimating μ_a compared with that of μ_s'. The system reproducibility (or precision), as measured by the coefficient of variation (or CV) with an absorption peak of 555 nm, was less than 10% for μ_a, and less than 4% for μ_s'. Additionally, the minimum detectable value of μ_a was 0.0117 cm^{-1}. The sensitivity of μ_s', as determined by the CV values, was always less than 3% because μ_s' was much larger than μ_a for the investigated range of $7.0 \leq \mu_s' \leq 39.9 \, cm^{-1}$. These results have shown that the OPA has achieved acceptable accuracies for measuring

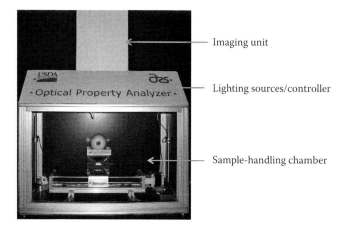

FIGURE 7.9 Photograph of the multipurpose OPA for measuring the optical absorption and scattering properties of food. (From Lu, R. and H. Cen. Non-destructive methods for food texture assessment. In *Instrumental Assessment of Food Sensory Quality*, 230–254. UK: Woodhead Publishing, 2013.)

the absorption and reduced scattering coefficients. The instrument has been used to measure the absorption and scattering spectra of peaches and apples, tomatoes, and sugar beets and also predict the quality of fruits (i.e., firmness, sucrose, and soluble solids) (Cen et al., 2012a,b; Zhu et al., 2015). Application examples are given in Section 7.4 of the use of OPA for measuring the optical properties of food products and studying the microstructural characteristics of apple tissues.

7.3.4 Spatial Frequency-Domain Imaging

SFDI, as a novel noncontact optical imaging method, was originally proposed by Cuccia et al. (2005). The technique uses spatially modulated illumination for optical characterization of investigated samples in the spatial frequency domain (SFD). The SFDI method quantifies absorption and reduced scattering coefficients by analyzing spatial frequency-dependent reflectance. The principle of this technique is based on the diffusion theory by introducing a spatially modulated source into the steady-state diffusion equation, and is related to the spatially-resolved measurement using Fourier transform. The SFDI technique integrates the determination of optical properties with depth-resolved imaging of sample heterogeneity using modulated illumination.

The general configuration of an SFDI platform includes a digital projector as the illumination source and a CCD camera as the detector. A schematic diagram of an SFDI system for multispectral imaging developed by Anderson et al. (2007) is shown in Figure 7.10. In this system, a broadband near-infrared (NIR) digital projector that consists of a 250-W quartz-halogen lamp, a 1025×768 binary digital micromirror device, commercial projector light engine, and a fixed focal length projection lens, is used to illuminate the sample with sinusoidal patterns of various spatial frequencies in one direction. A CCD camera system with a liquid crystal-tunable filter (LCTF)

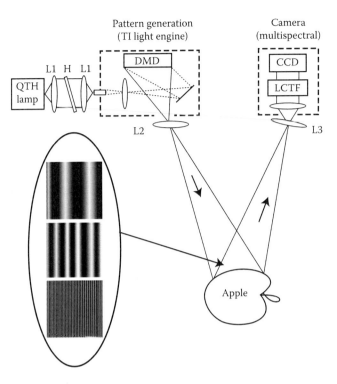

FIGURE 7.10 SFDI system: CCD camera, DMD—digital micromirror device, H—hybrid hot mirror, L1—aspheric condenser, L2—projection lens, L3—camera lens, LCTF—liquid crystal-tunable filter, QTH—quartz tungsten halogen. (From Anderson, E. R., D. J. Cuccia and A. J. Durkin. Detection of bruises on Golden Delicious apples using spatial-frequency-domain imaging. In *Advanced Biomedical and Clinical Diagnostic Systems V—Proceedings of SPIE*, 6430, 2007.)

is employed to image the diffusely reflected light at multiple wavelengths. In addition, specular reflection is eliminated by using crossed polarizers mounted in front of the illumination source and the camera. From the acquired reflectance image, the absorption and reduced scattering coefficients are obtained by fitting the SFD reflectance profile to a modified diffusion model.

Like other spatially-resolved techniques, several factors in instrumentation related to projection boundary effects, binning of SFD images, sample–camera distance, and calibration of spatial frequencies have to be considered to achieve an accurate measurement of optical properties. Bodenschatz et al. (2014) presented a thorough discussion on the errors caused by these four factors in the SFDI method for quantitative characterization of optical properties. Experimentally, the projection of spatial frequency is affected by a limited spatial projection area. This boundary effect causes a decrease in reflectance, and results in errors in computing absorption and reduced scattering coefficients. It is required to have a distance from the projection boundaries in millimeters to achieve a deviation in SFD reflectance of <1% and to keep the errors on the estimated bulk optical properties μ_a and μ_s' below 5%.

It was also found that μ_a is more sensitive to small deviations in reflectance than μ_s'. In addition, the spatial frequency also has an effect on the SFD reflectance close to the projection boundaries, and the relative error in SFD reflectance at the projection boundaries increases slightly with higher spatial frequencies. Hence, both sensitivity to short-range photons and the boundary effect should be considered in the selection of spatial frequency. Binning of the acquired images can reduce noise and increase the acquisition speed. However, strong binning of SFD images could raise errors in SFD reflectance. A proper binning of images should be determined and the spatial resolution of SFDI data should be large enough to ensure several sampling points within one spatial oscillation. Meanwhile, it is essential to have a precise distance between the sample and the detector for an accurate measurement of SFD reflectance. Bodenschatz et al. (2014) reported that an increase in SFD reflectance can lead to an underestimation of absorption and an apparent increase in scattering. An accurate calibration of spatial frequency is thus necessary to avoid the errors in the values of μ_a and μ_s' due to inaccuracy in the projected spatial frequency pattern.

Although SFDI is a relatively new method, it has shown the potential for optical characterization of biological materials. The technique has been tested on various tissue-simulating model samples such as homogenous siloxane TiO_2 phantoms (Cuccia et al., 2005), homogenous and layered Intralipid phantoms (Bodenschatz et al., 2014; Weber et al., 2011), and poly(dimethylsiloxane)-based multilayer phantom tissues (Greening et al., 2014). SFDI has also been used to nondestructively quantify absorption and scattering image maps of apple fruit to distinguish the bruise regions from the nonbruised regions in apples (Anderson et al., 2007). A considerable difference in the reduced scattering coefficients between the bruise and nonbruised regions for the different levels of bruising at the spectral region of 650–980 nm were observed.

7.4 MEASUREMENT OF OPTICAL PROPERTIES OF FOOD PRODUCTS

SRS has recently been used as a nondestructive and fast method for the optical characterization of food products including fruits, vegetables, meat, milk, and other foods. The technique not only provides an important means for analyzing light propagation inside the turbid medium, but also offers separate quantitative information on the absorption and scattering properties. This could be potentially useful for imaging and understanding the interior structures of food products, and eventually for developing effective optical techniques for quality evaluation and safety inspection of food. In this section, we offer several examples on measurement of the optical properties of food products by using SRS and its relationship with the structural properties of apple tissues.

7.4.1 OPTICAL PROPERTIES OF FOOD PRODUCTS

Light absorption and scattering properties carry the information related to the chemical composition and physical structure of food products. Optical absorption and reduced scattering coefficients for food products such as fruits, vegetables, and meat measured using a spatially-resolved spectroscopic technique were recently reported in the literature. In Figure 7.10, the typical absorption and scattering coefficient spectra are shown for five fruit and vegetable samples, three meat samples, and three liquid food samples. The μ_a spectra of "Golden Delicious" (GD), "Delicious," and "Granny Smith" (GS) apples and a "Redstar" peach fruit (Figure 7.11a) have absorption peaks at 675 nm, which corresponds to the chlorophyll-a absorption waveband, and the μ_a values range from 0.10 to 0.48 cm^{-1} at this wavelength. The "GS" apple has the highest chlorophyll absorption due to its greenish

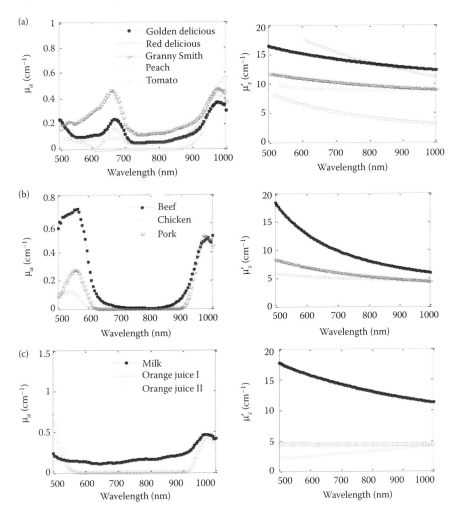

FIGURE 7.11 Absorption (left pane) and reduced scattering coefficient (right pane) spectra for (a) three apples of different varieties, a peach fruit, and a tomato fruit; (b) beef, chicken, and pork; and (c) a full-fat milk sample and two orange juice samples from two different sources. (From Lu, R. and H. Cen. Measurement of food optical properties. In *Hyperspectral Imaging Technology in Food and Agriculture*, 203–226, Springer, 2015.)

skin and flesh. Similar patterns in the absorption spectra of other fruits such as kiwifruits, pears, "Fuji" apples, and "Braeburn" apples were reported in the studies of Nguyen Do Trong et al. (2014a) and Qin and Lu (2008). Tomatoes at the "red" stage do not show an absorption peak at 675 nm, because chlorophyll-a in fully ripened tomatoes decreases greatly or even disappears completely. In fully ripened tomatoes, anthocyanin becomes the dominant pigment, which absorbs light at 535 nm, as shown in Figure 7.11a. Qin and Lu (2008) and Lu et al. (2011) also reported that absorption peaks due to chlorophyll-a absorption at 675 nm were not obvious for the samples of plum, cucumber, and zucchini squash. The absorption values of the fruit and vegetable samples in the 720–900 nm range are relatively small and consistent, but they increase dramatically above 900 nm and peak at 970 nm at the absorption peak of water. Compared with μ_a spectra, the reduced scattering coefficient spectra of fruit and vegetable samples are relatively flat and have fewer features. For most of the tested samples, their μ'_s values decrease steadily with the increasing wavelength (see plots on the right pane of Figure 7.11a). This pattern of changes is consistent with Mie-scattering theory and other reported studies that scattering is wavelength dependent (Keener et al., 2007; Michels et al., 2008). The apple samples have higher μ'_s values (9.0–17.0 cm^{-1}) over the entire spectral region from 500 to 1000 nm, while the tomato has the lowest μ'_s values (4.5–6.0 cm^{-1}) (Figure 7.11a). This can be explained by the fact that apple tissue is very porous with air between the cells, which creates larger refractive index mismatches than the water-filled tomato tissue.

For meat products, it has been known that the quality attributes such as tenderness, juiciness, and flavor are mostly affected by the fat, collagen, and myofibrillar proteins, which are associated with the light scattering and absorption properties of meat (Emerson et al., 2013; Lida et al., 2014; Maltin et al., 2003). In the absorption coefficient spectra of the beef, chicken, and pork samples as shown in Figure 7.11b, we observe two prominent absorption peaks at 560 and 970 nm. The absorption peak at 560 nm could be attributed to the combined effect of myoglobin, oxymyoglobin, metmyoglobin, and oxyhemoglobin (Xia et al., 2007), while the absorption at 970 nm is caused by water absorption in the meat samples. The three meat samples also have different μ'_s values over the entire wavelengths; the beef has the highest μ'_s values, while the chicken sample has the lowest (Figure 7.11b). Xia et al. (2007) measured absorption and reduced scattering coefficients at wavelengths in the range of 450–950 nm using a fiber-optic probe-based SRS method, and compared the absorbance spectra with the absorption and reduced scattering coefficient spectra of *semimembranosus* muscle and *psoas major* beef muscle samples. They found that there was not much difference in the absorption spectra for these two samples, while the reduced scattering coefficient spectra were significantly different for these two samples over the entire wavelength range. The result demonstrated that light scattering could be more effective to differentiate different beef samples than the conventional optical method. A detailed description on studying the structural characteristics of meat muscles during postmortem aging, and of meat analogs, using the SRS technique and other related techniques is presented in Chapter 9.

SRS technique was also used for measuring the optical properties of liquid food products such as fruit juice and milk. Figure 7.11c shows that the absorption spectra of orange juice samples over 550–900 nm are relatively flat, which is consistent with the study reported by Qin and Lu (2007). There is a steep rise toward 500 nm due to the carotenoids that make the orange juice orange, and one prominent absorption peak is also found at 970 nm due to absorption by water. For the full-fat milk sample obtained from a local supermarket, there are several small absorption peaks and one dominant peak at 970 nm caused by water absorption. In addition, the milk sample has distinctly higher μ_s' values than the orange juice samples over the spectral region of 500–1000 nm. Fat globules and casein micelles are the main, excellent scattering particles in milk. Qin and Lu (2007) measured the optical properties of five milk samples with different levels of fat contents from 0% to 3.25% by the HISRS technique. It was found that μ_a and μ_s' values are highly correlated with the fat content of milk. Aernouts et al. (2015) reported that the fat globule size of milk samples has a significant effect on the bulk optical properties of milk at the wavelength region of 500–1900 nm, and the bulk-scattering coefficient and scattering anisotropy factor reduce and become more dependent on the wavelength with the diameter of the fat globules decreasing. Considering a scattering particle as a sphere, the scattering ability depends on its density, size, and complex refractive indices of the particle and the medium; while the final reduced scattering coefficient μ_s' is related to the scattering efficiency factor Q_s and the probability distribution of scattering angles $p(\theta)$ based on Mie theory. Q_s and $p(\theta)$ are complicated, and the wavelength dependence of Q_s varies greatly with both size and refractive index of the scatterer. Therefore, the value of μ_s' and its pattern of change with wavelength can provide useful information on the structural and physical characteristics of food products. If we assume that the medium has a polydisperse system with a smooth particle size distribution and the particle size close to a certain region in which $r \geq \lambda$ but $(2\pi r n_m/\lambda) < 70$ (r is the particle radius and n_m is the refractive index of the medium), the wavelengths dependence of μ_s' at 350–950 nm can be approximated to an empirical function $\mu_s' = a\lambda^{-b}$, which describes the relationship between μ_s' and wavelength λ, where a is proportional to the density of the scattering particles, and b depends on the particle size (Mourant et al., 1997). The empirical equation simplifies the computation of μ_a and μ_s'; the potential errors caused by this approximation may need to be considered for some specific samples. It was noted that in Figure 7.11, the μ_s' value calculated by the empirical equation for an orange juice sample increases with the wavelength. This is probably due to the large particle size or the large size variation of the scatterers in orange juice (Aernouts et al., 2015; Betoret et al., 2009), which could violate the assumption when using the empirical exponential decay function to calculate the μ_s' values. Furthermore, the validation of this empirical equation needs to be investigated for the wavelength above 950 nm. Although a reasonable absorption spectral pattern for the milk sample was obtained, the μ_a value may be overestimated over the entire wavelength range due to the very high scattering, as the absorption coefficient of milk is considered to be similar to that of water as reported by Aernouts et al. (2015). More research is needed related to the validation of the empirical equation to obtain an accurate measurement of both μ_a and μ_s' values for different food products.

7.4.2 Relationship between Optical Properties and Structural Properties of Apple Fruit

It is generally accepted that optical absorption and scattering properties change with the structural properties inside the food product. However, it has not been well understood until very recently how these changes are related to the structural properties of food products. In this section, we present some recent findings from the study of Cen et al. (2013) on the relationship of optical absorption and scattering properties with the microstructural and mechanical properties of apple fruit. Applications of optical absorption and scattering properties for quality assessment of fruits and vegetables using the spatially-resolved spectroscopic technique as well as the time-resolved technique are presented in Chapters 8 and 9, respectively.

A study was conducted to determine the relationship between the optical properties measured by the HISRS method and the basic structural and mechanical properties of "GD" and "GS" apples (Cen et al., 2013) during an accelerated softening storage over a period of 30 days. The firmness of apples was measured using an acoustic/impact firmness sensor, whereas compression tests were performed to measure the fundamental mechanical properties of fruit tissue specimens. To quantify the structural characteristics of apple tissues, confocal laser-scanning microscopy (CLSM) was carried out to measure the microstructural changes in the flesh tissue specimens that were stored at 22°C and 95% relative humidity for 1–30 days. Image processing was subsequently performed to extract the quantitative information (cell area, diameter, etc.) of individual cells from the CLSM images of the apple tissue specimens. Finally, linear regression analysis was carried out to correlate acoustic/impact firmness, Young's modulus and the cell's area, and equivalent diameter with the absorption and scattering properties of apple fruit.

It was found that the changes or decreases in the optical absorption and scattering coefficients were accompanied with decreases in the acoustic/impact firmness and Young's modulus of apple fruit. Low to very-high correlations between the average values of the four optical parameters ($\mu_{a,675\,nm}$, $\mu'_{s,675\,nm}$, $a_{\mu_s'}$, and $b_{\mu_s'}$) and the average values of acoustic and impact firmness (*FI* and *IF*) and Young's modulus (*E*) for "GD" and "GS" apples for the five test dates were obtained (Table 7.1). $\mu_{a,675\,nm}$ and $\mu'_{s,675\,nm}$ represent the absorption and reduced scattering coefficients at a wavelength of 675 nm, which corresponds to the chlorophyll-*a* absorption waveband, and revealed the main characteristic of the absorption spectra. Since the chlorophyll content in the apple is related to fruit ripeness and senescence, it would be interesting to perform correlation analysis at this important wavelength. $a_{\mu_s'}$ and $b_{\mu_s'}$ are two parameters in the wavelength-dependent exponential function $\left(\mu'_s = a_{\mu_s'} \lambda^{-b_{\mu_s'}}\right)$ used to quantify the entire μ'_s spectra. Table 7.2 presents correlations between the average values of the optical parameters and cell size parameters. Results indicated that the optical absorption and reduced scattering coefficients were positively correlated with the cell's area and equivalent diameter, although the correlations varied greatly for the two apple cultivars. These findings suggest that the spatially-resolved spectroscopic technique is useful for determining the optical properties of food products,

TABLE 7.1
Correlations of Selected Optical Parameters with Acoustic and Impact Firmness and Young's Modulus for Golden Delicious "GD" and "GS" Apples, Using the Average Values of Each Test Date[a]

Optical Parameters	GD			GS		
	FI	IF	E	FI	IF	E
$\mu_{a,675\ nm}$	0.870	0.918	0.585	0.334	0.421	0.292
$\mu'_{s,675\ nm}$	0.903	0.932	0.766	0.993	0.992	0.694
$a_{\mu s'}$	0.948	0.939	0.947	0.902	0.941	0.584
$b_{\mu s'}$	0.924	0.938	0.804	0.974	0.974	0.620

Source: Adapted from Cen, H. Y. et al. *Postharvest Biology and Technology* 85, 2013: 30–38.

[a] $\mu_{a,675\ nm}$ and $\mu'_{s,675\ nm}$ are the absorption and reduced scattering coefficients at 675 nm, and $a_{\mu s'}$ and $b_{\mu s'}$ are the parameters related to the μ'_s spectra ($\mu'_s = a_{\mu s'}\lambda^{-b_{\mu s'}}$, where λ is the wavelength in nm); *FI* and *IF* are the acoustic firmness index and impact firmness index, respectively, and *E* is Young's modulus.

TABLE 7.2
Correlations between the Optical Parameters and Cell Size Parameters Extracted from the Confocal Laser-Scanning Microscopic Images of "GD" and "GS" Apples, Using the Average Values of Each Test Date

Optical Parameter	GD		GS	
	Area	Equivalent Diameter	Area	Equivalent Diameter
$\mu_{a,675\ nm}$	0.934	0.941	0.630	0.607
$\mu'_{s,675\ nm}$	0.777	0.657	0.581	0.572
$a_{\mu s}$	0.767	0.774	0.903	0.906
$b'_{\mu s}$	0.660	0.657	0.850	0.845

Source: Adapted from Cen, H. Y. et al. *Postharvest Biology and Technology* 85, 2013: 30–38.

and for gaining new knowledge on the complex relationship between the optical and structural properties of food products.

7.5 CONCLUSIONS AND FUTURE TRENDS

SRS provides an effective means for optical characterization of turbid or diffuse media. The technique, originally developed for measuring biological tissues in the biomedical area, has now been extended to food applications thanks to recent advances in the development of more cost-effective, practical measurement configurations. The technique allows separating the absorption and scattering properties, which cannot be achieved with conventional NIR point spectroscopic measurement.

Several different measurement configurations (i.e., fiber-optic probe, monochromatic imaging, hyperspectral imaging, and SFDI) have been recently developed for food and agricultural products. Each measurement configuration has shown its merits and shortcomings in determining the optical properties of food products. Considerable challenges for the technique still remain in accurate and reliable measurement of optical properties for a large class of food products. These challenges, however, also offer future research opportunities to make this new, emerging technique applicable for a wide range of food products.

Compared with conventional visible and NIR spectroscopy, SRS technique generally requires more sophisticated instrumentation and complex mathematical models and algorithms. Complex structures of food products, combined with unknown factors in the real experiment, make it challenging to evaluate the system performance, where many error sources in instrumentation and the sample are intertwined. Factors related to the instrumentation such as characteristics of the light source, dynamic range of the camera, and source–detector arrangement should be carefully considered in the design of a spatially-resolved system. Although a reference method and model samples (or phantom tissues, as commonly called in the biomedical research) are required for system calibration, optimization, and evaluation, no standard procedure has been established by the scientific community. This makes it difficult to compare and evaluate the performance of different instruments developed by different research groups. Other factors such as sample surface condition or geometry (i.e., roughness, unevenness, or irregularity) can introduce additional uncertainty and variability for the measurement of optical properties. There are still great difficulties in measuring the optical absorption and scattering coefficients for multilayer or heterogenous media, because of the complexity of solving the diffusion model and a large number of unknown parameters to be estimated (Cen and Lu, 2009). Further research is thus needed to address these critical issues in the development and application of a spatially-resolved spectroscopic technique for food and agricultural products.

Solving inverse radiative transfer equations is a critical step for the measurement of optical properties using the SRS technique. It is important to develop fast, efficient inverse algorithms. Analytical diffusion models are based on certain assumptions (e.g., scattering domination) and for simplified geometries (e.g., semi-infinite media). It is thus essential to evaluate the limitations and application scopes of these models when they are applied to food products. On the other hand, the inverse metamodeling approaches are limited to the range of optical properties for which they have been trained. Moreover, additional errors in the optical properties measurement could be introduced if the inverse algorithm is not optimized. The novel Monte Carlo simulation methods presented in Chapter 4 and the parameter estimation methods presented in Chapter 5 should thus be considered in conducting the sensitivity analysis and in developing the optimal inverse algorithm to estimate the optical parameters.

REFERENCES

Aernouts, B., R. Van Beers, R. Watté, T. Huybrechts, J. Jordens, D. Vermeulen, T. Van Gerven, J. Lammertyn and W. Saeys. Effect of ultrasonic homogenization on the VIS/NIR bulk optical properties of milk. *Colloids and Surfaces B: Biointerfaces* 126, 2015: 510–519.

Anderson, E. R., D. J. Cuccia and A. J. Durkin. Detection of bruises on Golden Delicious apples using spatial-frequency-domain imaging. In *Advanced Biomedical and Clinical Diagnostic Systems V—Proceedings of SPIE*, 6430, 2007, San Jose, CA.

Baranyai, L. and M. Zude. Analysis of laser light propagation in kiwi fruit using backscattering imaging and Monte Carlo simulation. *Computers and Electronics in Agriculture* 69(1), 2009: 33–39.

Betoret, E., N. Betoret, J. V. Carbonell and P. Fito. Effects of pressure homogenization on particle size and the functional properties of citrus juices. *Journal of Food Engineering* 92(1), 2009: 18–23.

Bodenschatz, N., A. Brandes, A. Liemert and A. Kienle. Sources of errors in spatial frequency domain imaging of scattering media. *Journal of Biomedical Optics* 19(7), 2014: 071405.

Cen, H. and R. Lu. Quantification of the optical properties of two-layer turbid materials using a hyperspectral imaging-based spatially-resolved technique. *Applied Optics* 48(29), 2009: 5612–5623.

Cen, H. and R. Lu. Optimization of the hyperspectral imaging-based spatially-resolved system for measuring the optical properties of biological materials. *Optics Express* 18(16), 2010: 17412–17432.

Cen, H., R. Lu, F. A. Mendoza and D. P. Ariana. Assessing multiple quality attributes of peaches using optical absorption and scattering properties. *Transactions of the ASABE* 55(2), 2012a: 647–657.

Cen, H. Y., R. F. Lu and F. A. Mendoza. Analysis of absorption and scattering spectra for assessing the internal quality of apple fruit. *Acta Horticulturae* 945, 2012b: 181–188.

Cen, H. Y., R. F. Lu, F. Mendoza and R. M. Beaudry. Relationship of the optical absorption and scattering properties with mechanical and structural properties of apple tissue. *Postharvest Biology and Technology* 85, 2013: 30–38.

Cletus, B., R. Kunnemeyer, P. Martinsen, A. McGlone and R. Jordan. Characterizing liquid turbid media by frequency-domain photon-migration spectroscopy. *Journal of Biomedical Optics* 14(2), 2009: 024041.

Cuccia, D. J., F. Bevilacqua, A. J. Durkin and B. J. Tromberg. Modulated imaging: Quantitative analysis and tomography of turbid media in the spatial-frequency domain. *Optics Letters* 30(11), 2005: 1354–1356.

Doornbos, R. M. P., R. Lang, M. C. Aalders, F. W. Cross and H. J. C. M. Sterenborg. The determination of *in vivo* human tissue optical properties and absolute chromophore concentrations using spatially resolved steady-state diffuse reflectance spectroscopy. *Physics in Medicine and Biology* 44(4), 1999: 967–981.

Du, H. and K. J. Voss. Effects of point-spread function on calibration and radiometric accuracy of CCD camera. *Applied Optics* 43(3), 2004: 665–670.

Emerson, M. R., D. R. Woerner, K. E. Belk and J. D. Tatum. Effectiveness of USDA instrument-based marbling measurements for categorizing beef carcasses according to differences in longissimus muscle sensory attributes. *Journal of Animal Science* 91(2), 2013: 1024–1034.

Erkinbaev, C., E. Herremans, N. Nguyen Do Trong, E. Jakubczyk, P. Verboven, B. Nicolaï and W. Saeys. Contactless and non-destructive differentiation of microstructures of sugar foams by hyperspectral scatter imaging. *Innovative Food Science and Emerging Technologies* 24, 2014: 131–137.

Fabbri, F., M. A. Franceschini and S. Fantini. Characterization of spatial and temporal variations in the optical properties of tissue-like media with diffuse reflectance imaging. *Applied Optics* 42(16), 2003: 3063–3072.

Falconet, J., A. Laidevant, R. Sablong, A. da Silva, M. Berger, F. Jaillon, E. Perrin, J. M. Dinten and H. Saint-Jalmes. Estimation of optical properties of turbid media: Experimental comparison of spatially and temporally resolved reflectance methods. *Applied Optics* 47(11), 2008: 1734–1739.

Farrell, T. J., M. S. Patterson and B. Wilson. A diffusion-theory model of spatially resolved, steady-state diffuse reflectance for the noninvasive determination of tissue optical-properties *in vivo*. *Medical Physics* 19(4), 1992: 879–888.

Foschum, F., M. Jäger and A. Kienle. Fully automated spatially resolved reflectance spectrometer for the determination of the absorption and scattering in turbid media. *Review of Scientific Instruments* 82(10), 2011: 103104.

González-Rodríguez, P. and A. D. Kim. Light propagation in two-layer tissues with an irregular interface. *Journal of the Optical Society of America A: Optics and Image Science, and Vision* 25(1), 2008: 64–73.

Greening, G. J., R. Istfan, L. M. Higgins, K. Balachandran, D. Roblyer, M. C. Pierce and T. J. Muldoon. Characterization of thin poly(dimethylsiloxane)-based tissue-simulating phantoms with tunable reduced scattering and absorption coefficients at visible and near-infrared wavelengths. *Journal of Biomedical Optics* 19(11), 2014: 115002.

Groenhuis, R. A. J., H. A. Ferwerda and J. J. Tenbosch. Scattering and absorption of turbid materials determined from reflection measurements.1. Theory. *Applied Optics* 22(16), 1983: 2456–2462.

Haskell, R. C., L. O. Svaasand, T. T. Tsay, T. C. Feng and M. S. McAdams. Boundary-conditions for the diffusion equation in radiative-transfer. *Journal of the Optical Society of America A—Optics Image Science and Vision* 11(10), 1994: 2727–2741.

Herremans, E., E. Bongaers, P. Estrade, E. Gondek, M. Hertog, E. Jakubczyk, N. Nguyen Do Trong et al. Microstructure–texture relationships of aerated sugar gels: Novel measurement techniques for analysis and control. *Innovative Food Science and Emerging Technologies* 18, 2013: 202–211.

Hollmann, J. L. and L. V. Wang. Multiple-source optical diffusion approximation for a multi-layer scattering medium. *Applied Optics* 46(23), 2007: 6004–6009.

Huang, C., J. R. G. Townshend, S. N. V. Kalluri and R. S. De Fries. Impact of sensor's point spread function on land cover characterization: Assessment and deconvolution. *Remote Sensing of Environment* 80, 2002: 203–212.

Keener, J. D., K. J. Chalut, J. W. Pyhtila and A. Wax. Application of Mie theory to determine the structure of spheroidal scatterers in biological materials. *Optics Letters* 32(10), 2007: 1326–1328.

Kienle, A. and M. S. Patterson. Improved solutions of the steady-state and the time-resolved diffusion equations for reflectance from a semi-infinite turbid medium. *Journal of the Optical Society of America A—Optics Image Science and Vision* 14(1), 1997: 246–254.

Kienle, A., M. S. Patterson, N. Dognitz, R. Bays, G. Wagnieres and H. van den Bergh. Noninvasive determination of the optical properties of two-layered turbid media. *Applied Optics* 37(4), 1998: 779–791.

Langerholc, J. Beam broadening in dense scattering media. *Applied Optics* 21(9), 1982: 1593–1598.

Liao, Y. K. and S. H. Tseng. Reliable recovery of the optical properties of multi-layer turbid media by iteratively using a layered diffusion model at multiple source–detector separations. *Biomedical Optics Express* 5(3), 2014: 975–989.

Lida, F., K. Saitou, T. Kawamura, S. Yamaguchi and T. Nishimura. Effect of fat content on sensory characteristics of marbled beef from Japanese black steers. *Animal Science Journal* 86(7), 2014: 707–715.

Lorente, D., M. Zude, C. Regen, L. Palou, J. Gómez-Sanchis and J. Blasco. Early decay detection in citrus fruit using laser-light backscattering imaging. *Postharvest Biology and Technology* 86, 2013: 424–430.

Lu, R., D. P. Ariana and H. Cen. Optical absorption and scattering properties of normal and defective pickling cucumbers for 700–1000 nm. *Sensing and Instrumentation for Food Quality and Safety* 5, 2011: 51–56.

Lu, R. and H. Cen. Non-destructive methods for food texture assessment. In *Instrumental Assessment of Food Sensory Quality*, 230–254. UK: Woodhead Publishing, 2013.

Lu, R. and H. Cen. Chapter 8. Measurement of food optical properties. In *Hyperspectral Imaging Technology in Food and Agriculture*, B. Park and R. Lu (ed.), 203–226. Springer, New York, 2015.

Malsan, J., R. Gurjar, D. Wolf and K. Vishwanath. Extracting optical properties of turbid media using radially and spectrally resolved diffuse reflectance. In *Progress in Biomedical Optics and Imaging—Proceedings of SPIE*, 8936, 2014, San Francisco, CA.

Maltin, C., D. Balcerzak, R. Tilley and M. Delday. Determinants of meat quality: Tenderness. *Proceedings of the Nutrition Society* 62, 2003: 337–347.

Marquet, P., F. Bevilacqua, C. Depeursinge and E. B. Dehaller. Determination of reduced scattering and absorption-coefficients by a single charge-coupled-device array measurement 1. Comparison between experiments and simulations. *Optical Engineering* 34(7), 1995: 2055–2063.

Michels, R., F. Foschum and A. Kienle. Optical properties of fat emulsions. *Optics Express* 16(8), 2008: 5907–5925.

Mourant, J. R., T. Fuselier, J. Boyer, T. M. Johnson and I. J. Bigio. Predictions and measurements of scattering and absorption over broad wavelength ranges in tissue phantoms. *Applied Optics* 36(4), 1997: 949–957.

Nguyen Do Trong, N., C. Erikinbaev, M. Tsuta, J. De Baerdemaeker, B. Nicolaï and W. Saeys. Spatially resolved diffuse reflectance in the visible and near-infrared wavelength range for non-destructive quality assessment of Braeburn apples. *Postharvest Biology and Technology* 91, 2014a: 39–48.

Nguyen Do Trong, N., A. Rizzolo, E. Herremans, M. Vanoli, G. Cortellino, C. Erkinbaev, M. Tsuta et al. Optical properties–microstructure–texture relationships of dried apple slices: Spatially resolved diffuse reflectance spectroscopy as a novel technique for analysis and process control. *Innovative Food Science and Emerging Technologies* 21, 2014b: 160–168.

Nguyen Do Trong, N., R. Watté, B. Aernouts, E. Verhoelst, M. Tsuta, E. Jakubczyk, E. Gondek, P. Verboven, B. M. Nicolaï and W. Saeys. Differentiation of microstructures of sugar foams by means of spatially resolved spectroscopy. In *Proceedings of SPIE—The International Society for Optical Engineering*, 8439, 2012, Brussels, Belgium.

Nichols, M. G., E. L. Hull and T. H. Foster. Design and testing of a white-light, steady-state diffuse reflectance spectrometer for determination of optical properties of highly scattering systems. *Applied Optics* 36(1), 1997: 93–104.

Nicolai, B. M., B. E. Verlinden, M. Desmet, S. Saevels, W. Saeys, K. Theron, R. Cubeddu, A. Pifferi and A. Torricelli. Time-resolved and continuous wave NIR reflectance spectroscopy to predict soluble solids content and firmness of pear. *Postharvest Biology and Technology* 47(1), 2008: 68–74.

Pham, T. H., F. Bevilacqua, T. Spott, J. S. Dam, B. J. Tromberg and S. Andersson-Engels. Quantifying the absorption and reduced scattering coefficients of tissue-like turbid media over a broad spectral range with noncontact Fourier-transform hyperspectral imaging. *Applied Optics* 39(34), 2000: 6487–6497.

Pilz, M., S. Honold and A. Kienle. Determination of the optical properties of turbid media by measurements of the spatially resolved reflectance considering the point-spread function of the camera system. *Journal of Biomedical Optics* 13(5), 2008: 054047.

Pilz, M. and A. Kienle. Determination of the optical properties of turbid media by measurement of the spatially resolved reflectance. In *Progress in Biomedical Optics and Imaging—Proceedings of SPIE*, 6629, 2007, Munich, Germany.

Qin, J. and R. Lu. Measurement of the absorption and scattering properties of turbid liquid foods using hyperspectral imaging. *Applied Spectroscopy* 61(4), 2007: 388–396.

Qin, J. and R. Lu. Measurement of the optical properties of fruits and vegetables using spatially resolved hyperspectral diffuse reflectance imaging technique. *Postharvest Biology and Technology* 49(3), 2008: 355–365.

Qin, J., R. Lu and Y. Peng. Prediction of apple internal quality using spectral absorption and scattering properties. *Transactions of the ASABE* 52(2), 2009: 499–507.

Raulot, V., P. Gérard, B. Serio, M. Flury, B. Kress and P. Meyrueis. Modeling of the angular tolerancing of an effective medium diffractive lens using combined finite difference time domain and radiation spectrum method algorithms. *Optics Express* 18(17), 2010: 17974–17982.

Reynolds, L., C. Johnson and A. Ishimaru. Diffuse reflectance from a finite blood medium—Applications to modeling of fiber optic catheters. *Applied Optics* 15(9), 1976: 2059–2067.

Shampine, L. F. Vectorized adaptive quadrature in MATLAB. *Journal of Computational and Applied Mathematics* 211, 2008: 131–140.

Tseng, S. H., A. Grant and A. J. Durkin. *In vivo* determination of skin near-infrared optical properties using diffuse optical spectroscopy. *Journal of Biomedical Optics* 13(1), 2008: 014016.

Watté, R., N. Nguyen Do Trong, B. Aernouts, C. Erkinbaey, J. De Baerdemaeker, B. Nicolaï and W. Saeys. Metamodeling approach for efficient estimation of optical properties of turbid media from spatially resolved diffuse reflectance measurements. *Optics Express* 21(23), 2013: 32630–32642.

Weber, J. R., D. J. Cuccia, W. R. Johnson, G. H. Bearman, A. J. Durkin, M. Hsu, A. Lin, D. K. Binder, D. Wilson and B. J. Tromberg. Multispectral imaging of tissue absorption and scattering using spatial frequency domain imaging and a computed-tomography imaging spectrometer. *Journal of Biomedical Optics* 16(1), 2011: 011015.

Xia, J. J., E. P. Berg, J. W. Lee and G. Yao. Characterizing beef muscles with optical scattering and absorption coefficients in VIS–NIR region. *Meat Science* 75(1), 2007: 78–83.

Zandhuis, J., D. Pycock, S. Quigley and P. Webb. Sub-pixel non-parametric PSF estimation for image enhancement. *IEE ProceedingsVision Image Signal Processing* 14, 1997: 285–292.

Zhang, L., Z. Wang and M. Zhou. Determination of the optical coefficients of biological tissue by neural network. *Journal of Modern Optics* 57(13), 2010: 1163–1170.

Zhu, Q., C. He, R. Lu, F. Mendoza and H. Cen. Ripeness evaluation of "sun bright" tomato using optical absorption and scattering properties. *Postharvest Biology and Technology* 103, 2015: 27–34.

8 Time-Resolved Technique for Measuring Optical Properties and Quality of Food

Anna Rizzolo and Maristella Vanoli

CONTENTS

8.1 Introduction 187
8.2 Instrumentation and Data Analysis 189
    8.2.1 TRS Data Analysis 191
8.3 Absorption and Scattering Spectra in Various Fruit Species 193
8.4 Fruit Maturity 198
    8.4.1 Modeling of Softening 200
    8.4.2 Absorption and Maturity 204
        8.4.2.1 Fresh Fruit 205
        8.4.2.2 Processed Fruit 207
    8.4.3 Scattering and Maturity 208
8.5 Texture 209
    8.5.1 Correlations..209
    8.5.2 Regression Models 210
    8.5.3 Classification Models... 212
8.6 Internal Disorders 213
    8.6.1 Absorption and Scattering Spectra and Internal Disorders.............. 213
    8.6.2 Absorption and Reduced Scattering Coefficients
        and Internal Disorders 214
    8.6.3 Correlations and Threshold Values.. 216
    8.6.4 Classification Models... 217
8.7 Conclusions 218
References 219

8.1 INTRODUCTION

Time-resolved reflectance spectroscopy (TRS) provides nondestructive assessment of diffusive media with the simultaneous measurement of the bulk optical properties (absorption and scattering) (Cubeddu et al., 2001a).

    Briefly, in TRS, a short pulse of monochromatic light is injected into the diffusive medium; whenever a photon strikes a scattering center, it changes its trajectory and keeps on propagating in the medium, until it eventually re-emits across the boundary, or it is captured by an absorbing center. The temporal distribution of the re-emitted photons at a distance from the injection point will be delayed, broadened, and attenuated. By using an appropriate theoretical model of light transfer for the analysis of light attenuation with time (see Chapter 3 for further details), it is possible to simultaneously measure, as a function of wavelength, the scattering coefficient (μ_s) and the absorption coefficient (μ_a), which indicate the scattering probability per unit length and the absorption probability per unit length, respectively. To account for nonisotropic propagation of photons, the reduced scattering coefficient μ'_s is commonly used. A detailed description of the time-resolved approach for the determination of the optical properties in turbid media is given in Torricelli (2009).

The applications of TRS to the nondestructive assessment of bulk optical properties of foods were mostly focused on fruits and vegetables (apple, pear, plum, peach, nectarine, mango, and potato). As a rule of thumb, TRS would require a sample volume larger than 10 cm^3; however, it is the combination of optical properties and sample geometry that determines the applicability of TRS: small but strongly diffusive samples, such as cherries, produce a large broadening of laser pulses, whereas for samples with weak diffusion, such as gels, or in the presence of voids, such as in cereal flakes, TRS cannot provide robust results (Torricelli et al., 2013).

Fruit are complex systems and many factors affect the optical properties, that is, the interactions between light and fruit tissue. To a first approximation, the absorption coefficient is mainly determined by the chemical constituents of the pulp (e.g., water and sugar) as well as by pigments (e.g., chlorophylls, carotenoids, and anthocyanins), while the scattering coefficient is mainly dependent on microscopic discontinuities in the dielectric properties present in the tissue of membranes, cell walls, air, vacuoles, starch granules, and organelles. The major advantages of using TRS in determining the optical properties of fruit include little or limited influence from the skin and penetration into the pulp to a depth of 1–2 cm (Cubeddu et al., 2001b; Torricelli et al., 2008), in contrast to continuous-wave spectrophotometers, which have a useful penetration depth of a few millimeters, depending on the wavelength (Lammertyn et al., 2000).

This chapter will focus on TRS applications for horticultural products that have been developed over the last 15 years. Section 8.2 introduces the various types of TRS instrumentation used in the food sector and TRS data analysis. Section 8.3 reports absorption and scattering spectra of various fruit species in relation to preharvest and postharvest factors, such as fruit development, ripening, and storage, impacting on quality parameters. Section 8.4 describes the use of the absorption coefficient measured at 670 nm (μ_a670) as a maturity index to segregate fruit in the same batch with different quality characteristics, and to model softening in nectarines and mangoes. Section 8.5 reports on the relationships between TRS optical properties and fruit texture, focusing on the various types of models studied to predict the mechanical properties, mainly firmness, as well as the sensory characteristics related to texture (sensory firmness, crispness, juiciness, and mealiness). Section 8.6 reviews the use of absorption and scattering optical properties to detect internal defects in various fruit species, such as apple (mealiness, watercore, and internal browning), peach (woolliness, internal browning, and bleeding), plum (internal browning and jelling), and kiwifruit (translucency). Section 8.7 gives some final remarks on the application of TRS techniques in food analysis.

8.2 INSTRUMENTATION AND DATA ANALYSIS

Three types of instrumentation for TRS measurements have been developed at Politecnico di Milano: a broadband TRS system for multiwavelength measurements (TRS–MW), a portable compact TRS system working at single wavelengths (TRS–SW), and a portable compact system working at discrete wavelengths (TRS–DW). Figure 8.1 shows the schematics of the three TRS setups.

The TRS–MW system is a fully automated laboratory setup for broadband measurements in the red and near-infrared (NIR) wavelength range (650–1000 nm) (Figure 8.1a). It utilizes a synchronously pumped mode-locked dye laser (CR-599, Coherent, Santa Clara, California) for excitation from 650 to 695 nm with a repetition rate of 76 MHz, and an actively mode-locked titanium:sapphire laser (Mod. 3900, Spectra Physics, Santa Clara, California) for excitation in the wavelength range 700–1000 nm with a repetition rate of 100 MHz. Two 1-mm optical fibers delivered light into the sample and collected the reflected photons. For measurement on fruit, the power density at the distal end of the illumination fiber is limited to less than 10 mW. A monochromator coupled with a double microchannel plate photomultiplier (R1564/U with S1 photocathode, Hamamatsu Photonics, Shizuoka, Japan), and an electronic chain for time-correlated single-photon counting (TCSPC) are used for detection. A small fraction of the incident beam, acting as a reference, is coupled directly to the photomultiplier in order to correct for any time drift of the instrumentation. Overall, the system transfer function was less than 120 ps and less than 180 ps (FWHM, or full width at half maximum) in the red and NIR regions, respectively. A personal computer controls the laser tuning and power, the monochromator scanning, and the optimization of the system transfer function. Typical measurement time for data acquisition and system adjustment is 8–10 s/wavelength. More details about the TRS–MW system can be found in Cubeddu et al. (2001b) and Pifferi et al. (2007).

The TRS–SW system is a portable prototype for TRS measurements at a single wavelength (Figure 8.1b). Briefly, the light source is a pulsed laser diode (model PDL800, Pico Quant GmbH, Berlin, Germany), with 80 MHz repetition frequency, 100 ps duration, and 1 mW average power. A compact photomultiplier tube (R5900U-L16, Hamamatsu Photonics, Shizuoka, Japan) and an integrated PC board for TCSPC (SPC130, Becker & Hickl GmbH, Berlin, Germany) are used to detect the distribution of time-of-flight of diffuse photons. An interference filter tuned at the laser wavelength is used to cut off the fluorescence signal due to chlorophyll. Light is delivered to and collected from the sample by two 1-mm optical fibers. The overall width (FWHM) of instrumental response function (IRF) is less than 160 ps, and the typical acquisition time is 1 s. The entire setup is controlled by an in-house developed software program written in C language in the LabWindows/CVI environment (National Instruments, Austin, Texas). The TRS–SW system can have more than one pulsed laser diode, each one operating at different wavelengths. In this case, a fiber optics switch (Eol 1 × 4, Piezojena GmbH, Germany) is used to sequentially

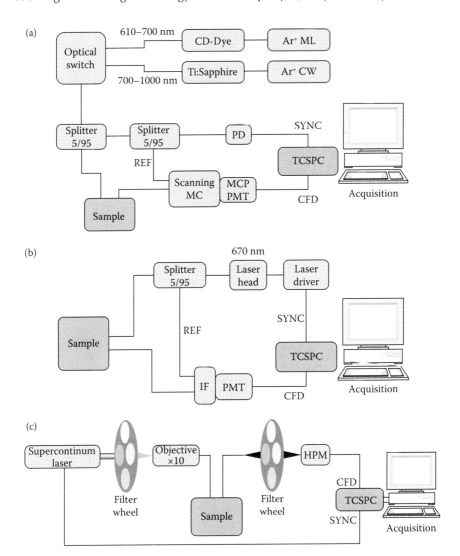

FIGURE 8.1 Schematics of three different TRS setups developed at Politecnico di Milano—Department of Physics: (a) broadband laboratory setup (TRS–MW); (b) single-wavelength portable system (TRS–SW); and (c) discrete wavelength portable system (TRS–DW). ML, mode-locked; CD, cavity dumped; CW, continuous wave; PD, photodiode; REF, optical reference signal; MC, monochromator; MCP, microchannel plate; PMT, photomultiplier tube; IF, interference filter; HPM, hybrid photomultiplier tube; TCSPC, time-correlated single-photon counting computer board; SYNC, synchronization signal; CFD, constant fraction discriminator. (Courtesy of Lorenzo Spinelli and Alessandro Torricelli, Politecnico di Milano, Department of Physics.)

select the operating wavelengths. More details about the TRS–SW system can be found in Torricelli et al. (2008).

Finally, the TRS–DW system is a portable compact setup working at discrete wavelengths (Figure 8.1c). The light source is a supercontinuum fiber laser (SC450-6W, Fianium, Southampton, UK) providing white-light picosecond pulses, with the duration of a few tens of picoseconds. A custom-made filter wheel loaded with 14 band-pass interference filters (NT-65 series, Edmund Optics, Barrington, New Jersey) is used for spectral selection in the range 540–940 nm. As in the previously described systems, light is delivered to and collected from the sample by means of optical fibers. A second filter wheel identical to the first one is used for cutting off the fluorescence signal originating from the fruit when it is illuminated in the visible spectral region. The light then is detected with a photomultiplier (HPM-100-50, Becker & Hickl, Berlin, Germany) and the photon time-of-flight distribution is measured by a TCSPC board (SPC-130, Becker & Hickl, Berlin, Germany). The IRF has a FWHM of about 260 ps and the typical acquisition time is 1 s per wavelength. More details about this system are described in Spinelli et al. (2012).

For all three systems described above, a custom-made holder is used to position the fibers 1.5 cm apart, parallel to each other, normal to and in contact with the sample surface.

Table 8.1 summarizes spectral regions and wavelengths for TRS measurements of horticultural products by TRS–MW, TRS–SW, and TRS–DW instrumentations.

8.2.1 TRS Data Analysis

The reduced scattering coefficient (μ_s') and the absorption coefficient (μ_a) of the diffusive medium are obtained by fitting the photon time-of-flight reflectance data at each wavelength with an analytical solution of the diffusion approximation to the radiative transfer equation. A complete description of available solutions for different geometries can be found in Martelli et al. (2009).

When TRS measurements are performed at multiple wavelengths with TRS–MW or TRS–DW setups, the absorption spectrum and the reduced scattering spectrum of the diffusive medium are obtained by iterating the fitting procedure for all wavelengths, using a spectrally constrained approach (D'Andrea et al., 2006). The spectral dependence of the optical properties is inserted in the analytical solution for light transport in diffusive media and best fit is obtained by simultaneously considering TRS data at multiple wavelengths and using chromophore concentrations (e.g., chlorophyll and water) and structure parameters (a and b of the Mie theory, see below) as free parameters.

The relationship between absorption coefficient and tissue constituents is given by the Beer law:

$$\mu_a(\lambda) = \sum_i c_i \varepsilon_i(\lambda) = c_{CHL}\varepsilon_{CHL}(\lambda) + c_{H_2O}\varepsilon_{H_2O}(\lambda) + bkg \tag{8.1}$$

where c_{CHL} and c_{H_2O} are the chlorophyll and water concentration, respectively, ε_{CHL} and ε_{H_2O} are the chlorophyll and water specific absorption coefficients, respectively, and bkg is a constant value to account for the contribution of other absorbers.

TABLE 8.1
Spectral Regions and Wavelengths Used for Horticultural Products for Broadband (TRS–MW), Single-Wavelength (TRS–SW), and Discrete Wavelength (TRS–DW) Setups

Setup	Product	Wavelength	References
TRS–MW (Cubeddu et al., 2001a)	Apple, peach, tomato, and kiwifruit	650–1000 nm every 5 nm 672, 750, 800, and 900–1000 nm every 10 nm	Cubeddu et al. (2001a) Valero et al. (2004b, 2005)
	Apple	670, 750, 790, and 912 nm	Vanoli et al. (2009c)
		600–700 nm every 5 nm	Cubeddu et al. (2001b)
	Kiwifruit	610–1010 nm every 5 nm	Valero et al. (2004a)
	Pear	710–850 nm every 10 nm	Eccher Zerbini et al. (2002)
		875–1030 nm every 5 nm	Nicolaï et al. (2008)
TRS–MW (Pifferi et al., 2007)	Apple	670–1100 nm every 10 nm	Vanoli et al. (2011a, c)
		740–1040 nm every 40 nm	Vanoli et al. (2014b)
		670 and 740–1100 nm at 20 nm intervals	Barzaghi et al. (2009)
TRS–SW (Torricelli et al., 2008)	Pear	690 nm	Eccher Zerbini et al. (2002, 2005c)
	Nectarine	670 nm	Eccher Zerbini et al. (2003, 2005a, 2006, 2009, 2011) Jacob et al. (2006) Rizzolo et al. (2009, 2010b, 2015) Tijskens et al. (2006, 2007a, b) Vanoli et al. (2005, 2008, 2009a)
		670 and 780 nm	Lurie et al. (2011)
	Peach	670 nm	Rizzolo et al. (2013a)
	Plum	670 and 780 nm	Vangdal et al. (2012)
		670 and 758 nm	Vangdal et al. (2010)
	Apple	630 nm	Eccher Zerbini et al. (2005b) Vanoli et al. (2005)
		670 nm	Rizzolo et al. (2011a, 2012, 2013b, 2014b)
		630, 670, 750, and 780 nm	Rizzolo et al. (2010a) Vanoli et al. (2007, 2009b)
		780 nm	Vanoli et al. (2014b)
	Mango	630 nm	Pereira et al. (2010) Vanoli et al. (2011b)
TRS–DW (Spinelli et al., 2012)	Mango	540, 580, 630, 650, 670, 690, 730, 780, 800, 830, 850, 880, and 900 nm	Eccher Zerbini et al. (2015) Spinelli et al. (2012, 2013) Vanoli et al. (2012a, 2013a)
		540, 630, and 690 nm	Vanoli et al. (2014a)
		540, 580, 630, 650, 670, 690, 730, and 780 nm	Vanoli et al. (2014a)

(Continued)

TABLE 8.1 (Continued)
Spectral Regions and Wavelengths Used for Horticultural Products for Broadband (TRS–MW), Single-Wavelength (TRS–SW), and Discrete Wavelength (TRS–DW) Setups

Setup	Product	Wavelength	References
	Apple, plum, potato, and peach	540, 580, 630, 650, 670, 690, 730, 780, 800, 830, 850, 880, and 900 nm	Attanasio (2012) Seifert et al. (2015) Vanoli et al. (2012b)
	Apple	630, 650, 670, 690, 730, 830, 850, and 900 nm	Rizzolo et al. (2014a) Vanoli et al. (2013b)
		670 nm	Vanoli et al. (2015) Zanella et al. (2013)

Mie theory predicts the wavelength dependence of the scattering coefficient and the relationship between scattering coefficient and sphere size. Under the hypothesis that scattering centers are homogenous spheres behaving individually, the relationship between μ'_s and wavelength can be empirically described by Equation 8.2:

$$\mu'_s = a\left(\frac{\lambda}{\lambda_0}\right)^{-b} \tag{8.2}$$

where λ is the wavelength, a is the scattering coefficient at the reference wavelength $\lambda_0 = 600$ nm proportional to the density of the scattering centers, and b is the parameter related to the equivalent size of the scattering centers (Mourant et al., 1997; Nilsson et al., 1998). Parameter a represents the scatter amplitude and parameter b represents the scatter power (D'Andrea et al., 2006). By interpreting the reduced scattering spectra with Mie theory, it is possible to obtain information about the density and size of the scattering centers and, consequently, about the structural properties of the sampled fruit.

The performances of TRS setups during each experiment are assessed by measuring over time calibrated solid tissue phantoms made of epoxy resin, black toner powder, and TiO_2 particles with known optical properties. Details of the protocol used can be found in Pifferi et al. (2005) and an example of TRS performance assessment during a whole experiment is described in Seifert et al. (2015). In the range of the measured values for the optical properties, the accuracy in the absolute estimate of both μ_a and μ'_s is usually better than 10% (Torricelli et al., 2013). However, the error incurred in the assessment of the absorption line shape is usually less than 2% (Cubeddu et al., 1996).

8.3 ABSORPTION AND SCATTERING SPECTRA IN VARIOUS FRUIT SPECIES

A typical absorption spectrum in the 540–1100 nm range presents maxima mainly at 540, 670, and 980 nm, corresponding to carotenoids/anthocyanins, chlorophyll a, and water absorption peaks, whereas scattering spectra show no particular spectral features, with scattering values generally decreasing as wavelength increases. The amplitude of the maxima of absorption spectra and the values and slopes of scattering spectra differ with species and cultivars. Examples of absorption and scattering spectra for peaches, apples, and mangoes are shown in Figure 8.2.

The absorption and scattering spectra obtained using the TRS–MW system for "Spring Belle" peaches are shown in the top panel of Figure 8.2 (each line in the figure represents one of the 30 fruit of each batch belonging to the same harvest date but stored for 1 month at different temperatures [0°C and 4°C]). The absorption spectra were dominated by a peak centered around 980 nm, corresponding to

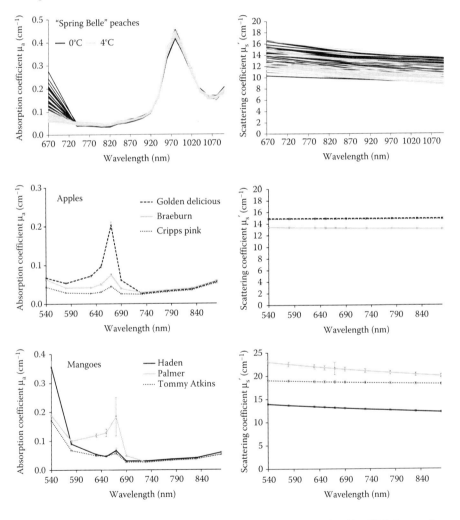

FIGURE 8.2 Absorption (left) and scattering (right) spectra obtained by TRS for peaches (unpublished data), apples (Barzaghi et al., 2009; Vanoli et al., 2011a, 2013b), and mangoes (Spinelli et al., 2012, 2013; Vanoli et al., 2013a). The continuous lines in the right graphs are the best fit of the scattering spectrum with the Mie theory approximation. Bars refer to standard error of the mean.

water, and by a significant absorption at 670 nm, with a great variability in the values (0.06–0.28 cm^{-1}), corresponding to chlorophyll a, along with minor absorption features detected from 740 to 840 nm. The scattering values in "Spring Belle" peaches decreased with increasing wavelength, with parameter a ranging from 10.58 to 17.66 cm^{-1}, and the spectra showed different slopes for each fruit, with the parameter b ranging from 0 to −0.83. Generally, high values of parameter a indicate a high density of the scattering centers, while high slope values reveal the presence of smaller particles. However, it is worth noting that these parameters do not assess the real size of the scattering centers in the tissue; rather, they are average equivalent parameters.

The values of absorption at 670 nm and in the 500–580 nm range vary among fruit species and cultivars, as highlighted by the comparison of absorption spectra shown in Figure 8.2 for apples (middle panel) and mangoes (bottom panel). After comparing the absorption spectra of apples, peaches, tomatoes, and kiwifruits, Cubeddu et al. (2001a) observed that the line shape of the absorption spectra was quite similar; however, for kiwifruits, the maximum chlorophyll a absorption value at 670 nm was up to two or three times the maximum water absorption value of about 0.4 cm^{-1}, centered near 970 nm, and was equivalent to a chlorophyll a content of about 7.0 μM versus 0.5–1.0 μM for the other species. "Cripps Pink" apples were characterized by a low content of chlorophyll a, compared to other cultivars (Cubeddu et al., 2001b; Rizzolo et al., 2014a; Vanoli et al., 2011a; Zanella et al., 2013). Similarly, "Ghiaccio" peaches, a white fleshed cultivar, showed the lowest absorption values at 670 nm in comparison with other 10 peach cultivars (Attanasio, 2012). The differences in the absorption spectra in the 500–580 nm range were due to the presence in the fruit pulp of carotenoids, as evidenced by the comparison of apples and mangoes spectra in Figure 8.2, and/or anthocyanins as found for plums (Seifert et al., 2015) and peaches (Attanasio, 2012). In fact, it has been reported that "Iride," a particularly red-colored peach cultivar, showed very high values of $\mu_a 580$ due to the large amount of anthocyanin in the pulp; in contrast, "Ghiaccio" peaches, being totally depigmented, showed very low values of $\mu_a 540$ and $\mu_a 580$ (Attanasio, 2012). Also, scattering spectra parameters are greatly influenced by fruit species and cultivar, and the differences in parameters a and b among fruit species and cultivars suggest that different fruit could have different densities and average dimensions of the scattering centers and, consequently, different structural characteristics. As shown in the middle panel of Figure 8.2, "Golden Delicious," "Braeburn," and "Cripps Pink" apples had flat scattering spectra and this feature was likely attributed to the fact that the equivalent size of scattering centers is much larger than the wavelength (Mourant et al., 1997; Nilsson et al., 1998). In contrast, the scattering spectra of "Spring Belle" peaches and of "Palmer" and "Haden" mangoes significantly decreased as wavelength increases, showing b values ranging from −0.270 ("Haden" mangoes) to −0.460 ("Spring Belle" peaches stored at 4°C). Likewise, the scattering values also vary with fruit species and cultivar; as shown in Figure 8.2, the highest μ'_s values were found in "Palmer" and "Tommy Atkins" mangoes and the lowest ones in "Braeburn" apples and in "Haden" mangoes. Similarly, Attanasio (2012), comparing different peach cultivars, observed high μ'_s values in "Big Top" fruit and the lowest ones in the "Glohaven" cultivar.

Within the same species and cultivar, the fruit optical properties also change according to preharvest and postharvest factors influencing fruit maturity, such as harvest date, storage conditions (temperature, atmosphere, and time), and shelf life. Seifert et al. (2015) studied the optical properties, measured by TRS, in developing "Elstar" apples and "Tophit plus" plums. They found that the absorption spectra in apples and plums showed a typical shape with the characteristic peak at 670 nm having the highest values at 60–80 days after full bloom and the lowest ones at 145 days after full bloom, owing to chlorophyll loss. The scattering spectra were much higher in apples than in plums, showing minimum changes during apple growth with the scattering power (parameter b) equal to zero at all dates; in contrast, in plums, both scattering power and scattering values (parameter a) decreased significantly during fruit development. This scenario reflects smaller sizes of scatterers in plums compared to apples, a "stable" cortex tissue in apples and a highly variable tissue in plums during fruit development. Similarly, in apples picked at different dates and stored for increasing times, the main change in absorption spectra during shelf life occurred in the chlorophyll peak, with a decrease in $\mu_a 670$ values as a consequence of chlorophyll breakdown due to fruit ripening as shown for "Braeburn" cultivar in Figure 8.3 and for other apple cultivars by Vanoli et al. (2011a, 2013b). The scenario for scattering spectra is more complex; in fact, parameters a and b can both change, or only one can vary, and when a change occurs, it does not always follow the same trend. For example, in "Braeburn" apples (Figure 8.3) scattering values decreased significantly with harvest date, showing the highest values in apples picked earlier and the lowest in those harvested later, while no changes were observed in the scattering power. In the same cultivar, parameter a increased with storage time in a controlled atmosphere as well as when fruit had been put in shelf life at 20°C, while parameter b did not change with time during the shelf life period, but showed the highest values after 3 months of storage and the lowest after 1 and 8 months of storage. In "Pink Lady®" apples (Vanoli et al., 2011a), scattering values did not change during storage while scattering power had the highest values after 7 days at 0°C, the lowest after 15 and 29 days and intermediate after 91 days.

In mangoes, the shelf life time influenced the absorption spectra not only at 670 nm (decrease in $\mu_a 670$ values) but also at 540 nm (increase of $\mu_a 540$ values), due to carotenoid accumulation typical of this fruit species, as revealed by the changes in the pulp color which turned from yellow to orange (Spinelli et al., 2012; Vanoli et al., 2013a). In "Tommy Atkins" mangoes, the scattering value decreased during the 5-day period of shelf life, whereas the scattering power increased only at the beginning of shelf life, from day 1 to day 2 (Vanoli et al., 2013a).

Absorption and scattering spectra also change with storage conditions: "Spring Belle" peaches stored at 4°C had lower $\mu_a 670$ and showed higher values of parameter b than peaches stored at 0°C, with no difference in the scattering values (Figure 8.2, top). "Golden Delicious" apples stored for 5 months at 1°C showed lower $\mu_a 670$ (0.109 ± 0.007 cm^{-1}, mean ± standard error) and lower μ'_s values (13.8 ± 0.1 cm^{-1}) when stored in normal atmosphere (air) than when stored under controlled atmosphere ($\mu_a 670$, 0.148 ± 0.007 cm^{-1}; μ'_s, 14.3 ± 0.1 cm^{-1}), while parameter b was 0 in both cold storage conditions (unpublished data). This scenario indicates that apples stored in controlled atmosphere differ from those stored in air for the density of scatterers, whereas the equivalent size of scatterers in both storage atmospheres was much larger than the wavelength as pointed out in other disciplines (Mourant et al., 1997; Nilsson et al., 1998).

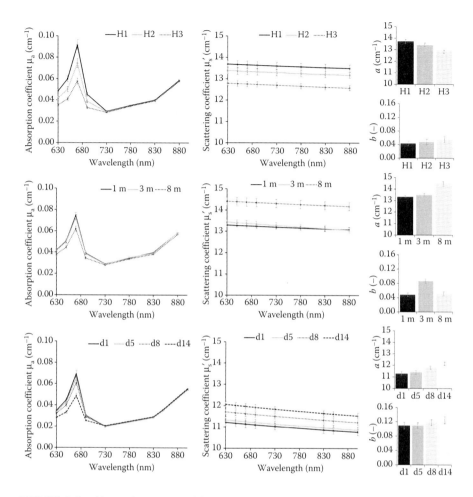

FIGURE 8.3 Absorption spectra (left), scattering spectra (center), density (parameter a), and size (parameter b) of the scattering centers (right) obtained by TRS for "Braeburn" apples in relation to harvest date (top), storage time (middle), and shelf life period (bottom). The continuous lines in the center graphs are the best fit of the scattering spectrum with the Mie theory approximation. Bars refer to standard error of the mean. (From Rizzolo, A., Vanoli, M., Bianchi, G. et al., *J. Hort. Res.* 22, 113–121, 2014a; Vanoli, M., Rizzolo, A., Zanella, A. et al., Apple texture in relation to optical, physical and sensory properties. *InsideFood Symposium: Book of Proceedings*, April 9–12, Leuven, Belgium, 2013b, http://www.insidefood.eu/INSIDEFOOD_WEB/UK/WORD/proceedings/032P.pdf.)

The changes in scattering parameters reflect the changes in the pulp structure due to different processes, such as fruit softening, water loss, and increase or decrease of intercellular spaces. The scattering decrease could be linked to fruit softening; this process is accompanied by the enzymatic cell wall breakdown, so the density of the scattering particles in the fruit flesh decreases. As a consequence of this phenomenon, the light scattering at the cell interfaces is reduced, leading to less scattering events in the tissue (Bobelyn et al., 2010). At the same time, the intercellular spaces could increase and some water is lost due to fruit transpiration, so the cells in the pulp tissue become smaller with more air-filled pores. The values of the scattering power decrease but the scattering values increase, as there is a higher refractive index mismatch leading to more and stronger scattering events (Bobelyn et al., 2010; Schotsmans et al., 2004). So an increase or a decrease in the scattering values or in the scattering power depends on which effect actually dominates the spectrum.

8.4 FRUIT MATURITY

Fruit quality during marketing and consumption mainly depends on fruit maturity at harvest, and the increasing consumer demand for better quality fruit has driven the development of nondestructive techniques for the assessment of fruit ripening. Among the nondestructive methods for assessing harvest maturity are the spectral indices, which allow the detection of changes in the content of pigments, such as chlorophyll, carotenoid, and anthocyanin, which are linked to fruit ripening (Vanoli and Buccheri, 2012). It has been shown that the absorption coefficient measured by TRS at a wavelength in the 630–690 nm range, near the chlorophyll peak, decreased with ripening. The μ_a measured at 670 nm in nectarines and at 630 nm in mangoes decreased during shelf life following a logistic curve occurring with the same rate in all the fruit, but shifted in time according to maturity, that is, to the individual age of the fruit on the tree. Tijskens et al. (2006), using the nonlinear mixed effects regression analysis, estimated for each fruit the biological time shift factor (BSF), which indicates the time required to reach the reference stage of maturity at the midpoint of the logistic curve of μ_a decay. Using this methodology, the biological variance that exists in the measured values of μ_a is transformed into variation in maturity, and expressed as time to ripen until the midpoint of the logistic curve (Tijskens et al., 2006).

Table 8.2 and Figure 8.4 summarize selected results from the literature of non-linear mixed effect regression analysis based on Equation 8.3 (Tijskens et al., 2006):

$$\mu_a = \frac{\mu_{a,max}}{1+e^{k_m \cdot (t+\Delta t)}} \qquad (8.3)$$

where
  μ_a is the absorption coefficient measured at 670 (nectarines) or at 630 (mangoes) nm
  $\mu_{a,max}$ is the absorption coefficient at minus infinite, that is, the maximum absorption ever possible
  k_m is the rate constant of μ_a decay in time
  t is the shelf life time

Δt is the biological shift factor computed according to Equation 8.4 (Tijskens et al., 2007a):

$$\Delta t = \frac{\log(\mu_{a,max}/\mu_a - 1)}{k_m} \qquad (8.4)$$

It was shown that the rate constant of μ_a decay in time is influenced by storage temperature ($k_{m,10°C}$ lower than $k_{m,20°C}$) (Tijskens et al., 2006) and cultivar ($k_{m,20°C}$

TABLE 8.2
Results of the Nonlinear Mixed Effects Regression Analysis Based on Equation 8.3 for Nectarines ($\mu_a 670$) and Mangoes ($\mu_a 630$) Stored at Different Temperatures

Fruit	Nectarines Spring Bright	Nectarines Spring Bright	Nectarines Morsiani 90	Mangoes Tommy Atkins
Temperature (°C)	10	20	20	20
k_m	0.134	0.226	0.065	0.222
Δt_mean	3.186	2.272	28.7	9.759
$\Delta t\_st\ dev$	1.664	1.625	3.9	2.938
$\mu_{a,max}$	0.6 (fixed)	0.6 (fixed)	0.6 (fixed)	0.4 (fixed)
N_{obs}	360	300	180	715
N_{fruit}	60	60	29	120
R^2_{adj}	0.850	0.972	0.94	0.78
References	Tijskens et al. (2007a)	Tijskens et al. (2007a)	Eccher Zerbini et al. (2011)	Pereira et al. (2010)

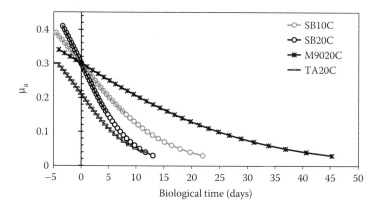

FIGURE 8.4 Predicted $\mu_a 670$ ("Spring Bright"—SB and "Morsiani 90"—M90 nectarines) and $\mu_a 630$ ("Tommy Atkins"—TA mangoes) according to the model (Equation 8.3) with parameters in Table 8.2, versus biological time during shelf life at 10°C and 20°C.

in the late-maturing nectarine was about 30% of that found in the early-maturing ones) (Eccher Zerbini et al., 2011). Figure 8.5 shows the changes of measured $\mu_a 670$ data with shelf life time at 10°C and 20°C for "Spring Bright" nectarines of increasing maturity at harvest shifted according to their BSF and the $\mu_a 670$ decay curves predicted according to the model (Equation 8.3) with parameters in Table 8.2. Less mature (LeM) fruit at harvest (R1–R9) are in the upper part of the curves and the more mature (MoM) ones (R22–R30) in the lower part, independently of shelf life temperature, showing that models take into account the different fruit maturity. A similar relationship between measured $\mu_a 670$ data with shelf life time

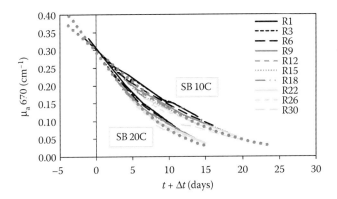

FIGURE 8.5 μ_a at 670 nm versus biological time ($t + \Delta t$) in "Spring Bright" nectarines of increasing maturity at harvest (R1, LeM, higher $\mu_a 670$ value; R30, MoM, lower $\mu_a 670$ value) during shelf life at 10°C and 20°C. Broken lines refer to measured $\mu_a 670$ during shelf life and are shifted in time according to the biological shift (Δt) of each fruit. Dotted curves are the $\mu_a 670$ predicted according to the model (Equation 8.3) with parameters in Table 8.2.

and predicted $\mu_a 670$ decay curve was reported by Eccher Zerbini et al. (2011) for "Morsiani 90" nectarines.

The $k_{m,20°C}$ in mangoes was similar to $k_{m,20°C}$ in "Spring Bright" nectarines, indicating a similar trend for the chlorophyll decay in the pulp (Table 8.2). However, using the nonlinear mixed effects regression analysis, in which k_m (fixed effect) is estimated in common for all fruit and the biological shift factor (random effects) is estimated separately for each individual fruit, only 78% of the overall variation for mangoes was explained versus 94%–97% for nectarines. This difference is due to the fact that during ripening in mango fruit flesh, there is a degradation of chlorophyll, accompanied with an accumulation of carotenoids, which caused an increase of the $\mu_a 630$ value with the shelf life time (Pereira et al., 2010; Rizzolo et al., 2011b). Hence, to account for this scenario specific to mango ripening, Pereira et al. (2010) corrected the $\mu_a 630$ by subtracting the estimated values for carotenoid development, resulting in an improvement of the $\mu_a 630$ decay model with $R^2 = 0.88$. In addition to considering $\mu_a 670$ for "Haden" mangoes, Eccher Zerbini et al. (2015) also took into consideration the absorption coefficient measured at 540 nm ($\mu_a 540$), which was shown to be positively correlated with total carotenoids content (Azzollini, 2012; Vanoli et al., 2014a), and follow a logistic but increasing trend.

8.4.1 Modeling of Softening

The most interesting results obtained so far from the use of $\mu_a 670$ alone or in combination with $\mu_a 540$ and scattering parameters for the evaluation of maturity were concerned with the modeling of softening in nectarines (Eccher Zerbini et al., 2006, 2011; Tijskens et al., 2007b) and mangoes (Eccher Zerbini et al., 2015; Pereira et al., 2010).

The model in nectarines was based on the fact that softening occurred earlier in MoM fruit (low $\mu_a 670$) and later in LeM ones (high $\mu_a 670$ at harvest), following the same sigmoid pattern in time (Eccher Zerbini et al., 2006). For firmness decay, a logistic model was developed by integrating the biological shift factor derived from the measurement of $\mu_a 670$ (Δt) with the kinetics of the fruit softening process (Tijskens et al., 2007b) according to Equation 8.5:

$$F = \frac{F_{max} - F_{min}}{1 + e^{k_f (F_{max} - F_{min})t + \Delta t*}} + F_{min} \tag{8.5}$$

where
F_{max} is the maximum firmness at minus infinite time
F_{min} is the minimum firmness at infinite time
t is time
k_f is the reaction rate constant of the softening reaction

Δt^* is the biological shift factor for firmness, which can be expressed as a function of the measured μ_a at harvest according to Equation 8.6, as firmness decay and chlorophyll decay are assumed to be synchronized and, hence, the biological shift factor for μ_a (Δt) and the biological shift factor for firmness (Δt^*) should be linearly related:

$$\Delta t^* = a \left(\log \left(\frac{\mu_{a,max}}{\mu_a} - 1 \right) + \beta \right) \tag{8.6}$$

where $\mu_{a,max}$ is the maximum μ_a value ever possible, and α and β are parameters to be estimated, with α being the synchronization factor between the ratio of chlorophyll (μ_a) and firmness (F) and β representing the commercial harvest date expressed as a shift factor relative to the midpoint of the logistic curve (Tijskens et al., 2007b).

It was shown that softening kinetics depend on the cultivar, and within the same cultivar, on harvest time, on shelf life temperature, and on storage time at 0°C (Table 8.3). Hence, μ_a at harvest was used to segregate fruit according to their softening capacity ("will never soften," "dangerously hard," "transportable," "ready to eat—firm," "ready to eat—ripe," and "overripe") (Rizzolo et al., 2009) and an export trial from Italy to the Netherlands, simulating, on a small scale (1000 fruit), the fruit supply chain from the packing house to the consumer, showed that the ripening classes had been correctly predicted (Eccher Zerbini et al., 2009).

Figure 8.6 shows the prediction of firmness decay during shelf life according to models reported in Table 8.3 for fruit having μ_a at harvest equal to the limits of "overripe" ($\mu_a = 0.09$ cm^{-1}) and "will never soften" ($\mu_a = 0.42$ cm^{-1}) classes for "Ambra" and "Spring Bright" early-maturing cultivars (Rizzolo et al., 2009), and "ready-to-eat very soft" ($\mu_a = 0.049$ cm^{-1}) and "hard" ($\mu_a = 0.25$ cm^{-1}) for "Morsiani 90" late-maturing cultivar (Rizzolo et al., 2015).

In "Spring Bright" nectarines, softening occurred earlier and faster than in "Ambra" and "Morsiani 90," with the latter being characterized by the slowest softening rate (Figure 8.6a); in fact, the midpoint of the firmness decay curve for the LeM class ("will never soften" and "hard") was reached at 8 days after a 5-day period of no changes

TABLE 8.3
Estimates of Parameters for the Nonlinear Regression Model of Softening (Equation 8.5) in Shelf Life for "Ambra," "Spring Bright" (SB), and "Morsiani 90" (M90) Nectarines

Cultivar	Ambra	SB	SB	SB	SB	SB	M90	M90
Harvest[a]	Comm	Comm	Comm	Early	Comm	Comm	Comm	Comm
Shelf life[b]	Harvest	Harvest	Harvest	Harvest	$6d_{cool}$	$17d_{cool}$	Harvest	$28d_{cool}$
Temperature (°C)	20	10	20	20	20	20	20	20
Season	2004	2005	2005	2005	2004	2004	2009	2009
k_f	0.000975	0.000315	0.001312	0.000677	0.000845	0.000693	0.000234	0.000227
α	3.423	1.940	1.595	2.909	2.79	2.80	1.40	1.03
β	−2.177	−1.470	−2.683	−1.470	−2.40	−2.05	−2.27	−2.72
F_{max}	58.22	78.97	72.27	78.97	79.64	78.27	85	85
F_{min}	2.16	4.50	3.05	4.50	5.53	4.84	4.7	4.7
N_{obs}	360	300	360	360	310	280	179	149
R^2_{adj}	0.76	0.91	0.92	0.81	0.87	0.81	0.78	0.59
Standard Error								
k_f	0.000128	0.000032	0.000113	0.000054	0.000127	0.000153	0.000068	0.000020
α	0.322	0.156	0.146	0.247	0.28	0.35	0.27	0.16
β	0.050	0.065	0.142	0.096	0.069	0.094	0.25	0.16
F_{max}	3.236	3.478	1.936	0.642	4.31	7.83	14	Fixed
F_{min}	Fixed	Fixed	Fixed	Fixed	0.87	0.81	Fixed	Fixed
References	(1)	(1)	(1)	(1)	(3)	(3)	(2)	(2)

[a] Harvest: comm: commercial harvest; early, 7 days before commercial harvest.
[b] Shelf life: harvest, after 1–2 days at 0°C; $6d_{cool}$, $17d_{cool}$, $28d_{cool}$, after 6, 17, and 28 days at 0°C
References: (1) Tijskens et al. (2007b), (2) Eccher Zerbini et al. (2011), and (3) Rizzolo et al. (2010b).

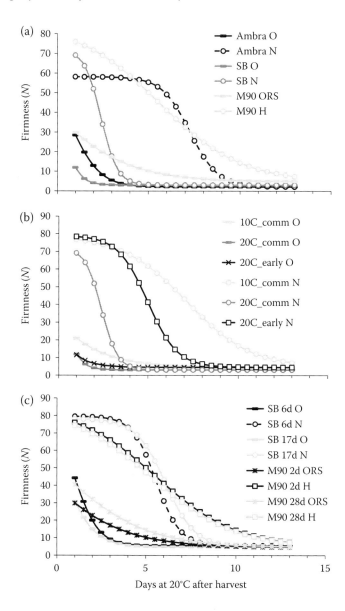

FIGURE 8.6 Prediction of firmness decay during shelf life according to models' parameters reported in Table 8.3. Lines represent firmness decay of fruit with μ_a values at harvest equal to the limits of classes of predicted firmness potential for handling and eating (overripe, O; never ripe, N; ready-to-eat very soft, ORS; hard, H). (a) Shelf life at harvest for "Ambra," "Spring Bright" (SB), and "Morsiani 90" (M90) nectarines; (b) shelf life at harvest at 10°C and 20°C for "Spring Bright" nectarines of commercial (comm) and early harvest; and (c) shelf life at 20°C after 6 and 17 days ("Spring Bright," SB) and 2 and 28 days ("Morsiani 90," M90) at 0°C.

in firmness in "Ambra," at 7 days without any initial plateau in "Morsiani 90," and after 3 days in "Spring Bright." In "Spring Bright" nectarines, softening occurred later and slower at 10°C than at 20°C, in fruit of early harvest than in those of commercial harvest (Figure 8.6b), and after a cold storage period (Figure 8.6c). In "Morsiani 90," the cold storage period at 0°C did not have any significant impact on k_f value (Table 8.3), but did influence the softening prediction of the MoM fruit in the batch ("ready-to-eat very soft" class) but not that of the LeM ones ("hard" class) (Figure 8.6c).

Applying this methodology at the time of harvest, Rizzolo et al. (2009) measured μ_a on all fruit and firmness on two samples of about 30 fruits, representative of all μ_a range, with the first as soon as possible after harvest and the second after 24 h at 20°C. They found that it was possible to estimate the parameters of the firmness decay model for the season and cultivar, and hence to compute the time required to reach the midpoint of the firmness decay curve of the μ_a values in each softening class. These time values (hours for fast-ripening cultivars and days for late-maturing ones) could be used to select fruit with different stages of maturity for different marketing segments, such as distant or close-by markets.

Pereira et al. (2010) showed that the $\mu_a 630$ value measured at harvest in "Tommy Atkins" mangoes could be used for the softening prediction in shelf life at 20°C, but the model explained only about 70% of the variation in firmness decay rate. To improve the softening prediction, Eccher Zerbini et al. (2015) assumed that the biological shift factor for firmness (Δt^*) is linearly related to the biological shift factors of chlorophyll and of carotenoids, and the firmness decay is parallel to chlorophyll degradation and carotenoid accumulation, and is also paralleled to changes in the scattering parameters. As a consequence, they expressed Δt^* as a function of the measured absorptions at 540 and 670 nm, as well as of the scattering value (a) and scattering power (b) according to Equation 8.7:

$$\Delta t^* = \alpha_{540}\left(\log\frac{\mu_{a,max}^{540} - \mu_{a,0}^{540}}{\mu_{a,0}^{540} - \mu_{a,min}^{540}}\right) + \alpha_{670}\left(\log\frac{\mu_{a,max}^{670} - \mu_{a,0}^{670}}{\mu_{a,0}^{670} - \mu_{a,min}^{670}}\right) + \beta + k_A a + k_B b \quad (8.7)$$

where α_{540}, α_{670}, β, k_A, and k_B are parameters to be estimated. The index 0 indicates the μ_a measured in each fruit, whereas the indices max and min indicate the maximum and minimum μ_a values at 540 and 670 nm ever possible, that is, which had been fixed based on experiments on mango fruits. For "Haden" cultivar, Eccher Zerbini et al. (2015) reported that the β and k_A parameters in Equation 8.7 were not significant and hence could be omitted, and the presence of the scattering parameter b in the model increased the coefficient of determination adjusted for the number of predictors (R^2_{adj}) to 0.80, while, without it, R^2_{adj} was 0.75.

8.4.2 Absorption and Maturity

The absorption coefficient μ_a in the 630–690 nm range has been proved to be an effective maturity index for various fruit species, and it was used to group fruit in a batch into TRS maturity classes, ranging from the LeM class with higher μ_a values, to the MoM class having lower μ_a values.

The selection of a specific wavelength in the 630–690 nm range to be used to assess maturity in different species and for different cultivars within the same species mainly depends on the signal-to-noise ratio (SNR) at each wavelength. When SNR at 670 nm (chlorophyll *a* peak) is too low (i.e., number of collected photons in 1 s less than 10,000) due to a high concentration of chlorophyll *a* in the fruit pulp (apples) or in the 2–3 mm green layer under the skin (mangoes), either 630 nm (chlorophyll *b* peak) (as in the case of "Jonagored" apples (Vanoli et al., 2005) and of "Tommy Atkins" mangoes (Pereira et al., 2010)), or 650 nm (shoulder of chlorophyll *a* peak) or 690 nm (tail of chlorophyll *a* peak) (as in the case of "Haden" and "Palmer" mangoes (Rizzolo et al., 2014c)), is chosen on the basis that all fruits in a batch might have a sufficient SNR at the selected wavelength in order to have a correct fruit classification at harvest.

The ability of μ_a in the 630–690 nm range in segregating fruit of the same batch into different maturity classes and, hence, different shelf life behaviors, is determined by the interplay of various factors such as fruit species, cultivar, maturity at harvest, and storage condition. Table 8.4 reports on the literature data of μ_a values for the LeM and MoM classes in the 630–690 nm range for apples, peaches, nectarines, tomatoes, kiwifruits, pears, plums, and mangoes.

8.4.2.1 Fresh Fruit

In evaluating the influence of TRS classification at harvest on pome fruit quality parameters, Vanoli et al. (2005) found that "Jonagored" apples classified as MoM according to $\mu_a 630$ had at harvest higher percent blush, lower titratable acidity, and a yellower skin than apples of LeM class; after 6 months of storage, MoM "Jonagored" apples also had higher soluble solids, and by sensory analyses were judged sweeter, more aromatic, and pleasant than LeM fruit. In a study on "Braeburn" and "Cripps Pink" apples classified according to $\mu_a 670$ measured after 6 months of storage, a different evolution of the pulp mechanical properties with shelf life as a function of harvest date and TRS maturity class was found (Vanoli et al., 2013b). LeM "Braeburn" apples showed the highest values of firmness, stiffness, and energy-to-rupture, and softened during the shelf life for all the harvests, whereas MoM "Braeburn" apples, characterized by lower values of pulp mechanical properties, softened during shelf life only in fruit of early harvest. In contrast, the classification of "Cripps Pink" apples was not so effective and firmness decreased with shelf life only in LeM apples from early harvest (Vanoli et al., 2013b). "Abate Fétel" pears classified as LeM according to $\mu_a 690$ (Eccher Zerbini et al., 2005c) had at harvest lower soluble solids and titratable acidity, after 4 months of storage in normal atmosphere showed higher firmness, and during poststorage shelf life produced less ethylene than MoM pears, with a delay in the climacteric peak. The TRS classification at harvest was most effective on stone fruit. "Spring Bright" nectarines with low $\mu_a 670$ (MoM class) had at harvest lower firmness and higher fruit mass (Eccher Zerbini et al., 2003, 2005b), as well as higher soluble solids content, lower acidity, and a redder hue, showing a more advanced ripening stage than LeM nectarines (Eccher Zerbini et al., 2005b). With shelf life, MoM "Spring Bright" nectarines softened earlier with higher percent juice, and at sensory analysis, they were perceived less firm, more juicy, sweet, pulpy, and aromatic (Eccher Zerbini et al., 2003). MoM "Ambra"

TABLE 8.4
Absorption Coefficient Values Measured in the 630–690 nm Range (Chlorophyll Peak) in Various Fruit Species According to TRS Maturity Class

Fruit	Cultivar	Sample	λ (nm)	μ_a (cm^{-1})	References
Apple	Golden Delicious	Retail	675	0.12	(1)
		At harvest	670	0.187 (LeM)–0.035 (MoM)	(2)
	Granny Smith	Retail	675	0.18	(1)
	Starking Delicious	Retail	675	0.18	(1)
	Cripps Pink	After storage	670	0.042 (H1)–0.040 (H3)	(3)
		At harvest	670	0.076–0.035	(4)
		At harvest	670	0.049 (LeM)–0.035 (MoM)	(17)
	Jonagored	At harvest	630	0.05–0.20 (H1) 0.05–0.16 (H2)	(5,6)
	Braeburn	After storage	670	0.109 (H1)–0.063 (H3)	(3)
Peach	Yellow-fleshed	Retail	675	0.08	(1)
	Spring Belle	At harvest	670	0.247 (LeM)–0.127 (MoM)	(7)
Nectarine	Ambra	At harvest	670	0.148 (LeM)–0.068 (MoM)	(8)
	Spring Bright	At harvest	670	0.35–0.03	(9)
		At harvest	670	0.22 (LeM)–0.012 (MoM)	(10)
	Morsiani 90	At harvest	670	0.19–0.03	(11)
Tomato	Daniela	Retail	675	0.15	(1)
Kiwifruit	Hayward	Retail	675	>1.8	(1)
Pear	Abbé Fétel	At harvest	690	0.075 (LeM)–0.047 (MoM)	(12)
		At harvest	670	0.070 (LeM)–0.044 (MoM)	(13)
Plum	Jubileum	At harvest	670	0.362 (LeM)–0.097 (MoM)	(2)
Mango	Tommy Atkins	At harvest	630	0.259 (LeM)–0.044 (MoM)	(2)
		At harvest	630	0.16 (LeM)–0.06 (MoM)	(14)
	Haden	At harvest	650	0.086 (LeM)–0.032 (MoM)	(15)
	Palmer	At harvest	690	0.074 (LeM)–0.021 (MoM)	(16)

Note: LeM = less mature; MoM = more mature; and harvest date (H1 = early harvest; H2 = commercial harvest; H3 = late harvest).

References: (1) Cubeddu et al. (2001a), (2) Rizzolo et al. (2011b), (3) Zanella et al. (2013), (4) Vanoli et al. (2011a), (5) Vanoli et al. (2005), (6) Vanoli et al. (2009b), (7) Rizzolo et al. (2013a), (8) Vanoli et al. (2009a), (9) Tijskens et al. (2007b), (10) Eccher Zerbini et al. (2005a), (11) Eccher Zerbini et al. (2011), (12) Eccher Zerbini et al. (2005b), (13) Eccher Zerbini et al. (2005c), (14) Vanoli et al. (2011b), (15) Spinelli et al. (2013), (16) Vanoli et al. (2014a), (17) Rizzolo et al. (2011a).

nectarines had a higher total sugars content with a higher proportion of sucrose, higher malic and quinic acid contents, and lower total acids, and citric acid content than fruit having higher $\mu_a 670$ value (LeM) (Jacob et al., 2006). After 3 days of shelf life at 20°C, LeM "Ambra" nectarines had higher titratable acidity and were much more firm than the MoM ones and, at sensory analysis, were firmer, sourer, and less juicy, and less sweet and aromatic (Vanoli et al., 2009a). LeM "Spring Belle" peaches at harvest were firmer, had lower expressible juice, lower glucose content, lower ratio of total sugars to total acids, and higher malic acid and citric acid contents than MoM fruit (Rizzolo et al., 2013a).

The studies carried out on mangoes showed that it was possible to use μ_a in the 630–690 nm range measured at harvest to distinguish mango fruit differing in the Yellowness Index (a parameter used for determining the amount of "yellow" color in the pulp), which is a commonly used criterion for assessing maturity in mangoes. "Tommy Atkins" mangoes sorted at harvest as LeM according to $\mu_a 630$ value had the lowest value of Yellowness Index at harvest; with shelf life, the Yellowness Index increased steeply in the LeM class, while it showed only slight changes in the MoM one (Rizzolo et al., 2011b; Vanoli et al., 2011b). Similarly, LeM "Haden" mangoes, sorted according to $\mu_a 650$ value at harvest, were characterized by lower values for pulp color parameters a^* and b^* (parameters that indicate color directions in the *CIELab* color space: $+a^*$ is the red direction, $-a^*$ is the green direction, $+b^*$ is the yellow direction, and $-b^*$ is the blue direction), lower chroma and Yellowness Index, and higher hue and firmness than MoM fruit (Spinelli et al., 2013).

8.4.2.2 Processed Fruit

Studies were carried out on the relationships between fruit maturity at harvest assessed by $\mu_a 670$ and quality characteristics of air-dried apple rings and fresh-cut apples.

Rizzolo et al. (2011a, 2012) showed that the classification of "Cripps Pink" and "Golden Delicious" apples at harvest based on the $\mu_a 670$ value was able to segregate raw and air-dried rings of different quality. For LeM "Cripps Pink" apples, fresh rings had lower soluble solids content, dry matter, and a^* value, and higher b^* value (Rizzolo et al., 2011a). LeM "Cripps Pink" and "Golden Delicious" fruit resulted in air-dried rings with higher shrinkage and greater browning, coupled with higher ring hardness and higher crispness index (Rizzolo et al., 2011a, 2012, 2013b). It was also shown that TRS maturity class influences the microstructure features evaluated by x-ray microtomography and crispness characteristics: air-dried apple rings prepared from LeM "Golden Delicious" apples stored in air for 5 months had higher porosity and higher average level of sound peaks (lower than 60 dB) than those prepared from MoM apples (Rizzolo et al., 2014b). After an osmodehydration pretreatment in 60% sucrose syrup, osmo-air-dried rings prepared from LeM apples had a more connected solid structure, with lower degree of tissue and pore anisotropy, which are a measure of preferential alignment of the structure, and were less crispy, as determined by acoustic parameters, than osmo-air-dried rings produced from MoM fruit (Rizzolo et al., 2014b).

The classification of fruit at harvest based on $\mu_a 670$ value was effective also in segregating fresh-cut apples packed in air and sealed in polypropylene bowls for ethylene evolution and pulp mechanical properties during storage at 4°C up to 15 days (Cortellino, personal communication): fresh-cut apples classified MoM at harvest had an earlier ethylene climacteric peak than LeM ones, and, similarly to what was found for dried apple rings, had lower firmness and stiffness than those classified LeM at harvest.

8.4.3 Scattering and Maturity

It has been shown that it is possible to use the reduced scattering coefficient measured at selected wavelengths to differentiate fruit at different stages of maturity.

Generally, the μ'_s values decreased with maturity and ripening. By measuring $\mu'_s 630$, it was possible to distinguish unripe (8.2 ± 0.6 cm^{-1}) from ripe (4.9 ± 0.5 cm^{-1}) kiwifruits (Eccher Zerbini et al., 2008). Values of $\mu'_s 720$ in the range 22–24 cm^{-1} were characteristic for "Conference" pears at removal from storage, whereas $\mu'_s 720$ values lower than 20 cm^{-1} were found for overripe pears (Eccher Zerbini et al., 2002). In peaches and nectarines, a decrease of $\mu'_s 750$ was observed when the harvest was delayed and after shelf life: for example, in "Big Top" nectarines, $\mu'_s 750$ values decreased from 20 ± 0.2 cm^{-1} at commercial harvest to 16 ± 0.2 cm^{-1} in fruit harvested 1 week later and further decreased, respectively, to 17.8 ± 0.3 and 14.4 ± 0.3 cm^{-1} after 3 days of shelf life (unpublished data). Likewise, in "Jonagored" apples, $\mu'_s 750$ values decreased from 14.2 ± 0.1 cm^{-1} at commercial harvest to 13.7 ± 0.1 cm^{-1} in fruit harvested 2 weeks later, and further decreased to 12.2 ± 0.1 and 10.9 ± 0.1 cm^{-1}, respectively, after 6 months of storage in a controlled atmosphere (Vanoli et al., 2007). In "Jubileum" plums, $\mu'_s 758$ decreased from 6.7 ± 0.15 to 4.3 ± 0.09 cm^{-1} after 4 days at 20°C (Vangdal et al., 2010), whereas in "Palmer" mangoes, $\mu'_s 780$ decreased from 20.7 ± 0.4 cm^{-1} at the beginning of shelf life to 15.1 ± 1.7 cm^{-1} after 4 days and to 12.1 ± 1.4 cm^{-1} after 11 days at 20°C (Spinelli et al., 2012).

The scattering parameters related to the density (a) and size (b) of scatterers may also be influenced by fruit maturity. Vanoli et al. (2012a) reported that for "Haden" mangoes with firmness values ranging from 5 to 108 N, the scattering values ranged from 10.31 to 17.66 cm^{-1} and scattering power from −0.256 to −0.098, and that both parameters were higher in the LeM maturity class ($a = 4.02 \pm 0.36$ cm^{-1} and $b = -0.265 \pm 0.017$) with respect to the MoM one ($a = 13.34 \pm 0.36$ cm^{-1} and $b = -0.256 \pm 0.015$). In apples, Vanoli et al. (2013b) found that the density and the size of the scattering centers changed in a different way depending on apple cultivars, fruit maturity, harvest time, and shelf life, as shown in Figure 8.7. In "Braeburn" apples, picked at different times (H1, early harvest; H2, commercial harvest; H3, late harvest) after 1 day at 20°C the scattering values (a) were highest in MoM apples picked at commercial harvest and lowest in LeM H1 apples and in MoM H3 ones; after 7 days of shelf life parameter a was higher in MoM fruit than in LeM ones, regardless of harvest date. In contrast, in "Cripps Pink" apples the scattering values were not significantly affected by fruit maturity class. As for the scattering power (b), no consistent trend was observed with fruit maturity, harvest time, and shelf life period in both cultivars.

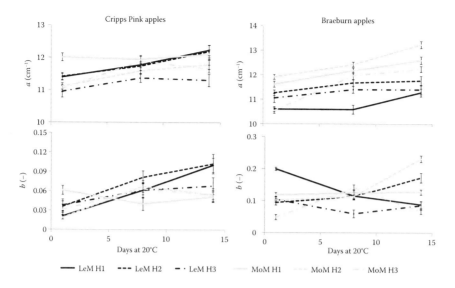

FIGURE 8.7 Density (parameter a, cm^{-1}) and size (parameter b, –) of the scattering centers for the LeM and MoM classes of "Cripps Pink" and "Braeburn" apples as a function of harvest time (i.e., H1, H2, and H3) and days of shelf life at 20°C. Bars refer to standard error of the mean. (From Vanoli, M., Rizzolo, A., Zanella, A. et al., Apple texture in relation to optical, physical and sensory properties. *InsideFood Symposium: Book of Proceedings*, April 9–12, Leuven, Belgium, 2013b, http://www.insidefood.eu/INSIDEFOOD_WEB/UK/WORD/proceedings/032P.pdf.)

8.5 TEXTURE

The scattering and absorption optical properties measured by TRS reflect the fruit texture characteristics. Close relationships of the sensory texture profiles (i.e., firm, crispy, juicy, and mealy) and the pulp mechanical characteristics (i.e., firmness, stiffness, and energy-to-rupture) with TRS scattering parameters in the 630–900 nm were found for "Braeburn" apples (Rizzolo et al., 2014a); scattering value (parameter a) increased from 11.10 ± 0.23 cm^{-1} for very firm and very crispy apples, to 11.56 ± 0.14 cm^{-1} for firm/crispy/juicy fruit, and to 11.93 ± 0.35 cm^{-1} and 12.12 ± 0.20 cm^{-1} for very mealy and mealy apples. Vanoli et al. (2015) reported that "Braeburn" apples with the lowest values of firmness and stiffness were scored as the most mealy and the highest scattering values (15.11 ± 0.28 cm^{-1}), with respect to apples characterized by the highest values of firmness and stiffness, and scored as the most crispy and juicy (13.36 ± 0.16 cm^{-1}).

8.5.1 CORRELATIONS

Studies have been carried out on the correlations between the optical properties measured by TRS and the texture characteristics of fruit measured by instrumental methods or by sensory tests. In "Jonagored" apples (Vanoli et al., 2007), the reduced scattering coefficients at 750 and 780 nm measured at harvest positively correlated with relative internal space volume (RISV) and titratable acidity, but not with firmness and stiffness. In contrast, μ'_s750 and μ'_s780 measured after storage were negatively correlated to titratable acidity, firmness, stiffness, percent juice, and sensory crispness, and positively to mealiness. Values of μ'_s780 lower than 11 cm^{-1} characterized crispy, not mealy "Jonagored" apples, having firmness values higher than 50 N and percent juice higher than 30%. After storage, μ'_s750 and μ'_s780 showed a high and positive correlation with galacturonic acid content in water-soluble pectin, and a negative correlation with residue insoluble pectin and protopectin index (Vanoli et al., 2009b). In "Braeburn" apples, μ'_s790 and μ'_s912 were positively correlated to sensory mealiness and to the RISV, and negatively correlated to sensory crispness, juiciness, and firmness, and to percent juice (Vanoli et al., 2009c). In contrast for "Cripps Pink" apples stored at 1°C up to 91 days, Vanoli et al. (2011a) found low correlation coefficients ($r \leq 0.4$) between μ'_s measured in the wavelength range 740–1100 nm and firmness and RISV. The μ'_s900 and parameter a in "Breaburn" apples were negatively correlated to sensory firmness, crispness, and juiciness and positively correlated to mealiness, and, with higher coefficients, were negatively correlated to firmness, stiffness, and energy-to-rupture (Vanoli et al., 2013b). In contrast, in "Cripps Pink" cultivar μ'_s900 and parameter a did not significantly correlate with sensory attributes, and positively correlated, with lower r values, with firmness, stiffness, and energy-to-rupture (Vanoli et al., 2013b). In "Jubineum" plums, Vangdal et al. (2010) found correlations between μ_a670 and firmness (positive), total acidity (positive), and soluble solids (negative) as well as positive correlations between μ'_s670 and μ'_s758 and soluble solids content.

The results of the correlation between μ'_s and textural parameters after storage in apples are consistent between "Jonagored" and "Braeburn" cultivars which underwent dramatic changes in texture with ripening. In the case of "Cripps Pink" apple cultivar, whose texture showed only little changes with ripening, only poor correlations were found.

8.5.2 Regression Models

Selected reduced scattering coefficients, as well as scattering spectra alone, in conjunction with selected absorption coefficients or absorption spectra, were used to develop models for the prediction of mechanical properties of fruit pulp, mainly firmness, and of sensory profiles in various fruit species. In kiwifruit, Valero et al. (2004a) built multiple linear regression models to predict firmness, measured as the ratio of force/deformation with a steel ball-headed probe, which is related to Young's module, using either only scattering (μ'_s675 and μ'_s750) or absorption (μ_a675, μ_a750, and μ_a800) plus scattering (μ'_s675 and μ'_s750). The models performed better when scattering and absorption coefficients were combined ($R^2_{adj} = 0.81$, SEE = 0.94 N/mm) than when including only scattering coefficients ($R^2_{adj} = 0.6$, SEE = 1.37 N/mm). Also by using partial least squares (PLS) regression models, Vanoli et al. (2011a, 2012a, 2013a) found better prediction for firmness in apples and mangoes when combining absorption and scattering coefficients, compared to using either scattering or absorption. Table 8.5 summarizes the literature data on the performance of PLS

TABLE 8.5
Performance of Different Regression Models on Original TRS Spectral Data for Prediction of Firmness, Relative Intercellular Space Volume (RISV), Streif Index in Apples, and Firmness and Pulp Color in Mangoes

Fruit	TRS Parameter	Parameter	LV	R^2_{CV}	RMSECV	References
Apple Cripps Pink	$\mu_a 740 - 900$	Firmness	11	0.87	7.44	Barzaghi et al. (2009)
	$\mu'_s 740 - 900$	Firmness	4	0.65	9.80	
Apple Golden Delicious	$\mu_a 740 - 900$	Firmness	3	0.74	5.60	Barzaghi et al. (2009)
	$\mu'_s 740 - 900$	Firmness	4	0.82	4.87	
Apple Cripps Pink storage	$\mu_a 670 - 1100$	Firmness	1	0.82	5.56	Vanoli et al. (2011a)
	$\mu_a 670 - 1100 + \mu'_s 670 - 1100$	Firmness	4	0.83	5.44	
	$\mu_a 670 - 1100$	RISV	1	0.77	1.30	
	$\mu_a 670 - 1100 + \mu'_s 670 - 1100$	RISV	4	0.82	1.19	
	$\mu_a 670 - 1100$	Streif Index	1	0.70	0.0073	
	$\mu_a 670 - 1100 + \mu'_s 670 - 1100$	Streif Index	4	0.75	0.0068	
Mango Haden	$\mu_a 540 - 690$	Firmness	2	0.43	16.19	Vanoli et al. (2012a)
		ln firmness	2	0.67	0.47	
		a^* pulp	2	0.90	1.98	
		$H°$ pulp	2	0.88	1.83	
	$\mu_a 540 - 690 +$ Mie $a\ b$	Firmness	2	0.50	14.77	
		ln firmness	4	0.78	0.39	
		a^* pulp	2	0.92	1.78	
		$H°$ pulp	2	0.91	1.58	
Mango Tommy Atkins	$\mu_a 540 - 880$	Firmness	7	0.65	16.5	Vanoli et al. (2013a)
		a^* pulp	3	0.89	1.13	
		$H°$ pulp	3	0.90	1.07	
	$\mu_a 540 - 880 +$ Mie $a\ b$	Firmness	4	0.73	14.5	
		a^* pulp	5	0.90	1.07	
		$H°$ pulp	7	0.93	0.91	

Note: LV, number of latent variables; R^2_{CV}, coefficient of determination in cross validation; and RMSECV, root mean square error in cross validation.

models to predict firmness and other texture parameters (RISV) and maturity index (Streif Index in apples and pulp color in mangoes). Barzaghi et al. (2009) reported better regression models for firmness prediction using scattering for "Golden Delicious" apples and absorption for "Cripps Pink" fruit. On the other hand, Vanoli et al. (2011a) reported: poor correlations between scattering spectra and firmness, RISV, and Streif Index in "Cripps Pink" apples after storage; good predictions when using absorption spectra; and better predictions, especially for RISV and Streif Index, when scattering and absorption spectra were combined (Table 8.5). A similar prediction trend was also observed for mango of the "Haden" cultivar (Vanoli et al., 2012a)—poor correlations ($R^2_{pred} < 0.3$) between scattering spectra and firmness, $a*$ and $H°$ pulp color parameters, and good correlations when using absorption spectra, and better predictions when either scattering spectra or scattering value and power (parameters a and b) were added (Table 8.5). Also for "Tommy Atkins" mangoes low correlations ($R^2_{pred} < 0.4$) between scattering spectra in the wavelength 540–880 nm range and firmness, $a*$ and $H°$ pulp color parameters were found, while good correlations were obtained when using absorption spectra, with further improvements when either scattering spectra or scattering value and power (parameters a and b) were added (Table 8.5).

8.5.3 Classification Models

TRS optical properties were used as explanatory variables to build classification models for firmness and texture sensory profiles. Valero et al. (2004b) used absorption and reduced scattering coefficients measured at 650, 750, and 800 nm to cluster fruit of different species according to three levels of firmness, and reported 81% correct classification rate for tomatoes, 77% for peaches, and 76% for apples. The study also confirmed that the models for firmness were slightly better when using both scattering and absorption coefficients, than when using only scattering. For pears, Nicolaï et al. (2008) found a highly nonlinear relationship between scattering spectra in the 875–1030 nm range and firmness, which prevented the construction of significant PLS calibration models for firmness based on μ'_s. Hence they attempted to develop nonlinear models based on kernel PLS, but with little success.

To discriminate mealy from nonmealy apples, with mealiness being measured by means of a tenderometric test, Valero et al. (2005) used 15 TRS parameters, nine of which were absorption coefficients, plus six scattering coefficients at 670, 818, 900, 930, 960, and 980 nm. The model correctly classified 98% of fruit. However, when the fruit were classified into the fresh, mealy, and not-mealy stages, the model performance decreased to 71%. For "Jonagored" apples, Rizzolo et al. (2010a) used both absorption and scattering coefficients measured at 650, 670, 750, and 780 nm to build models to classify apples into three classes (low, medium, and high) for firm, crispy, mealy, and juicy sensory attributes. All models showed better classification performance when absorption coefficients were added to scattering variables. The correct classification rates were 57% for juicy, 67% for crispy, 71% for firm, and 72% for mealy. Furthermore, considering all sensory attributes as a whole, Rizzolo et al. (2010a) clustered apples according to five sensory profiles differing for texture attributes (firm, crispy, mealy, and juicy) and sour, sweet, and aromatic intensities.

By sorting apples into the five sensory profiles using $\mu'_s 670$, $\mu'_s 750$, $\mu'_s 780$, and $\mu_a 630$, the model showed only 56% classification accuracy, while it was able to correctly classify almost 70% of fruit into either a mealy-dry texture without flavor or a not mealy-juicy texture with a moderate flavor.

8.6 INTERNAL DISORDERS

The development of internal disorders in fruit induces changes in tissue optical properties. The internal browning phenomena can be associated with membrane damage caused by stresses such as low temperatures and low oxygen and/or high carbon dioxide concentration in controlled atmosphere storage. The loss of integrity of membranes can lead to leakage of cellular liquid into the cellular spaces with a release of phenolic compounds from the vacuoles and of polyphenoloxidase from the plastides, resulting in the enzymatic oxidation of phenols to o-quinones and the formation of brown-colored pigments (Tomás-Barberán and Espín, 2001). Browning is mainly detected by absorption coefficients measured in the 720–780 nm range in coincidence of the red area of the spectrum (650–750 nm); the absorbance at these wavelengths increases with the increase of browning severity (Clark et al., 2003; Han et al., 2006). Furthermore, Clark et al. (2003) also reported that browned tissue was more saturated with free water than healthy tissue, and that more intercellular water reduces refractive index changes at cellular boundaries, resulting in less light scattering.

The internal disorders mainly affecting tissue structure can be revealed by scattering parameters. Mealiness is a negative characteristic of sensory texture that combines the sensation of a disaggregated tissue with the sensation of a lack of juiciness (Barreiro et al., 2000). In peaches and nectarines, the mealy texture is known as woolliness when the lack of juiciness is caused by gel structures that retain water molecules, and as leatheriness when there are hard-textured fruit with no juice (Lurie and Crisosto, 2005). In apples, the lack of juiciness in mealy fruit depends on the type of softening process, which mainly involves cell separation rather than cell breakages due to the weakness of the middle lamella, and the intact cells are responsible for the mealy texture (Harker et al., 2002). Watercore develops in some apple cultivars at harvest; the affected areas look glassy due to the presence of water instead of air in the intercellular spaces. The water-soaked areas are usually located around the vascular bundles or the core area (Marlow and Loescher, 1984) and the elevated sorbitol levels found in the intercellular spaces of the affected tissue change the microstructure (Gao et al., 2005). Jelling in plums manifests itself as a translucent gelatinous breakdown of the mesocarp tissue around the stone and it is related to changes in membrane permeability and to the presence of water-soluble pectins (Candan et al., 2008).

8.6.1 ABSORPTION AND SCATTERING SPECTRA AND INTERNAL DISORDERS

It has been reported that TRS optical properties change in response to internal disorder occurrence. In "Braeburn" apples affected by internal browning (Vanoli et al., 2011c, 2014b), healthy and browned fruit measured in the 670–1040 nm range produced a similar pattern of absorption spectra; however, browned fruit showed significantly higher μ_a values in the 670–940 nm spectral range than the healthy ones, similar to what were found in pears affected by brown heart (Eccher Zerbini et al., 2002) and in potatoes affected by internal brown spot (Vanoli et al., 2012b). The scattering spectra were rather flat in the 740–900 nm range for both healthy and browned apples; however, healthy tissues showed higher μ_s' values at 740, 780, 820, 860, and 900 nm than browned ones (Vanoli et al., 2014b). In contrast, Eccher Zerbini et al. (2002) found minimal differences between the scattering values of healthy pears and of those affected by brown heart. The highest differences between the absorption spectra of browned and healthy tissues were found in the 670–780 nm range (Eccher Zerbini et al., 2002; Vanoli et al., 2011c, 2012b), and, hence, these wavelengths were selected to distinguish healthy fruit from those affected by internal disorders, as shown in Table 8.6.

8.6.2 Absorption and Reduced Scattering Coefficients and Internal Disorders

Generally, healthy fruit showed lower μ_a values at the selected wavelengths than those affected by internal browning, internal bleeding, and watercore, while Vanoli et al. (2009c) found no difference at $\mu_a 790$ between mealy and nonmealy apples, but lower $\mu_a 912$ values in mealy fruit with respect to the nonmealy ones. Healthy fruits were also characterized by higher μ_s' values at the selected wavelengths than fruit affected by internal defects (Table 8.6), with some exceptions such as nectarines affected by internal bleeding, where no difference was found between healthy and internal bled fruit (Lurie et al., 2011); "Braeburn" apples affected by internal browning, where $\mu_s'670$ values were higher in browned tissues (Vanoli et al., 2014b); and mealy apples, which showed higher values of $\mu_s'790$ and $\mu_s'912$ than nonmealy fruit (Vanoli et al., 2009c). Furthermore, it has been observed in apples (Vanoli et al., 2009c, 2011c, 2014b), plums (Vangdal et al., 2012), and nectarines (Lurie et al., 2011) that the values of $\mu_a 740$, $\mu_a 750$, and $\mu_a 780$ increased as browning severity increased. Vanoli et al. (2009c, 2011c, 2014b) found also that the localization of the internal disorder had an impact on $\mu_a 740$, $\mu_a 750$, and $\mu_a 780$ values; they observed higher values when the internal browning affected the pulp tissues rather than the inner part of a fruit (core), which could be, depending on fruit size, at a depth greater than 25 mm, the estimated maximum depth that would be achieved by the TRS setup.

Moreover, $\mu_a 670$ changed with browning development, but it was not able to discriminate healthy fruit from browned ones because, as described in Section 8.4, its value was also affected by the chlorophyll content of the pulp. So, high $\mu_a 670$ values can be due to either high chlorophyll content or the presence of browned tissues (Vanoli et al., 2014b). Similarly, Eccher Zerbini et al. (2002) observed in pears that $\mu_a 690$ value was simultaneously influenced by fruit ripening and by brown heart development and hence concluded that $\mu_a 690$ cannot be used to discriminate healthy from browned pears. Lurie et al. (2011) found in nectarines that $\mu_a 670$ value was able to discriminate healthy fruit from those simultaneously affected by bleeding and browning, but not from those affected by either bleeding alone or browning alone, whereas $\mu_a 780$ was able to discriminate between healthy fruit and those affected by

TABLE 8.6
TRS μ_a and μ_s' Values at Selected Wavelengths for Sound Fruit (Healthy) and Fruit Affected by Internal Disorder (Defect) in Apples, Nectarines, and Plums, and Number of Measurement Points/Fruit (N_{points})

Type of Defect	Fruit	Cultivar	TRS Properties	Healthy	Defect	N_{points}	Reference
Internal browning	Apple	Granny Smith	$\mu_a 750$	0.029 b	0.037 a	4	Vanoli et al. (2009c)
			$\mu_s' 750$	12.22 a	11.24 b	4	
		Braeburn	$\mu_a 670$	0.091 b	0.140 a	8	Vanoli et al. (2014b)
			$\mu_a 780$	0.047 b	0.068 a	8	
			$\mu_s' 670$	10.82 b	10.97 a	8	
			$\mu_s' 780$	10.93 a	10.60 b	8	
	Nectarine	Morsiani 90	$\mu_a 670$	0.065 b	0.083 a	2	Lurie et al. (2011)
			$\mu_a 780$	0.046 b	0.066 a	2	
			$\mu_s' 670$	19.87 a	18.73 b	2	
			$\mu_s' 780$	18.32 a	16.76 b	2	
Internal bleeding	Nectarine	Morsiani 90	$\mu_a 670$	0.065 b	0.066 a	2	Lurie et al. (2011)
			$\mu_a 780$	0.046 b	0.051 a	2	
			$\mu_s' 670$	19.87 a	19.67 a	2	
			$\mu_s' 780$	18.32 a	17.74 a	2	
Internal browning and jelling	Plum	Jubileum	$\mu_a 670$	0.151 b	0.252 a	2	Vangdal et al. (2012)
			$\mu_a 780$	0.106 b	0.165 a	2	
			$\mu_s' 670$	5.02 a	4.69 b	2	
			$\mu_s' 780$	4.46 a	4.24 b	2	
Watercore	Apple	Fuji	$\mu_a 670$	0.079 b	0.096 a	4	Vanoli et al. (2009c)
			$\mu_a 790$	0.049 b	0.059 a	4	
			$\mu_s' 670$	12.29 a	9.52 b	4	
			$\mu_s' 790$	8.99 a	8.21 b	4	
Mealiness	Apple	Braeburn	$\mu_a 790$	0.037 a	0.037 a	2	Vanoli et al. (2009c)
			$\mu_a 912$	0.091 a	0.084 b	2	
			$\mu_s' 790$	16.41 b	20.13 a	2	
			$\mu_s' 912$	16.26 b	19.47 a	2	

Note: Means followed by different letters are statistically different (Bonferroni's test at $P < 0.05$).

chilling injuries. In contrast, Vangdal et al. (2012) found in plums that both $\mu_a 670$ and $\mu_a 780$ were able to distinguish healthy fruit from those affected by internal disorders. In addition, Eccher Zerbini et al. (2002) found in pears that μ_s' value decreased when fruit tissue became translucent, as in overripe fruit, as well as in the presence of mechanical damage. In kiwifruit, by using $\mu_s'630$, it was possible to distinguish the sound region of the fruit from that affected by *Botrytis*, which was characterized by a higher translucency, corresponding to lower $\mu_s'630$ (Eccher Zerbini et al., 2008).

The $\mu_s'670$, $\mu_s'650$, and $\mu_s'680$ showed different trends in relation to browning severity and localization in the fruit. Vanoli et al. (2009c) found in "Granny Smith" apples that $\mu_s'750$ allowed to distinguish healthy fruit from browned ones only when browning affected the pulp region. In "Braeburn" apples, $\mu_s'780$ was higher in healthy fruit with respect to those affected by internal browning in two subsequent seasons, but different trends in $\mu_s'780$ value in relation to internal browning position and severity were found: in 2009, $\mu_s'780$ did not change with browning severity and position, while in 2010, $\mu_s'780$ was lower for slight severity and when internal browning affected the pulp (Vanoli et al., 2014b). In nectarines, $\mu_s'670$ failed to differentiate between healthy and disordered tissues, whereas $\mu_s'780$ could differentiate between fruit with high expressible juice and those having low values of expressible juice, which are fruit that are affected by some woolliness (Lurie et al., 2011). In plums, $\mu_s'670$ and $\mu_s'780$ were not influenced by chilling injury development (Vangdal et al., 2012).

8.6.3 Correlations and Threshold Values

The correlations among TRS optical properties and internal disorders have been studied for various fruit species. Positive correlations between μ_s' measured at 630, 670, 750, 780, 790, and 912 nm and sensory mealiness in apples have been reported by Rizzolo et al. (2010a) and Vanoli et al. (2009c), with correlation coefficient ranging from 0.51 to 0.72; in contrast, very low or no correlations were found among the μ_a measured at 630, 670, 750, 780, and 790, and mealiness, while $\mu_a 912$ showed a negative correlation with mealiness ($r = -0.72$). On the contrary, Valero et al. (2005) found low correlations ($r < 0.4$) between μ_s' measured at 672, 750, 818, 900, and 950 nm as well as μ_a measured at 900 and 950 nm and the descriptive tests used to measure mealiness, while Vangdal et al. (2012) did not find any correlation between $\mu_s'670$ and $\mu_s'780$ and jelling and browned area in plums.

Threshold values for absorption and reduced scattering coefficients have been established for mealy apples; for "Braeburn" cultivar, mealy, noncrispy, nonjuicy apples had $\mu_s'790$ and $\mu_s'912$ values above 19 cm^{-1} and $\mu_a 912$ values below 0.09 cm^{-1} (Vanoli et al., 2009c), whereas for "Jonagored" cultivar, nonmealy and crispy apples had $\mu_s'780$ values lower than 11 cm^{-1} (Vanoli et al., 2007).

For internal browning, high and positive correlations were found among $\mu_a 670$ and $\mu_a 780$ and browning score in plums and nectarines (Lurie et al., 2011; Vangdal et al., 2012). Changes in absorption coefficient revealed changes in the pulp color due to internal browning development. High negative or positive correlations were found between $\mu_a 740$ and L^* ($r = -0.95$), a^* ($r = 0.88$), and $H°$ ($r = -0.88$) in "Braeburn" apples (Vanoli et al., 2011c) and between $\mu_a 750$ and L^* ($r = -0.66$), a^* ($r = 0.87$), and

$H°$ ($r = -0.86$) in "Granny Smith" apples (Vanoli et al., 2009c). Correlations were also found between μ'_s and pulp color in "Granny Smith" cultivar, but they were low, ranging from 0.54 (L^*) to -0.57 (a^*) (Vanoli et al., 2009c). It was possible to use $\mu_a 750$ to distinguish healthy "Granny Smith" apple fruit from browned ones, with the former being characterized by $\mu_a 750 < 0.030$ cm^{-1} (Vanoli et al., 2009c). Similarly, in "Braeburn" apples, $\mu_a 740$ values below 0.038 cm^{-1} indicated only healthy pulp (Vanoli et al., 2011c). Eccher Zerbini et al. (2002) found in "Conference" pears that sound tissue had $\mu_a 720 \leq 0.04$ cm^{-1}, while tissues affected by brown heart showed $\mu_a 720 > 0.04$ cm^{-1}.

8.6.4 CLASSIFICATION MODELS

Absorption and scattering coefficients measured at different wavelengths have been used as explanatory variables to build discriminant models to distinguish healthy fruit from those affected by internal disorders, such as mealiness (Rizzolo et al., 2010a; Valero et al., 2005) and internal browning (Vanoli et al., 2014b). Valero et al. (2005) used 15 TRS variables measured in the 670–980 nm range to discriminate between mealy and nonmealy apples belonging to two cultivars ("Cox" and "Golden Delicious") achieving 80% correct classification rate. When the models were used to estimate three (fresh, nonmealy, and mealy) or four (fresh, dry, soft, and mealy) textural stages, the performance decreased to 51% and 47% in the validation set, respectively, even though the mealy category was better predicted than the other groups. By reducing the wavelengths used for the model to 670 nm combined with 960–980 nm for the absorption coefficient only, the segregation between mealy and nonmealy fruit reached 88% in validation. This fact confirms the increased robustness for the classification model with a lower number of variables, which avoided the potential risk of overfitting for the model. When separate models for each variety were developed, the classification was best for "Cox" apples, while for "Golden Delicious," the model was unable to obtain a better score than the pooled data.

Rizzolo et al. (2010a) used the scattering coefficients measured at 630, 670, and 750 nm and $\mu_a 630$ for the classification of apples into three classes of mealiness (low, medium, and high), and the model achieved 72% overall correct classification rate (cv "Jonagored"); it well classified the low class (82%) and the medium class (72%) but not the high class (47%).

In a two-year experiment on the internal browning of "Braeburn" apples, the scattering coefficient measured at 780 nm plus $\mu_a 780$ or plus μ_a measured in the 670–1000 nm range gave good classification models for discriminating healthy apples from those affected by internal browning (Vanoli et al., 2014b). The best classification was obtained in 2010 by using $\mu'_s 780$ plus $\mu_a 780$ when the region affected by internal browning (pulp or core) was considered, the model allowed to discriminate 71% of browned fruit and 90% of healthy ones. In 2009, only 71% of healthy fruit were correctly classified together with the same percentage of browned ones. Vanoli et al. (2014b) suggested that the better classification result obtained in 2010 was due to the increase in the number of TRS measurement points/fruit that allowed better probing of the fruit tissues. However, the asymmetric development of this disorder made its difficult, especially when the disorder was localized in the core region and when it occurred in spots. A different TRS setup (position and distance of fibers, time resolution) should be studied in order to reach the deeper tissue within the fruit, thus improving the browning detection.

8.7 CONCLUSIONS

We have reviewed the use of TRS as a tool for nondestructively sensing food quality. The applicability of TRS to the nondestructive assessment of food depends on the combination of optical properties and sample geometry; usually TRS requires a sample volume larger than 10 cm^3, but it could be carried out on strongly diffusive samples having smaller dimensions. On the other hand, for samples with weak diffusion, such as gels, or in the presence of voids, such as cereal flakes, TRS cannot provide robust results.

The major advantage of using TRS in determining the optical properties of fruit is a penetration into the pulp to a depth of 1–2 cm. This feature allows (1) to measure the pigment (chlorophyll and carotenoids) content in fruit pulp which is related to fruit maturity, (2) to discriminate fruit having different texture characteristics, and (3) to detect internal disorders causing changes in pulp color and texture.

As a rule of thumb, the chlorophyll content is measured by using the absorption coefficient at 670 nm. However, when the content of chlorophyll is too high, the signal at 670 nm could be too low, and hence another wavelength in the 630–690 nm range has to be selected in order to have a sufficient SNR for all the fruit, allowing correct fruit classification. In order to measure the carotenoid content, as for example to assess maturity in mangoes, it is advisable to use the absorption coefficient at 540 nm.

Spectral measurements give better performances in discriminating fruit having different texture characteristics than the absorption and reduced scattering coefficients at single wavelengths. Better predictions are obtained by adding the a and b scattering parameters to the absorption spectra.

Using TRS at selected wavelengths, it is possible to detect internal disorders related to changes in color or texture of fruit pulp. To ensure the detection of the disorder, the number of measurement points has to be suited to the localization and distribution of the affected tissue. To detect disorders involving all the pulp, two measurement points at a distance of 180° are sufficient, while to detect disorder involving only part of the pulp, eight points at a distance of 45° are advisable, even if the measurement time/fruit greatly increases.

Over the last few years, efforts have been made in developing a scaled-down version of TRS instrumentation from the first ones, which occupied the space equivalent to a laboratory room, as described here. The new TRS instrumentations are compact desktop or rack-based systems, with lower cost. Notwithstanding this, TRS technology nowadays is still expensive and not commercially available. However, the use of semiconductor lasers and detectors is under study to further scale down TRS systems to the level of hand-held devices, with a further reduction of cost.

REFERENCES

Attanasio, G. 2012. Assessment of flesh texture in peach (*Prunus persica* L. Batsch). PhD thesis, Graduate School in Molecular Sciences and Plant, Food and Environmental Biotechnology, Università degli Studi di Milano, Milano.

Azzollini, S. 2012. Valutazione delle caratteristiche nutraceutiche di frutti di mango (*Mangifera indica* L.) in relazione allo stato di maturazione alla raccolta e in shelf life. Diss. Thesis in Biology Applied to Nutrition Science, Università degli Studi di Milano, Milano.

Barreiro, P., C. Ortiz, M. Ruiz-Altisent et al. 2000. Mealiness assessment in apples and peaches using MRI techniques. *Magn. Reson. Imag.* 18: 1175–1181.

Barzaghi, S., M. Vanoli, K. Cremonesi et al. 2009. Outer product analysis applied to time-resolved reflectance (TRS) and NIR reflectance spectra of apples. In Saranwong S., S. Kasemsumran, W. Thanapase, and P. Williams (eds), *Proceedings NIRS2009—The 14th International Conference on Near Infrared Spectroscopy*, 213–218, Chichester: IM Publications LLP.

Bobelyn, E., A. S. Serban, M. Nicu, J. Lammertyn, B. M. Nicolaï, and W. Saeys. 2010. Postharvest quality of apple predicted by NIR-spectroscopy: Study on the effect of biological variability on spectra and model performance. *Postharvest Biol. Technol.* 55: 133–143.

Candan, A. P., J. Graell, and C. Larrigaudière. 2008. Role of climacteric ethylene in the development of chilling injury in plums. *Postharvest Biol. Technol.* 47: 107–112.

Clark, C. J., V. A. McGlone, and R. B. Jordan. 2003. Detection of brownheart in "Braeburn" apple by transmission NIR spectroscopy. *Postharvest Biol. Technol.* 28: 87–96.

Cubeddu, R., C. D'Andrea, A. Pifferi et al. 2001a. Non-destructive quantification of chemical and physical properties of fruits by time-resolved reflectance spectroscopy in the wavelength range 650–1000 nm. *Appl. Opt.* 40: 538–543.

Cubeddu, R., C. D'Andrea, A. Pifferi et al. 2001b. Non-destructive measurements of the optical properties of apples by means of time-resolved reflectance spectroscopy. *Appl. Spectrosc.* 55: 1368–1374.

Cubeddu, R., A. Pifferi, P. Taroni, A. Torricelli, and G. Valentini. 1996. Experimental test of theoretical models for time-resolved reflectance. *Med. Phys.* 23: 1625–1633.

D'Andrea, C., L. Spinelli, A. Bassi et al. 2006. Time-resolved spectrally constrained method for quantification of chromophore concentrations and scattering parameters in diffusing media. *Opt. Express* 14: 1888–1898.

Eccher Zerbini, P., M. Grassi, R. Cubeddu, A. Pifferi, and A. Torricelli. 2002. Nondestructive detection of brown heart in pears by time-resolved reflectance spectroscopy. *Postharvest Biol. Technol.* 25: 87–97.

Eccher Zerbini, P., M. Grassi, M. Fibiani et al. 2003. Selection of "Spring Bright" nectarines by time-resolved reflectance spectroscopy (TRS) to predict fruit quality in the marketing chain. *Acta Hortic.* 604: 171–177.

Eccher Zerbini, P., M. Vanoli, M. Grassi et al. 2005a. Time-resolved reflectance spectroscopy measurements as a non-destructive tool to assess the maturity at harvest and to model the softening of nectarines. *Acta Hortic.* 682: 1459–1464.

Eccher Zerbini, P., M. Vanoli, M. Grassi et al. 2005b. Una nuova tecnica per la valutazione non distruttiva della qualità interna dei frutti: la spettroscopia risolta nel tempo. *Atti VII Giornate Scientifiche SOI*, 136–138, Firenze: Società Ortoflorofrutticola Italiana.

Eccher Zerbini, P., P. Cambiaghi, M. Grassi et al. 2005c. Effect of 1-MCP on "Abbé Fétel" pears sorted at harvest by time-resolved reflectance spectroscopy. *Acta Hortic.* 682: 965–971.

Eccher Zerbini, P., M. Vanoli, M. Grassi et al. 2006. A model for the softening of nectarines based on sorting fruit at harvest by time-resolved reflectance spectroscopy. *Postharvest Biol. Technol.* 39: 223–232.

Eccher Zerbini, P., M. Vanoli, F. Lovati et al. 2011. Maturity assessment at harvest and prediction of softening in a late maturing nectarine cultivar after cold storage. *Postharvest Biol. Technol.* 62: 275–281.

Eccher Zerbini, P., M. Vanoli, A. Rizzolo, R. Cubeddu, L. Spinelli, and A. Torricelli. 2008. Time-resolved reflectance spectroscopy: A non-destructive method for the measurement of internal quality of fruit. *Acta Hortic.* 768: 399–406.

Eccher Zerbini, P., M. Vanoli, A. Rizzolo et al. 2009. Time-resolved reflectance spectroscopy as a management tool in the fruit supply chain: An export trial with nectarines. *Biosyst. Eng.* 102: 360–363.

Eccher Zerbini, P., M. Vanoli, A. Rizzolo et al. 2015. Optical properties, ethylene production and softening in mango fruit. *Postharvest Biol. Technol.* 101: 58–65.

Gao, Z., S. Jayanty, R. Beaudry, and W. Loescher. 2005. Sorbitol transporter expression in apple sink tissues: Implications for fruit sugar accumulation and watercore development. *J. Am. Soc. Hort. Sci.* 130: 261–268.

Han, D., R. Tu, C. Lu, X. Liu, and Z. Wen. 2006. Nondestructive detection of brown core in the Chinese pear "Yali" by transmission visible-NIR spectroscopy. *Food Control* 17: 604–608.

Harker, F. R., J. Maindonald, S. H. Murray, F. A. Gunson, and S. B. Walker. 2002. Sensory interpretation of instrumental measurements 1: Texture of apple fruit. *Postharvest Biol. Technol.* 24: 225–239.

Jacob, S., M. Vanoli, M. Grassi et al. 2006. Changes in sugar and acid composition of "Ambra" nectarines during shelf life based on non-destructive assessment of maturity by time-resolved reflectance spectroscopy. *J. Fruit Ornam. Plant Res.* 14(Suppl. 2): 183–194.

Lammertyn, J., A. Peirs, J. De Baedemaeker, and B. M. Nicolaï. 2000. Light penetration properties of NIR radiation in fruit with respect to non-destructive quality assessment. *Postharvest Biol. Technol.* 18: 121–132.

Lurie, S. and C. H. Crisosto. 2005. Chilling injury in peach and nectarine. *Postharvest Biol. Technol.* 37: 195–208.

Lurie, S., M. Vanoli, A. Dagar et al. 2011. Chilling injury in stored nectarines and its detection by time-resolved reflectance spectroscopy. *Postharvest Biol. Technol.* 59: 211–218.

Marlow, G. and W. H. Loescher. 1984. Watercore. *Hort. Rev.* 6: 189–251.

Martelli, F., S. Del Bianco, A. Ismaelli, and G. Zaccanti. 2009. *Light Propagation through Biological Tissue and Other Diffusive Media: Theory, Solution, and Software*, Washington DC, SPIE Press.

Mourant, J. R., T. Fuselier, J. Boyer, T. M. Johnson, and I. J. Bigio. 1997. Predictions and measurements of scattering and absorption over broad wavelength ranges in tissue phantoms. *Appl. Opt.* 36: 949–957.

Nicolaï, B. M., B. E. Verlinden, M. Desmet et al. 2008. Time-resolved and continuous wave NIR reflectance spectroscopy to predict soluble solids content and firmness of pear. *Postharvest Biol. Technol.* 47: 68–74.

Nilsson, M. K., C. Sturesson, D. L. Liu, and S. Andersson-Engels. 1998. Changes in spectral shape of tissue optical properties in conjunction with laser-induced thermotherapy. *Appl. Opt.* 37: 1256–1267.

Pereira, T., L. M. M. Tijskens, M. Vanoli et al. 2010. Assessing the harvest maturity of Brazilian mangoes. *Acta Hortic.* 880: 269–276.

Pifferi, A., A. Torricelli, A. Bassi et al. 2005. Performance assessment of photon migration instruments: The MEDPHOT protocol. *Appl. Opt.* 44: 2104–2114.

Pifferi, A., A. Torricelli, P. Taroni, D. Comelli, A. Bassi, and R. Cubeddu. 2007. Fully automated time domain spectrometer for the absorption and scattering characterization of diffusive media. *Rev. Sci. Instrum.* 78: 053103.

Rizzolo, A., G. Bianchi, M. Vanoli, S. Lurie, L. Spinelli, and A. Torricelli. 2013a. Electronic nose to detect volatile compound profile and quality changes in "Spring Belle" peaches

(*Prunus persica* L.) during cold storage in relation to fruit optical properties measured by time-resolved reflectance spectroscopy. *J. Agric. Food Chem.* 61: 1671–1685.

Rizzolo, A., M. Vanoli, G. Bianchi et al. 2014a. Relationship between texture sensory profiles and optical properties measured by time-resolved reflectance spectroscopy during post storage shelf life of "Braeburn" apples. *J. Hort. Res.* 22: 113–121.

Rizzolo, A., M. Vanoli, G. Cortellino, L. Spinelli, and A. Torricelli. 2011a. Quality characteristics of air-dried apple rings: Influence of storage time and fruit maturity measured by time-resolved reflectance spectroscopy. *Procedia Food Sci.* 1: 216–223.

Rizzolo, A., M. Vanoli, G. Cortellino, L. Spinelli, and A. Torricelli. 2012. Potenzialità della spettroscopia di riflettanza risolta nel tempo per l'ottenimento di rondelle di mele essiccate con elevate caratteristiche organolettiche. In Porretta, S. (ed.), *Ricerche e Innovazioni nell'industria alimentare*, vol. X, 283–288, Pinerolo: Chiriotti Editori.

Rizzolo, A., M. Vanoli, G. Cortellino, L. Spinelli, and A. Torricelli. 2013b. Crispness of air-dried rings in relation to osmosis time and fruit maturity measured by time-resolved reflectance spectroscopy. InsideFood Symposium: Book of Proceedings, April 9–12, Leuven, Belgium, http://www.insidefood.eu/INSIDEFOOD_WEB/UK/WORD/proceedings/030P.pdf.

Rizzolo, A., M. Vanoli, G. Cortellino et al. 2014b. Characterizing the tissue of apple air-dried and osmo-air-dried rings by X-CT and OCT and relationship with ring crispness and fruit maturity at harvest measured by TRS. *Inn. Food Sci. Emerg. Technol.* 24: 121–130.

Rizzolo, A., M. Vanoli, P. Eccher Zerbini et al. 2009. Prediction ability of firmness decay models of nectarines based on the biological shift factor measured by time-resolved reflectance spectroscopy. *Postharvest Biol. Technol.* 54: 131–140.

Rizzolo, A., M. Vanoli, M. Grassi, L. Spinelli, and A. Torricelli. 2015. Time-resolved reflectance spectroscopy as a management tool for late-maturing nectarine supply chain. In Guidetti, R., L. Bodria, and S. Bist (eds) *Chemical Engineering Transactions*, vol. 44, 7–12, Milano: AIDIC Servizi S.r.L.

Rizzolo, A., M. Vanoli, L. Spinelli, and A. Torricelli. 2010a. Sensory characteristics, quality and optical properties measured by time-resolved reflectance spectroscopy in stored apples. *Postharvest Biol. Technol.* 58: 1–12.

Rizzolo, A., M. Vanoli, L. Spinelli, and A. Torricelli. 2011b. Relationship between continuous wave reflectance measurement of pulp colour and the optical properties measured by time-resolved reflectance spectroscopy in various fruit species. In Rossi, M. (ed.), *Color and Colorimetry. Multidisciplinary Contributions*, vol. VII B, 328–335, Santarcangelo di Romagna: Maggioli Editore.

Rizzolo, A., M. Vanoli, L. Spinelli, and A. Torricelli. 2014c. Non destructive assessment of pulp colour in mangoes by time-resolved reflectance spectroscopy; problems and solutions. *Lecture Held at the Non-Destructive Assessment of Fruit Attributes Symposium, 29th International Horticultural Congress*, August 17–21, Brisbane Australia.

Rizzolo, A., M. Vanoli, P. E. Zerbini, L. Spinelli, and A. Torricelli. 2010b. Influence of cold storage time on the softening prediction in "Spring Bright" nectarines. *Acta Hortic.* 877: 1395–1402.

Schotsmans, W., B. E. Verlinden, J. Lammertyn, and B. M. Nicolaï. 2004. The relationship between gas transport properties and the histology of apple. *J. Sci. Food Agric.* 84: 1131–1140.

Seifert, B., M. Zude, L. Spinelli, and A. Torricelli. 2015. Optical properties of developing pip and stone fruit reveal underlying structural changes. *Physiol. Plant.* 153: 327–336.

Spinelli, L., A. Rizzolo, M. Vanoli et al. 2012. Optical properties of pulp and skin in Brazilian mangoes in the 540–900 nm spectral range: Implication for non-destructive maturity assessment by time-resolved reflectance spectroscopy. *Proceedings of the 3rd CIGR*

International Conference of Agricultural Engineering (CIGR-AgEng2012), July 8–12, Valencia, Spain, ISBN 84-615-9928-4.

Spinelli, L., A. Rizzolo, M. Vanoli et al. 2013. Nondestructive assessment of fruit biological age in Brazilian mangoes by time-resolved reflectance spectroscopy in the 540–900 nm spectral range. *InsideFood Symposium: Book of Proceedings*, April 9–12, Leuven, Belgium, http://www.insidefood.eu/INSIDEFOOD_WEB/UK/WORD/proceedings/027P.pdf.

Tijskens, L. M. M., P. Eccher Zerbini, and R. E. Schouten. 2007a. Biological variation in ripening of nectarines. *Veg. Crops Res. Bull.* 66: 205–212.

Tijskens, L. M. M., P. Eccher Zerbini, R. E. Schouten et al. 2007b. Assessing harvest maturity in nectarines. *Postharvest Biol. Technol.* 43: 204–213.

Tijskens, L. M. M., P. Eccher Zerbini, M. Vanoli et al. 2006. Effects of maturity on chlorophyll-related absorption in nectarines, measured by non-destructive time-resolved reflectance spectroscopy. *Int. J. Postharvest Technol. Innov.* 1: 178–188.

Tomás-Barberán, F. A. and J. C. Espín. 2001. Phenolic compounds and related enzymes as determinants of quality in fruits and vegetables. *J. Sci. Food Agric.* 81: 853–876.

Torricelli, A. 2009. Optical sensing—Determination of optical properties in turbid media: Time-resolved approach. In Zude M. (ed.), *Optical Monitoring of Fresh and Processed Agricultural Crops*, 55–81, Boca Raton, Florida: CRC Press.

Torricelli, A., L. Spinelli, D. Contini et al. 2008. Time-resolved reflectance spectroscopy for non-destructive assessment of food quality. *Sens. Instrum. Food Qual. Safety* 2: 82–89.

Torricelli, A., L. Spinelli, M. Vanoli et al. 2013. Optical coherence tomography (OCT), space-resolved reflectance spectroscopy (SRS) and time-resolved reflectance spectroscopy (TRS): Principles and applications to food microstructures. In Morris, V. J., and K. Groves (eds), *Food Microstructures: Microscopy, Measurement and Modelling*, 132–162, Cambridge: Woodhead Publishing Limited.

Valero, C., P. Barreiro, M. Ruiz-Altisent, R. Cubeddu, A. Pifferi, and P. Taroni. 2005. Mealiness detection in apples using time-resolved reflectance spectroscopy. *J. Texture Stud.* 36: 439–458.

Valero, C., M. Ruiz-Altisent, R. Cubeddu et al. 2004a. Detection of internal quality of kiwi with time-domain diffuse reflectance spectroscopy. *Appl. Eng. Agric.* 20: 223–230.

Valero, C., M. Ruiz-Altisent, R. Cubeddu et al. 2004b. Selection models for the internal quality of fruit, based on time domain laser reflectance spectroscopy. *Biosyst. Eng.* 88: 313–323.

Vangdal, E., M. Vanoli, P. Eccher Zerbini, S. Jacob, A. Torricelli, and L. Spinelli. 2010. TRS measurements as a nondestructive method assessing stage of maturity and ripening in plum (*Prunus domestica* L.). *Acta Hortic.* 858: 443–448.

Vangdal, E., M. Vanoli, A. Rizzolo, P. Eccher Zerbini, L. Spinelli, and A. Torricelli. 2012. Detecting internal physiological disorders in stored plums (*Prunus domestica* L.) by time-resolved reflectance spectroscopy. *Acta Hortic.* 945: 197–203.

Vanoli, M. and M. Buccheri. 2012. Overview of the methods for assessing harvest maturity. *Stewart Postharvest Rev.* 1: 4.

Vanoli, M., P. Eccher Zerbini, M. Grassi et al. 2005. The quality and storability of apples cv "Jonagored" selected at harvest by time-resolved reflectance spectroscopy. *Acta Hortic.* 682: 1481–1488.

Vanoli, M., P. Eccher Zerbini, A. Rizzolo, M. Grassi, A. Torricelli, and L. Spinelli. 2009a. Influenza della selezione alla raccolta con spettroscopia in riflettanza risolta nel tempo (TRS) sulle proprietà sensoriali di nettarine al consumo. In Bertuccioli, M., and E. Monteleone (eds.), *Secondo Convegno Nazionale della Società Italiana di Scienze Sensoriali, atti dei Lavori*, 121–126, Firenze: University Press.

Vanoli, M., P. Eccher Zerbini, L. Spinelli, A. Torricelli, and A. Rizzolo. 2009b. Polyuronide content and correlation to optical properties measured by time-resolved reflectance spectroscopy in "Jonagored" apples stored in normal and controlled atmosphere. *Food Chem.* 115: 1450–1457.

Vanoli, M., S. Jacob, L. Spinelli, A. Torricelli, P. Eccher Zerbini, and A. Rizzolo. 2008. Time-resolved reflectance spectroscopy as a tool for selecting at harvest "Ambra" nectarines for aroma quality. *Acta Hortic.* 796: 231–235.

Vanoli, M., A. Rizzolo, S. Azzollini, L. Spinelli, and A. Torricelli. 2014a. Carotenoid content and pulp colour non destructively measured by time-resolved reflectance spectroscopy in different cultivars of Brazilian mangoes. *Poster Code 2284 Presented at the Non-Destructive Assessment of Fruit Attributes Symposium, 29th International Horticultural Congress*, August 17–21, Brisbane Australia.

Vanoli, M., A. Rizzolo, P. Eccher Zerbini, L. Spinelli, and A. Torricelli. 2009c. Non-destructive detection of internal defects in apple fruit by time-resolved reflectance spectroscopy. In Nunes C. (ed.), *Environmentally Friendly and Safe Technologies for Quality of Fruits and Vegetables*, 20–26, Faro: Universidade do Algarve.

Vanoli, M., A. Rizzolo, M. Grassi et al. 2007. Relationship between scattering properties as measured by time-resolved reflectance spectroscopy and quality in apple fruit. *CD-ROM Proceedings of the Third International Symposium CIGR Section VI "Food and Agricultural Products: Processing and Innovation,"* September 24–26, Naples, Italy, 13pp.

Vanoli, M., A. Rizzolo, M. Grassi et al. 2011a. Time-resolved reflectance spectroscopy non-destructively reveals structural changes in Pink Lady® apples during storage. *Procedia Food Sci.* 1: 81–89.

Vanoli, M., T. Pereira, M. Grassi et al. 2011b. Changes in pulp colour during postharvest ripening of Tommy Atkins mangoes and relationship with optical properties measured by time-resolved reflectance spectroscopy. *CD-ROM Proceedings of the 6th CIGR Section VI International Symposium "Towards a Sustainable Food Chain: Food Process, Bioprocessing and Food Quality Management,"* April 18–20, Nantes, France, 4pp.

Vanoli, M., A. Rizzolo, M. Grassi et al. 2011c. Non destructive detection of brown heart in Braeburn apples by time-resolved reflectance spectroscopy. *Procedia Food Sci.* 1: 413–420.

Vanoli, M., A. Rizzolo, M. Grassi et al. 2012a. Valutazione non distruttiva dell'età biologica di mango brasiliani mediante spettroscopia VIS/NIR risolta nel tempo. In Cattaneo T.M.P. and P. Berzaghi (eds.), *Atti 5° Simposio Italiano di Spettroscopia NIR*, 113–118, Lodi Milano: SISNIR – Società Italiana Spettroscopia NIR.

Vanoli, M., A. Rizzolo, M. Grassi et al. 2013a. Quality of Brazilian mango fruit in relation to optical properties non-destructively measured by time-resolved reflectance spectroscopy. In Bellon-Maurel, V., P. Williams, and G. Downey (eds), *NIR2013 Proceedings. A1 – Agriculture and Environment*, 177–181, Montpellier: IRSTEA-France Institut National de recherche en sciences et technologies pour l'environnement et l'agriculture.

Vanoli, M., A. Rizzolo, M. Grassi, L. Spinelli, B. E. Verlinden, and A. Torricelli. 2014b. Studies on classification models to discriminate "Braeburn" apples affected by internal browning using the optical properties measured by time-resolved reflectance spectroscopy. *Postharvest Biol. Technol.* 91: 112–121.

Vanoli, M., A. Rizzolo, M. Grassi, L. Spinelli, A. Zanella, and A. Torricelli. 2015. Characterizing apple texture during storage through mechanical, sensory and optical properties. *Acta Hortic.* 1079: 383–390.

Vanoli, M., A. Rizzolo, L. Spinelli, B. Parisi, and A. Torricelli. 2012b. Non destructive detection of Internal Brown Spot in potato tubers by time-resolved reflectance spectroscopy: Preliminary results on a susceptible cultivar. *Proceedings of the 3rd CIGR International Conference of Agricultural Engineering (CIGR-AgEng2012)*, July 8–12, Valencia, Spain, ISBN 84-615-9928-4.

Vanoli, M., A. Rizzolo, A. Zanella et al. 2013b. Apple texture in relation to optical, physical and sensory properties. *InsideFood Symposium: Book of Proceedings*, April 9–12, Leuven, Belgium, http://www.insidefood.eu/INSIDEFOOD_WEB/UK/WORD/proceedings/032P.pdf.

Zanella, A., M. Vanoli, A. Rizzolo et al. 2013. Correlating optical maturity indices and firmness in stored "Braeburn" and "Cripps Pink" apples. *Acta Hortic.* 1012: 1173–1180.

9 Spectral Scattering for Assessing the Quality of Fruits and Vegetables

Yibin Ying, Lijuan Xie, and Xiaping Fu

CONTENTS

9.1 Introduction 225
9.2 Optical Properties of Fruits and Vegetables 226
    9.2.1 Absorption and Scattering 227
    9.2.2 Anisotropy Factor 233
9.3 Quality Assessment of Fruits and Vegetables 234
    9.3.1 Ripeness and Quality 236
    9.3.2 Defects 243
    9.3.3 Other Applications 245
9.4 Conclusion 246
References 247

9.1 INTRODUCTION

Fruits and vegetables are a major category of food products in the human diet for their nutritional value and health benefits. Quality assessment of fruits and vegetables is of increasing importance in the ever more competitive global and domestic markets. A considerable amount of effort has been invested in developing methods to measure the quality and composition of food in general, and fruits and vegetables in particular. Except the external quality of fruits and vegetables, internal quality is commonly assessed by destructive techniques, which includes chemical, physical, and mechanical properties. However, there is an increasing demand for nondestructive detection at both research and commercial application levels, because of the importance of determining the optimum time for harvest, monitoring the changes of postharvest quality and composition during storage, and sorting and grading the internal quality of individual pieces of fruits and vegetables at the packinghouse.

Different nondestructive techniques, such as computer vision, x-ray, nuclear magnetic resonance or magnetic resonant imaging, fluorescence, biosensing, and wireless sensing, have been investigated and/or developed for evaluating quality-related attributes in fruits and vegetables over the past decades (Ruiz-Altisent et al., 2010). Most of these techniques have not been adopted for commercial use because of their limited capability, unsatisfactory performance, or high cost in instrumentation. Hence, researchers have been continuing to investigate or develop new sensing techniques for more effective and efficient quality detection of fruits and vegetables. Spectral scattering is a new, emerging technique that has been investigated in recent years as a nondestructive tool for quality evaluation of fruits and vegetables through simultaneous estimation of the absorption and scattering coefficients or by extracting the critical scattering features from one-dimensional (1D) or two-dimensional (2D) scattering profiles or images.

This chapter reviews the measurement of optical properties of fruits and vegetables, including absorption, scattering, and anisotropy factor. It provides an overview of applications of light scattering technique for quality assessment, monitoring of ripening, identification of defects, and others on fruits and vegetables.

9.2 OPTICAL PROPERTIES OF FRUITS AND VEGETABLES

Biological tissues are optically inhomogeneous and absorbing media; the average refractive index of which is higher than that of air. This refractive index mismatch is responsible for partial reflection of the radiation at the tissue/air interface (Fresnel reflection), while the remaining radiation penetrates the tissue. Multiple scattering and absorption are responsible for light broadening and eventual decay as it travels through a tissue (Tuchin, 2007). The shape, size, and spatial distribution of the cellulosic structures, together with the composition of the tissue, will affect how light propagates in the tissue. The fundamental optical parameters that are used to characterize light propagation in the tissue are the absorption coefficient μ_a, the scattering coefficient μ_s, and the anisotropy factor g (Lorenzo, 2012).

Direct and indirect methods can be used to determine the optical parameters of tissues (Tuchin, 2007). Direct methods include those based on some fundamental concepts and rules such as the Beer–Lambert law and the single scattering phase function for thin samples, or the effective light penetration depth for slabs. Indirect methods, divided into iterative and noniterative models, rely on the solutions of the inverse scattering problem using a theoretical model of light propagation. A detailed description of *ex vivo* (invasive or destructive) and *in vivo* (nondestructive) measurement methods and instrumentation for the optical parameters is presented in Chapters 6 through 8. This section is focused on the optical properties of fruits and vegetables that have been measured and reported by researchers using different techniques, both destructive and nondestructive.

The first study about measuring the optical properties of fruits and vegetables was reported by Birth (1978). The distribution of 632 nm radiation, generated by a laser source, transmitted through samples of high-moisture nonpigmented slices of raw white potato tissue was measured. The rate of change in the radiometric measurements with respect to the distance from the point of incidence was found to be related to the Kubelka–Munk (K–M) scattering coefficient of the tissue. Birth and colleagues also investigated the potential of light scattering to measure the quality of agricultural products (pork and grain) (Birth et al., 1978; Birth, 1986). Chen and Nattuvetty (1980) utilized optical fiber to measure light transmittance through a region of an intact fruit in order to estimate the depth through which the detected light penetrated into the fruit. They reported the light penetration depths at 680 nm of 1.5–2.5 cm for oranges, 1–1.5 cm for apples, and 0.7–1.1 cm for green tomatoes.

Since these early studies, there was a hiatus in research for food and agricultural products, until the arrival of the twenty-first century because of the advent of, and advances in, new optical measurement techniques. On the contrary, great progress has been made since the 1980s in both theories and measurement techniques for optical characterization of biological tissues in the biomedical field (Tuchin, 2007). This has inspired renewed interest from agricultural and food researchers over the past decade in the measurement of light scattering and absorption properties of food products.

9.2.1 Absorption and Scattering

Light absorption is the process by which the incident radiant flux is converted to another form of energy, usually heat, or into photons with a much lower frequency (e.g., fluorescence) (Bass et al., 2001). Light scattering is a physical interaction where photons are forced to deviate from a straight trajectory by one or more paths due to localized nonuniformities in the medium through which they pass. The absorption coefficient is the probability, per unit distance traveled by a photon, of being absorbed by pigments (e.g., chlorophylls and carotenoids) and by the principal chemical constituents of the pulp (e.g., water and sugars), while the scattering coefficient is the probability, per unit distance traveled by a photon, of changing direction due to the refractive index mismatch caused by cellular structure, such as membranes, cell walls, intercellular spaces, starch granules, vacuoles, or organelles (Rizzolo et al., 2010). The reduced (transport) scattering coefficient μ'_s is often used instead of μ_s. The relationship between μ'_s and μ_s is $\mu'_s = \mu_s(1-g)$, where g is the anisotropy factor.

Different techniques, such as time-resolved reflectance spectroscopy (TRS), spatial-frequency domain imaging (SFDI), integrating sphere (IS), spatially-resolved reflectance spectroscopy (SRS) (including optical fiber based and hyperspectral imaging based), were used for nondestructive determination of tissue absorption and scattering properties of fruits and vegetables. Data presented in Table 9.1 reflect the current situation in the measurement of optical properties based on the above techniques. It should be pointed out that some data in Table 9.1 were estimated from the figures of the reported studies. Apparently, much attention was paid to the optical properties of apples perhaps because of the importance of this crop worldwide. Other fruits (peach, pear, kiwifruit, plum, etc.) and vegetables (tomato, cucumber, zucchini squash, onion, etc.) were also studied but not as extensively as apple. Research has been reported on the changes or specificities of absorption and scattering properties for fruits and vegetables during ripening, for different tissue parts such as skin and flesh, for defects (e.g., bruise and internal defect), under different storage environments or different pretreatment conditions, etc.

Cubeddu et al. (2001a, 2001b) first measured the optical parameters of μ_a and μ'_s of apples by using TRS in the wavelength range of 650–1000 nm. They reported that the measured optical properties were those of the pulp because no major variations were found in their study when the fruit was peeled, and the change in chlorophyll absorption associated with the ripening process during storage and ripening could be tracked. Besides, they found that photons traveled more than 2 cm deep

TABLE 9.1
Optical Absorption (μ_a) and Reduced Scattering (μ_s') Coefficients for Fruits and Vegetables

Product	λ (nm)	μ_a (cm^{-1})	μ_s' (cm^{-1})	Measuring Technique	References
Apple	600–700	0.02–0.25	8–22	TRS	Cubeddu et al. (2001a)
	650–1000	0.02–0.4	18–24.5	TRS	Cubeddu et al. (2001b)
	630	0.077–0.112	18.94–21.54	TRS	Rizzolo et al. (2010)
	670	0.141–0.179	17.90–21.07		
	750	0.034–0.036	10.72–12.91		
	780	0.025–0.026	10.41–12.53		
	670	0.069–0.173	9.99–15.76	TRS	Vanoli et al. (2014)
	780	0.032–0.078			
	980	0.405–0.458			
	500–1000	0–0.62	9–13	HISR	Qin and Lu (2008)
	500–1000	0.04–2.52	1.02–12.61	HISR	Qin and Lu (2009)
	500–1000	Normal: 0.1–0.9	Normal: 7.5–9	HISR	Lu et al. (2010)
		Bruised: 0–1.1	Bruised: 4.2–7.5		
	500–1000	Dried slices: 0–1.2	Dried slices: 60–140	SRS	Trong et al. (2014)
	350–2,200	Skin: 1–70	Skin: 35–100	IS	Saeys et al., 2008
	350–1,900	Flesh: 1–28	Flesh: 12–15		
	650–980	Bruised: 0.01–1.42	Bruised: 4.2–9	SFDI	Anderson et al. (2007)
	400–1050	0.1–1.45	8–16	IS	Rowe et al. (2014)
	540–940	0.02–0.58	15.2–18.6	TRS	Seifert et al. (2015)
Peach	650–1000	0.02–0.45	21–23	TRS	Cubeddu et al. (2001b)
	500–1000	0.12–0.3	13–15.5	HISR	Qin and Lu (2008)
	515–1000	0.01–0.45	6–17.5	HISR	Cen et al. (2012)
Pear	710–850	0.025–0.088	23–26	TRS	Zerbini et al. (2002)
	500–1000	0.02–0.95	8–9	HISR	Qin and Lu (2008)
Kiwifruit	650–1000	0.05–0.45	10–16	TRS	Cubeddu et al. (2001b)
	500–1000	0.2–1.0	7–8	HISR	Qin and Lu (2008)
	785	0.9	40	SRS	Baranyai and Zude (2009)

(Continued)

TABLE 9.1 *(Continued)*
Optical Absorption (μ_a) and Reduced Scattering (μ'_s) Coefficients for Fruits and Vegetables

Product	λ (nm)	μ_a (cm⁻¹)	μ'_s (cm⁻¹)	Measuring Technique	References
Nectarine	670	0.05–0.35	/	TRS	Tijskens et al. (2007)
Plum	500–1000	0.04–1.18	7.8–8.1	HISR	Qin and Lu (2008)
	540–940	0.02–1.42	4.8–9.2	TRS	Seifert et al. (2015)
Tomato	650–1000	0.04–0.48	6–9	TRS	Cubeddu et al. (2001b)
	500–950	0.0007–0.275	3.59–9.32	HISR	Zhu et al. (2015)
Cucumber	500–1000	0.02–0.55	9–10	HISR	Qin and Lu (2008)
Zucchini squash	500–1000	0–0.45	10.5–11.5	HISR	Qin and Lu (2008)
Onion	632.8	Outermost dried skin: 7.5, 20 Outer skin: 3, 6 First flesh: 0.5, 1.5 Second flesh: 0.5, 1	Outermost dried skin: 190, 200 Outer skin: 40, 20 First flesh: 1, 5 Second flesh: 2, 5	IS	Wang and Li (2012)
	633	Dry skin: 5.01–19.74 Wet skin: 2.84–5.99 Flesh: 0.33–1.05	Dry skin: 184.8–224.6 Wet skin: 19.9–55.7 Flesh: 2.0–6.7	IS	Wang and Li (2013)
	550–880 950–1650	Healthy dry skin: 1.8–39 Healthy flesh: 0.2–27.8 Sour skin dry skin: 1.2–52 Neck rot dry skin: 2.5–59 Sour skin flesh: 0.25–29 Neck rot flesh: 0.76–27	Healthy dry skin: 160–190 Healthy flesh: 9.5–14.5 Sour skin dry skin: 180–205 Neck rot dry skin: 180–220 Sour skin flesh: 2.5–6.5 Neck rot flesh: 5–10	IS	Wang et al. (2014)

Note: Some of the data in the table were estimated from the figures reported in these studies. HISR, hyperspectral imaging-based spatially-resolved spectroscopy; IS, integrating sphere; SFDI, spatial frequency domain imaging; SRS, spatially-resolved reflectance spectroscopy; TRS, time-resolved reflectance spectroscopy.

beneath the skin of the tested "Granny Smith" apple at 675 nm with $\mu_a = 0.07$ cm^{-1} and $\mu'_s = 18$ cm^{-1}. However, for the scattering spectra of the same fruit, no particular features were found. Cubeddu et al. (2001b) also used a fully automated TRS system to measure the optical properties of other fruits, including peach, kiwifruit, and tomato in the wavelength range of 650–1000 nm. Seifert et al. (2015) measured the absorption and reduced scattering coefficients of apple and plum harvested four times (65–145 days after full bloom) in the spectral region of 540–940 nm using the TRS technique and found that apples were more isotropic and uniform in structure compared with plums.

The research group at CRA-IAA (Consiglio per la Ricerca e la Sperimentazione in Agricoltura, Unità di ricerca per i processi dell'industria agroalimentare) in Italy (Zerbini et al., 2002; Vanoli et al., 2009, 2014; Rizzolo et al., 2010) used the TRS to study the optical properties of pear and apple. Zerbini et al. (2002) investigated the potential of the TRS for detecting brown heart in intact pears. An increase in μ_a at 720 nm was observed for brown heart, which was significantly higher than that in sound tissues. Sound tissues had μ_a values at 720 nm ≤0.04 cm^{-1}. At 690 nm, the value of μ_a increased in the fruit with brown heart and decreased with the ripening in sound fruit. The reduced scattering coefficient apparently was not affected by brown heart. Later, TRS measurements at 630, 670, 750, and 780 nm were performed to determine the μ_a and μ'_s of "Jonagored" apples at harvest, after 6 months' storage in normal and controlled atmosphere, and after 7 days of poststorage shelf life at 20°C (Rizzolo et al., 2010). Their study showed higher values for μ_a at 630 nm and μ_a at 670 nm, while μ_a values at 750 and 780 nm were, on average, as low as 0.025–0.036 cm^{-1} for apples of different harvest time and storage atmosphere conditions. The μ_a at 630 nm and μ_a at 670 nm values were significantly influenced by harvest time. The values of μ'_s decreased with increasing wavelength, mainly from 21.54 to 10.41 cm^{-1} independent of harvest. Compared with apples stored in normal atmosphere, apples stored under controlled atmosphere at the end of shelf life were characterized by significantly higher values of μ_a at 630 nm, and μ_a at 750 nm, and significantly lower values of μ'_s at the four detected wavelengths of 630, 670, 750, and 780 nm. Vanoli et al. (2014) reported their more recent work that was carried out in two seasons: in 2009, apples were measured by TRS at 670 nm and in the 740–1040 nm spectral range on four equidistant points around the equator, whereas in 2010 apples were measured by TRS at 670 nm and at 780 nm on eight equidistant points. They found that μ_a increased with internal browning development of apples in 670–940 nm, while μ'_s at 780 nm was higher in healthy apples than in internal browning ones. A complete review on using TRS to assess fruit maturity or ripeness, texture, and internal defect is given in Chapter 8.

The research group at East Lansing, Michigan, USA (Qin and Lu, 2008, 2009; Lu et al., 2010) developed a hyperspectral imaging-based spatially-resolved (HISR) system for the measurement of the optical properties of fresh fruits and vegetables over the visible and short-wave near-infrared region (500–1000 nm). They used a hyperspectral imaging system (in line scan mode) to acquire spatially-resolved diffuse reflectance images from apples of different varieties as well as normal and bruised apples (Lu et al., 2010). When quantifying the changes in the absorption and scattering properties of bruised apple tissues over time, the results indicated that bruising had different effects on the absorption and reduced scattering coefficients. While no consistent pattern of changes for absorption between the normal and bruised tissues and between the bruised tissues of different ages was ascertained, a pattern of consistent decrease in the reduced scattering coefficient for the bruised tissue over time was observed. Qin and Lu (2008) also studied the optical properties of other fruits and vegetables such as peach, pear, kiwifruit, plum, cucumber, and zucchini squash by using the HISR. Cen et al. (2012) reported an improved optical property measuring instrument for acquiring hyperspectral reflectance images of 500 "Redstar" peaches. More recently, the instrument was used to measure the optical absorption and scattering properties of tomatoes harvested at six ripeness grades (i.e., "green," "breaker," "turning," "pink," "light red," and "red") (Zhu et al., 2015). It was shown that absorption peak around 675 nm decreased consistently with the progress of ripeness, while the value of μ_s' decreased from "green" to "turning," and an opposite trend was found from "pink" to "red."

Trong et al. (2014) investigated the potential of spatially-resolved diffuse reflectance spectroscopy in the 500–1000 nm range by means of a fiber-array probe for measuring the scattering and absorption properties of air-dried apple rings subjected to different pretreatment conditions. There were no significant differences between the scattering properties of the samples with osmo-dehydration for 1 h (OSMO1) and 3 h (OSMO2). However, the samples without osmo-dehydration were found to have significantly higher scattering properties than OSMO1 and OSMO2 apple rings.

SFDI is a noncontact optical imaging technology, first developed by Cuccia et al. (2005) for biomedical applications. Anderson et al. (2007) applied SFDI to detect bruising on "Golden Delicious" apples. They obtained the quantitative absorption and scattering image maps for two levels of bruising severity from 650 to 980 nm. The average scattering and absorption spectra were calculated for the two levels of bruising and compared with the adjacent nonbruised regions. There was a considerable difference in the average reduced scattering coefficients between the bruised and nonbruised regions for the two levels of bruising.

IS method, including single IS and double IS configurations, is commonly used to measure total diffuse reflectance and transmittance of biological tissues, from which it is possible to estimate the optical parameters using the inverse adding–doubling (IAD) method (Phral, 2011). A detailed description on IS and IAD methods can be found in Chapter 6. The reported optical property measurements of fruits and vegetables were based on the single IS configuration. Saeys et al. (2008) used single IS measurements combined with IAD method to estimate the optical properties of apple skin and flesh tissue in the range of 350–2200 nm for three cultivars. The skin tissue was more highly scattering than the flesh tissue. Wang and Li (2012) measured the optical properties of dry skin and fleshy tissues of onions at 632.8 nm. The results indicated that dry skins had much higher absorption and reduced scattering coefficients than onion flesh tissues. Later, Wang and Li (2013) estimated the optical properties of dry skin, wet skin, and flesh of red, Vidalia sweet, white, and yellow onions at a wavelength of 633 nm using the same method. The results indicated that both dry and wet skins had significantly higher μ_a and μ_s' values than the flesh at 633 nm. Wang et al. (2014) further investigated the feasibility of detecting diseased onions based on the optical characteristics. The optical properties of dry skin and flesh tissues of healthy onions, *Burkholderia cepacia*-infected onions, and *Botrytis aclada*-infected onions were estimated in the spectral region of 550–1650 nm. The results indicated that the optical properties were related with the color change, moisture content, and level of decomposition, which could be applied in food safety and quality inspection. Rowe et al. (2014) measured the absorption and reduced scattering coefficients between 400 and 1050 nm for "Royal Gala" apples using a single IS and IAD method. The absorption coefficient was found to correlate with penetrometer firmness, with the correlation coefficients of −0.69 at 500 nm and 0.52 centered on 680 nm. The reduced scattering coefficient across all wavelengths increased as the fruit softened with an average correlation of −0.68 between the reduced scattering coefficient (550–900 nm) and penetrometer firmness.

It is evident from previous studies that different fruit and vegetable tissues have specific optical properties. For the same tissue, the scattering coefficients are much higher than absorption coefficients, which confirms that fruits and vegetables are highly scattering objects. For fruits and vegetables, the absorption spectra are featured by major pigments (chlorophyll, anthocyanin, and carotenoid) in the visible region and by water in the near-infrared region, as shown in Figure 9.1 (Qin and Lu, 2008), typically marked with absorption peaks around 525 nm for anthocyanin, 620–630 nm for chlorophyll b, 670–675 nm for chlorophyll a, 750–780 nm, and 970 nm for water. Saeys et al. (2008) estimated the apple optical properties in a broader range of 350–2200 nm. The curves of the absorption and scattering coefficients had prominent

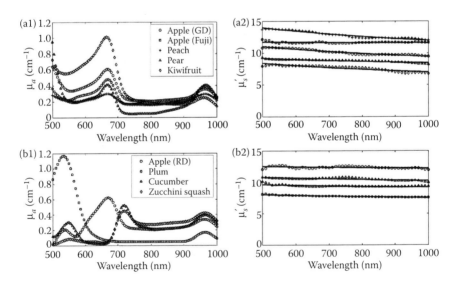

FIGURE 9.1 Optical properties of fruit and vegetable samples: (a1) absorption and (a2) reduced scattering coefficients of "Golden Delicious" and "Fuji" apple, peach, pear, and kiwifruit; (b1) absorption and (b2) reduced scattering coefficients of "Delicious" apple, plum, cucumber, and zucchini squash. Solid lines in (a2) and (b2) are the best fits of μ'_s using the wavelength-dependent function $\mu'_s = a\lambda^{-b}$, where λ is the wavelength in nm and a and b are constants. (From Qin, J. and R. Lu. *Postharvest Biology and Technology* 49 (3), 2008: 355–365. With permission.)

water absorption peaks at 1450 and 1900 nm. The change in chlorophyll absorption (around 675 nm) could be used to track the ripening process during ripening or storage (Cubeddu et al., 2001a; Zhu et al., 2015). Moreover, the reduced scattering coefficients decreased progressively with increasing wavelength (500–1000 nm) for most of the tested samples. This phenomenon was verified in many studies using different measuring methods and for different fruits and vegetables.

Although the trends of μ_a and μ'_s for different types of fruit and vegetable tissues are similar, their values and the trend of changes are specific because of varietal difference, difference in tissue parts (e.g., skin vs. flesh), the presence or absence of bruise or internal defect, etc. Tissues with less moisture content, such as dried apple slices (Trong et al., 2014) and apple skin (Saeys et al., 2008), had much larger μ'_s values than tissues with higher moisture content (like apple flesh). This phenomenon was also reported by Wang and Li (2012, 2013) on the optical properties of dried onion skin, wet onion skin, and flesh. Besides, the absorption coefficients of onion skin were also much higher than those of onion flesh. In addition, internal defect, bruise, and disease were all found to have effects on the optical properties. Other factors such as storage environment and pretreatment condition also had significant effects on the optical properties.

However, inconsistencies in the measured results for the same type of fruit were reported in previous studies. For example, the measured μ'_s values for apples in the studies of Qin and Lu (2009), Rowe et al. (2014), and Seifert et al. (2015) were significantly different. At present, it is impossible to compare the relative accuracy of these results, since the system validation methods are different for these studies. It is difficult to distinguish the main reason for the differences in the reported optical properties for apples. Were the differences due to samples species, detecting parts, detecting methods, or other factors? Therefore, it is necessary to establish a consistent method or adopt a common standard for validating different measuring systems. This is an important area for research in the future.

9.2.2 Anisotropy Factor

The anisotropy factor g, sometimes called the anisotropy coefficient, represents the average of the cosine between the incident and scattered fields, which tells us how much scattering occurs in the forward direction compared with the backward direction. When the scattering phase function $p(\theta)$ is available from goniophotometry, g can be calculated (Tuchin, 2007). In more practical terms, this can be understood as how "transparent" a particle is, at least when dealing with tissue. The more transparent the particle, the more the light is distributed in the forward direction (Lorenzo, 2012). According to this relaxed definition, transparency is only affected by scattering, not absorption. The anisotropy factor approaching 1, 0, and −1 describes extremely forward, isotropic, and highly backward scattering, respectively. In human tissue optics, typical values of g have been shown to vary approximately between 0.8 and 0.9 (Cheong et al., 1990; Tuchin, 2007). Sensitivity analysis has shown that μ_a and μ'_s estimations are not very sensitive to the value of g (Tuchin, 2007).

For fruit and vegetable tissues, only a few studies have been reported on the measurement of g directly, which were carried out based on the IS method. IAD method can be used to estimate the anisotropy factor by using the collimated transmittance (Phral, 2011). The g values were about 0.45–0.75 for apple skin in the range of 350–2200 nm and 0.55–0.75 for apple flesh in the range of 350–1900 nm (Saeys et al., 2008). For onions (Wang and Li, 2013), the g values for the dry skin, wet skin, and flesh at 633 nm were 0.371–0.589, 0.421–0.72, and 0.427–0.688, respectively. The g values of apple and onion reported in the literature are smaller than those reported for human tissues.

The measured μ_a, μ_s', and g can be used not only for characterizing the absorption and scattering features of fruit and vegetable tissues, but also for investigations and applications such as (1) simulating light propagation inside tissues; (2) assessing quality of fruits and vegetables; and (3) providing guides for process optimization for fruits and vegetables. The following section highlights the research on quality assessment of fruits and vegetables based on both optical property parameters and empirical models or image processing methods for spectral scattering image data.

9.3 QUALITY ASSESSMENT OF FRUITS AND VEGETABLES

As mentioned before, biological materials such as fruits and vegetables are optically inhomogeneous or opaque. The light that enters a fruit will be attenuated through absorption and multiple scattering, which are affected by the compositional and structural characteristics of the tissue (Cen et al., 2012). The study of light interaction, primarily of photon absorption and scattering events, in biological materials is useful for quality assessment of biological materials such as fruits and vegetables. Knowledge of the optical properties will enable us to gain insight into the interaction of light with fruit and vegetable tissue, which would be valuable in developing an optical technique for evaluating the properties and characteristics that are indicative of specific quality attributes. In contrast to conventional *ex vivo* or *in vivo* optical methods, technologies such as Monte Carlo (MC) simulation, laser backscattering imaging, TRS, fiber-optic-based or HISR spectroscopy, and SFDI have been used for nondestructive detection of properties of fruits and vegetables. Birth and colleagues explored the potential to assess the quality of agricultural products using light scattering (Birth, 1978). The fruit flesh scatters light due to abrupt changes in the refractive index in the microstructural composition of the tissue. In applying light scattering techniques to assess fruits and vegetables, there are two main approaches to scattering analysis. The first one is based on the fundamental absorption and scattering coefficients to directly assess fruits and vegetables. The technique is not easy to implement and may not be suitable for many food products in practical applications. Another is applying empirical models or image processing methods to quantify the 1D or 2D scattering profiles or images. Table 9.2 summarizes the research on quality assessment of fruits and vegetables based on light scattering techniques, mainly including laser backscattering imaging, multispectral scattering imaging, HISR, and TRS. Quality indices, including maturity, defect, and other properties such as the structural and mechanical properties, have been measured. Since Chapter 8 has provided an extensive coverage on the time-resolved technique for evaluating defects, ripeness, texture, a full review on the applications of TRS for quality assessment of fruits and vegetables will not be given here.

TABLE 9.2
Summary of Research on Quality Assessment of Fruits and Vegetables Using Various Light Scattering Techniques

Measuring Technique	Product (Index)	References
Laser beam	Pear (firmness)	Budiastra et al. (1992)
Laser scattering imaging	Tomato (ripeness and firmness)	Tu et al. (1995, 2000)
Near-infrared light scattering	Kiwifruit (firmness)	McGlone et al. (1997)
Kubelka–Munk theory	Apple and pear (maturity and SC)	Budiastra et al. (1998)
Laser backscatter imaging	Apple (ripeness and firmness)	Tu et al. (1995)
	Apple (SSC and flesh firmness)	Qing et al. (2007, 2008)
	Banana (moisture content)	Romano et al. (2008)
	Apple (optical properties)	Baranyai et al. (2009)
	Apple (moisture content)	Romano et al. (2010)
	Apple (moisture content, SSC, and hardness)	Romano et al. (2011)
	Apple (firmness), tomato (elasticity), plum (firmness), and mushroom (elasticity)	Mollazade et al. (2013)
	Banana (chilling injury)	Hashim et al. (2013, 2014)
	Papaya (moisture content, shrinkage, lightness, chroma, and hue)	Udomkun et al. (2014)
	Pear (maturity and color change) and sweet pepper (maturity and color change)	Zsom et al. (2014)
	Citrus (decay)	Lorente et al. (2015)
Multispectral scattering imaging	Apple (firmness and SSC)	Lu (2003, 2004a, 2004b)
	Apple (firmness)	Peng and Lu (2005, 2006a, 2006b, 2006c)
Hyperspectral imaging-based spatially-resolved	Peach (firmness)	Lu and Peng (2006)
	Apple (firmness and SSC)	Lu (2007)
	Tomato (optical properties, ripeness, and firmness)	Qin and Lu (2008)
	Apple (bruise)	Lu et al. (2010)
	Peach (firmness, SSC, and skin and flesh color parameters)	Cen et al. (2012)
	Apple (mechanical properties and microstructural changes)	Cen et al. (2013)
Spatial frequency domain imaging	Apple (bruise)	Anderson et al. (2007)
MC simulation	Apple (bruise)	Baranyai and Zude (2008)
Inverse adding–doubling technique	Apple (penetrometer firmness and optical properties)	Rowe et al. (2014)

(*Continued*)

TABLE 9.2 (*Continued*)
Summary of Research on Quality Assessment of Fruits and Vegetables Using Various Light Scattering Techniques

Measuring Technique	Product (Index)	References
Time-resolved reflectance spectroscopy	Apple, peach, tomato, and nectarine (firmness, SSC, and acidity)	Valero et al. (2004a)
	Apple (mealiness)	Valero et al. (2001)
	Kiwifruit and tomato (firmness, SSC, and acidity)	Valero et al. (2004b)
Total internal reflectance, continuous-wave, and time-resolved spectroscopy	Cherry (anthocyanins)	Zude et al. (2009)
Lambert–Beer and multivariate regression	Cherry (anthocyanins)	Zude et al. (2011)

9.3.1 Ripeness and Quality

The ripeness of fruits and vegetables is determined by multiple indices, including internal quality attributes such as firmness, sweetness, and acidity and external characteristics such as color. Firmness is an important indicator of ripeness, textural properties, shelf life, and consumer acceptance for many fruits and vegetables. Firmness may be assessed by visual inspection according to the appearance on whether the fruit surface is shriveled or flaccid. It may also be evaluated from the resistance to light manual pressure, which is subjective. The method commonly used for firmness assessment is based on the penetrometric method such as the Magness–Taylor (MT) firmness tester or the Effegi penetrometer (Sinha et al., 2011). Sweetness and acidity determine the flavor. Sweetness is an important quality parameter for fruits. It is a good indication of the state of ripeness. For most fresh fruits, sweetness is normally measured in terms of sugar content (SC) or soluble solids content (SSC) in °Brix, which is commonly measured using Brix refractometry on the basis of the refraction of light by juice samples. For acidity measure, titration with a suitable alkaline solution is generally used. Brix-to-acid ratios or both SSC and acidity are sometimes used as a maturity/ripeness indicator. Many fruits and vegetables undergo color changes during the ripening process. Changes in the skin/flesh color during fruit and vegetable ripening are mainly related to the breakdown of chlorophylls and the increase of other pigments (Cen et al., 2012). For some fruits, unripe fruit are green and the green color becomes lighter during maturation and ripening. Measurement of color could be indicative of eating quality and thus is useful for grading and shelf-life assessment. Zsom et al. (2014) detected two pear and five sweet pepper varieties during cold storage and shelf life by nondestructive optical methods, including laser backscattering imaging, chlorophyll fluorescence analysis, and surface color measurement. Among them, laser scattering parameters showed significant and cultivar-dependent changes with time during cold storage and shelf life.

Studies on the measurement of fruit firmness by light scattering have been conducted for more than 20 years. Budiastra et al. (1992) reported the relationships of the scattering coefficient with the firmness of apples and Japanese pears. On the basis of the observation that the firmness and the scattering coefficient of apples and Japanese pears decreased with storage time (absorption was small at a wavelength of 783 nm), they drew a conclusion that the scattering coefficient could be used to monitor the quality changes of fruits during storage. During softening, the fruit tissue becomes flimsy with weakened cell walls and hydrated cellular matter, which would result in a lower density of scatters with longer scatter free paths between scattering events (McGlone et al., 1997). Based on the underlying idea that as fruit soften, there would be an increase in the intensity of the scattered light exiting at points from the surface surrounding the irradiation area, McGlone et al. (1997) estimated kiwifruit firmness by measuring the scattering of 864 nm laser light in the fruit at exit angles of 20°–55° around the circumference of the fruit from the incident beam direction (Figure 9.2). The samples were stored at 10°C for up to 15 days. During storage, the fruit were subsampled on four occasions and softer kiwifruit of the same size class were added to extend the measurements to a lower firmness limit. An NIR Systems 6500 spectrometer with a fiber optic interactance probe was used to obtain the spectra. Scattered light measurements were taken at 5° intervals. Correlations of the scattered light intensity at each angle with the stiffness and rupture force measurements were calculated. And the highest correlations occurred at the largest scattering angle of 55° with the determination coefficient (r^2) of 72% and 67% for the stiffness and rupture force measurements, respectively. Besides, the scattered light intensity decreased with increasing scattering angle. McGlone et al. (1997) reasoned that the poor correlations obtained at the low scattering angles might be due to the near-surface regions constituting a great proportion of the total light path at these lower angles. A simple two-parameter phenomenological model was developed for the intensity decrease by assuming that the intensity is strongly and negatively related to the path length of the light (D) (defined as $\sin(\theta/2)$) in the fruit, which is given in the following equation:

$$I(D) = S / (\sin(\theta/2))^b$$

where θ is the exit angle. With increasing firmness, the scatter constant S increased and the power coefficient D and the path length index $-b$ decreased. The scatter

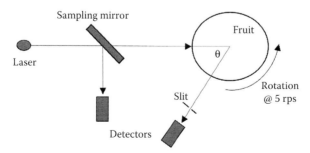

FIGURE 9.2 Schematic of the laser scattering experimental arrangement. θ is the scattering angle. (From McGlone, V. A., Abe Kawano, and S. Kawano, *J. Near Infrared Spectrosc.*, 5 (2), 83–89, 1997. With permission.)

constant S depends on the type, density, and distribution of the scattering elements within the fruit. For firmness prediction, the best r^2 value was around 80%. The error increased in proportion to the level of the firmness measurement, which, they suspected, could be due to lower signal-to-noise ratio for harder fruit as well as the increased error in the penetrometer firmness measurements.

The Kubelka–Munk theory is one of the several theories that could determine the scattering and absorption properties of turbid or translucent materials (Budiastra et al., 1998). Budiastra et al. (1998) studied the optical properties of apple and Japanese pear on the wavelengths from 240 to 2600 nm using the K–M theory. The result showed that K–M absorption coefficient and scattering coefficient were influenced by maturity. Moreover, the presence of the skin yielded an increase in the scattering coefficient in the region of 400–2000 nm.

Tu et al. (1995) used a 670 nm, 3 mW He–Ne laser diode as a light source to generate scattering images at the surface of tomato and apple fruit and obtained the scattering images using a color charge-coupled device (CCD) camera (Figure 9.3). The total number of pixels in the image of scattered light was used as a texture indicator and related to fruit ripeness and firmness. Tu et al. (2000) further used a 670 nm, 3 mW solid-state laser diode module as the light source to monitor quality changes of tomatoes stored at room temperature for 8 days. The total number of pixels in the scattering images was taken as an indicator of fruit ripeness. The results showed that there were significant differences in the image size for tomatoes at different ripeness stages. Fruit firmness had a negative correlation with the image size (Tu et al., 2000).

Quantification of the absorption and scattering properties can be used to assess plant products' ripeness. Since the physiological process of fruit ripeness is associated with simultaneous changes in both chemical composition (related to μ_a) and physical structure (related to μ_s'), the combinations of μ_a and μ_s' could improve the prediction of maturity of fruits such as peach (Cen et al., 2012). Large variation of μ_a values existed due to different maturity levels.

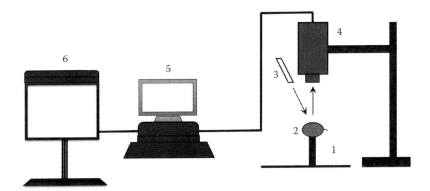

FIGURE 9.3 Laser scattering image acquisition system. (1) Sample holder, (2) sample (tomato and apple), (3) laser light source, (4) camera, (5) computer with frame grabber, and (6) monitor. (From Tu, K. et al. *Proceedings of the Food Processing Automation IV Conference* 1995: 528–536. With permission.)

Lu (2003, 2004a) proposed a new concept of using a rotating filter wheel-based optical system for acquiring multispectral images to quantify light backscattering profiles in the visible and near-infrared region from apple fruit for predicting firmness and SSC (Figure 9.4). A circular broadband light beam was used to generate light backscattering at the surface of apple fruit and scattering images were acquired at four discrete wavelengths between 680 and 1060 nm. Scattering images were then radially averaged to produce 1D spectral scattering profiles, which were input into a backpropagation neural network (NN) for predicting apple fruit firmness and SSC. The three ratio combinations with four spectral bands (680, 880, 905, and 940 nm) gave the best predictions of apple firmness, with a correlation coefficient (r) of 0.87 and a standard error for validation (SEV) of 5.8 N. For SSC, two ratio combinations with the three wavelengths of 880, 905, and 940 nm gave the best predictions with an r value of 0.277 and an SEV of 0.78%. An improved multispectral imaging system was developed (Lu, 2004b), which used a common-aperture multispectral imaging spectrograph (Optical Insights, LLC, Santa Fe, New Mexico) to split the beam passing through the focusing lens into four separate, equal beams without losing the original spatial information. As a result, spectral images at four discrete wavelengths or bands were acquired simultaneously, which allowed rapid, real-time sensing of fruit internal quality. In addition, improvements in image processing algorithms were made to enhance the NN for fruit firmness prediction. With three ratios of spectral profiles involving all four wavelengths, the NN gave apple firmness predictions with an r of 0.76 and SEV of 6.2 N. These results are not nearly as good as those reported in his previous study. The author reasoned that the improved multispectral imaging system was less efficient in receiving light due to the use of the multispectral imaging spectrograph, which would negatively affect the scattering size and the signal-to-noise ratio.

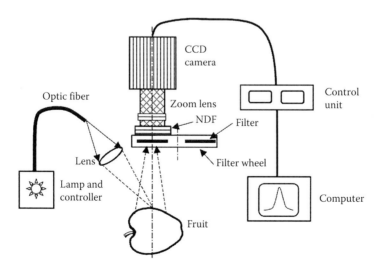

FIGURE 9.4 Schematic of the multispectral imaging system for measuring scattering images from apple fruit. NDF = neutral density filter. (From Lu, R. *Postharvest Biology and Technology* 31 (2), 2004a: 147–157. With permission.)

Peng and Lu (2005) proposed a Lorentzian distribution (LD) function with three independent profile parameters to characterize the scattering profiles of multispectral scattering images. Firmness prediction models were constructed using multilinear regression against 12 Lorentzian parameters for four wavelengths. The prediction models gave firmness predictions with $r = 0.82$ and SEV = 6.39 N for one set of apple samples, and $r = 0.76$ and SEV = 6.01 N for another set. Later, Peng and Lu (2006a, 2006b) proposed a modified LD (MLD) function with four independent parameters to describe the entire spectral scattering profiles, including saturation area, of spectral scattering images acquired from a compact multispectral imaging system with a liquid crystal tunable filter (LCTF). The MLD function described the scattering profiles accurately for wavelengths between 650 and 1000 nm with an average r of 0.999. Using this LCTF-based multispectral imaging system, an optimal set of wavelengths between 650 and 1000 nm was determined for predicting apple fruit firmness. A linear prediction model was developed to describe the relationship between the MT firmness of "Delicious" apples and their MLD parameters at seven wavelengths with $r = 0.82$ and SEV = 6.64 N. Peng and Lu (2006c) further improved the multispectral imaging system to quantify light backscattering profiles from apple fruit for measuring firmness and SSC. To increase the stability of the light source, they equipped the system with a light intensity controller to measure light scattering from "Delicious" apples at seven wavelengths and "Golden Delicious" apples at eight wavelengths. To reduce noisy signals in the scattering images during radial averaging of image pixels and improve prediction, they proposed methods to filter out pixels with unusually high or low intensities due to abnormal tissue spots on the scattering images and correct the effect of fruit shape/size on the scattering intensity and distance. The MLD function with four parameters was used to fit the scattering profiles. Multilinear regression against MLD parameters was applied to develop firmness prediction models. Better firmness predictions were obtained with $r = 0.898$ and SEV = 6.41 N for "Delicious" apples and $r = 0.897$ and SEV = 6.14 N for "Golden Delicious" apples. To further improve firmness prediction, Lu and Peng (2006) used hyperspectral imaging technique to simultaneously acquire scattering profiles for all 153 wavebands over 500–1000 nm for peach firmness prediction. Best firmness predictions were obtained with $r^2 = 0.77$ and SEV = 14.2 N for "Red Haven" peaches with 10 wavelengths (603, 616, 629, 642, 648, 664, 671, 677, 690, and 707 nm) when two Lorentzian parameters representing peak scattering value and the full width of the scattering profile at one-half of the peak value were used as independent variables. A hybrid method combining the backpropagation feedforward NN with principal component analysis (PCA) was developed relating hyperspectral scattering characteristics to fruit firmness and SSC (Lu, 2007). The NN models were able to predict fruit firmness with $r^2 = 0.76$ and the standard error of prediction (SEP) of 6.2 N for "Golden Delicious" apples and $r^2 = 0.55$ and SEP = 6.1 N for red "Delicious" apples. Better SSC predictions were obtained with $r^2 = 0.79$ and 0.64 and SEP = 0.72% and 0.81% for "Golden Delicious" and "Delicious" apples, respectively.

Qin and Lu (2008) developed a hyperspectral imaging configuration for acquiring spatially-resolved reflectance profiles from intact samples of fruits and

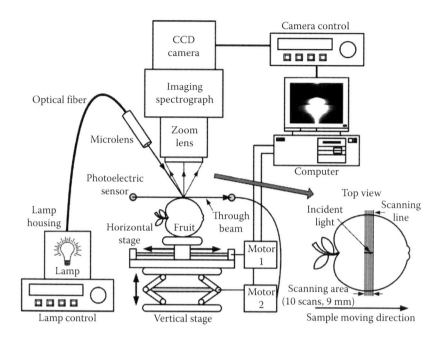

FIGURE 9.5 Hyperspectral imaging system for acquiring spatially-resolved scattering images from fruit sample. (From Qin, J. and R. Lu. *Postharvest Biology and Technology* 49 (3), 2008: 355–365. With permission.)

vegetables (Figure 9.5). Spectra of the absorption and reduced scattering coefficients for the samples of apple (three varieties), peach, pear, kiwifruit, plum, cucumber, zucchini squash, and tomato (at three ripeness stages) over the spectral range of 500–1000 nm were obtained using an inverse algorithm for a diffusion equation, coupled with a method for correcting the effect of the curved sample surface. Values of μ_a and μ'_s varied greatly among the test samples. Large differences in the absorption spectra were observed for tomatoes of three ripeness stages (green, pink, and red), and their ripeness was correctly classified using the ratio of μ_a at 675 nm (for chlorophyll) to that at 535 nm (for anthocyanin). Values of the μ'_s positively correlated with the firmness of tomatoes at individual wavelengths of 500–1000 nm, with a maximum correlation of 0.66 being obtained at 790 nm. Cen et al. (2012) further improved the HISR technique through optimization of the optical designs and algorithms. They measured the spectral absorption and reduced scattering coefficients of peaches for maturity/quality assessment. The optical property measuring instrument was used for acquiring hyperspectral reflectance images of 500 "Redstar" peaches. The results show that both μ_a and μ'_s were correlated with peach firmness, SSC, and skin and flesh color parameters. Better correlation results for partial least squares models were obtained using the combined values of μ_a and μ'_s than using μ_a or μ'_s. Improved results for firmness, SSC, skin lightness, and flesh lightness were obtained using least squares support vector machine models.

To gain a quantitative understanding of the relationship between the optical and mechanical/structural properties of apple fruit during the softening process, Cen et al. (2013) measured μ_a and μ'_s for the wavelengths of 500–1000 nm, acoustic/impact firmness and tissue elasticity of "Golden Delicious" and "Granny Smith" apples stored at high temperature (~22°C)/high humidity (~95%) for five times over a 30-day period of storage. They also quantified the morphological features of apple cells (i.e., area and equivalent diameter) and their changes during the softening process, using confocal laser scanning microscopic technique. The absorption and scattering properties of apple fruit were correlated with acoustic/impact firmness ($r = 0.870$–0.948 for "Golden Delicious" and $r = 0.334$–0.993 for "Granny Smith") and with Young's modulus ($r = 0.585$–0.947 for "Golden Delicious" and $r = 0.292$–0.694 for "Granny Smith"). Preliminary analysis also showed that optical absorption and scattering parameters were positively correlated with the cell's area and equivalent diameter. These findings suggest that the optical properties could be used to study the mechanical properties of apples and possibly their microstructural changes during ripening and postharvest storage.

Valero et al. (2001, 2004a, 2004b, 2007) applied time-domain laser reflectance spectroscopy to evaluate internal fruit quality. They developed models for absorption and scattering coefficients, using statistical techniques such as the PCA, multiple stepwise linear regression, clustering, and discriminant analysis, to estimate firmness and soluble solids and acid contents in kiwifruit, tomato, apple, peach, and nectarine. Different quality grades of fruits could be sorted, according to their firmness, soluble solids, and acidity. The research demonstrated that time-domain laser reflectance spectroscopy technique has potential for assessing the internal properties of fruits and other food products.

The research group led by Manuela Zude at Leibniz-Institute for Agricultural Engineering in Potsdam, Germany has done much work on laser backscattering for quality assessment. Backscattering images at five wavelength bands (680, 780, 880, 940, and 980 nm) were obtained from "Elstar" and "Pinova" apples (Qing et al., 2007, 2008), using the setup shown in Figure 9.6. Partial least squares regression models for "Elstar" apple gave the highest correlation coefficients of $r = 0.89$ and 0.90 and the lowest standard errors of cross-validation (SECV) of 0.73 °Brix and 5.44 N for SSC and flesh firmness, respectively. An intercultivar validation on "Pinova" apples resulted in an SEP < 13% (calculated with error value divided by mean values of measured fruit parameters), indicating that fruit firmness can be measured in parallel with fruit SSC (Qing et al., 2007). For apples grown in sites with different plant water availability, calibration models gave an SECV < 5% for analyzing flesh firmness and an SECV < 8% for predicting SSC for both "Elstar" and "Pinova" apple cultivars, which implied that backscattering imaging gave relevant information for fruit monitoring in the production (Qing et al., 2008). The selected optical properties of "Elstar" and "Pinova" apples during controlled atmosphere storage (2% CO_2, 1.5% O_2) at 2°C were monitored using laser backscattering imaging at 785 nm (Baranyai et al., 2009). The total interaction coefficient μ_t increased significantly within the first 81 days and decreased afterwards, and it might have been affected by other factors, including storage time, position in the orchard, and cultivar in addition to firmness. The anisotropy factor showed almost a monotonous decrease, and

Spectral Scattering for Assessing the Quality of Fruits and Vegetables

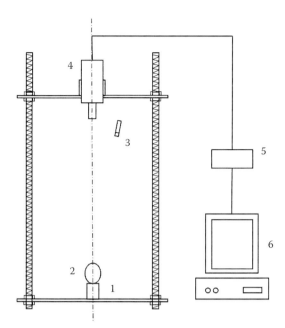

FIGURE 9.6 Schematic view of a multilaser scatter imaging system for measuring backscattering images from fruit. (1) Sample holder, (2) sample (apple), (3) laser source with lens, (4) CCD camera with lens, (5) video I/O card, and (6) computer for data processing. (From Qing, Z., B. Ji, and M. Zude, *J. Food Eng.*, 82 (1), 58–67, 2007. With permission.)

its change of 2.1% was less than the 15% change for the μ_t. Backscattering imaging at 660 nm was used to monitor fruit flesh firmness and elastic modulus of apple, plum, tomato, and mushroom (Mollazade et al., 2013). The results from adaptive neuro-fuzzy inference system models showed that integrating selected real-time features of both texture analysis and space-domain techniques provided the best prediction results. Satisfactory performance was obtained for apple ($r = 0.887$), tomato ($r = 0.919$), and mushroom ($r = 0.896$), while the performance was poor for plum fruit ($r = 0.790$). In high-value sweet cherry (*Prunus avium*), the red coloration determined by the anthocyanin content is correlated with the fruit ripeness stage and market value. Zude et al. (2009) used total internal reflectance, continuous-wave, and time-resolved spectroscopy in the anthocyanins content study of cherry. Zude et al. (2011) also used the Beer–Lambert law and multivariate regression based on visible and near-infrared spectroscopy and scatter correction by the effective path length in the fruit tissue obtained from time-resolved readings of the distribution of time of flight for nondestructive analysis of anthocyanins in cherries. The scatter correction could improve the performance of pigment analysis.

9.3.2 Defects

Defects reduce the market value of fruits and vegetables and cause economic loss. The assessment of external defects is largely carried out by human inspectors or computer imaging technique. Manual methods for determining the presence of internal defects would require cutting open samples. Encouraging results have been reported on applying spectral scattering for noninvasive detection of defects in fruits and vegetables, including bruise, mealiness, brown heart (Cubeddu et al., 2002; Zerbini et al., 2002), decay, and chilling injury.

Bruises are primarily caused by two types of mechanical load, that is, static and dynamic (Mohsenin, 1970). Most bruises are caused by dynamic load in the form of vibration or impact (Lu et al., 2010). Impact bruising is the most common, and it occurs frequently during harvest, transport, and postharvest handling (Baritelle and Hyde, 2001). Bruising can occur from dropping or it could occur if the produce is stacked too high or overfilled. Bruising causes cell destruction and reduces intercellular air spaces, which are initially filled by water released from the ruptured cells (Lu et al., 2010). As time progresses, the bruised tissue starts to lose moisture and eventually becomes desiccated (Lu et al., 2010). Bruise can alter the color, flavor, and texture of the fruit. The shelf life of many fresh fruits and vegetables is considerably reduced by bruising damage. The detection of bruises is dependent on fruit variety, age of bruise, bruise type, and severity. Lu et al. (2010) used a hyperspectral imaging technique for acquiring spatially-resolved scattering images from normal and bruised apples. The scattering profiles were corrected first for instrument nonuniformity effect to estimate μ_a and μ'_s. The reduced scattering coefficient for normal apples was much higher than that for bruised apples and it decreased consistently with time after bruising. The results suggested that bruising has a greater impact on scattering than on absorption (Lu et al., 2010). SFDI was also used to obtain quantitative absorption and scattering image maps from 650 to 980 nm for the detection of bruises and quantification of levels of bruising severity on "Golden Delicious" apples (Anderson et al., 2007). The average scattering and absorption spectra were calculated for the two levels of bruise severity and compared with those obtained for adjacent nonbruised regions. Bruises were induced using impact energies of 0.06 and 0.314 J and they were distinguishable by comparing the scattering pixel count for the bruised region to the nonbruised region. Bruised regions were clearly distinguishable from the surrounding nonbruised region, though the ratio of the average scattering coefficients for the bruise region to the average scattering coefficients for the nonbruise region using the different energies of 0.06 and 0.314 J is approximately equal.

Mealiness is a negative texture attribute common in some apple and potato varieties as they age. It gives rise to the sensation of a desegregated tissue with lack of juiciness (Valero et al., 2007). Mealiness reduces shelf life with respect to eating quality. Valero et al. (2001) used classification models based on time-domain laser reflectance spectroscopy technique at several wavelengths (672, 750, and 818 nm and from 900 to 1000 nm) to identify the mealiness in apples. The classification accuracy ranged from 47% to 100% for mealy versus nonmealy apples.

The suitability of MC method for estimating the valuable information regarding the mechanical impact damage of "Golden Delicious" and "Idared" apple tissue (Baranyai and Zude, 2008), and the optical properties of "Hayward" kiwifruit tissue (Baranyai and Zude, 2009) was investigated, and the simulated results were compared with experimental data produced by backscattering imaging. Significant changes of the scattering profiles on the boundary of a semi-infinite homogeneous media were found ($p < 0.05$), even though the optical measurements were repeated after only 1 day of storage (Baranyai and Zude, 2008).

Laser backscattering imaging was tried for automatically detecting early symptoms of decay in citrus fruit after infection with the pathogen *Penicillium digitatum* at five wavelength bands (680, 780, 880, 940, and 980 nm) (Lorente et al., 2013, 2015). Gaussian–Lorentzian cross product distribution function described radial profiles accurately with average r^2 values higher than or equal to 0.998 and average root mean squared errors lower than or equal to 2.54 (Lorente et al., 2013). A diffusion theory model by Farrell et al. (1992) (see Chapters 3 and 7 for further details) was also used, and both models described backscattering profiles accurately with slightly better curve-fitting results ($r^2 > 0.996$) for the Gaussian–Lorentzian model compared with the diffusion theory model ($r^2 > 0.982$) (Lorente et al., 2015). The best classification results were achieved using a reduced set of Gaussian–Lorentzian parameters, yielding the best overall classification accuracy of 93.4%, with 92.5% and 94.3% for sound and decaying samples, respectively (Lorente et al., 2015).

The potential of laser backscattering imaging for predicting chilling injury in bananas was studied (Hashim et al., 2013, 2014). Bananas were stored for 2 days at 13°C (control), 6°C (chilling temperature), and subsequently 1 day at ambient temperature to allow the symptom to develop. Classification of the control and chill-injured samples in ripe fruit measured at 660 and 785 nm resulted in misclassification errors as low as 6% and 8% for early detection, and 0.67% and 1.33% for after-storage detection, respectively (Hashim et al., 2013). All parameters obtained from the backscattering profiles, such as inflection point, slope after inflection point, full width at half maximum, and saturation radius were statistically significant to storage temperature, ripening stage, and treatment time. In addition, backscattering parameters were strongly influenced by pigment changes and textural properties (Hashim et al., 2014).

9.3.3 OTHER APPLICATIONS

Variations in the moisture content of fruit slices subjected to different drying conditions were evaluated by laser backscattering imaging (Romano et al., 2008, 2010, 2011; Udomkun et al., 2014). Backscattering images of banana slices were acquired at 670 nm each hour during the drying process at three different temperatures: 53°C, 58°C, and 63°C (Romano et al., 2008). Significant relationship was found between changes in backscattering area and moisture content, especially at lower temperatures when no tissue browning occurred ($r = 0.76$ for 53°C). At higher temperatures, correlations were observed between the parameters extracted by image processing and a* standard color index mainly due to tissue browning. Backscattering images of apple slices dried in a high-precision through-flow laboratory dryer at air temperatures of 60°C and 70°C were measured every 30 min over 3 h of drying (Romano et al., 2010). Backscattering area and gray values, corresponding to the light luminescence, ranging from 0 (dark) to 255 (white), were highly correlated with moisture content with Pearson coefficients of 0.95 and 0.96 at 635 nm, and of 0.74 and 0.87 at 785 nm. Simultaneous prediction of the variations in hardness, moisture content, and SSC of "Gala" apple slices during drying was also done with backscattering imaging at 635 nm (Romano et al., 2011). Calibration models showed high correlation coefficients ($r = 0.8$ and 0.89) with low SECV (SECV = 11.6% and 9.8%) between moisture content and backscattering area and light luminescence, respectively. However, predictions obtained for SSC were slightly lower. A laser backscattering analysis at three wavelengths (532, 650, and 780 nm) was also reported for predicting the moisture content, shrinkage, lightness, chroma, and hue changes of papaya during drying at four different temperatures (50°C, 60°C, 70°C, and 80°C) (Udomkun et al., 2014). The results showed that moisture content, lightness, and chroma values decreased, whereas hue and shrinkage values increased with the increasing drying temperature. In addition, multivariate regressions of measured illuminated area and light intensity parameters at 650 nm were found to yield the best moisture content, lightness, and hue predictions ($r > 0.92$). The study showed that backscattering imaging provided a useful tool for quality control in the fruit drying process.

9.4 CONCLUSION

In the past decade, an increasing interest in optical characterization of fruits and vegetables has been springing up for nondestructive quality and safety evaluation. Optical properties of fruit and vegetable tissues provide abundant information about the chemical and physical characteristics. Spectral scattering has been investigated as a useful tool for quality evaluation of fruits and vegetables through simultaneous estimation of the absorption and scattering properties or by extracting critical scattering features from 1D or 2D scattering profiles or images. Commonly used spectral scattering techniques for measuring optical properties of fruits and vegetables include TRS, SFDI, IS method, fiber optic-based SRS, and HISR spectroscopy. Specificity of the optical properties for different fruit and vegetable tissues could be detected for different varieties or skin/flesh parts. Internal defect, bruise, and disease as well as storage environment and pretreatment condition were all found to have effects on the optical properties compared with sound/normal tissue. However, since different results were reported in different studies using diverse optical property measurement techniques, it is necessary to establish a standard method for validating and comparing the different measuring systems.

So far, optical property measurements have been reported for apple, peach, pear, kiwifruit, nectarine, cucumber, plum, zucchini squash, tomato, onion, banana, cherry, mushroom, papaya, citrus, and sweet pepper, which only account for a small portion of this large category of food. The structural complexity, diversity, and inhomogeneity of fruits and vegetables present critical challenges in the accurate measurement of optical properties as well as their qualities. Spectral scattering technique has been only used for research and commercial applications have not been developed. Many technical issues and limitations still need to be addressed so that the technique can be eventually used for quality and ripening assessment or monitoring of quality changes in horticultural and food products during long-term storage. A compact, ease-of-use, portable, and low-cost instrument needs to be developed before industrial applications can take place. Recent advances in biomedical tissue optics can offer new research opportunities in optical property measurement and quality assessment of fruits and vegetables.

REFERENCES

Anderson, E. R., D. J. Cuccia, and A. J. Durkin. Detection of bruises on golden delicious apples using spatial-frequency-domain imaging. *Proceedings of SPIE* 6430, 2007: 64301O.

Baranyai, L., C. Regen, and M. Zude. Monitoring optical properties of apple tissue during cool storage. *CIGR Workshop on Image Analyses in Agriculture*, 2009, Potsdam, Germany.

Baranyai, L. and M. Zude. Analysis of laser light migration in apple tissue by Monte Carlo simulation. *Progress in Agricultural Engineering Sciences* 4 (1), 2008: 45–59.

Baranyai, L. and M. Zude. Analysis of laser light propagation in kiwifruit using backscattering imaging and Monte Carlo simulation. *Computers and Electronics in Agriculture* 69 (1), 2009: 33–39.

Baritelle, A. L. and G. M. Hyde. Commodity conditioning to reduce impact bruising. *Postharvest Biology and Technology* 21 (3), 2001: 331–339.

Bass, M., E. W. Van Stryland, D. R. Williams, and W. L. Wolfe. *Handbook of Optics* (2nd edn, Vol. III). New York: McGraw-Hill, 2001.

Birth, G. S. The light scattering properties of foods. *Journal of Food Science* 43, 1978: 916–925.

Birth, G. S. The light scattering characteristics of ground grains. *International Agrophysics* 2 (1), 1986: 59–67.

Birth, G. S., C. E. Davis, and E. Townsend. The scattering coefficient as a measure of pork quality. *Journal of Animal Science* 46 (3), 1978: 639–645.

Budiastra, I. W., Y. Ikeda, and T. Nishizu. Optical methods for quality evaluation of fruits, 1: Optical properties of selected fruits using the Kubelka–Munk theory and their relationships with fruit maturity and sugar content. *Journal of the Japanese Society of Agricultural Machinery (Japan)* 60 (2), 1998: 117–128.

Budiastra, I. W., Y. Ikeda, T. Nishizu, and T. Kataoka. The scatter coefficients as quality indices of the fruits. *Journal of the Japanese Society of Agricultural Machinery* 54 (Supplement), 1992: 301–302.

Cen, H., R. Lu, F. Mendoza, and D. Ariana. Assessing multiple quality attributes of peaches using optical absorption and scattering properties. *Transactions of the ASABE* 55 (2), 2012: 647–657.

Cen, H., R. Lu, F. Mendoza, and R. M. Beaudry. Relationship of the optical absorption and scattering properties with mechanical and structural properties of apple tissue. *Postharvest Biology and Technology* 85, 2013: 30–38.

Chen, P. and V. Nattuvetty. Light transmittance through a region of an intact fruit. *Transactions of the ASABE* 23 (2), 1980: 519–522.

Cheong, W. F., S. A. Prahl, and A. J. Welch. A review of the optical properties of biological tissues. *IEEE Journal of Quantum Electronics* 26 (12), 1990: 2166–2185.

Cubeddu, R., C. D'andrea, A. Pifferi, P. Taroni, A. Torricelli, G. Valentini, C. Dover, D. Johnson, M. Ruiz-Altisent, and C. Valero. Nondestructive quantification of chemical and physical properties of fruits by time-resolved reflectance spectroscopy in the wavelength range 650–1000 nm. *Applied Optics* 40 (4), 2001b: 538–543.

Cubeddu, R., C. D'andrea, A. Pifferi, P. Taroni, A. Torricelli, G. Valentini, M. Ruiz-Altisent et al. Time-resolved reflectance spectroscopy applied to the nondestructive monitoring of the internal optical properties in apples. *Applied Spectroscopy* 55 (10), 2001a: 1368–1374.

Cubeddu, R., A. Pifferi, P. Taroni, and A. Torricelli. Measuring fresh fruit and vegetable quality: Advanced optical methods. In *Fruit and Vegetable Processing: Improving Quality*, ed. Jongen, W. Cambridge: Woodhead Publishing Ltd; Boca Raton, Florida: CRC Press LLC, 2002, 150–169.

Cuccia, D. J., F. Bevilacqua, A. J. Durkin, and B. J. Tromberg. Modulated imaging: Quantitative analysis and tomography of turbid media in the spatial-frequency domain. *Optics Letters* 30 (11), 2005: 1354–1356.

Farrell, T. J., M. S. Patterson, and B. Wilson. A diffusion-theory model of spatially resolved steady-state diffuse reflectance for the noninvasive determination of tissue optical-properties *in vivo*. *Medical Physics* 19, 1992: 879–888.

Hashim, N., R. B. Janius, R. Abdul, A. O. Rahman, M. Shitan, and M. Zude. Changes of backscattering parameters during chilling injury in bananas. *Journal of Engineering Science and Technology* 9 (3), 2014: 314–325.

Hashim, N., M. Pflanz, C. Regen, R. B. Janius, R. A. Rahman, A. Osman, M. Shitan, and M. Zude. An approach for monitoring the chilling injury appearance in bananas by means of backscattering imaging. *Journal of Food Engineering* 116 (1), 2013: 28–36.

Lorente, D., M. Zude, C. Idler, J. Gómez-Sanchis, and J. Blasco. Laser-light backscattering imaging for early decay detection in citrus fruit using both a statistical and a physical model. *Journal of Food Engineering* 154, 2015: 76–85.

Lorente, D., M. Zude, C. Regen, L. Palou, J. Gómez-Sanchis, and J. Blasco. Early decay detection in citrus fruit using laser-light backscattering imaging. *Postharvest Biology and Technology* 86, 2013: 424–430.

Lorenzo, J. R. *Principles of Diffuse Light Propagation: Light Propagation in Tissues with Applications in Biology and Medicine*. Singapore: World Scientific Printers, 2012.

Lu, R. Near-infrared multispectral scattering for assessing internal quality of apple fruit. *Proceedings of SPIE* 8369, 2003: 313–320.

Lu, R. Nondestructive measurement of firmness and soluble solids content for apple fruit using hyperspectral scattering images. *Sensing and Instrumentation for Food Quality and Safety* 1 (1), 2007: 19–27.

Lu, R. Multispectral imaging for predicting firmness and soluble solids content of apple fruit. *Postharvest Biology and Technology* 31 (2), 2004a: 147–157.

Lu, R. Prediction of apple fruit firmness by near-infrared multispectral scattering. *Journal of Texture Studies* 35 (3), 2004b: 263–276.

Lu, R., H. Cen, M. Huang, and D. P. Ariana. Spectral absorption and scattering properties of normal and bruised apple tissue. *Transactions of the ASABE* 53 (1), 2010: 263–269.

Lu, R. and Y. Peng. Hyperspectral scattering for assessing peach fruit firmness. *Biosystems Engineering* 93 (2), 2006: 161–171.

McGlone, V., H. Abe Kawano, and S. Kawano. Kiwifruit firmness by near infrared light scattering. *Journal of Near Infrared Spectroscopy* 5 (2), 1997: 83–89.

Mohsenin, N. N. *Physical Properties of Plant and Animal Materials*. New York: Gordon & Breach Science Publishers Inc., 1970.

Mollazade, K., M. Omid, F. A. Tab, Y. R. Kalaj, S. S. Mohtasebi, and M. Zude. Analysis of texture-based features for predicting mechanical properties of horticultural products by laser light backscattering imaging. *Computers and Electronics in Agriculture* 98, 2013: 34–45.

Peng, Y. and R. Lu. Modeling multispectral scattering profiles for prediction of apple fruit firmness. *Transactions of the ASAE* 48 (1), 2005: 235–242.

Peng, Y. and R. Lu. An LCTF-based multispectral imaging system for estimation of apple fruit firmness: Part I. Acquisition and characterization of scattering images. *Transactions of the ASAE* 49 (1), 2006a: 259–267.

Peng, Y. and R. Lu. An LCTF-based multispectral imaging system for estimation of apple fruit firmness: Part II. Selection of optimal wavelengths and development of prediction models. *Transactions of the ASAE* 49 (1), 2006b: 269–275.

Peng, Y. and R. Lu. Improving apple fruit firmness predictions by effective correction of multispectral scattering images. *Postharvest Biology and Technology* 41 (3), 2006c: 266–274.

Prahl, S. Everything I Think You Should Know about Inverse Adding–Doubling. *Oregon Medical Laser Center, Manual of the Inverse Adding–Doubling Program*, 2011, http://omlc.ogi.edu/software/iad/. (Accessed October 1, 2012.)

Qin, J. and R. Lu. Measurement of the optical properties of fruits and vegetables using spatially resolved hyperspectral diffuse reflectance imaging technique. *Postharvest Biology and Technology* 49 (3), 2008: 355–365.

Qin, J. and R. Lu. Monte Carlo simulation for quantification of light transport features in apples. *Computers and Electronics in Agriculture* 68 (1), 2009: 44–51.

Qing, Z., B. Ji, and M. Zude. Predicting soluble solid content and firmness in apple fruit by means of laser light backscattering image analysis. *Journal of Food Engineering* 82 (1), 2007: 58–67.

Qing, Z., B. Ji, and M. Zude. Non-destructive analyses of apple quality parameters by means of laser-induced light backscattering imaging. *Postharvest Biology and Technology* 48 (2), 2008: 215–222.

Rizzolo, A., M. Vanoli, L. Spinelli, and A. Torricelli. Sensory characteristics, quality and optical properties measured by time-resolved reflectance spectroscopy in stored apples. *Postharvest Biology and Technology* 58 (1), 2010: 1–12.

Romano, G., D. Argyropoulos, and J. Müller. Laser light backscattering for monitoring changes in moisture content during drying of apples. *XVIIth World Congress of the International Commission of Agricultural and Biosystems Engineering (CIGR)*, 2010, Quebec, Canada.

Romano, G., L. Baranyai, K. Gottschalk, and M. Zude. An approach for monitoring the moisture content changes of drying banana slices with laser light backscattering imaging. *Food and Bioprocess Technology* 1 (4), 2008: 410–414.

Romano, G., M. Nagle, D. Argyropoulos, and J. Müller. Laser light backscattering to monitor moisture content, soluble solid content and hardness of apple tissue during drying. *Journal of Food Engineering* 104 (4), 2011: 657–662.

Rowe, P. I., R. Künnemeyer, A. McGlone, S. Talele, P. Martinsen, and R. Seelye. Relationship between tissue firmness and optical properties of "Royal Gala" apples from 400 to 1050 nm. *Postharvest Biology and Technology* 94, 2014: 89–96.

Ruiz-Altisent, M., L. Ruiz-Garcia, G. P. Moreda, R. Lu, N. Hernandez-Sanchez, E. C. Correa, B. Diezma, B. Nicolaï, and J. Gaicía-Ramos. Sensors for product characterization and quality of specialty crops—A review. *Computers and Electronics in Agriculture* 74 (2), 2010: 176–194.

Saeys, W., M. A. Velazco-Roa, S. N. Thennadil, H. Ramon, and B. M. Nicolaï. Optical properties of apple skin and flesh in the wavelength range from 350 to 2200 nm. *Applied Optics* 47 (7), 2008: 908–919.

Sinha, N. K., Y. H. Hui, E. Özgül Evranuz, M. Siddiq, and J. Ahmed. Tomato processing, quality, and nutrition. In *Handbook of Vegetables and Vegetable Processing*, ed. Sinha, N. K. Oxford: Wiley-Blackwell Publishing Ltd., 2011, 753.

Seifert, B., M. Zude, L. Spinelli, and A. Torricelli. Optical properties of developing pip and stone fruit reveal underlying structural changes. *Physiologia Plantarum* 153, 2015: 327–336.

Tijskens, L. M. M., P. E. Zerbini, R. E. Schouten, M. Vanoli, S. Jacob, M. Grassi, R. Cubeddu, L. Spinelli, and A. Torricelli. Assessing harvest maturity in nectarines. *Postharvest Biology and Technology* 45 (2), 2007: 204–213.

Trong, N. N. D., A. Rizzolo, E. Herremans, M. Vanoli, G. Cortellino, C. Erkinbaev, M. Tsuta, L. Spinelli, D. Contini, and A. Torricelli. Optical properties-microstructure-texture relationships of dried apple slices: Spatially resolved diffuse reflectance spectroscopy as a novel technique for analysis and process control. *Innovative Food Science and Emerging Technologies* 21, 2014: 160–168.

Tu, K., R. De Busscher, J. De Baerdemaeker, and E. Schrevens. Using laser beam as light source to study tomato and apple quality non-destructively. *Proceedings of the Food Processing Automation IV Conference*, 1995, 528–536, Chicago.

Tu, K., P. Jancsók, B. Nicolaï, and J. D. Baerdemaeker. Use of laser-scattering imaging to study tomato-fruit quality in relation to acoustic and compression measurements. *International Journal of Food Science & Technology* 35 (5), 2000: 503–510.

Tuchin, V. *Tissue Optics-Light Scattering Methods and Instruments for Medical Diagnosis* (2nd edn). Bellingham, Washington: SPIE Press, 2007.

Udomkun, P., M. Nagle, B. Mahayothee, and J. Müller. Laser-based imaging system for non-invasive monitoring of quality changes of papaya during drying. *Food Control* 42, 2014: 225–233.

Valero, C., P. Barreiro Elorza, and C. Ortiz. Optical detection of mealiness in apples by laser TDRS. *Acta Horticulturae* 553 (2), 2001: 513–518.

Valero, C., P. Barreiro, C. Ortiz, M. Ruiz-Altisent, R. Cubeddu, A. Pifferi et al. Optical detection of mealiness in apples by laser TDRS. *Proceedings of SPIE* 6430, 2007: 64301O.

Valero, C., M. Ruiz-Altisent, R. Cubeddu, A. Pifferi, P. Taroni, A. Torricelli, G. Velentini, D. S. Johnson, and C. J. Dover. Selection models for the internal quality of fruit, based on time domain laser reflectance spectroscopy. *Biosystems Engineering* 88 (3), 2004a: 313–323.

Valero, C., M. Ruiz-Altisent, R. Cubeddu, A. Pifferi, P. Taroni, A. Torricelli, G. Velentini, D. S. Johnson, and C. J. Dover. Detection of internal quality in kiwi with time-domain diffuse reflectance spectroscopy. *Applied Engineering in Agriculture* 20 (2), 2004b: 223–230.

Vanoli, M., A. Rizzolo, M. Grassi, L. Spinelli, B. E. Verlinden, and A. Torricelli. Studies on classification models to discriminate "Braeburn" apples affected by internal browning using the optical properties measured by time-resolved reflectance spectroscopy. *Postharvest Biology and Technology* 91, 2014: 112–121.

Vanoli, M., P. E. Zerbini, L. Spinelli, A. Torricelli, and A. Rizzolo. Polyuronide content and correlation to optical properties measured by time-resolved reflectance spectroscopy in "Jonagored" apples stored in normal and controlled atmosphere. *Food Chemistry* 115 (4), 2009: 1450–1457.

Wang, W. and C. Li. The optical properties of onion dry skin and flesh at the wavelength 632.8 nm. *Proceedings of SPIE* 8369, 2012: 83690G.

Wang, W. and C. Li. Measurement of the light absorption and scattering properties of onion skin and flesh at 633 nm. *Postharvest Biology and Technology* 86, 2013: 494–501.

Wang, W., C. Li, and R. D. Gitaitis. Optical properties of healthy and diseased onion tissues in the visible and near-infrared spectral region. *Transactions of the ASABE* 57 (6), 2014: 1771–1782.

Zerbini, P. E., M. Grassi, R. Cubeddu, A. Pifferi, and A. Torricelli. Nondestructive detection of brown heart in pears by time-resolved reflectance spectroscopy. *Postharvest Biology and Technology* 25 (1), 2002: 87–97.

Zhu, Q., C. He, R. Lu, F. Mendoza, and H. Cen. Ripeness evaluation of "Sun Bright" tomato using optical absorption and scattering properties. *Postharvest Biology and Technology* 103, 2015: 27–34.

Zsom, T., V. Zsom-Muha, D. L. Dénes, G. Hitka, L. P. Nguyen, and J. Felföldi. Non-destructive postharvest quality monitoring of different pear and sweet pepper cultivars. *Acta Alimentaria* 43, 2014: 206–214.

Zude, M., M. Pflanz, L. Spinelli, C. Dosche, and A. Torricelli. Non-destructive analysis of anthocyanins in cherries by means of lambert–beer and multivariate regression based on spectroscopy and scatter correction using time-resolved analysis. *Journal of Food Engineering* 103 (1), 2011: 68–75.

Zude, M., L. Spinelli, C. Dosche, and A. Torricelli. *In-situ* analysis of fruit anthocyanins by means of total internal reflectance, continuous wave and time-resolved spectroscopy. *SPIE Optical Engineering + Applications*, International Society for Optics and Photonics, 2009, 74320H, San Diego.

10 Light Propagation in Meat and Meat Analog
Theory and Applications

Gang Yao

CONTENTS

10.1 Meat, Meat Analog, and Optical Inspection for Quality Control 251
    10.1.1 Meat and Meat Tenderness 251
    10.1.2 Meat Analog and Fiber Formation 253
    10.1.3 Light Interaction with Meat and Meat Analog 254
10.2 Light Propagation in Meat Analogs 256
    10.2.1 Anisotropic Diffusion Theory 256
    10.2.2 Monte Carlo Simulation of Light Propagation in Fibrous Tissue 258
    10.2.3 Conclusion 260
10.3 Light Propagation in Skeletal Muscle 261
    10.3.1 Optical Diffraction by a Single Muscle Fiber 262
    10.3.2 Optical Reflectance in Whole Muscle 264
    10.3.3 Conclusion 266
10.4 Optical Characterization of Fiber Formation in Meat Analog 266
10.5 Optical Characterization of Beef Tenderness 271
10.6 Conclusion 276
References 277

10.1 MEAT, MEAT ANALOG, AND OPTICAL INSPECTION FOR QUALITY CONTROL

10.1.1 MEAT AND MEAT TENDERNESS

Meat, in the form of animal muscles, poultry, and fish are the most common sources of high-protein food. Specifically, the U.S. beef industry is the largest segment of the nation's food and fiber industry and produces nearly 25% of the world's beef supply.

    Whole muscle consists of fascicles, which are collections of muscle fibers (muscle cells) encapsulated in an elaborate connective tissue matrix (Gerrard and Grant, 2003). Morphologically, each muscle fiber consists of many smaller myofibrils. Each myofibril is in the form of a cylindrical structure organized into repetitive units named "sarcomeres." The major proteins in the highly organized sarcomeres include actin-containing thin filaments, myosin-containing thick filaments, the lateral boundary Z-discs, and elastic filaments known as titin. Both the thin filaments and titin are anchored to and extend in opposing directions from the Z-disc. Sarcomeres are aligned precisely in muscle fibers and are readily observed using light microscopy as alternating light and dark bands called I- and A-bands. The thin filaments, with parallel arrays, span the whole I-band and overlap with thick filaments in the A-band (Clark et al., 2002). Force generation and contraction of muscle depend on the actin–myosin interactions in the overlap area of the thick and thin filaments driven by the cycling cross-bridges (Huxley, 1957; Huxley and Simmons, 1971). The normal mechanical and physiological functions of striated muscle are maintained and realized by sarcomeres (Bonnemann and Laing, 2004). The length of a sarcomere (or "sarcomere length") can vary significantly depending on species, anatomical locations, and functional states (Burkholder and Lieber, 2001).

The sophisticated muscle structure directly affects muscle functions as well as meat quality, in particular meat tenderness. Tenderness is essentially a sensory characterization of meat hardness. Albeit somewhat ambiguous in nature, it can be defined as the sum of mechanical strength needed to break down and fragment meat during the chewing process. Variation in meat tenderness remains one of the most critical quality problems facing the meat industry, especially the beef industry, aside from food safety concerns (Huffman et al., 1996; Shackelford et al., 2001). Even with the arrival of the genetic era, variations in tenderness continue to exist among those animals genetically selected for superior meat tenderness because handling and processing can adversely affect tenderness.

Sarcomere length has long been recognized as the major effect that changes meat tenderness (Herring et al., 1965; Marsh and Carse, 1974; Smulders et al., 1990). Sarcomeres retain the ability to contract for several hours after slaughter. The mean length of sarcomeres is related to the contractile state of muscle and thus influences meat tenderness (Herring et al., 1967). Bendall (1951) first described the muscle shortening that accompanies rigor mortis development and showed that it led to meat toughening. Subsequent studies (Marsh and Carse, 1974; Smulders et al., 1990) further supported sarcomere length as a primary factor that determines meat tenderness. Empirical practice (Hostetler et al., 1970; Wang et al., 1994) has been used in the meat industry to alter sarcomere length in order to improve meat tenderness. Several *in vivo* methods exist to artificially induce changes in sarcomere length by carcass hanging (Hostetler et al., 1970) or mechanically stretching muscle (Buege and Stouffer, 1974; Marsh and Carse, 1974). Regardless of the procedure utilized, all methods are capable of altering beef tenderness by lengthening sarcomeres, which argues that fresh meat tenderness, in part, is a function of sarcomere length, or the inherent ultrastructure of muscle cells.

In addition to sarcomere length, the connective tissue content and proteolysis of myofibril proteins may also affect meat tenderness (Koohmaraie et al., 2002). The collagenous structures serve to distribute compressive and tensional forces through the muscle and maintain muscle fiber alignment (Borg and Caulfield, 1980). Type I collagen predominates in perimysial connective tissue sheets, while type IV collagen is the major component in the endomysial connective tissue. Collagen content is a contributor to meat toughness (Ramsbottom and Strandine, 1948), though its contribution may be through interactions with other components of the muscle. However, collagen content alone offers limited value as a predictor of toughness especially for young animals (Cross et al., 1973; Dransfield, 1977). The maturity of the structural substance is likely more important as mature collagen becomes highly crosslinked and therefore, more thermally stable and resistant to solubilization (Horgan, 1991; McCormick, 1994). This is consistent with the observation that collagen solubility during cooking decreases with animal age (Shorthose and Harris, 1991). Regardless of the exact mechanism, the quantity and quality of connective tissue potentially contribute to changes in meat tenderness.

In addition to the effects of sarcomere and collagen, it is important to note that meat has the potential to become tender upon storage at refrigerated temperatures because of the proteolysis of various structural proteins (Taylor et al., 1995). Degradation of these proteins results in the weakening of myofibrils and ultimately improves meat tenderness (Koohmaraie et al., 1996). Proteolysis degrades various inter- and intrasarcomeric and costameric proteins that maintain the muscle fibers' highly organized state, and results in fragmentation of myofibrils. Tenderness at any given time depends both on sarcomere length and extent of proteolysis occurring during the ageing process (Koohmaraie et al., 1996; Wheeler and Koohmaraie, 1994). In beef, ageing reduces toughness in all cases but those beef carcasses with extremely short sarcomeres (Harris and Shorthose, 1998). These data show the importance of proteolysis for meat tenderization and underscore the significance of the underlying subcellular structure in the process.

Because meat tenderness is a critical issue especially in the beef industry, it is highly desirable to have a technology that can *segregate* beef carcasses, or cuts, on the production line based on predicted tenderness. The current industrial beef grading system based on marbling and carcass maturity has been implemented for over 80 years, and it has limited capability in determining the variation in beef tenderness (Campion et al., 1975; Wheeler et al., 1994). Many mechanical tests are used to measure tenderness objectively (Bratzler, 1932; Shackelford et al., 1999; Stephens et al., 2004). Among them is the Warner–Bratzler shear force (WBSF) (Bratzler, 1932), which is the most frequently used method and has the best correlation with sensory panel evaluations of meat tenderness (Shackelford et al., 1995; Tornberg, 1996). Unfortunately, mechanical tests often require specialized instrumentation and are time consuming, costly, and destructive. As a result, nondestructive, rapid, and cost-effective alternatives have been pursued, many of which are optical based, including computer vision (Tan, 2004), optical fluorescence (Swatland, 1995; Egelandsdal et al., 2002), Raman spectroscopy (Beattie et al., 2004), and optical reflectance spectroscopy (Park et al., 1998; Liu et al., 2003; Shackelford et al., 2005).

10.1.2 MEAT ANALOG AND FIBER FORMATION

Although meat is the most important and common source of protein for human consumption, it may not be the ideal choice for certain populations due to various reasons such as health issues, cost, lifestyle, or religious restrictions. In such cases, vegetable proteins become an important alternative source for protein food. Among various sources, soy protein is one of the major vegetable proteins and is abundantly available with relatively low cost (Wolf et al., 1971; Liu and Hsieh, 2007). In addition, soy protein has many proven health benefits, including low saturated fat and no cholesterol. As a result, soy-based foods have become very popular in the Western world in the last 30 years.

"Meat analog" is a unique vegetable protein-based food product. It improves the consumer's eating experience by emulating the chewing sensation of real meat. To achieve meat-like chewing experience, special manufacturing technologies are used to produce fibrous texture in such "meat analog" products to mimic muscle fibers. Thermoplastic expansion is a common method used in producing texturized vegetable proteins (Lin et al., 2002). This method has become popular because it is less sophisticated and low in cost, and produces no waste. In particular, high-moisture extrusion technology has been very successful in texturizing vegetable proteins into fibrous products that resemble chicken or turkey breast meat in appearance and taste sensation (Lin et al., 2000; Liu and Hsieh, 2008).

Both single- and twin-screw extruders can be used in the extrusion process. Twin-screw extruders have more mixing capacity and higher throughput than single-screw extruders. The extrusion process generates high shear force and pressure in the barrel (Lin et al., 2000). As a result, the protein melts into a viscoelastic material and forms new bonds and cross-links. In the event of moisture insertion, this plasticized material expands and obtains the shape of the die. Bonded proteins form fibers aligning with the direction of extrusion while passing through the die. The amount of moisture and the temperature of heating units can change the fiber formation in high-moisture soy extrudates. Under high-moisture conditions (typically 50%–80%), a twin-screw food extruder with a smooth barrel can be used to convert soy proteins into meat analog products, resembling chicken or turkey breast meat (Liu and Hsieh, 2008).

Well-formed fibrous structures are the key factor affecting meat analogs' visual appearance and palatability. Therefore, assessing fiber formation in high-moisture extrusion products becomes an important quality control requirement. Textural profile analysis (Breene and Barker, 1975) and microstructure examination (Lin et al., 2000, 2002) are conventionally used to investigate extruded protein products. However, these methods are not successful in describing the fibrous characteristics of protein products extruded under high moisture. Results obtained from the textural profile analysis are mainly influenced by sample moisture content and have poor correlation with actual fiber formation (Yao et al., 2004). Although microstructure examination provides microscopic information in small sections, it does not accurately describe the global details of fiber formation. For quality control purposes during the manufacturing process, it is important to develop nondestructive techniques to quantify the textural properties of extrudates. Only recently have optical imaging-based techniques (Yao et al., 2004; Ranasinghesagara et al., 2006, 2009) been applied and shown great promise for assessing fiber formation in meat analogs.

10.1.3 LIGHT INTERACTION WITH MEAT AND MEAT ANALOG

Optical spectroscopy and imaging techniques have been increasingly applied for examining the quality of meat and meat analog products. Optical-based methods are attractive because they are rapid and noninvasive in nature. In addition, optical devices are becoming rather inexpensive. This makes optical detection very cost effective. As additional benefits, optical devices are usually quite portable and flexible.

All optical-based methods require the use of one or more light sources to illuminate the sample and detect the optical intensity at the sample surface. According to light transport theory (Wang and Wu, 2012), once the light enters a sample, it is absorbed by various chromophores and scattered by various muscle structural components while propagating inside the tissue. Absorption attenuates the light intensity, whereas scattering changes the light's trajectory. The amount of absorption depends on the concentration of the chromophores as well as the light's pathlength inside the sample. Optical scattering refers to any processes that alter light propagation directions. Based on the classic electromagnetic theory, tissue scattering is induced by the inhomogeneous distribution of refractive index inside the sample (Tuchin, 1997). Different tissue components have different refractive indices and morphological profiles. The total scattering potential of each tissue component depends on its concentration as well as the scattering power of individual scatterers.

The sample scattering and absorption properties can be represented using two physical parameters: the *scattering coefficient* and the *absorption coefficient* (Wang and Wu, 2012). The scattering coefficient (μ_s) is defined as the probability of the light being scattered within an infinitely small distance. And the absorption coefficient (μ_a) is defined as the probability of the light being absorbed within an infinitely small distance. The physical unit for both coefficients is the inverse of a distance (such as cm^{-1}).

It is important to note that most biological tissues are turbid, meaning their scattering coefficients are high. The light usually experiences a huge number of scattering events inside a turbid sample. Although the absorption and scattering process by a single component can be perfectly solved using classic electromagnetic theory, such multiple scattering processes greatly complicate the problem. Conventionally, the light–tissue interaction can be described using the radiative transport equation (RTE) (Chandrasekhar, 1960). When scattering is significantly stronger than optical absorption, the diffusive approximation (Prahl, 1995) can be applied to simplify the transport equation into the diffusion equation. The diffusion equation can be solved analytically and is widely applied in many *in vivo* experimental configurations (Farrell et al., 1992). However, numerical methods, such as Monte Carlo techniques, are necessary to solve the RTE under general conditions. The Monte Carlo method has been considered the most accurate way to solve the RTE and has achieved tremendous success in our understanding of light–tissue interactions (Wang et al., 1995; Yao and Wang, 2000).

To develop effective optical methods for inspection of meat and meat analogs, it is important to understand how light interacts with them. In particular, most optical methods operate in the reflectance detection mode, that is, measuring the light backscattered or back reflected from the sample. Therefore, we need to understand what kind of information about sample optical properties is carried by which features of the optical reflectance. In the next few sections, we will introduce the theoretical and experimental techniques that are currently used to understand this very complicated process. In our discussion, the meat or meat analog samples will be treated as optical materials characterized with scattering coefficient (μ_s), the absorption coefficient (μ_a), and optical refractive index n. The scattering process by a single scatterer is also characterized by the scattering phase function, which represents the probability a photon is scattered into a particular direction. The average cosine of the scattering angle θ is defined as the anisotropy factor $g = \cos(\theta)$. Under diffusion approximation, the scattering coefficient and anisotropy factor are combined into the "reduced scattering coefficient" $\mu'_s = \mu_s(1-g)$.

10.2 LIGHT PROPAGATION IN MEAT ANALOGS

High-moisture (40%–80%) extrusion technology has achieved great success in texturizing vegetable proteins into fibrous products that resemble chicken or turkey breast meat in appearance and eating sensation. The key feature of such "meat analogs" is their abundant fibrous content although not all ingredients have been converted into a fibrous structure. To this regard, a simple model for a meat analog can be constructed as a mixture of fibers and nonfibers (Figure 10.1). Cylindrical objects can be conveniently used to simulate fibers, whereas spherical objects can be used to simulate nonfibrous scatterers such as nontexturized soy proteins.

Light has a different scattering probability at different incident angles in relation to a cylinder. In other words, the scattering coefficient of a cylinder is different when the light is incident in parallel and perpendicular to the cylinders. Such incident angle-dependent scattering process of the cylindrical objects is the major difference between fibrous samples and other "isotropic" samples where the light scattering probability is independent of the incident angle.

10.2.1 ANISOTROPIC DIFFUSION THEORY

The diffusion approximation (Wang and Wu, 2012) becomes valid if the reduced scattering coefficient μ'_s is much larger than the absorption coefficient μ_a and the distance between the incident light and the point of measurement is greater than the transport mean free path $1/\mu'_s$. The first requirement is true for most biological tissues in the red and near-infrared (NIR) wavelength region. Under diffusion approximation, the RTE can be simplified into the diffuse theory:

$$\frac{1}{c}\frac{\partial \Phi(\mathbf{r},t)}{\partial t} - D\nabla^2 \Phi(\mathbf{r},t) + \mu_a \Phi(\mathbf{r},t) = S(\mathbf{r},t), \qquad (10.1)$$

FIGURE 10.1 Optical model for simulating light propagation in a meat analog sample. Cylindrical objects are used to simulate fibers, whereas spherical objects are used to simulate nonfibrous components.

where c is the light speed in the turbid medium; $\Phi(\mathbf{r}, t)$ is the optical fluence rate at location $\mathbf{r}$ and time t; μ_a is the absorption coefficient; and S is the light source. D is the diffuse coefficient given by

$$D = \frac{1}{3\mu'_t} = \frac{1}{3(\mu_a + \mu'_s)}. \tag{10.2}$$

Under steady state, the time-resolved Equation 10.1 can be simplified as

$$-D\nabla^2 \Phi(\mathbf{r}) + \mu_a \Phi(\mathbf{r}) = S(\mathbf{r}). \tag{10.3}$$

For anisotropic samples shown in Figure 10.1, the diffuse coefficient is direction dependent. It can be conveniently represented using a tensor format along the three axes:

$$D = \begin{pmatrix} D_x & 0 & 0 \\ 0 & D_y & 0 \\ 0 & 0 & D_z \end{pmatrix}. \tag{10.4}$$

In the specific arrangement shown in Figure 10.1 where the fibers are all aligned in the y-axis, we have $D_z = D_x \ne D_y$. Therefore, the "isotropic" diffusion equation in Equation 10.3 becomes the "anisotropic" diffusion equation:

$$-\left(D_x \frac{\partial^2}{\partial x^2} + D_y \frac{\partial^2}{\partial y^2} + D_z \frac{\partial^2}{\partial z^2}\right)\Phi(\mathbf{r}) + \mu_a \Phi(\mathbf{r}) = S(\mathbf{r}). \tag{10.5}$$

This equation can be transformed back to the "isotropic" equation if we perform a coordinate transformation using (Johnson et al., 2002; Heino et al., 2003)

$$y' = y\sqrt{\frac{D_x}{D_y}}. \tag{10.6}$$

Note that the reflectance derived based on Fick's law (Wang and Wu, 2012) from the "isotropic" diffusion equation is a function of distance $r = \sqrt{x^2 + y^2}$ at the sample surface. The "equal-intensity" contour of the optical reflectance can be calculated by linking all surface locations with the same reflectance intensity. Therefore, the "equi-intensity" profile from the "isotropic" diffusion equation is $r = \sqrt{x^2 + y^2} = \text{constant}$, which represents a circle. However, such profile from the "anisotropic" diffusion equation becomes

$$x^2 + y^2 \frac{D_x}{D_y} = \text{constant}. \tag{10.7}$$

Equation 10.7 represents an ellipse. The ratio of the two elliptical axes is the square root of the ratio of the diffuse coefficients. If the optical absorption coefficient is much smaller than scattering, the shape of this elliptical "equi-intensity" profile is determined approximately by $\sqrt{\mu'_{s\_x}/\mu'_{s\_y}}$. Because the scattering coefficient along the fiber μ'_{s_y} is less than that perpendicular to the fiber (the μ'_{s_x}), the elliptical equi-intensity profile is elongated along the fibers.

10.2.2 Monte Carlo Simulation of Light Propagation in Fibrous Tissue

If the conditions for diffusion approximation are not met, the diffuse theory becomes invalid for the description of light propagation in tissue. However, the Monte Carlo method can still be applied. To apply Monte Carlo, we need to consider first the scattering process from a single scatterer. Since there are two kinds of scatterers in a fibrous sample (Figure 10.1), both cylindrical and spherical particles need to be considered. The conventional Henyey–Greenstein function (Wang and Wu, 2012) can be used as the scattering phase function for background spherical particles. The scattering coefficient of the background isotropic scatterers is denoted as $\mu_{s,b}$.

The scattering coefficient $\mu_{s,c}$ of the cylinders can be solved using electromagnetic theory (Yousif and Boutros, 1992). Without losing generality, we assume the incident light is within the x–z plane. As illustrated in Figure 10.2, ξ is the incident angle with the fiber (along the z-axis); ϕ is the scattering angle within the x–y plane; and θ is the scattering angle between an incident light and the corresponding scattered light.

The scattering coefficient of the cylinder at an incident angle ξ can be calculated as

$$\mu_{s,c}(\xi) = 2r * c_A * Q_s(\xi), \tag{10.8}$$

where r is the cylinder radius, c_A is the concentration of cylinders (mm^{-2}), and $Q_s(\xi)$ is the scattering efficiency and can be calculated using the algorithm described by Yousif and Boutros (1992). The volume density of the cylinders can be calculated as $c_A \times \pi r^2$. Because the incident angle ξ is defined as the angle formed by the

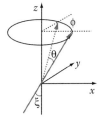

FIGURE 10.2 Illustration of various angles involved in light scattering by a cylinder. The cylinder is aligned with the z-axis. The incident light has an angle ξ in relation to the cylinder. The light is scattered to the direction (θ, ϕ) as illustrated.

incident light and the cylinder axis, $\xi = 0°$ indicates light incident along the cylinder and $\xi = 90°$ indicates light incident perpendicularly to the cylinder. The scattering anisotropy g_c can be derived as

$$g_c(\xi) = \langle \cos\theta \rangle = \cos^2\xi + \sin^2\xi \frac{\int_0^\pi \cos\varphi\, p(\varphi,\xi)\sin\varphi\, d\varphi}{\int_0^\pi p(\varphi,\xi)\sin\varphi\, d\varphi}. \qquad (10.9)$$

The scattering phase function $p(\phi,\xi)$ can be calculated using Yousif and Boutros's method (Yousif and Boutros, 1992). The reduced scattering coefficient of the cylinders is calculated as $\mu'_{s,c} = \mu_{s,c}(1 - g_c)$. The total reduced scattering coefficient of the sample is $\mu'_s = \mu'_{s,c} + \mu'_{s,b}$. Figure 10.3 shows examples of reduced scattering coefficients obtained in cylinders of different sizes.

To simulate light propagation in the whole sample, we need to trace the photons in all scattering events (Wang et al., 1995; Shuaib and Yao, 2010). The movement of a photon packet inside the sample can be determined based on the local scattering and absorption coefficients. At each scattering event, a sampling method is applied to determine whether the photon is scattered by the cylinders or the background spherical particles based on their corresponding scattering coefficients $\mu_{s,c}$ and $\mu_{s,b}$ (Shuaib and Yao, 2010). If the photon is scattered by spherical particles, the standard procedure (Wang et al., 1995) using the Henyey–Greenstein phase function is applied to determine the scattering direction.

If the photon is scattered by cylinders, the new photon direction is determined by sampling the cylindrical scattering phase function $p(\varphi,\xi)$. The cylindrical scattering phase function can be precalculated at different incident and scattering angles and read into the Monte Carlo program at the beginning of the simulation. The incident direction of a photon is then used as an index to retrieve the stored cylinder phase function. The new photon scattering angle ξ can then be sampled from the phase function. The simulation process continues until a photon is completely absorbed in

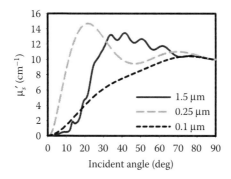

FIGURE 10.3 Example reduced scattering coefficients obtained from cylinders of different sizes. The densities of these aligned infinite-long cylinders are 2.3×10^3, 8.2×10^5, and 4.4×10^3 mm^{-2} for cylinders with radii of 1.5, 0.25, and 0.1 µm, respectively.

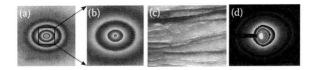

FIGURE 10.4 (a) Sample reflectance images obtained in anisotropic media. All image pixels with the same reflectance are shown in the same grayscale color to represent the equi-intensity contours. The cylinders are aligned with the horizontal direction. (b) A zoom-in image of the central portion in (a). (c) A meat analog sample produced from high-moisture soy extrusion. (d) The experimentally measured reflectance of the sample shown in (c).

the sample or exits the sample from the surface. The physical location and weight of the back-reflected photon are then stored in arrays to construct the images of the diffuse reflectance.

Figure 10.4a shows an example image of the diffuse reflectance in a fibrous sample. In the simulation, we assume a pencil beam is incident perpendicularly upon the semi-infinite sample. The backscattered light at the sample surface is directly imaged by a charge-coupled device (CCD) camera. The cylinders in the simulation have a radius of 1.5 µm, a refractive index of 1.46, and a volume density of $\rho = 9.78\%$. The scatter coefficient and anisotropy for the background spherical scatterers are 30.0 cm^{-1} and 0.8, respectively. The sample had an absorption coefficient of 0.01 cm^{-1}. As expected, the equi-intensity profiles in Figure 10.4a is in elliptical shape and the long axis of the ellipse is aligned with the fiber orientation.

Figure 10.4b shows the magnified view of the center portion of the reflectance image in Figure 10.4a. The equi-intensity profiles at the center deviate from the elliptical shape and have different orientations (Shuaib and Yao, 2010) due to single scattering events close to the incidence where the diffusion approximation becomes invalid. Figure 10.4c shows a meat analog sample obtained using high-moisture soy extrusion (Ranasinghesagara et al., 2006) and Figure 10.4d shows the measured reflectance. The black horizontal object in Figure 10.4d is the shadow of an optical fiber used to deliver the incident light. The equi-intensity contours of the optical reflectance in this meat analog are in elliptical shape at a distance far away from the light incident. Their orientations are visually aligned with the fiber formation shown in Figure 10.4c.

10.2.3 Conclusion

In summary, light propagation in meat analogs can be modeled using anisotropic diffusion theory and Monte Carlo simulation. Both methods indicate that the equi-intensity distribution of the optical reflectance is in elliptical shape when a point light source is used. Monte Carlo simulation further indicates that the elliptical shape is only valid when the measurement location is far away from the light incident point. Such optical reflectance patterns have been confirmed in real meat analog samples. The exact shape of the ellipse is related to the ratio of optical scattering coefficients parallel and perpendicular to the muscle fibers. Because the scattering coefficient parallel to the fiber is equivalent to the background nonfibrous components, the shape of the equi-intensity profile can be used as an indicator of the relative amount of fiber formation in a meat analog sample.

10.3 LIGHT PROPAGATION IN SKELETAL MUSCLE

Skeletal muscles are also fibrous tissue as in meat analogs. However, muscle fibers have the periodic sarcomere structure that does not exist in meat analogs. As an illustration, Figure 10.5 shows an optical model for a meat sample. In addition to muscle fibers, the cellular and subcellular organelles, myofibrils, connective tissues (epimysium, perimysium, and endomysium), and adipose tissues also contribute to optical scattering in meat. Light scattering by connective tissues or muscle fibers has been studied intensively in the past. Collagen scattering can be well modeled using the classic Mie scattering theory (Van de Hulst, 1981; Saidi et al., 1995). Optical absorption occurring at visible (VIS) to NIR (<1 µm) wavelengths is mainly due to myoglobin and its derivatives. For wavelengths greater than approximately 1 µm, water and adipose tissues become significant for light absorption.

As the individual sarcomere shows significant microscopic complexity, the scattering from muscle fibers is much more complicated (Yeh et al., 1980). Figure 10.5b shows the sarcomere model proposed by Thornhill et al. (1991). Two major proteins known as actin and myosin are contained in a sarcomere. The myosin filaments are located in the A-band and actin is located in both the I- and A-bands. The segment of the A-band with only myosin proteins is referred to as the H-zone. Z- and M-lines are located in the middle of the I-band and middle of the H-zone, respectively. The sarcomere length is usually defined as the distance between two consecutive Z-lines. The "overlap region" refers to the part of the A band where both actin and myosin are present.

Because the alternating I- and A-bands have different optical refractive indices, the periodic sarcomere structure resembles an optical phase grating that is used widely in spectroscopy and will diffract incident light into different directions. Such diffraction is sensitive to the sarcomere length. Since its first discovery in frog and rabbit skeletal muscles more than 100 years ago, light diffraction technique has been improved significantly by incorporating modern imaging and electronic processing techniques and has been a standard tool for measuring sarcomere length (Lieber et al., 1984; Cutts, 1988).

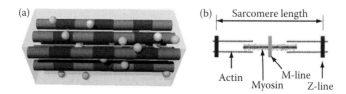

FIGURE 10.5 (a) An optical model for simulating light propagation in meat. Muscle fibers have the periodic sarcomere structure. Spherical objects represent nonfibrous scattering components such as cellular structures, collagen, adipose tissues, and others. (b) A microscopic sarcomere model proposed by Thornhill et al. (1991).

10.3.1 OPTICAL DIFFRACTION BY A SINGLE MUSCLE FIBER

As no simple analytic solution is available for studying light interaction with muscles, Monte Carlo simulation remains a practical solution. However, the light interaction with a single muscle fiber needs to be solved before a Monte Carlo model can be developed to simulate light propagation in a whole muscle. In particular, the scattering phase function has to be computed first. The three-dimensional (3D) coupled wave theory (Moharam and Gaylord, 1983) has been applied to study light diffraction in muscle fibers (Ranasinghesagara and Yao, 2008). For a muscle fiber with a thickness of d, the optical fields can be constructed within the following three regions. Region 1 represents the space in the incident side; region 2 is within the muscle fiber; and region 3 represents the space in the transmission side. The optical field in region 1 is the summation of incident light and back-reflected light:

$$\mathbf{E}_1 = \sum_i \mathbf{R}_i \exp[-j\mathbf{k}_{1i} \cdot \mathbf{r}] + \mathbf{E}_0 \exp(-j\mathbf{k}_1 \cdot \mathbf{r}), \tag{10.10}$$

where $\mathbf{E}_0$ is the incident amplitude; $\mathbf{R}_i$ is the electrical field of the ith reflected light; and $\mathbf{k}_{1i}$ is the wave vector. Similarly, the transmitted optical field $\mathbf{E}_3$ is the summation at all diffraction orders:

$$\mathbf{E}_3 = \sum_i \mathbf{T}_i \exp[-j\mathbf{k}_{3i} \cdot (\mathbf{r} - \mathbf{d})]. \tag{10.11}$$

The electrical field $\mathbf{E}_2$ inside the muscle fiber is the summation of space harmonics $\mathbf{S}$:

$$\mathbf{E}_2 = \sum_i [S_{xi}(z)\hat{x} + S_{yi}(z)\hat{y}]\exp[-j\sigma_i \cdot \mathbf{r}], \tag{10.12}$$

where $\sigma_i = k_{xi}\hat{x} + k_{yi}\hat{y}$. The corresponding magnetic field can be expressed in space harmonics $\mathbf{U}$:

$$\mathbf{H}_2 = \sqrt{\frac{\varepsilon_0}{\mu_0}} \sum_i [U_{xi}(z)\hat{x} + U_{yi}(z)\hat{y}]\exp[-j\sigma_i \cdot \mathbf{r}], \tag{10.13}$$

where ε_0 and μ_0 are the permittivity and permeability of the free space, respectively. The "coupled wave" equations in terms of $\mathbf{S}$ and $\mathbf{U}$ can be derived by substituting the above equations in Maxwell's equations and applying the permittivity distribution for the specific sarcomere structure: $\varepsilon(x) = n^2(x)$. The $\varepsilon(x)$ can be conveniently represented in the Fourier series as

$$\varepsilon(x) = \sum_n \hat{\varepsilon}_n \exp(jhKx), \tag{10.14}$$

where $\mathbf{K} = 2\pi/\Lambda$ is the grating vector with a grating period of Λ. The nth Fourier coefficient is given by

$$\hat{\varepsilon}_n = \frac{1}{\Lambda} \int_0^\Lambda \varepsilon(x) \exp(-jn\mathbf{K}x) dx. \tag{10.15}$$

The derived coupled wave equation can be solved by applying the Floquet condition and continuous boundary conditions as shown by Ranasinghesagara and Yao (2008). The final diffraction efficiencies of each diffraction order can be calculated as

$$DE_i = \text{Re}\left(\frac{k_{z3i}}{k_z}\right) |T_i|^2, \tag{10.16}$$

where k_z and k_{z3i} are the z components of the incident wave vector and the ith diffracted wave vector. Both the transverse electric (TE) or transverse magnetic (TM) polarized light can be calculated and the two diffraction efficiencies can be averaged to obtained diffraction efficiencies for an unpolarized incident light.

Ranasinghesagara and Yao (2008) showed that the 3D coupled wave theory can accurately simulate light diffraction by a single muscle fiber under various conditions. Figure 10.6 shows an example of the sarcomere length effect on diffraction efficiency. The experimental data are extracted from the study published by Baskin et al. (1979) in single frog *semitendinosus* muscle fibers. In the simulation, the laser wavelength is 633 nm and the muscle fiber has a thickness of 80 µm. The refractive indices of the sarcomere used are from Thornhill et al. (1991). The experimental results are in arbitrary unit because no absolute values are available. As shown in Figure 10.6, the simulation results have an excellent agreement with the experimental results.

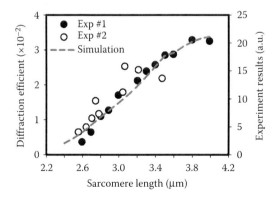

FIGURE 10.6 Effect of sarcomere length on the first-order optical diffraction from a single muscle fiber. The experimental data are extracted from Baskin et al. (1979) in a single frog *semitendinosus* muscle fiber.

10.3.2 Optical Reflectance in Whole Muscle

The calculated diffraction efficiency of a single muscle fiber can then be incorporated into a Monte Carlo model to simulate the light propagation in a whole muscle sample (Ranasinghesagara and Yao, 2007). Similar to the procedures described in Section 10.2.2, the diffraction coefficients can be precalculated and then used in Monte Carlo simulation. A photon packet subject to regular background scattering is processed using the Henyey–Greenstein phase function as in the conventional Monte Carlo model (Wang and Wu, 2012). Between two consecutive scatterings, the photon interacts with the muscle fibers and may be diffracted instead of following a ballistic trajectory. Using the precalculated diffraction data, the diffraction efficiency and angle can be computed based on the photon's incident angle and the pathlength between two regular scattering events. The actual diffraction angle can be sampled based on the relative efficiencies of all diffraction orders. In addition, different background scattering probabilities can be assigned depending on whether the photon propagation is parallel or perpendicular to the muscle fibers. The above procedure is repeated until a photon is fully absorbed or exits from the sample. The backscattering photons are recorded to construct the reflectance image as in Section 10.2.2.

Figure 10.7a shows an optical reflectance image obtained in a piece of *Sternomandibularis* beef muscle (Ranasinghesagara and Yao, 2007). The incident light is a pencil beam at normal incidence to the muscle sample surface. The image is displayed using a grayscale colormap to reveal the equi-intensity profiles, which clearly are not elliptical as in the meat analog samples. Figure 10.7b shows the reflectance image obtained using the Monte Carlo simulation as described above. The Thornhill sarcomere model was used in the simulation. The scattering coefficient along muscle fibers is $\mu_s = 13.4$ cm^{-1}, whereas the scattering coefficient perpendicular to the muscle fibers is $\mu_s = 30$ cm^{-1}. The other optical properties used are anisotropy $g = 0.94$ and absorption coefficient $\mu_a = 0.1$ cm^{-1}.

The Monte Carlo results are in excellent agreement with the experimental observation (Figure 10.7). Ranasinghesagara and Yao (2007) discovered that the nonelliptical equi-intensity profile can be well fitted using the following curve:

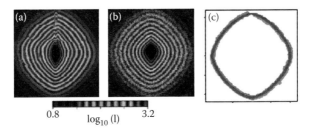

FIGURE 10.7 (a) The optical reflectance image obtained in a piece of beef *Sternomandibularis* muscle. (b) The reflectance image simulated using Monte Carlo simulation. (c) The curve-fitting result (the thick line) using Equation 10.17 on the equi-intensity profile (the dots) extracted at 10 mm from the light incidence.

$$f(x,y) = \left(\frac{|x|}{a}\right)^q + \left(\frac{|y|}{b}\right)^q - 1 = 0. \tag{10.17}$$

As shown in Figure 10.7c, Equation 10.17 can perfectly fit the equi-intensity data in the muscle sample. It is interesting to note that Equation 10.17 becomes an ellipse when the parameter q is equal to 2. The ratio of a/b describes the ratio of the two axes parallel and perpendicular to the muscle fibers. A "bias" parameter (Ranasinghesagara and Yao, 2007) can be defined as $B = (b/a)^2$ to help quantify the shape described in Equation 10.17.

The different equi-intensity profile of the optical reflectance obtained in muscle (or meat) can be attributed to the unique sarcomere structure, which is the major difference between true meat and meat analog. It is important to note that the exact equi-intensity profile depends on the relative scattering contribution from the background "isotropic" scattering and the organized muscle diffraction. Figure 10.8 shows that the equi-intensity contours are very different among the three muscles: *M. longissimus dorsi* (LD), *M. psoas major* (PM), and *M. semitendinosus* (ST) muscles (Ranasinghesagara et al., 2010). Such a significant difference is most likely due to the different muscle structure profiles. The PM muscle is known to have the longest sarcomere length leading to a strong optical diffraction away from the fiber orientation. The ST muscle has an equi-intensity profile that is most close to an ellipse, which suggests the sarcomere diffraction is probably overshadowed by strong background scatterings.

It is also important to note that the shape of the equi-intensity profiles change during the rigor process (Figure 10.9). The initial increase in the bias parameter implies sarcomere contracting (Ranasinghesagara and Yao, 2007). Eventually, proteolysis process becomes significant after approximately 10 h and the autolytic enzymes start to break down the sarcomere structure. The disintegration of sarcomere structures weakens the optical diffraction. This process is coincident with the increase of the q parameter after approximately 8 h. At the beginning of rigor mortis, the small q

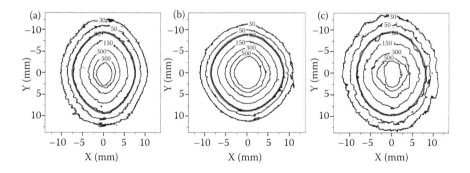

FIGURE 10.8 Equi-intensity contours obtained from (a) *M. longissimus dorsi* (LD), (b) *M. psoas major* (PM), and (c) M. *semitendinosus* (ST) muscles from a *Bos taurus* steer at 1-d postmortem. The fitted parameters are (a) LD muscle: $q = 1.70$, $B = 1.55$; (b) PM muscle: $q = 1.77$, $B = 1.03$; and (c) ST muscle: $q = 1.85$, $B = 1.36$. The y-axis is along the muscle fibers.

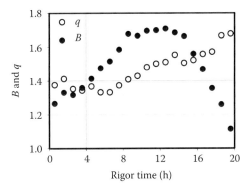

FIGURE 10.9 Change of the equi-intensity profile during the rigor process. The fitting parameters (i.e., bias parameter B and ellipse parameter q) are extracted from the equi-intensity reflectance at 10 mm from the incident light.

indicates a rhombus-like profile. After 20 h, the equi-intensity becomes closer to an ellipse as a result of the destruction of the sarcomere structure.

10.3.3 CONCLUSION

The sarcomere structure in muscle greatly affects the light scattering processing due to its strong diffraction effect. As explained in the theoretical model and demonstrated in experiment, such a unique structure creates a very different reflectance profile in true muscle from the elliptical pattern observed in meat analogs. In addition, the specific reflectance pattern is also influenced by other muscle structure properties such as the collagen content. The optical reflectance can be quantified using two parameters q and B. These two parameters can be used to monitor the muscle structure changes during the rigor process. Because meat tenderness is mainly attributed to the same muscle structure properties, the optical scattering properties of the meat and the resulting reflectance profiles may be used to characterize meat tenderness.

10.4 OPTICAL CHARACTERIZATION OF FIBER FORMATION IN MEAT ANALOG

As discussed in Section 10.2, the elliptical optical reflectance in meat analogs is related to the different optical scattering coefficients parallel and perpendicular to the fiber orientation. The scattering coefficient along the fiber direction is smaller than that perpendicular to the fiber. Photons can travel longer distances along fibers, leading to an elliptical reflectance with the major axis oriented along the fiber formation direction. For samples with less fiber formation, the scattering probabilities are nearly isotropic, which leads to a circular reflectance profile. Therefore, analyzing the elliptical shape may be used to quantify the fiber formation in a meat analog sample.

To implement this method, we need to ensure a small incident point using a laser or fiber optical delivery. In their first experimental demonstration, Ranasinghesagara et al. (2006) used a red LED (λ = 680 nm) coupled into a 400 µm optical fiber as the light source. The incident light was delivered at a 45° angle toward the sample without touching the sample surface. Such an oblique incident scheme minimized the image obstruction from the incident optical fiber. Its effect on light reflectance was insignificant when analyzing reflectance far away from the incident point. The incident light was approximately parallel to the extrusion direction. The optical reflectance image was acquired by using a CCD camera mounted above the sample. Multiple images of the same sample can be acquired and averaged to improve signal-to-noise ratio.

To quantify the reflectance profiles, an image processing algorithm needs to be developed to analyze the reflectance images and derive the characteristic parameters of the elliptical equi-intensity profiles. In Ranasinghesagara et al.'s study (Ranasinghesagara et al., 2006), the pixel locations in each reflectance image were transformed to the Cartesian coordinate system with the incident location chosen as the origin. The small area of the image obstructed by the incident optical fiber was removed from all images. A point on the x-axis was selected and its pixel intensity was used to find all image pixels with the same intensity. In order to compensate for intensity variations caused by noise, a ±1% margin was allowed when searching for these pixels. The corresponding equi-intensity pixels were identified and their locations, $(x_1, y_1), (x_2, y_2) \ldots (x_n, y_n)$ were then used to find the best-fit ellipse.

The direct least square error ellipse fit technique (Fitzgibbon et al., 1999) is a reliable method to find the best-fitting ellipse. A general polynomial equation can be used to describe an ellipse:

$$G(\mathbf{P},\mathbf{U}) = \mathbf{P} \cdot \mathbf{U} = ax^2 + bxy + cy^2 + dx + ey + f = 0, \quad (10.18)$$

where vectors $\mathbf{P} = [a, b, c, d, e, f]^T$ and $\mathbf{U} = [x^2, xy, y^2, x, y, 1]$ are conic parameters. The constraint $4ac - b^2 = 1$ is required for an ellipse. The quantity $G(\mathbf{P}, \mathbf{U})$ represents the algebraic distance from a point (x, y) to the ellipse. The summation of algebraic distances from all equi-intensity points $\sum_{i=1}^{n} G_i(\mathbf{P},\mathbf{U})^2$ can be minimized to find the best-fitted ellipse. After the conic parameter $\mathbf{P}$ is obtained from the fitting, the azimuth angle (θ) or tilt angle of the ellipse is derived as

$$\theta = \begin{cases} \dfrac{\pi}{4} & \text{if } a = c \\ 0.5\tan^{-1}\left(\dfrac{b}{a-c}\right) & \text{otherwise} \end{cases}. \quad (10.19)$$

And the "bias parameter" B is defined as the square of the long axis L_L divided by the short axis L_S and can be calculated by using the following equation:

$$B = \left(\frac{L_L}{L_S}\right)^2 = \frac{K(d\cos\theta + e\sin\theta)^2 + (-d\cos\theta + e\sin\theta)^2 - 4f}{(d\cos\theta + e\sin\theta)^2 + (1/K)(-d\cos\theta + e\sin\theta)^2 - 4f}, \quad (10.20)$$

where the parameter K is calculated from the conic parameters:

$$K = \frac{a\sin^2\theta - b\sin\theta\cos\theta + c\cos^2\theta}{a\cos^2\theta + b\sin\theta\cos\theta + c\sin^2\theta}. \tag{10.21}$$

Figure 10.10 shows the results obtained in an extrudate of poor fiber formation and one with good fiber formation. The meat analog samples were produced using a twin-screw food extruder (MPF 50/25, APV Baker Inc., Grand Rapids, Michigan). The ingredients used were 60% soy protein isolate (Profam 974, ADM, Decatur, Illinois), 35% wheat gluten (MGP Ingredients, Atchison, Kansas), and 5% wheat starch (MGP Ingredients, Atchison, Kansas). Sample no. 1 was extruded at 65% moisture content and 165°C temperature, whereas sample no. 2 was extruded at 55% moisture content and 182°C temperature. Digital pictures of the extrudates (surface peeled) are also displayed for reference. The raw reflectance images were shown in grayscale pseudo color (in log scale) to reveal the equi-intensity profiles. Elliptical fitting was obtained for equi-intensity contours at 5.0 and 6.0 mm from the light incidence.

The calculated bias parameter B has a good match with visual inspections. To provide a quantitative assessment of this technique, we compared the B values with the fiber index measured directly from peeled samples using the image processing algorithm reported previously (Ranasinghesagara et al., 2005). After reflectance imaging measurements, the extrusion samples were hand peeled to reveal the internal fibrous structures. Digital images of the peeled samples were then acquired and analyzed using the Hough transform-based processing method that can quantify fiber formations. Figure 10.11 shows that a good correlation exists between the bias parameter and the fiber index.

The major advantage of the "photon migration"-based method is noncontact and nondestructive. It does not need sample preparation and all measurements are conducted directly from the sample surface. Therefore, it is ideally positioned for real-time online monitoring. Ranasinghesagara et al. (2009) recently implemented this

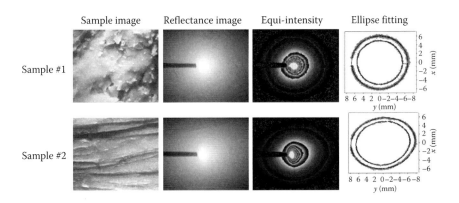

FIGURE 10.10 Images of extrusion samples with limited fiber formation (sample no. 1, bias parameter $B = 1.22$) and with good fiber formation (sample no. 2, bias parameter $B = 1.77$).

Light Propagation in Meat and Meat Analog

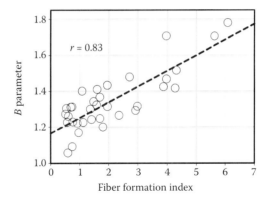

FIGURE 10.11 Correlation between the bias parameter B and the fiber index measured in meat analogs.

method using a laser scanning system and demonstrated its capability for real-time measurements of fiber formation in meat analogs.

Figure 10.12 shows a schematic diagram of the laser scanning system used by Ranasinghesagara et al. (2009). The light source was a common nonpolarized He–Ne laser at 633 nm and with a 10 mW output laser power. A $f = 450$ mm lens was used to slightly focus the laser to a small beam size on the sample surface. A reflection mirror was used to redirect the laser to a mirror with a center hole. The incident

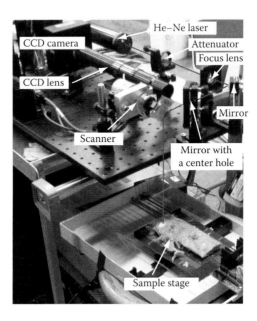

FIGURE 10.12 Digital picture of the real-time laser scanning system. The system is mounted on top of a simulated conveying belt moving at a speed of 10 mm/s.

light can pass through the center hole, and the mirror surface can reflect the backscattered light from the sample to the imaging CCD camera. A two-dimensional (2D) galvanometer scanner (M2, General Scanning, Billerica, Massachusetts) was used to scan the incident light across the extrusion sample by rotating the galvanometer. The beam diameter of the incident laser beam was approximately 1.0 mm at the sample surface. The CCD camera captured the reflectance image at a resolution of 512 × 512 pixels. To eliminate the room light effect, a band-pass optical filter at 633 nm (bandwidth = 2.4 nm, N47-494, Edmund optics, Barrington, New Jersey) was inserted into the imaging lens of the CCD camera.

The CCD image acquisition was synchronized with the 2D scanner and different scanning patterns can be achieved. For example, the scanner can start at one side of the extrudates and scan the laser toward to other side. The scanner can then reverse the direction when reaching the sample boundary, and scan backward toward the initial position. This process can be repeated until stopped by the operator via a software interface. At every scanning location, the reflectance image can be acquired and the image analysis procedure can be carried out as described earlier. The obtained bias parameter B represents the fiber formation at this particular scanning position. Because the extrudate is continuously moving on the extrusion machine, the aforementioned scanning mode creates a "zig-zag" pattern on the sample surface. However, if the image is acquired only during the forward scanning period, the scanning forms a series of parallel lines at the sample surface. Either way, a map of the fiber formation in the entire sample can be constructed by visualizing all B-parameters at the sample surface at all scanning locations. In addition, since the long axis of the ellipse is aligned with the fiber orientation, an orientation map can be simultaneously obtained to examine the fiber orientation change in the entire meat analog sample.

Figure 10.13 shows example measurement results of a freshly extruded meat analog sample (Ranasinghesagara et al., 2009). The image represents the fiber formation (the bias parameter B) using different grayscale colors and the fiber orientation using vector plot. The meat analog samples were produced using a co-rotating, twin-screw food extruder (MPF 50/25, APV Baker Inc., Grand Rapids, Michigan) with a smooth barrel and a length/diameter ratio of 15:1 (Ranasinghesagara et al., 2009). The ingredients used were 57% soy protein isolate (Profam 974, ADM,

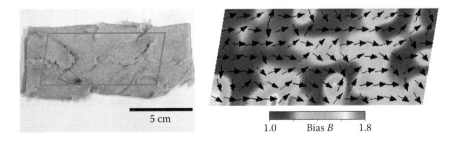

FIGURE 10.13 Bias parameter B and orientation maps of a fresh extrusion sample.

Decatur, Illinois), 38% wheat gluten (MGP Ingredients, Atchison, Kansas), and 5% wheat starch (MGP Ingredients, Atchison, Kansas). The samples were extruded at 65% moisture content and 165°C temperature. During the imaging, the sample was moving at 10 mm/s on a simulated convey belt. Figure 10.13 shows that variations in fiber formation and fiber orientation can be clearly visualized.

In conclusion, the backscattered reflectance images from extruded samples can be used to assess fiber formation in meat analog products. By numerically fitting the equi-intensity reflectance profiles, the calculated bias parameter B showed a good correlation with the degree of fiber formation in extrudates. This method is noncontact, nondestructive, and simple to implement, and requires no sample preparation. By implementing in a laser scanning system, this method can be used for online real-time measurements. Ranasinghesagara et al. (2009) demonstrated in a prototype system that can achieve a high speed of 60 measurements per second. By scanning over the sample surface, the system provides an accurate assessment of an overall fiber formation and orientation distribution in the entire meat analog product.

10.5 OPTICAL CHARACTERIZATION OF BEEF TENDERNESS

As described in Section 10.1.1, meat tenderness mainly depends on meat structural properties such as sarcomere length, collagen, and processes that alter these properties such as proteolysis. Meanwhile these structural traits also determine tissue scattering properties. Therefore, it is worth exploring the potential of using optical scattering characteristics to predict meat tenderness.

Many techniques have been developed to measure optical properties in bulk tissue. Oblique-incidence reflectometry (Wang and Jacques, 1995) is a simple way to measure the absorption coefficient and reduced scattering coefficient. Xia et al. (2007a) implemented this method using a low-cost spectroscopy system. They used two 400 μm optical fibers, one to deliver light to the meat sample at an oblique incident angle of 40° and the other fiber was used to detect backscattering light at 90° to the sample surface. A 20 W broadband halogen light (HL-2000-FHSA-HP, Ocean Optics Inc., Dunedin, Florida) was used as the light source. The detection fiber was connected to a spectrometer (USB2000, Ocean Optics Inc., Dunedin, Florida) to measure the reflectance spectra from 450 to 950 nm. The detection fiber scanned over the sample via a translational stage to measure reflectance at 13 positions. These scanning positions were from 9.0 to 6.5 mm on one side of the incident fiber and 4.0 to 7.0 mm on the other side of the incident fiber with an interval of 0.5 mm. As discussed in the previous section, the light scattering by a fibrous object is stronger perpendicular to the fiber orientation. Measuring optical scattering at an orientation perpendicular to the muscle fibers may better capture the fiber properties (Xia et al., 2006, 2007a, 2008a, b).

According to the theory of oblique-incidence reflectometry (Wang and Jacques, 1995; Xia et al., 2007a), two parameters (Δx and μ_{eff}) can be derived by fitting the spatial resolved measurements with the optical diffuse equation. μ_{eff} is referred to as the "effective attenuation coefficient": $\mu_{eff} = \sqrt{3\mu_a(\mu_a + \mu'_s)}$. And Δx is the lateral offset between the diffuse center and the light incident position. The absorption coefficient (μ_a) and reduced scattering coefficients (μ'_s) can then be derived as

$$\mu_a = \frac{\mu_{\text{eff}}^2 \Delta x}{3 \sin \theta_t}, \quad (10.22)$$

$$\mu'_s = \frac{\sin \theta_t}{\Delta x} - 0.35 \mu_a, \quad (10.23)$$

where θ_t is the incident angle (28°) of light inside the muscle sample and is calculated from the oblique incident angle, the refractive index of the cover glass (1.52), and the meat sample (1.37). The absorption and scattering coefficients represent the probabilities of a photon being absorbed and scattered inside the sample. The reduced scattering coefficient μ'_s includes the effect of scattering anisotropy $\mu'_s = \mu_s(1-g)$. The absorption and scattering coefficients of a sample are usually independent of each other.

Figure 10.14 shows an example measurement in a *Semimembranosus* meat sample at 721 nm wavelength. The zero distance in the figure represents the light incident point. As expected, the reflectance was higher when measured at positions closer to light incidence. All measurements appeared to be symmetric around the "diffuse center," which is the effective location inside the sample derived from the diffusion theory (Wang and Jacques, 1995). Owing to the oblique incident, the diffuse reflectance center deviated from the incident point by Δx. The dashed line represents a diffusion fit (Xia et al., 2007a), which fitted very well to the experimental data.

As discussed in Section 10.3, sarcomere structure plays an important role in modulating light propagation in muscle. Figure 10.15 shows the scattering coefficients measured from *psoas major* muscles of different sarcomere lengths (Xia et al., 2006). To make samples of different sarcomere lengths, pre-rigor muscles

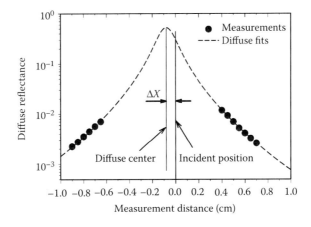

FIGURE 10.14 Example diffusion fitting of the measured optical reflectance at 721 nm wavelength obtained in a *Semimembranosus* meat sample.

Light Propagation in Meat and Meat Analog 273

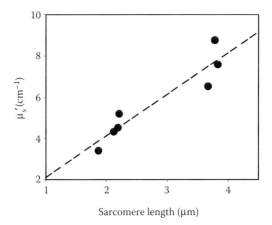

FIGURE 10.15 Effect of sarcomere length on optical scattering coefficients.

were either left on carcasses to induce longer sarcomere lengths, or were removed and subjected to reduced ambient temperatures ("cold shortening") to induce shorter sarcomere lengths. After completion of rigor mortis (~24 h), the reduced scattering coefficients of these samples were measured using oblique-incidence reflectometry (Xia et al., 2006). The actual sarcomere lengths were determined using histology analysis. This experiment (Figure 10.15) indicated optical scattering properties of whole muscle are strongly correlated to sarcomere length.

To further investigate the effect of sarcomere length on light propagation, the reduced scattering coefficients in a whole pre-rigor *sternomandibularis* muscle were measured while stretching the muscle. The measurement location was held constant by stretching the sample equally from both ends. Stretching pre-rigor whole muscles can induce sarcomere length changes. Figure 10.16a shows the μ'_s obtained when the sample was stretched 40% over the original length at the resting state and then released to recover to the original length. The measured reduced scattering

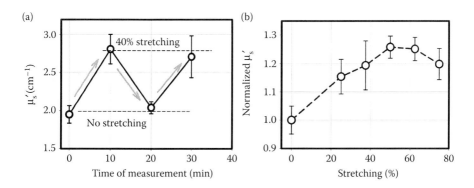

FIGURE 10.16 (a) Effect of repeated muscle stretching and releasing on optical scattering. (b) Optical scattering measured in pre-rigor muscles with a different amount of stretching.

coefficient can reliably track the muscle stretching and releasing process. All measurements in Figure 10.16a were concluded within 2 h of slaughter.

Figure 10.16b shows that the reduced scattering coefficient changes with the amount of muscle stretching. The measured reduced scattering coefficients were normalized to initial measurement at 0% stretching. It is interesting to note that μ'_s increased with stretching up to 50% stretching. As stretching became greater than 50%, the reduced scattering coefficient decreased slightly. This observation is likely caused by the disintegration of sarcomere structures or the increase of the sarcomere disorder under excessive stretching. The damage in sarcomere structure reduced its diffraction power (Leung and Cheung, 1988) and therefore led to a decrease in optical scattering.

Figure 10.17a shows the change of the reduced scattering coefficients and pH value during the rigor process in bovine *M. sternomandibularis* (Xia et al., 2007b). The temporal change of pH was very consistent: it decreased rapidly at the beginning, and then transited to a relatively flat stage. The reduced scattering coefficient of beef muscle decreased with time and slowly reached a steady state. However, the similar trend between pH and optical scattering stopped at approximately 10 h postmortem. The measured reduced scattering coefficient started to increase slowly after 10 h and showed a dramatic rapid increase after approximately 15 h postmortem. The changes in optical scattering agreed very well with the passive tension measured using a force sensor (Xia et al., 2007b) in the same sample during rigor (Figure 10.17b).

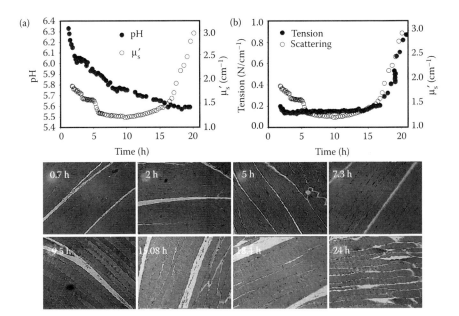

FIGURE 10.17 (a) pH value and reduced scattering coefficient change with time during rigor development in bovine *M. sternomandibularis*. (b) The temporal profiles of passive tension and optical scattering during the rigor process in bovine *M. sternomandibularis*. The lower panel shows a series of histology images obtained from the sample during the test.

During rigor mortis, temporary cross-bridges form between the actin and myosin filaments in the overlapping area and break due to early postmortem adenosine triphosphate (ATP) hydrolysis and subsequent ATP binding. At rigor completion, cross-bridges become "locked" due to the depletion of ATP. Such sarcomere changes may lead to a reduced refractive index difference between the A-band and I-band (Baskin et al., 1986), which results in a smaller scattering efficiency. However, this does not explain the subsequent increase in the optical scattering coefficient.

A careful examination of the histology samples obtained from the same sample (Figure 10.17) revealed significant muscle morphological changes during the rigor process. Muscle fibers appeared well integrated as big bundles during the first several hours until 9.5 h postmortem when fiber separation started to appear. At 15.08 h postmortem, the separation between muscle fibers became very significant. The fiber separation continued to develop and at 24 h postmortem, only threaded fibers can be observed.

It appears that the increase of optical scattering was coincident with the fiber separation revealed in the histology analysis. The separation of muscle fibers is caused by the lateral shrinkage of the myofibrils, which expels water out of the myofibrillar matrix and results in the redistribution of water within the bulk muscle (Bertram et al., 2004). Such structural changes increase the inhomogeneity of refractive index distribution in muscle and therefore increase optical scattering.

The studies discussed above clearly demonstrate that optical scattering is affected by the muscle sarcomere structure as well as the structural change during rigor when the muscle is eventually transformed into meat. Therefore, optical scattering properties have the potential to predict the meat tenderness (Xia et al., 2007a, 2008b). However, optical scattering coefficient depends significantly on measurement orientation due to the strong "anisotropic" structural components in muscle (Xia and Yao, 2007; Fan et al., 2011). To completely evaluate the optical scattering in muscle or meat, measurements at various orientations may be necessary. An alternative approach is to evaluate the entire optical reflectance, including the equi-intensity profiles as described in Section 10.3.

Ranasinghesagara et al. (2010) conducted a comprehensive study to investigate the optical reflectance and its implication for tenderness assessment. In addition to the q and B parameters used to describe the equi-intensity profile (Section 10.3), several other parameters were also evaluated. The total reflectance intensity excluding the surface specular reflection was referred to as "scattering intensity." This parameter is affected by both absorption and scattering properties. In general, higher tissue absorption reduces the total scattering intensity; while a higher scattering increases the scattering intensity. If the tissue absorption remains relatively stable, the "scattering intensity" is a reliable indicator of overall scattering properties in the sample. The spatial decay rate of the reflectance describes how quickly the reflectance is attenuated over the distance from the light incident. This parameter is related to the effective attenuation coefficient μ_{eff} (Equation 10.22). This parameter can be measured both parallel and perpendicular to the muscle fibers.

Different optical reflectance parameters had different correlations with tenderness in different muscles. To measure tenderness, WBSF in cooked meat was measured, following the standard procedure (Ranasinghesagara et al., 2010).

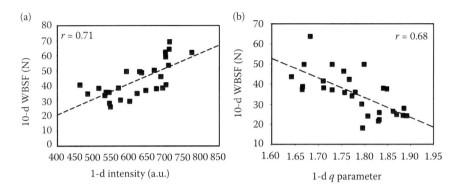

FIGURE 10.18 Correlation between WBSF measured at 10-d and (a) the scattering intensity measured at 1-d in *M. longissimus dorsi* muscles and (b) the q parameter measured at 1-d in *M. psoas major* muscles.

As shown in Figure 10.18, the 1-d scattering intensity was positively correlated with the 10-d WBSF in *M. longissimus dorsi* (LD) muscle ($R^2 = 0.50$). Meanwhile, the 1-d q-parameter showed a negative correlation with 10-d WBSF in *M. psoas major* (PM) muscle ($R^2 = 0.46$). For LD muscle, a tougher meat sample tends to have a higher scattering intensity (Figure 10.18a). Since a higher optical scattering coefficient can increase the scattering intensity, this result is consistent with previous results that WBSF increases with scattering coefficient in LD muscle (Xia et al., 2008a). In PM muscle, a lower WBSF was associated with a higher q-value, which indicates a longer sarcomere. This result was consistent with previous findings that sarcomere length is the dominant factor regulating tenderness in PM muscle (Xia et al., 2008a).

In conclusion, the reduced optical scattering coefficient measured from meat samples is greatly affected by the muscle structural properties such as the sarcomere as well as structural changes during the rigor process. Because optical scattering and meat tenderness are controlled by the same meat structure, meat scattering can be used as a nondestructive means to assess meat tenderness. In addition, the optical reflectance images obtained using a point-incident light provides a more comprehensive evaluation of the overall meat optical scattering. Multiple optical parameters can be extracted to quantitatively describe each reflectance image. The current experimental study suggested that the q parameter and scattering intensity measured at 1-d postmortem may be used to predict the 10-d WBSF in *M. psoas major* and *M. longissimus dorsi* muscles, respectively.

10.6 CONCLUSION

Owing to the large thickness of chunk meat and meat analog products, most optical measurements are limited to the optical reflectance signals at the sample surface. The optical reflectance actually has two components. The first part is from the light directly reflected from the tissue–air surface, usually denoted as the "specular" reflection. Because specularly reflected light does not enter the sample, it carries limited information regarding the internal sample structure. The second part is from that backscattered light that is multiply scattered inside the sample, survives the absorption, and eventually exits from the surface. These photons carry all the information related to the sample optical absorption and scattering properties. In particular, the results described in this chapter illustrate that light scattering in tissue is significantly affected by the sample structures, which are also the main contributors to the eating quality of meat and meat analog products. Ample theoretical and experimental studies support that tissue scattering properties can be assessed by analyzing optical reflectance based on optical scattering theories.

By using a point-like incident light, imaging experiments and light transport theories indicate that distinct reflectance profiles exist in meat and meat analogs. These different optical reflectance profiles are directly related to the different scattering structures inside the sample. Fiber formation is a key quality parameter in meat analogs. The fibrous structures in meat analogs induce elliptical-shaped equi-intensity contours in optical reflectance. The elongation of such elliptical shapes provides a good measure of the degree of fiber formation in meat analogs. In addition to the fibrous structures, muscle fibers have unique repetitive sarcomere units, which produce a strong optical diffraction effect. The overall scattering by the isotropic, fibrous, and diffraction tissue components in meat leads to the unique rhombus-like equi-intensity reflectance patterns, which provide useful information about meat tenderness.

The technologies presented in this chapter can be further extended by incorporating and analyzing the spectral information that can be acquired using a multispectral system. The spectral data not only provide specific information on the chemical composition of the sample, but also carry important information regarding the morphological properties of the scattering components. A challenging issue in optical characterization of meat tenderness is to decipher the relative contributions from various tissue structural components in meat. Both sarcomere and collagen components can affect the optical scattering process, which in turn changes the optical reflectance. However, their effects may be counteractive and intertwined. For example, higher collagen content and a longer sarcomere length can both lead to stronger optical reflectance, whereas their contributions to meat tenderness are very different. Therefore, different optical measures may be used for different types of muscle depending on their specific structural properties. A sophisticated multimodality technology that combines multiple measures, including spatial, temporal, and spectral measurements, may provide a more comprehensive evaluation of meat quality. Nevertheless, optical technologies developed based on advanced light–tissue interaction theories have shown great promise for real-time online assessment of meat and meat analog products. Future technology development will be able to further realize their potential.

REFERENCES

Baskin, R. J., K. P. Roos, and Y. Yeh. Light diffraction study of single skeletal muscle fibres. *Biophysical Journal* 28, no. 1, 1979: 45–64.

Baskin, R. J., Y. Yeh, K. Burton, J. S. Chen, and M. Jones. Optical depolarization changes in single, skinned muscle fibers. Evidence for cross-bridge involvement. *Biophysical Journal* 50, no. 1, 1986: 63–74.

Beattie, R. J., S. J. Bell, L. J. Farmer, B. W. Moss, and D. Patterson. Preliminary investigation of the application of Raman spectroscopy to the prediction of the sensory quality of beef silverside. *Meat Science* 66, no. 4, 2004: 903–913.

Bendall, J. R. The shortening of rabbit muscles during rigor mortis: Its relation to the breakdown of adenosine triphosphate and creatine phosphate and to muscular contraction. *The Journal of Physiology* 114, no. 1–2, 1951: 71–88.

Bertram, H. C., A. Schäfer, K. Rosenvold, and H. J. Andersen. Physical changes of significance for early post mortem water distribution in porcine *M. longissimus*. *Meat Science* 66, no. 4, 2004: 915–924.

Bonnemann, C. G. and N. G. Laing. Myopathies resulting from mutations in sarcomeric proteins. *Current Opinion in Neurology* 17, no. 5, 2004: 529–537.

Borg, T. K. and J. B. Caulfield. Morphology of connective tissue in skeletal muscle. *Tissue and Cell* 12, no. 1, 1980: 197–207.

Bratzler, L. J. Measuring the tenderness of meat by use of the Warner–Bratzler method. MS thesis, Kansas State College, Manhattan, Kansas, 1932.

Breene, W. M. and T. G. Barker. Development and application of a texture measurement procedure for textured vegetable protein. *Journal of Texture Studies* 6, no. 4, 1975: 459–472.

Buege, D. R. and J. R. Stouffer. Effects of pre-rigor tension on tenderness of intact bovine and ovine muscle. *Journal of Food Science* 39, no. 2, 1974: 396–401.

Burkholder, T. J. and R. L. Lieber. Sarcomere length operating range of vertebrate muscles during movement. *Journal of Experimental Biology* 204, no. 9, 2001: 1529–1536.

Campion, D. R., J. D. Crouse, and M. E. Dikeman. Predictive value of USDA beef quality grade factors for cooked meat palatability. *Journal of Food Science* 40, no. 6, 1975: 1225–1228.

Chandrasekhar, S. *Radiative Transfer*. Dover Publications, New York, 1960.

Clark, K. A., A. S. McElhinny, M. C. Beckerle, and C. C. Gregorio. Striated muscle cytoarchitecture: An intricate web of form and function. *Annual Review of Cell and Developmental Biology* 18, no. 1, 2002: 637–706.

Cross, H. R., G. C. Smith, and Z. L. Carpenter. Pork carcass cutability equations incorporating some new indices of muscling and fatness. *Journal of Animal Science* 37, no. 2, 1973: 423–429.

Cutts, A. The range of sarcomere lengths in the muscles of the human lower limb. *Journal of Anatomy* 160, 1988: 79.

Dransfield, E. Intramuscular composition and texture of beef muscles. *Journal of the Science of Food and Agriculture* 28, no. 9, 1977: 833–842.

Egelandsdal, B., J. P. Wold, A. Sponnich, S. Neegard, and K. I. Hildrum. On attempts to measure the tenderness of *longissimus dorsi* muscles using fluorescence emission spectra. *Meat Science* 60, no. 2, 2002: 187–202.

Fan, C., A. Shuaib, and G. Yao. Path-length resolved reflectance in tendon and muscle. *Optics Express* 19, no. 9, 2011: 8879–8887.

Farrell, T. J., M. S. Patterson, and B. Wilson. A diffusion theory model of spatially resolved, steady-state diffuse reflectance for the noninvasive determination of tissue optical properties *in vivo*. *Medical Physics* 19, no. 4, 1992: 879–888.

Fitzgibbon, A., M. Pilu, and R. B. Fisher. Direct least square fitting of ellipses. *IEEE Transactions on Pattern Analysis and Machine Intelligence* 21, no. 5, 1999: 476–480.

Gerrard, D. E. and A. L. Grant. *Principles of Animal Growth and Development*. Kendall Hunt, Dubuque, IA, 2003.

Harris, P. V. and W. R. Shorthose. Meat texture. In R. Lawrie (ed.), *Developments in Meat Science*, Vol. 4, pp. 245–296. Elsevier Applied Science, London, 1998.

Heino, J., S. Arridge, J. Sikora, and E. Somersalo. Anisotropic effects in highly scattering media. *Physical Review E* 68, no. 3, 2003: 031908.

Herring, H. K., R. G. Cassens, and E. J. Rriskey. Further studies on bovine muscle tenderness as influenced by carcass position, sarcomere length, and fiber diameter. *Journal of Food Science* 30, no. 6, 1965: 1049–1054.

Herring, H. K., R. G. Cassens, G. G. Suess, V. H. Brungardt, and E. J. Briskey. Tenderness and associated characteristics of stretched and contracted bovine muscles. *Journal of Food Science* 32, no. 3, 1967: 317–323.

Horgan, D. J. The estimation of the age of cattle by the measurement of thermal stability of tendon collagen. *Meat Science* 29, no. 3, 1991: 243–249.

Hostetler, R. L., W. A. Landmann, B. A. Link, and H. A. Fitzhugh. Influence of carcass position during rigor mortis on tenderness of beef muscles: Comparison of two treatments. *Journal of Animal Science* 31, no. 1, 1970: 47–50.

Huffman, K. L., M. F. Miller, L. C. Hoover, C. K. Wu, H. C. Brittin, and C. B. Ramsey. Effect of beef tenderness on consumer satisfaction with steaks consumed in the home and restaurant. *Journal of Animal Science* 74, no. 1, 1996: 91–97.

Huxley, A. F. Muscle structure and theories of contraction. *Progress in Biophysics and Biophysical Chemistry* 7, 1957: 255–318.

Huxley, A. F. and R. M. Simmons. Proposed mechanism of force generation in striated muscle. *Nature* 233, no. 5321, 1971: 533–538.

Johnson, P. M., B. P. Bret, J. G. Rivas, J. J. Kelly, and A. Lagendijk. Anisotropic diffusion of light in a strongly scattering material. *Physical Review Letters* 89, no. 24, 2002: 243901.

Koohmaraie, M., M. E. Doumit, and T. L. Wheeler. Meat toughening does not occur when rigor shortening is prevented. *Journal of Animal Science* 74, no. 12, 1996: 2935–2942.

Koohmaraie, M., M. P. Kent, S. D. Shackelford, E. Veiseth, and T. L. Wheeler. Meat tenderness and muscle growth: Is there any relationship? *Meat Science* 62, no. 3, 2002: 345–352.

Leung, A. F. and M. K. Cheung. Decrease in light diffraction intensity of contracting muscle fibres. *European Biophysics Journal* 15, no. 6, 1988: 359–368.

Lieber, R. L., Y. Yeh, and R. J. Baskin. Sarcomere length determination using laser diffraction. Effect of beam and fiber diameter. *Biophysical Journal* 45, no. 5, 1984: 1007–1016.

Lin, S., H. E. Huff, and F. Hsieh. Texture and chemical characteristics of soy protein meat analog extruded at high moisture. *Journal of Food Science* 65, no. 2, 2000: 264–269.

Lin, S., H. E. Huff, and F. Hsieh. Extrusion process parameters, sensory characteristics, and structural properties of a high moisture soy protein meat analog. *Journal of Food Science* 67, no. 3, 2002: 1066–1072.

Liu, K. S. and F.-H. Hsieh. Protein–protein interactions in high moisture-extruded meat analogs and heat-induced soy protein gels. *Journal of the American Oil Chemists' Society* 84, no. 8, 2007: 741–748.

Liu, K. S. and F.-H. Hsieh. Protein–protein interactions during high-moisture extrusion for fibrous meat analogues and comparison of protein solubility methods using different solvent systems. *Journal of Agricultural and Food Chemistry* 56, no. 8, 2008: 2681–2687.

Liu, Y., B. G. Lyon, W. R. Windham, C. E. Realini, T. D. D. Pringle, and S. Duckett. Prediction of color, texture, and sensory characteristics of beef steaks by visible and near infrared reflectance spectroscopy. A feasibility study. *Meat Science* 65, no. 3, 2003: 1107–1115.

Marsh, B. B. and W. A. Carse. Meat tenderness and the sliding-filament hypothesis. *International Journal of Food Science & Technology* 9, no. 2, 1974: 129–139.

McCormick, R. J. The flexibility of the collagen compartment of muscle. *Meat Science* 36, no. 1, 1994: 79–91.

Moharam, M. G. and T. K. Gaylord. Three-dimensional vector coupled-wave analysis of planar-grating diffraction. *JOSA* 73, no. 9, 1983: 1105–1112.

Park, B., Y. R. Chen, W. R. Hruschka, S. D. Shackelford, and M. Koohmaraie. Near-infrared reflectance analysis for predicting beef *longissimus* tenderness. *Journal of Animal Science* 76, no. 8, 1998: 2115–2120.

Prahl, S. A. The diffusion approximation in three dimensions. In A. J. Welch and M. J. C. van Gemert (eds.), *Optical-Thermal Response of Laser-Irradiated Tissue*, pp. 207–231. Springer, Berlin, 1995.

Ramsbottom, J. M. and E. J. Strandine. Comparative tenderness and identification of muscles in wholesale beef cuts. *Journal of Food Science* 13, no. 4, 1948: 315–330.

Ranasinghesagara, J., F. H. Hsieh, and G. Yao. An image processing method for quantifying fiber formation in meat analogs under high moisture extrusion. *Journal of Food Science* 70, no. 8, 2005: e450–e454.

Ranasinghesagara, J., F. Hsieh, and G. Yao. A photon migration method for characterizing fiber formation in meat analogs. *Journal of Food Science* 71, no. 5, 2006: E227–E231.

Ranasinghesagara, J., F-H. Hsieh, H. Huff, and G. Yao. Laser scanning system for real-time mapping of fiber formations in meat analogues. *Journal of Food Science* 74, no. 2, 2009: E39–E45.

Ranasinghesagara, J. and G. Yao. Imaging 2D optical diffuse reflectance in skeletal muscle. *Optics Express* 15, no. 7, 2007: 3998–4007.

Ranasinghesagara, J. and G. Yao. Effects of inhomogeneous myofibril morphology on optical diffraction in single muscle fibers. *JOSA A* 25, no. 12, 2008: 3051–3058.

Ranasinghesagara, J., T. M. Nath, S. J. Wells, A. D. Weaver, D. E. Gerrard, and G. Yao. Imaging optical diffuse reflectance in beef muscles for tenderness prediction. *Meat Science* 84, no. 3, 2010: 413–421.

Saidi, I. S., S. L. Jacques, and F. K. Tittel. Mie and Rayleigh modeling of visible-light scattering in neonatal skin. *Applied Optics* 34, no. 31, 1995: 7410–7418.

Shackelford, S. D., T. L. Wheeler, and M. Koohmaraie. Relationship between shear force and trained sensory panel tenderness ratings of 10 major muscles from *Bos indicus* and *Bos taurus* cattle. *Journal of Animal Science* 73, no. 11, 1995: 3333–3340.

Shackelford, S. D., T. L. Wheeler, and M. Koohmaraie. Tenderness classification of beef: II. Design and analysis of a system to measure beef *longissimus* shear force under commercial processing conditions. *Journal of Animal Science* 77, no. 6, 1999: 1474–1481.

Shackelford, S. D., T. L. Wheeler, and M. Koohmaraie. On-line classification of US Select beef carcasses for *longissimus* tenderness using visible and near-infrared reflectance spectroscopy. *Meat Science* 69, no. 3, 2005: 409–415.

Shackelford, S. D., T. L. Wheeler, M. K. Meade, J. O. Reagan, B. L. Byrnes, and M. Koohmaraie. Consumer impressions of tender select beef. *Journal of Animal Science* 79, no. 10, 2001: 2605–2614.

Shorthose, W. R. and P. V. Harris. Effects of growth and composition on meat quality. In A. M. Pearson and T. R. Dutson (eds.), *Growth Regulations in Farm Animals: Advances in Meat Research*, Vol. 7, pp. 515–554. Elsevier Applied Science, London, 1991.

Shuaib, A. and G. Yao. Equi-intensity distribution of optical reflectance in a fibrous turbid medium. *Applied Optics* 49, no. 5, 2010: 838–844.

Smulders, F. J. M., B. B. Marsh, D. R. Swartz, R. L. Russell, and M. E. Hoenecke. Beef tenderness and sarcomere length. *Meat Science* 28, no. 4, 1990: 349–363.

Stephens, J. W., J. A. Unruh, M. E. Dikeman, M. C. Hunt, T. E. Lawrence, and T. M. Loughin. Mechanical probes can predict tenderness of cooked beef *longissimus* using uncooked measurements. *Journal of Animal Science* 82, no. 7, 2004: 2077–2086.

Swatland, H. J. Microscope spectrofluorometry of bovine connective tissue using a photodiode array. *Journal of Computer-Assisted Microscopy* 7, no. 3, 1995: 165–170.

Tan, J. Meat quality evaluation by computer vision. *Journal of Food Engineering* 61, no. 1, 2004: 27–35.

Taylor, R. G., G. H. Geesink, V. F. Thompson, M. Koohmaraie, and D. E. Goll. Is Z-disk degradation responsible for postmortem tenderization? *Journal of Animal Science* 73, no. 5, 1995: 1351–1367.

Thornhill, R. A., N. Thomas, and N. Berovic. Optical diffraction by well-ordered muscle fibres. *European Biophysics Journal* 20, no. 2, 1991: 87–99.

Tornberg, E. Biophysical aspects of meat tenderness. *Meat Science* 43, 1996: 175–191.

Tuchin, V. V. Light scattering study of tissues. *Physics-Uspekhi* 40, no. 5, 1997: 495.

Van de Hulst, H. C. *Light Scattering by Small Particles.* Dover Publications, Inc., New York, 1981.

Wang, H., J. R. Claus, and N. G. Marriott. Selected skeletal alterations to improve tenderness of beef round muscles. *Journal of Muscle Foods* 5, no. 2, 1994: 137–147.

Wang, L. and S. L. Jacques. Use of a laser beam with an oblique angle of incidence to measure the reduced scattering coefficient of a turbid medium. *Applied Optics* 34, no. 13, 1995: 2362–2366.

Wang, L., S. L. Jacques, and L. Zheng. MCML—Monte Carlo modeling of light transport in multi-layered tissues. *Computer Methods and Programs in Biomedicine* 47, no. 2, 1995: 131–146.

Wang, L. V. and H-I. Wu. *Biomedical Optics: Principles and Imaging.* John Wiley & Sons, New Jersey, 2012.

Wheeler, T. L., L. V. Cundiff, and R. M. Koch. Effect of marbling degree on beef palatability in *Bos taurus* and *Bos indicus* cattle. *Journal of Animal Science* 72, no. 12, 1994: 3145–3151.

Wheeler, T. L. and M. Koohmaraie. Prerigor and postrigor changes in tenderness of ovine *longissimus* muscle. *Journal of Animal Science* 72, no. 5, 1994: 1232–1238.

Wolf, W. J., J. C. Cowan, and H. Wolff. Soybeans as a food source. *Critical Reviews in Food Science & Nutrition* 2, no. 1, 1971: 81–158.

Xia, J. J., E. P. Berg, J. W. Lee, and G. Yao. Characterizing beef muscles with optical scattering and absorption coefficients in VIS-NIR region. *Meat Science* 75, no. 1, 2007a: 78–83.

Xia, J. J., J. Ranasinghesagara, C. W. Ku, and G. Yao. Monitoring muscle optical scattering properties during rigor mortis. In Y. R. Chen, G. E. Meyer, and S. I. Tu (eds.), *Optics for Natural Resources, Agriculture, and Foods II*, pp. 67610H–67610H. International Society for Optics and Photonics, Bellingham, WA, 2007b.

Xia, J. J., A. Weaver, D. E. Gerrard, and G. Yao. Monitoring sarcomere structure changes in whole muscle using diffuse light reflectance. *Journal of Biomedical Optics* 11, no. 4, 2006: 040504–040504.

Xia, J. J., A. Weaver, D. E. Gerrard, and G. Yao. Distribution of optical scattering properties in four beef muscles. *Sensing and Instrumentation for Food Quality and Safety* 2, no. 2, 2008a: 75–81.

Xia, J. J., A. Weaver, D. E. Gerrard, and G. Yao. Heating induced optical property changes in beef muscle. *Journal of Food Engineering* 84, no. 1, 2008b: 75–81.

Xia, J. J. and G. Yao. Angular distribution of diffuse reflectance in biological tissue. *Applied Optics* 46, no. 26, 2007: 6552–6560.

Yao, G., K. S. Liu, and F. Hsieh. A new method for characterizing fiber formation in meat analogs during high-moisture extrusion. *Journal of Food Science* 69, no. 7, 2004: 303–307.

Yao, G. and L. Wang. Propagation of polarized light in turbid media: Simulated animation sequences. *Optics Express* 7, no. 5, 2000: 198–203.

Yeh, Y., R. J. Baskin, R. L. Lieber, and K. P. Roos. Theory of light diffraction by single skeletal muscle fibers. *Biophysical Journal* 29, no. 3, 1980: 509–522.

Yousif, H. A. and E. Boutros. A FORTRAN code for the scattering of EM plane waves by an infinitely long cylinder at oblique incidence. *Computer Physics Communications* 69, no. 2, 1992: 406–414.

11 Spectral Scattering for Assessing Quality and Safety of Meat

Yankun Peng

CONTENTS

11.1 Introduction 284
    11.1.1 Spectral Scattering Mechanism in Meat 284
    11.1.2 Spectral Scattering Systems 284
        11.1.2.1 Light Sources 285
        11.1.2.2 Wavelength Dispersion Devices..286
        11.1.2.3 Area Detectors 288
    11.1.3 Scattering Image Acquisition Methods 288
        11.1.3.1 Hyperspectral Scattering Imaging......................................288
        11.1.3.2 Multispectral Scattering Imaging 289
    11.1.4 Extraction and Analysis of Scattering Characteristics......................290
        11.1.4.1 Modified Lorentzian Functions 290
        11.1.4.2 Modified Gompertz Functions 291
        11.1.4.3 Boltzmann Function..293
11.2 Assessment of Meat Quality Attributes...294
    11.2.1 Color 294
    11.2.2 Tenderness 295
    11.2.3 Water-Holding Capacity 297
    11.2.4 pH Value 298
    11.2.5 Detection of Multi Quality Attributes 299
11.3 Assessment of Meat Safety Attributes...300
    11.3.1 Total Viable Count 300
    11.3.2 Total Volatile Basic Nitrogen 302
    11.3.3 *E. coli* Contamination..304
    11.3.4 Shelf-Life Estimation..306
11.4 Portable and Movable Prototype Devices...309
    11.4.1 Freshness Detection Device...309
    11.4.2 TVC Detection Device..309
11.5 Conclusions 312
Acknowledgments..312
References..313

11.1 INTRODUCTION

11.1.1 Spectral Scattering Mechanism in Meat

Spectral scattering is a promising technique for nondestructive and rapid assessment of meat quality and safety. Meat quality can be defined in terms of consumers' appreciation of texture and flavor, and safety, which includes the health implications of both composition and microbiological contamination. The main parameters of meat quality and safety are lean-to-fat ratio, protein, marbling, tenderness, water content, drip loss, pH, juiciness, etc. The content and distribution of fat, water, protein, and other compositional parameters affect the structural characteristics of meat. The scattering characteristics vary across different regions of animal carcasses, depending on the density and structural properties of meat, such as sarcomere length and collagen. Spectral scattering in the muscle tissue is related to the morphology and refractive index of the muscle tissue composition. These muscle tissue structures are the primary contributor to meat texture, and they also influence the light scattering characteristics of meat. The pattern and amount of light scattering and propagation depends on factors such as wavelength of the incident light and physical and chemical properties of meat samples. Chapter 11 gives a detailed description on how light propagation is related to the meat muscle structure and its change during postmortem aging. The spatial profile of diffusely reflected light depends on the scattering of light inside the meat sample as shown in Figure 11.1. Generally, meat quality and safety parameters are mainly affected by the muscle tissue structures and their chemical composition. Therefore, spectral scattering techniques have attracted significant attention due to the great potential for meat quality and safety detection.

11.1.2 Spectral Scattering Systems

A spectral scattering system, which is mainly based on hyperspectral or multispectral imaging technique, is used to collect scattering profiles with high spatial and spectral resolutions in a short acquisition time. Hyperspectral imaging techniques for the detection of single quality parameters as well as for simultaneous detection of

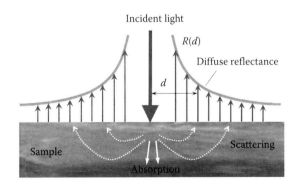

FIGURE 11.1 Spatial reflectance profile depends on the scattering of light inside the meat sample.

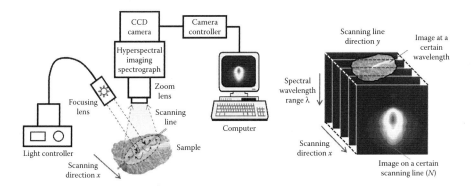

FIGURE 11.2 Schematic diagram of a line-scanning hyperspectral scattering imaging system.

multiple quality parameters have been extensively reported in the literature. The techniques have been applied for the detection of nutritional content, water content, pH, tenderness, microbiological spoilage, water-holding capacity (WHC), marbling, drip loss, and color. A typical hyperspectral scattering imaging system, as shown in Figure 11.2, consists of a charge-coupled device (CCD) camera, an imaging spectrograph covering the spectral range of 400–1100 nm, a light unit, a desktop computer, and a positioning sensor. The imaging spectrograph acquires spectral information for each pixel of the scanned line on the surface of meat samples. Each sample is line scanned multiple times at a given position to obtain multiple scattering images. Typically, a total of 20 images obtained from five different positions are needed to take the structural variation of meat samples into consideration. The acquired images are then averaged to obtain the final scattering information. To ensure that the hyperspectral imaging system performs satisfactorily, each component in the system should be carefully designed or selected. A detailed description about each optical component of a hyperspectral scattering imaging system is given in the following subsections.

11.1.2.1 Light Sources
The proper selection of a light source is critical to the acquisition of scattering images. Typical light sources include halogen lights, light-emitting diodes (LEDs), lasers, and tunable light sources (Wu and Sun, 2013).

11.1.2.1.1 Halogen Lights
Halogen lamps, also known as tungsten halogen, quartz halogen, or quartz iodine lamps, are a common broadband illumination to generate a continuous spectral distribution of light. As a highly reliable source, halogen lamps have been widely employed for the illumination of visible and near-infrared (VIS/NIR) spectral regions. The halogen light source provides adequate illumination with excellent color rendering, making it perfect for various applications. Tungsten halogen lamps are popularly used as an illumination unit in hyperspectral reflectance measurement for meat inspection. The illumination unit is integrated with optical fibers to deliver broadband light to the tissue and receive diffusely reflected light. Different forms of

illumination, such as point, line, and ring lights, are available for different application purposes. The point light source is commonly used to obtain the scattering images. Compared to other light sources, tungsten lamps have comparative advantages such as a long lifetime, small size, lightweight, high efficiency, ease of replacement, and constant lumen maintenance.

11.1.2.1.2 Light-Emitting Diodes
An LED is a two-lead semiconductor light source. As a reliable light source, LEDs have gained popularity in practical applications with the advantages of a long operational life, fast response, high shock resistance, low heat generation, low power consumption, and insensitivity to vibration. Compared to halogen lamps, LEDs are sensitive to voltage fluctuations and junction temperature, with low light intensities. LEDs can be assembled in different arrangements such as spot, line, and ring lights. Currently, as one of the promising light sources, LED technology has gained importance in meat quality and safety inspection.

11.1.2.1.3 Lasers
Unlike broadband illumination sources, lasers are powerful monochromatic sources, which are commonly used as excitation sources. Recently, lasers have received increasing interest in research and industrial applications such as Raman imaging and hyperspectral fluorescence imaging. Generally, lasers are regarded as an ideal monochromatic light source. They have the advantages of good brightness, good directionality, highly concentrated energy, efficient irradiance, and true monochromaticity. However, high-power lasers are expensive, and most lasers can only provide single wavelengths or narrow wavebands.

11.1.2.1.4 Tunable Light Sources
To overcome the limitations with lasers, an economic, broadband alternative method has been exploited by a combination of wavelength dispersion and broadband illumination, which leads to a tunable light source. Tunable light sources are used to illuminate materials with only a small specific range of wavelengths. Combined with the wavelength dispersion device, a tunable light source is created from white light. For automatic image collection, it is necessary to keep the wavelength dispersion device synchronized with the detector, thus simultaneously acquiring the spectral and spatial information by performing area scanning. The major shortcoming of tunable light sources is that they cannot be used in practical production, especially for conveyor belt systems.

11.1.2.2 Wavelength Dispersion Devices
Wavelength dispersion devices are considered as an integral component of the spectral scattering system. It is essential to use a wavelength dispersion device for dispersing broadband light into different wavelengths and projecting the dispersed light to the area detectors. Typical dispersive devices are imaging spectrographs, filter wheels, tunable band-pass filters, and beam splitters, which are discussed below.

11.1.2.2.1 Imaging Spectrographs

Compared with filter- and color camera-based imaging systems, an imaging spectrograph can produce full, contiguous spectral information for each line of pixels with high spectral and spatial resolutions. Coupled with a monochromatic area camera, the imaging spectrograph can become a line-scanning spectral imaging camera. It can also be built for simultaneous measurement of several points of a sample by combining with multichannel optic fibers. Each image contains the line pixels in one dimension (spatial axis) and the spectral pixels in the other dimension (spectral axis), providing full spectral information for each line pixel. Most imaging spectrographs are developed with two main design types, including transmission gratings and reflection gratings. The transmission-grating imaging spectrograph utilizes a prism-grating-prism component, where the collimated beam is dispersed into different wavelengths. On the other hand, the reflection-grating imaging spectrograph is composed of an entrance slit, two concentric spherical mirrors, an aberration-corrected convex reflection grating, and a detector. It is coupled with a pair of spherical mirrors to form a continuous spectrum after its entrance through the input slit. These two types of imaging spectrographs have been widely used in various line-scan spectral imaging systems for practical applications.

11.1.2.2.2 Filter Wheels

A filter wheel provides automated and efficient optical filter changing. To meet different demands, a broad range of filters is designed from ultraviolet (UV), VIS to NIR wavelength specifications. Furthermore, it is widely supported by almost any available imaging application software. The disadvantage of using a filter wheel is that the wavelength switching is slow. Accurate filter positioning is essential to eliminate imaging errors that can affect the optical performance of the imaging system.

11.1.2.2.3 Tunable Band-Pass Filters

A tunable band-pass filter changes the band-pass wavelength electronically. There are two types of tunable band-pass filters: acousto-optic tunable filters (AOTFs) and liquid crystal-tunable filters (LCTFs). AOTFs can simultaneously modulate the intensity and wavelength of multiple laser lines from one or more sources, while LCTFs possess controlled liquid crystal elements to transmit light with a selectable wavelength. Because of a high image quality and rapid tuning over a broad spectral range, LCTFs are commonly used in hyperspectral or multispectral imaging systems. However, these tunable filters have a long exposure time and low light collection efficiency.

11.1.2.2.4 Beam Splitters

A beam splitter is an optical device that divides an incident light into eight, four, or two beams of equivalent power with only minor loss in total power. Beam-splitting cubes can be used not only for simple light beams, but also for beams carrying images. Fiber-optic splitters are required for fiber-optic interferometers, whose splitting ratio depends on the wavelength and polarization of the input. Beam-splitting devices are utilized in the multispectral imaging system with two or three bands for application in real-time and online meat inspection.

11.1.2.3 Area Detectors

In a front-illuminated device, photons falling on the CCD must travel through the region of the gate electrode structures. Both CCD and complementary metal–oxide–semiconductor (CMOS) image sensors start at the same point where they convert light into electrons in the form of accumulated charge. The next step is to read out this charge of each cell and to digitize this information to make it computer readable.

11.1.2.3.1 CCD Detectors

CCD detectors are silicon-based multichannel array detectors of UV, VIS, and NIR light. They produce image information by converting light (photons) into electrical current (electrons). Owing to the extreme sensitivity to light, they are the most commonly used detectors for spectral scattering imaging systems. CCD detectors have many advantages, including a high signal-to-noise ratio, good uniformity, large dynamic range, linear response, and high detection sensitivity. Typical CCD detectors have one-dimensional (linear) or two-dimensional (2-D) (area) arrays of thousands or even millions of individual detector elements (also known as pixels). They have been in production for a long period of time; so, the technology has become more mature in their features of high quality and more pixel resolution. They are perfectly suited for scattering imaging applications.

11.1.2.3.2 CMOS Detectors

CMOS consumes low power. Compared with CCD, CMOS cameras are usually less expensive and have greater battery life. The efficiency of CMOS cameras has been in steady improvement to meet the needs of agricultural and industrial applications. Owing to these advantages, CMOS cameras are especially suited for high-speed imaging for online and real-time meat inspection in industrial-processing lines.

11.1.3 SCATTERING IMAGE ACQUISITION METHODS

A typical scattering image is a three-dimensional (3-D) data cube, which contains 2-D spatial information (x, y) and one-dimensional spectral information (λ). On the basis of whether the collected scattering image data are continuous in the wavelength domain, there are generally two types of scattering imaging systems: hyperspectral and multispectral.

11.1.3.1 Hyperspectral Scattering Imaging

There are two main approaches, that is, line and area scanning, to acquire a 3D hyperspectral scattering image cube (hypercube (x, y, λ)), as illustrated in Figure 11.3 (Qin et al., 2013). The line-scanning method (with a push broom scanner) acquires a slit of spatial information as well as full spectral information for each spatial pixel in the linear field of view, as illustrated in Figure 11.3a. The second spatial dimension is achieved through the platform or sample movement. A complete hypercube is thus obtained from the composite set of spatial line scans. Line scanning is ideal for online applications of individual or a continuous stream of meat samples (Kim et al., 2011). A line-scan system, integrated with an imaging spectrograph (i.e., VIS/NIR, UV, Raman, etc.) can be used to scan meat samples moving on a conveyor belt.

Spectral Scattering for Assessing Quality and Safety of Meat 289

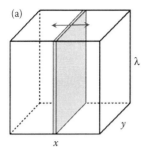
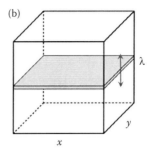

FIGURE 11.3 Methods for acquisition of 3-D hyperspectral image cubes in two spatial (x, y) and one spectral (λ) dimensions: (a) line scanning and (b) area scanning.

The other acquisition approach, illustrated in Figure 11.3b, is referred to as area or plane scanning, where the imaging system uses a band-pass filter (e.g., LCTF or filter wheel) placed or installed in front of the CCD camera. This approach acquires a 2-D monochrome image (x, y) with full spatial information at one wavelength or a narrow waveband after another in a time sequence.

11.1.3.2 Multispectral Scattering Imaging

Multispectral scattering imaging is usually implemented in the field or industrial application to acquire spatial information from meat samples at selected wavelengths in real time. A spectrophotometer equipped with a point detector can be used to acquire a single spectrum for each pixel in the scene. However, to scan a complete scene, either the sample or the detector has to be moved along two spatial dimensions. This method is time consuming, as it needs to scan along two spatial dimensions. The other two methods (i.e., line scan and area scan), as described earlier, can be adapted to satisfy the requirements of rapid multispectral image acquisition. As illustrated in Figure 11.4, both line-scan and area-scan methods can be implemented to collect images at fewer wavelengths (normally <10).

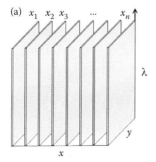
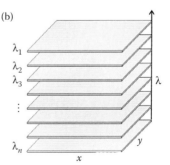

FIGURE 11.4 Methods for acquisition of multispectral images at specific wavelengths of $\lambda_1, \lambda_2, \lambda_3, \ldots, \lambda_n$, where n is normally less than 10: (a) line scanning and (b) area scanning.

For the line-scan method, as illustrated in Figure 11.4a, multispectral imaging can be achieved by specifying the positions of all useful tracks along the spectral dimension of the CCD detector. Only the data from the selected tracks are acquired, which reduces the amount of data for each line-scan image (y, λ) and consequently shortens the acquisition time. In comparison to the line-scan method, the area-scan method for multispectral imaging can simultaneously collect single-band images for multiple selected wavelengths. Light from the spatial scene is divided into several parts of the same spatial scene by an optical separation device (e.g., a common aperture multispectral imaging spectrograph). Divided scenes pass through preset band-pass filters separately. Narrowband images are then formed on multiple cameras or on one camera with a large CCD sensor (Qin et al., 2013). Area scanning-based multispectral scattering imaging can greatly reduce the scattering image acquisition time so as to overcome the long sequence-scanning issue that is commonly associated with this type of scanning method.

11.1.4 Extraction and Analysis of Scattering Characteristics

Nonlinear curve-fitting algorithms, including Lorentzian and Gompertz functions, have been proposed to quantify the optical scattering characteristics of meat from the spatially-resolved reflectance profiles of the scattering images. Figure 11.5 presents an original 2-D hyperspectral scattering image and a median-filtered image shown in three dimensions (with the third, vertical axis for the intensity).

11.1.4.1 Modified Lorentzian Functions

The Lorentzian distribution (LD) function is commonly used to describe the laser profiles and light distribution patterns in optics research (Davis, 1996). Peng and Lu (2005) proposed a three-parameter LD function to describe the spatial scattering profiles of apples (Equation 11.1) and concluded that the function gives an excellent fitting result

$$R_{w_i} = a_{w_i} + \frac{b_{w_i}}{1 + (x/c_{w_i})^2} \tag{11.1}$$

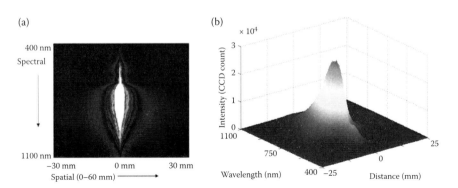

FIGURE 11.5 Hyperspectral scattering image: (a) original image and (b) median-filtered 3-D intensity image.

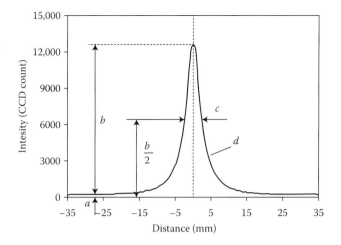

FIGURE 11.6 Modified LD function with four parameters.

where R is the average light intensity of each radial circular band; a is the asymptotic value of light intensity; b is the peak value of the estimated light intensity at the light incident point; c is the full scattering width at half maximal (FWHM) value; x is the scattering distance (mm); and the subscript w_i denotes a specific wavelength with $i = 1, 2, 3,..., N$, where N is the total number of wavelengths. The Lorentzian function is a single-peaked function that can be normalized so that its maximum values are equal to one at its center. The modified Lorentzian function with four parameters is given by Equation 11.2 and also shown in Figure 11.6

$$R_{w_i} = a_{w_i} + \frac{b_{w_i}}{1+(x/c_{w_i})^d} \tag{11.2}$$

where d is the slope around the inflection point.

Figure 11.7 shows the three Lorentzian parameters of a, b, and c collected from beef samples for the spectral range of 500–1000 nm (Tao, 2013). The parameter a (Figure 11.7a) had dramatic changes in its value over the spectral region between 570 and 645 nm. There was a prominent peak around 700 nm for parameter b spectra (Figure 11.7b), which corresponds to the pigment-absorbing band. The spectra of parameter c had greater changes in its value over the spectral region between 550 and 950 nm (Figure 11.7c). A peak or valley around 540 nm was observed in all three-parameter spectra, which was attributed to the chemical changes of myoglobin, hemoglobin, and their isomerides during meat storage (Tao et al., 2012b).

11.1.4.2 Modified Gompertz Functions

The Gompertz function is used for describing the ascending gradient profiles (Fekedulegn et al., 1999; Khamis et al., 2005; Peng and Lu, 2007). Its modified version with three or four parameters, termed as the modified Gompertz function, is

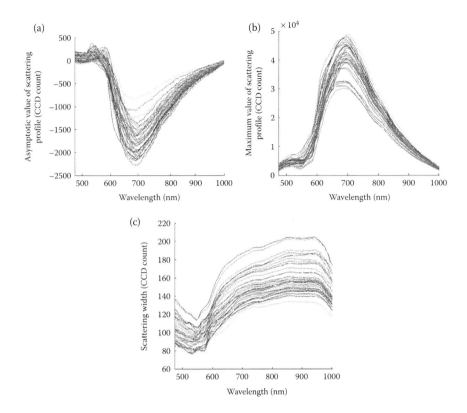

FIGURE 11.7 Spectra of Lorentzian parameters for selected pork samples: (a) Lorentzian parameter *a*, (b) Lorentzian parameter *b*, and (c) Lorentzian parameter *c*. (From Tao, F. 2013. Study on the rapid and nondestructive detection methods and antimicrobial delivery for the bacterial contamination of pork. The PhD thesis of China Agricultural University, May, 2013, No. B10209185, Beijing, China.)

appropriate to describe the scattering profiles accurately at individual wavelengths, which are given in Equations 11.3 and 11.4, respectively

$$R_{w_i} = \alpha_{w_i} + \beta_{w_i}[1 - \exp\{-\exp(\varepsilon_{wi} - x)\}] \quad (11.3)$$

$$R_{w_i} = \alpha_{w_i} + \beta_{w_i}[1 - \exp\{-\exp(\varepsilon_{wi} - \delta_{wi}x)\}] \quad (11.4)$$

where R is the average light intensity of each radial circular band; α is the asymptotic value of light intensity; β is the peak value of the estimated light intensity at the light incident point; ε is the full scattering width at the inflection point; δ is the slope around the inflection point; x is the horizontal linear distance (mm); and the subscript w_i denotes a specific wavelength with $i = 1, 2, 3, ..., N$, where N is the total number of wavelengths.

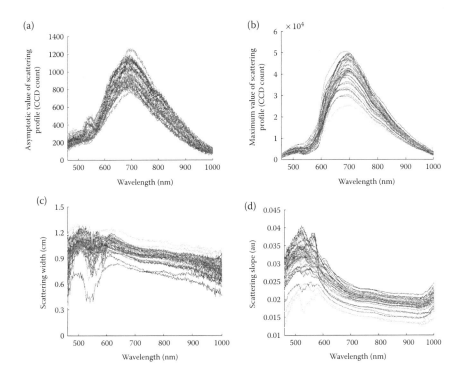

FIGURE 11.8 Spectra of Gompertz parameters (Equation 11.4) for selected pork samples: (a) parameter α, (b) parameter β, (c) parameter ε, and (d) parameter δ.

The four Gompertz parameters of α, β, ε, and δ collected from pork samples are depicted in Figure 11.8. The figure clearly shows that the four Gompertz parameters of chilled pork at different storage times generally had larger variations in the spectral range of 500–600 nm, compared to the other spectral range, which were clearly observed at the wavelength of 550 nm. As the meat is being spoiled by bacteria, it is partly oxidized into the metmyoglobin as a result of protein degradation. Oxymyoglobin has a stronger characteristic absorption peak at 540–550 nm (Zhang et al., 2013b).

11.1.4.3 Boltzmann Function

The diffusion approximation theory model has been effectively applied to determine the optical absorption coefficient (μ_a) and reduced scattering coefficient (μ_s') (Cen et al., 2010, 2013). The Boltzmann equation (Tuchin, 2000), also known as the radiation transport equation, describes light scattering and absorption in turbid materials. The equation may be simplified to a diffusion approximation equation for a large class of biological materials, in which scattering is dominant (i.e., $\mu_s' \gg \mu_a$). An analytical equation is available for describing diffuse reflectance profiles at the surface of a semiinfinite medium under the illumination of a vertically incident beam (Farrell et al., 1992). The equation has been used to estimate the absorption and reduced scattering coefficients of fruits, vegetables, and meats (Qin and Lu, 2007,

2008; Tao et al., 2012c). A detailed description of the theory of light transfer and spatially-resolved spectroscopy technique is given in Chapters 3 and 7.

11.2 ASSESSMENT OF MEAT QUALITY ATTRIBUTES

Meat quality is often referred to as a comprehensive evaluation of various quality traits (Elmasry et al., 2012a; Dale et al., 2013; Wu and Sun, 2013). Meat quality mostly depends on the interaction of various chemical components of the muscle with the palatability factors such as visual appearance, flavor, and tenderness. It is the physical, chemical, and organoleptic attributes of muscles that largely determine meat quality traits. Quality assurance and control is among the main tasks in meat production and processing, which includes physical qualities, chemical compositions, and nutritional contents. Physical quality attributes of meat may include marbling, WHC, etc. Protein, fat, calcium, and moisture content represent the chemical quality attributes of meat (Kamruzzaman et al., 2012). Chemical attributes are also indicative of the nutritional value of meat. Sensory attributes such as color, smell, flavor, and tenderness influence the purchase decision of consumers, as these parameters give the first impression regarding meat quality. The most important quality attributes to be evaluated for pork and beef include color, tenderness, WHC, cooking loss, and pH value (Qiao et al., 2007; ElMasry et al., 2012b). These quality attributes can directly or indirectly reflect the structural characteristics of meat, and determine the optical scattering characteristics in meat. Therefore, the spectral scattering technique could be of great practical significance for the rapid, accurate, and nondestructive detection of meat quality.

11.2.1 COLOR

As one of the most important quality attributes, the color of meat directly affects consumers' purchasing decision. Beef and lamb meat in bright red and pork in pink are desirable and attractive for consumers. The discoloration of fresh meat is known to be determined by three redox forms of myoglobin derivatives (deoxyhemoglobin, oxymyoglobin, and metmyoglobin). Color is usually measured based on one channel for luminance (lightness) (L^*) and two color channels (a^* and b^*), where L^* denotes the brightness, a^* axis extends from green to red, and b^* axis extends from blue to yellow.

Color measurement is not difficult, and it can be done using simple techniques. As long as the spectral range of the scattering imaging system covers the visible light, the meat color can be simultaneously detected, along with other quality parameters.

Wu et al. (2012) applied the hyperspectral scattering technique to predict all three color parameters (L^*, a^*, and b^*) in fresh beef. The scattering profiles were first derived from the hyperspectral images and then fitted to the LD function. In the study, the reference meat color was measured on the basis of L^*, a^*, and b^* values using a calibrated portable color reader (model HP-200, Shanghai Chinaspec Optoelectronics Technology Co. Ltd, China). Finally, multilinear regression (MLR) models were established using the LD parameters in conjunction with the optimal wavelengths determined by the stepwise regression method. Hyperspectral

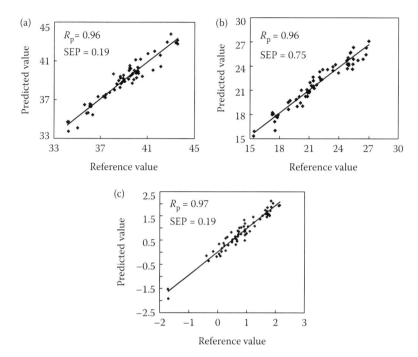

FIGURE 11.9 Cross-validation results of the MLR model for color parameters: (a) color parameter L^*, (b) color parameter a^*, and (c) color parameter b^*.

scattering technique predicted the color parameters (L^*, a^*, and b^*) in beef with a high degree of accuracy. As shown in Figure 11.9, the models gave the correlation coefficient of prediction or $R_P = 0.96$ and standard error of prediction or SEP = 0.61 for L^* value, $R_P = 0.96$ and SEP = 0.75 for a^* value, and $R_P = 0.97$ and SEP = 0.19 for b^* value (Wu et al., 2012). These results also provide support for the development of a multispectral system for online quality detection of beef steaks.

11.2.2 TENDERNESS

Tenderness is one of the most important quality attributes, because it directly relates to the edible quality of meat and is one key factor for consumer acceptance (Wu et al., 2010). Warner–Bratzler shear force (WBSF) and slice shear force (SSF) are the two most common methods for evaluating the tenderness of red meats. Studies have shown that variations in meat tenderness highly affect consumers' decision for repeat purchase.

Light scattering can be potentially used as an indicator of beef tenderness (Xia et al., 2007). Hyperspectral or multispectral imaging techniques have been applied to predict the tenderness of fresh meat. An NIR hyperspectral imaging system (900–1700 nm) with mathematical modeling algorithms was reported for measuring the tenderness of beef that had been aged for 14 days (Naganathan et al., 2008). The authors reported that the partial least-squares regression (PLSR) modeling had the accuracy of 74% for classification of beef samples as "tender" (SSF ≤ 205.8 N), "intermediate" (205.8 N ≤ SSF ≤ 254.8 N), and "tough" (SSF ≥ 254.8 N). In a similar research, a hyperspectral imaging system over 496–1036 nm was used to measure the tenderness of beef (Cluff et al., 2008). Furthermore, with a hyperspectral imaging system (922–1739 nm), Cluff et al. (2013) used a larger incident beam of 1 cm in diameter to illuminate the entire length of a steak to avoid the effect of fat flecks on tenderness prediction. The predicted WBSF values were used to classify the samples into "tender" (WBSF ≤ 58.8 N) and "tough" (WBSF > 58.8 N) groups, as described by Zhao et al. (2006), with the accuracy of 96.9% for the tender group and 90.9% for the tough group. The weighted regression coefficients resulting from the best PLSR prediction model were used to identify the most important wavelengths and to reduce the high dimensionality of the hyperspectral data, and only the selected feature wavelengths were used for a multispectral imaging system to predict quality traits (ElMasry et al., 2012b).

Compared with the aforementioned hyperspectral imaging technique with a uniform, broad-area illumination configuration, spectral scattering imaging technique offers a better solution to the detection of meat tenderness. A hyperspectral imaging system, as illustrated in Figure 11.2, was used to collect hyperspectral scattering images from beef samples. A digital meat tenderness meter with Warner–Bratzler shear accessory (model C-LM3B, Northeast Agricultural University, China) was used to measure samples' tenderness as reference values. The original hyperspectral scattering images of the meat samples were fitted by the modified three-parameter Lorentzian function. Figure 11.10 shows the tenderness calibration and fullcross-validation results for the MLR model based on LD parameter combinations of the eight optimal wavelengths (485, 524, 541, 645, 700, 720, 780, and 820 nm). The MLR model yielded good predictions for calibration with $R_C = 0.95$ and SEC = 7.95 N and for validation with $R_P = 0.91$ and SEP = 9.93 N, respectively (Wu et al., 2012).

Further works were reported for the simultaneous detection of different quality traits by the scattering imaging technique. For instance, Tao et al. (2012a) reported on the hyperspectral scattering technique for the simultaneous determination of tenderness and *Escherichia coli* bacteria in pork, and the prediction models established with MLR method gave high R^2 values, ranging from 0.831 to 0.930.

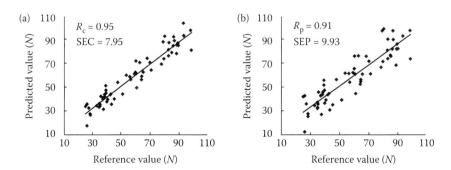

FIGURE 11.10 Calibration and full cross-validation results of the MLR model for tenderness: (a) calibration set and (b) prediction set.

11.2.3 WATER-HOLDING CAPACITY

For fresh meat, WHC is one of the most important quality parameters as it determines the juiciness of meat. During slaughtering, storage, and processing, it is easy to lose moisture in muscles. For the meat-processing industry, predicting the WHC of meat is essential because it is an indication of weight loss in raw and cooked meat as well as processed meat.

In the late 1990s and the early 2000s, NIR reflectance method (900–1800 nm) was reported for the measurement of WHC and drip loss in fresh pork with the correlation coefficient of greater than 80% (Forrest et al., 2000). Later, Pedersen et al. (2003) reported that infrared spectroscopy in the region of 1800–900 cm^{-1} contained the best predictive information for the WHC of pork. The authors used infrared spectroscopy with PLSR to measure the WHC in pork with $R = 0.89$, and the result was also confirmed by online measurement of the pork WHC by using Fourier transform-infrared (FT-IR) spectroscopy.

A recent study showed NIR spectroscopy as a potential tool for the quality assessment of chicken (Barbin et al., 2015). NIR reflectance spectroscopy (400–2500 nm) was used for the assessment of chicken quality based on pale, soft, and exudative (PSE), L^*, pH, and WHC. High accuracy and satisfactory results were reported for WHC; however, the system was not successful in completely discriminating PSE and pale-only muscles. ElMasry et al. (2011) developed an NIR hyperspectral imaging system for the nondestructive prediction of WHC in fresh beef. Six-feature wavelengths were selected for establishing a PLSR prediction model. The model predicted WHC with R_P^2 of 0.87 and SEP of 0.28%.

Raman spectral fingerprint was used as the basis for classification of the samples, and the spectral peak height was the basis for quantitative analysis of samples. A recent work was reported on the application of Raman spectroscopy (excitation by 785-nm laser) for the evaluation of WHC in broilers according to the growth rate (Phongpa-Ngan et al., 2014). A significant correlation between WHC and wave numbers was observed at 538, 691, 1367, and 1743 cm^{-1}. The relationships between ratios of peak intensities (538/1849, 691/1849, 1367/1849, and 1743/1849) and WHC were statistically significant with correlation coefficients $(R) = -0.85, 0.89, 0.73,$ and -0.72.

WHC is one of the structural parameters for meat, so that it could be detected by the spectral scattering imaging technique. A hyperspectral imaging system as illustrated in Figure 11.2 was used to collect the hyperspectral scattering images from beef samples. The Lorentzian function with three parameters fitted the scattering profiles accurately with the correlation coefficients of up to 0.99 for each of the four selected wavelengths at 495, 525, 980, and 990 nm (Figure 11.11).

MLR was performed between the reference WHC and the relative reflectance intensity of the calibration set for individual wavelengths. For the 30 calibration samples of pork, the following prediction model was established:

$$F = -0.46372 + 0.034875 a_{495} + 0.002459 b_{495} - 0.01971 c_{495} + 0.007869 a_{525}$$
$$- 0.00117 b_{525} + 0.21274 c_{525} - 0.04412 a_{980} - 0.00803 b_{980} - 0.02072 c_{980}$$
$$+ 0.00499 a_{990} + 0.008464 b_{990} - 0.07402 c_{990} \tag{11.5}$$

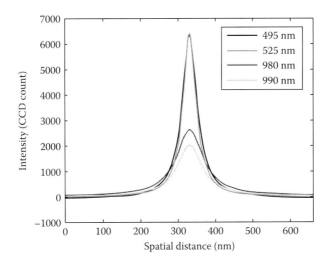

FIGURE 11.11 Fitting results from the Lorentzian function with three parameters (Equation 11.1) at four selected wavelengths.

where a, b, and c denote the three Lorentzian parameters (Equation 11.1) and the subscripts denote the wavelengths.

An independent set of samples were used to validate the model, which yielded $R_P^2 = 0.85$ and SEP = 0.44. Figure 11.12a and b represents the measured and predicted values for the 30 samples from the calibration set and 10 samples from the validation or prediction set, respectively (Wang et al., 2010a).

11.2.4 pH Value

pH is a chemical measure, which refers to the concentration of hydrogen ion in an aqueous solution, and has a great influence on the storage and quality of red meats by affecting their WHC and color. pH values for normal muscles are between 7.1 and 7.3, which change greatly in postmortem. The rate and extent of pH decline

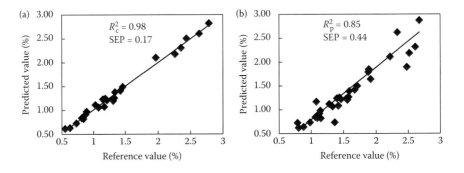

FIGURE 11.12 WHC prediction results: (a) calibration samples and (b) prediction samples.

have a great impact on the shelf life of red meats and their eating quality (Xiong et al., 2014).

Traditionally, pH is measured by inserting a pH meter into the muscle directly after its incision. Nowadays, there are several methods to nondestructively predict the pH value. Qiao et al. (2007) developed a hyperspectral imaging system, and selected six wavelengths as feature wavelengths for predicting pH values in pork. With these feature wavelengths, a feed-forward neural network model was established, which gave R_P^2 of 0.67. Similarly, Kamruzzaman et al. (2012) developed an NIR hyperspectral imaging system in conjunction with multivariate analysis for pH prediction of lamb meat. In that study, a prediction model was established with PLSR method, which gave $R_P^2 = 0.65$ and SEP = 0.085.

Liao et al. (2012) developed an online VIS/NIR spectroscopy system for the nondestructive measurement of pH values in fresh pork. VIS/NIR spectra (350–1100 nm) were collected from samples moving on a conveyor belt. The authors applied spectral denoising methods with efficient variable selection methods to achieve rapid, accurate, and efficient measurement of sample quality traits. They eliminated 85% of variables in the PLSR model using the uninformative variable elimination method, and the system was able to detect pH in the pork samples with R_C and R_P of 0.915 and 0.890, respectively.

Peng et al. (2009a) explored the hyperspectral scattering imaging technique for pH prediction of beef, and reported good performance with $R_P = 0.86$, SEP = 0.07. A hyperspectral imaging system, as illustrated in Figure 11.2, was used to collect the hyperspectral scattering images from pork samples. The pH values of meat were measured by a portable pH meter combined with pH electrodes (Mettler Toledo, China). The values were obtained at eight different locations from each pork sample and were averaged as the ultimate value of pH. The original hyperspectral scattering images of meat samples were accurately fitted by the three-parameter-modified Lorentzian function. The optimal wavelengths were determined for establishing MLR models for the LD function parameters. A portable multispectral imaging system was developed by using a filter wheel with the optimal wavelengths (Li et al., 2013). For pH, the MLR model yielded the best prediction with $R_P = 0.88$, SEP = 0.07.

11.2.5 Detection of Multi Quality Attributes

For rapid, nondestructive, and real-time inspection of meat quality traits in the industrial processing line, research needs to be focused on the simultaneous detection of quality attributes by a single system. Detection of each quality trait by multiple systems not only makes it difficult to implement them at the processing line, but is also expensive. Hence, current research is focused on the development of a single system capable of detecting multiple quality traits in meat samples. This would allow a comprehensive evaluation of various aspects of meat quality.

Only a few works have been reported for the simultaneous detection of different quality traits by one system. An NIR hyperspectral imaging system in the spectral range of 900–1700 nm was applied for the detection of moisture, color, and pH in cooked turkey hams (Iqbal et al., 2013). The research demonstrated the ability of NIR hyperspectral imaging to detect multiple quality attributes simultaneously.

TABLE 11.1
Two-Day pH Values and Color Parameters and WBSF Aged for 7 Days

Quality Attribute	Number	Mean (%)	Minimum (%)	Maximum (%)	SD
pH	33	5.06	5.56	6.16	1.30
L^*	33	39.08	34.25	43.49	2.29
a^*	33	21.85	15.26	26.88	3.05
b^*	33	0.83	−1.77	2.20	0.80
WBSF (N)	33	50.91	20.82	86.05	20.57

Note: N, number of steaks; SD, standard deviation.

For the simultaneous detection of multiple quality parameters, the hyperspectral scattering imaging system, as illustrated in Figure 11.2, was used to collect hyperspectral scattering images for beef samples from 33 beef cattle in a commercial slaughtering house in Beijing, China. Two-day pH values and hyperspectral scattering image of the samples were first measured, and the samples were then vacuum packed immediately and aged for 7 days postmortem in a refrigerator at 4°C before being subjected to color and tenderness measurements. Table 11.1 summarizes the descriptive statistics (mean, ranges, and standard deviations) of pH values, color parameters, WBSF, and cooking loss measured with traditional standard methods as reference values (Peng et al., 2009a).

The combinations of 7, 8, and 5 optimal wavelengths were determined for predicting pH value, tenderness, and WHC, respectively. Three combinations with seven wavelengths were determined to predict the color parameters. Table 11.2 shows the calibration and full cross-validation results of beef quality attributes for the MLR models based on the LD parameter combinations of the optimal wavelengths (Peng et al., 2009a).

11.3 ASSESSMENT OF MEAT SAFETY ATTRIBUTES

Owing to high moisture content and abundant nutritional contents, raw meat is recognized as one of the most perishable foods and is susceptible to contamination during storage and marketing, thus leading to food-borne diseases. Among the different causes for the spoilage of fresh meat, microorganisms are the most common, which may lead to food-borne diseases (Barbin et al., 2013; Dissing et al., 2013; Cheng et al., 2014). To ensure the safety of meat, it is essential to not only control and detect microbes effectively, but also measure the microbial species and population accurately.

11.3.1 TOTAL VIABLE COUNT

Meat is a highly perishable food product that, unless properly stored, processed, packaged, and distributed, spoils quickly due to microbial growth and becomes hazardous (McDonald and Sun, 1999). Spoilage in meat is caused by the growth and

TABLE 11.2
Calibration and Prediction Results for Quality Attribute of Beef Samples Based on the Optimal Wavelength Combinations for Lorentzian Parameters

Quality Attributes	Calibration Set		Cross-Validation	
	R_C	SEC	R_P	SEP
pH	0.85	0.07	0.82	0.08
L^*	0.94	0.84	0.92	0.90
a^*	0.91	1.32	0.89	1.51
b^*	0.92	0.33	0.88	0.41
WBSF (N)	0.92	9.40	0.87	11.0

Source: Adapted from Peng, Y., J. Wu, and J. Chen. 2009a. Prediction of beef quality attributes using hyperspectral scattering imaging technique. *ASABE Annual International Meeting*, Paper No. 096424, Reno, NV.

enzymatic activity of microorganisms, which result in the decomposition of nutritional matter and the formation of metabolites. Meat with excessive bacteria causes harm to human health, and it is thus critical to guarantee the safety of meat supplied to the market. However, the traditional methods that currently are still in use for bacterial spoilage detection, such as plate count, enumeration based on microscopy, adenosine triphosphate (ATP) bioluminescence, and measurement of electrical phenomena, cannot achieve rapid, accurate, and nondestructive detection of bacterially contaminated meat (Xiong et al., 2014).

Detection of microbial contamination, particularly total viable count (TVC) is a common microbiological test conducted in large scale for food, medical, and biological samples. TVC is a common and essential indicator in assessing the quality and safety of meat. It not only reflects the status of bacteriological contamination, but also determines whether the meat is fresh or putrid (Wang et al., 2010b, 2011; Peng et al., 2011; Tao et al., 2011; Zhang et al., 2013c).

The application of multispectral imaging technology using UV to NIR wavelengths (405–970 nm) was reported for the estimation of TVC in the determination of the microbiological quality of beef filets during aerobic storage (Panagou et al., 2014). In addition, hyperspectral imaging in the VIS/NIR spectral range was also reported for the determination of TVC in meats (Feng and Sun, 2013; Lin et al., 2013). The acquired reflectance hyperspectral images were transformed into absorbance and Kubelka–Munk (K–M) units. The absorbance model had superior results compared with the other models. The best full-wavelength PLSR model had the correlation coefficient of 0.97 and 0.93 for calibration and validation sets, respectively. To enhance the efficiency and effectiveness of the models, a stepwise regression method was used to select fewer-feature wavelengths (954, 957, 1138, 1148, and 1328 nm). The results showed that the K–M model had a superior result when the feature wavelengths were selected. The correlation coefficient and root mean-squared error were 0.96 and 0.40 $\log_{10}$CFU g^{-1}, respectively, for the calibration set and 0.94 and 0.50 $\log_{10}$CFU g^{-1}, respectively, for the validation set.

Meat spoilage causes changes in the structural characteristics. Spectral scattering imaging can thus be an effective method for the detection of meat TVC. Hyperspectral scattering imaging has been intensively researched for predicting microbiological attributes of red meats in recent studies (Peng and Wang, 2008; Peng et al., 2009b; Tao et al., 2010). The majority of these studies used linear regression methods for prediction modeling. For example, Tao et al. (2010) used hyperspectral reflectance imaging for assessing the total plate count on a chilled pork surface, and two prediction models were established using MLR and PLSR, with encouraging results of $R_P = 0.886$ and 0.863, respectively. Furthermore, hyperspectral scattering technique was also used to detect the TVC of beef (Peng et al., 2009b), in which an MLR prediction model was developed based on individual Lorentzian parameters and their combinations at different wavelengths. The best predictions were acquired with $R_P^2 = 0.96$ and SEP $= 0.23$ for $\log_{10}$ (TVC). Although PLSR or MLR is promising, neither of them is able to address the possible nonlinear problems. Hence, some researchers used nonlinear modeling methods such as artificial neural networks, spectral angle mapper, and support vector machine (SVM). For example, Peng and Wang (2008) successfully developed a hyperspectral scattering imaging system with SVM for detecting the TVC of bacteria in pork with $R_P = 0.87$, which was better than that of the MLR method. To improve the accuracy of prediction models, Wang et al. (2011) used a hyperspectral scattering imaging system coupled with the least-squares support vector machines (LS-SVMs) for predicting the TVC of pork. Eight optimal wavelengths (477, 509, 540, 552, 560, 609, 720, and 772 nm) were selected to construct the TVC prediction model using the average reflectance intensity of the hyperspectral scattering images at these wavelengths. The ultimate model with $R_P^2 = 0.924$ and SEP $= 0.33$ indicated that the hyperspectral scattering imaging system with LS-SVM was more effective to predict the TVC of pork.

In a further study, a hyperspectral imaging system was used to collect the hyperspectral scattering images from pork samples and the Boltzmann equation was applied to fit the scattering profiles of pork to extract the optical absorption and reduced scattering coefficients (μ_a and μ_s') (Tao et al., 2012c). The stepwise discrimination method was performed to determine the optimal wavelengths based on μ_a and μ_s', respectively. MLR models were established using the parameters μ_a and μ_s' for the determined wavelengths, and full cross-validation was conducted to evaluate the model performance. The prediction results of pork TVC using μ_a and μ_s' are shown in Figure 11.13, with $R_P = 0.86$ and 0.80, respectively.

11.3.2 Total Volatile Basic Nitrogen

Freshness is an essential index of meat quality, and directly influences storage and logistics. Total volatile basic nitrogen (TVB-N), pH value, and color are most important for meat freshness assessment, and among the three parameters, TVB-N is the most important index (Li et al., 2011; Zhang et al., 2012). Although the external

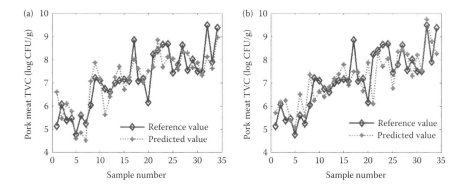

FIGURE 11.13 Prediction results for TVC based on (a) the absorption coefficient μ_a and (b) the reduced scattering coefficient μ_s'.

and sensory qualities can be determined by human perception, it is impossible for humans to visually detect freshness attributes, including pH value and TVB-N content.

Short-wavelength NIR spectroscopy (400–1000 nm) was reported for the detection of pH, TVB-N, mesophilic bacteria, and among others, as a means of predicting freshness in chicken (Grau et al., 2011). The wavelengths of 413, 426, 449, 460, 473, 480, 499, 638, 942, 946, 967, 970, and 982 nm were selected for prediction of the freshness parameters.

With regard to spectral scattering imaging, Li et al. (2013) developed a multi-spectral imaging system, which was used to collect scattering images from 64 pork samples. The TVB-N of the pork samples was measured using a semiautomatic nitrogen analyzer. The scattering images at the wavelengths of 517, 550, 560, 580, 600, 760, 810, and 910 nm were used for TVB-N modeling. Figure 11.14a shows an original scattering image of a sample at the wavelength of 560 nm, and Figure 11.14b is the 3-D intensity plot for the image shown in Figure 11.14a. In Figure 11.14b, the

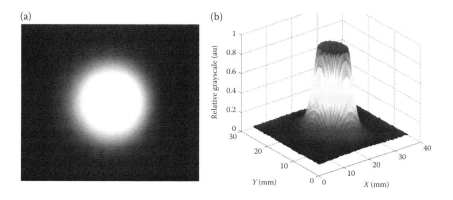

FIGURE 11.14 Raw scattering image and its 3-D curved surface plot at 560 nm: (a) 2-D raw scattering image and (b) 3-D scattering intensity plot.

relative grayscale value denotes the reflected light intensity of the scattering image area for the sample.

For obtaining the scattering profiles, image processing algorithms such as binarization, corrosion, and expansion were utilized to find the gravity center. Then, the scattering image was divided into concentric rings with the same bandwidth of 1–2 pixels. Taking the radius of the concentric ring as the horizontal ordinate (*x*-coordinate), and the grayscale value of the concentric band as the vertical coordinate (*y*-coordinate), the scattering profile of each image was finally obtained. The scattering profile at each wavelength was fitted by the four-parameter LD function mentioned above. The TVB-N prediction model was developed on the basis of the four fitting parameters of the LD function combined with the average grayscale values at the selected wavelengths. The correlation coefficient between the measured TVB-N values and predicted TVB-N values was 0.93, and the standard error of prediction was 1.68 mg/100 g (Figure 11.15).

11.3.3 *E. coli* Contamination

Hyperspectral scattering technique was applied in detecting *E. coli* contamination in pork (Tao et al., 2012a). In that study, the scattering profiles were fitted by the LD function with three parameters. The results showed that MLR models based on the parameter *a* gave a high R_p of 0.877.

In a further study, Tao and Peng (2014) showed that the modified four-parameter Gompertz distribution function (Equation 11.4) gave an accurate fit to the hyperspectral scattering profiles of pork in the spectral range of 400–1100 nm. A comparison

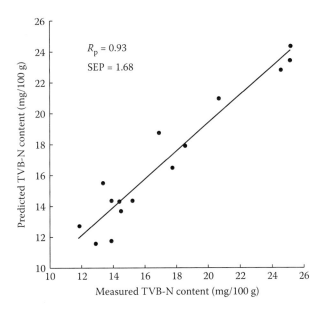

FIGURE 11.15 Correlation between the measured values and the estimated values of TVB-N.

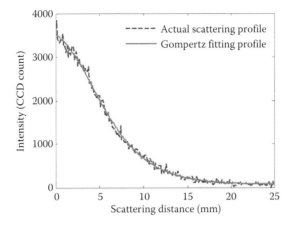

FIGURE 11.16 Fitting result by the modified Gompertz function (Equation 11.4) at 550 nm.

between the actual scattering profile and Gompertz fitting profile at 550 nm is shown in Figure 11.16.

Stepwise discrimination was performed to determine the optimal variable combination for parameters α, β, ε, and δ. The optimal wavelengths for establishing MLR models based on Gompertz distribution function parameters were determined. MLR models for each parameter of α, ε, and δ were developed using the determined wavelengths for *E. coli* contamination. In addition, the model for the combination of four parameters was also developed for *E. coli* prediction. The wavelengths of 625, 754, 791, 829, 865, and 868 nm were determined based on the descending order of the correlation coefficient. The results for *E. coli* prediction for calibration and validation sets of samples are summarized in Table 11.3. The calibration and validation results from the model for the combined four Gompertz parameters α, β, ε, and δ are shown in Figure 11.17 (Tao and Peng, 2014).

TABLE 11.3
Results of Calibration and Prediction from MLR with Gompertz Parameters for Prediction of *E. coli* Contamination

Gompertz Parameters	Calibration Set		Prediction Set	
	R_C	SEC	R_P	SEP
α	0.89	0.84	0.83	1.02
ε	0.92	0.73	0.86	0.88
δ	0.95	0.58	0.92	0.66
[α, β, ε, δ]	0.99	0.27	0.94	0.64

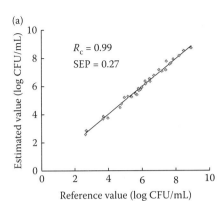
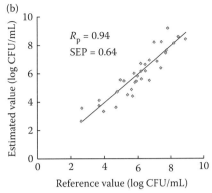

FIGURE 11.17 Prediction results of the MLR model with combined Gompertz parameters [α, β, ε, δ] (Equation 11.4) for *E. coli* contamination: (a) calibration samples and (b) prediction samples.

11.3.4 Shelf-Life Estimation

Predictive modeling based on variables in meat processing for predicting shelf life has gained interest in the meat industry (McDonald and Sun, 1999). The end of shelf life for fresh meat products is influenced by many factors, including breed difference and muscle fiber type, external influences such as diet and stress, and postharvest storage conditions including time, temperature, and packaging atmosphere. The characteristics that indicate the end of shelf life for fresh meat products include color deterioration due to myoglobin oxidation, rancidity due to lipid oxidation, and microbial spoilage (Antoniewski and Barringer, 2010). The pork industry uses pH to differentiate products of varying quality; thus, the effect of pH on shelf life is important during long transportation (Holmer et al., 2009). The rate and extent of pH decline have a great impact on the shelf life of meat and its eating quality. The moisture content of meat also affects shelf life. Both microbial growth and texture loss influence the shelf life of salted meat and products (Xiong et al., 2014).

TVC can also be a useful indicator to predict the shelf life of raw meat and distinguish meat spoilage during storage (Stanbridge and Davies, 1998). TVC, *Pseudomonas* spp., and *Brochothrix thermosphacta* are important parameters for determination of the microbiological quality and fresh level of meat during aerobic storage (Papadopoulou et al., 2011; Panagou et al., 2014).

On the basis of food microbiology growth law, a natural variation trend model of the bacterial contamination in meat was determined by an experiment, so that the remaining shelf life of a meat sample could be estimated from the bacterial contamination prediction value of the meat sample according to the trend model (Zhang et al., 2013a, b).

Spectral scattering imaging is effective for predicting the shelf life of meat. In a recent study, hyperspectral scattering images were collected from 54 pork samples (Zhang et al., 2013b). The bacteriological enumeration was traditionally performed according to the standard plate count method. The scattering profiles

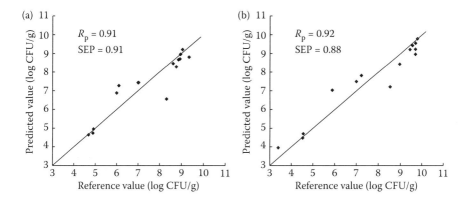

FIGURE 11.18 Predictions by the SVM model with combined Gompertz parameters [α, β, ε, δ] for pork spoilage attributes: (a) TVC and (b) *Pseudomonas* spp.

were accurately fitted by the four-parameter Gompertz function at individual wavelengths. The SVM provides an excellent generalization ability for limited samples. The SVM was applied to develop prediction models for two microbial indexes. The total samples were randomly divided into two groups: 40 samples were selected randomly as the training set and the remaining 14 samples were selected as the validation set. The SVM model was validated using the samples from the validation group. The Gompertz parameter δ yielded better prediction results than the other three parameters for TVC and *Pseudomonas* spp., with R_P = 0.88 and 0.92 and SEP = 0.85 and 0.92, respectively. Comparative analysis showed that a multiparameter combination [α, β, ε, δ] produced better prediction results for both TVC and *Pseudomonas* spp. than using individual Gompertz parameters, with R_P = 0.91 and 0.92 and SEP = 0.91 and 0.88, respectively (Zhang et al., 2013b). Figure 11.18 shows the prediction of SVM models for the two shelf-life parameters.

Gompertz function is widely used for describing the biological growth (Bhowmick and Bhattacharya, 2014). As an empirical bacterial growth function, the Gompertz model (Equation 11.6) was applied to describe the microbial growth for shelf-life prediction as a function of storage time (McDonald and Sun, 1999; Raab et al., 2008; Kreyenschmidt et al., 2010)

$$B = p + q\, e^{-e^{-k(t-m)}} \qquad (11.6)$$

where B is the bacterial count (TVC or *Pseudomonas* spp.) of meat; t is the storage time (days); p is the initial number of bacteria; q is the asymptotic value of bacteria growth; k is the maximum rate of bacteria growth; and m is the storage time at the maximum growth rate; $k \times m$ indicates the capacity of bacteria growth.

The experimental microbiological growth data of TVC and *Pseudomonas* spp. were fitted using nonlinear regression analysis, and the results are shown in Figure 11.19a and b, respectively. Usually, there is a good correlation between TVC and *Pseudomonas* spp. (Zhang et al., 2013b). Either of them can be used to determine the shelf life of meat.

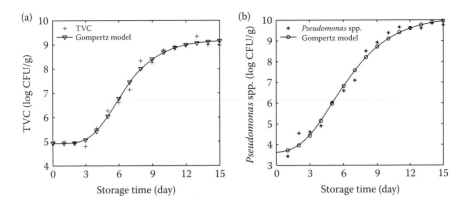

FIGURE 11.19 Growth curves of (a) TVC and (b) *Pseudomonas* spp. at 4°C fitted with Gompertz model (Equation 11.6).

If the TVC of a meat sample is obtained by the optical scattering method, the remaining shelf life of the meat can be predicted (Figure 11.20). For instance, the safety limit value of TVC (expressed as B_e) for the shelf life of a certain kind of pork, depends on the industrial meat safety standard. The end of shelf life is determined as the time t_e based on B_e for the TVC of the meat can be calculated from Equation 11.6. If the TVC of the i pork meat sample is B_i predicted by the optical scattering method, the storage time t_i for the pork meat can be obtained by inversion from Equation 11.6. Furthermore, the remaining shelf-life T of the meat can be predicted from the following equation:

$$T = t_e - t_i = t_e - \left(m - \frac{1}{k}\ln\left(-\ln\frac{1}{q}(B_i - p)\right) \right) \quad (11.7)$$

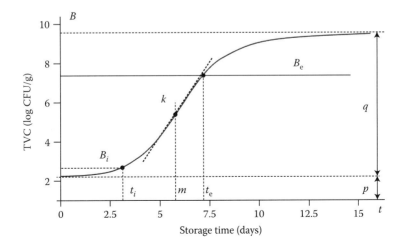

FIGURE 11.20 Prediction of the remaining shelf life of meat by TVC.

11.4 PORTABLE AND MOVABLE PROTOTYPE DEVICES

Advances in optical nondestructive detection technology over recent years have led to the development and use of a number of commercially viable detection devices for the meat industry. An online NIR system was recently developed by Analytical Spectral Devices (Boulder, CO), in partnership with the U.S. Department of Agriculture (USDA), to predict the tenderness of meat at the processing plant in real time (Shackelford et al., 2012). In addition, a hyperspectral and multispectral imaging system has been developed by the USDA Agricultural Research Service for online detection of contaminants on poultry carcasses (Lawrence et al., 2003; Chao et al., 2008). Our laboratory recently developed a nondestructive, real-time detection system for simultaneously assessing multiple quality parameters of pork, which is now being used by a meat-processing company in China (Zhang et al., 2013). The following sections present the latest progress on the development of portable and movable spectral scattering imaging instruments for practical use.

11.4.1 Freshness Detection Device

A portable device based on multispectral scattering imaging technique was developed for meat freshness detection (Li et al., 2012). The device consists of a light unit, an image acquisition unit, a display unit, and a computer data processing unit. The image acquisition unit contains a high-performance VIS/NIR CCD camera, a data-acquisition card, and eight narrowband filters with FWHM between 10 and 15 nm. The light source unit with a regulated power supply provides stable VIS/NIR light to the system. The data processing unit is used to extract effective information from multispectral images and predict meat freshness in real time.

The device is small in size, and can be easily brought to the scene. A schematic drawing of the portable freshness detection device is shown in Figure 11.21. When a sample is placed on the loading table or holder, the table height is automatically adjusted for keeping a constant distance between the sample surface and lens of the camera. The freshness is predicted by three important parameters including TVB-N, color (L^*, a^*, and b^*), and pH value. The detection speed is 4 s per sample; the detection accuracy is greater than 92% (Li et al., 2012, 2013). By comparison, traditional detection methods such as the Kjeldahl nitrogen determination methods would take about 30 min.

The portable freshness detection device provides stable and reliable performance, and it is now being used by Xinjiang Yurun Food Company in China. It is fast, easy to operate, and affordable and flexible in instrumentation, and also saves manpower.

11.4.2 TVC Detection Device

As an important microbiology indicator for the sanitary quality and safety evaluation of meat, TVC must be tested in meat production and processing (Zhang et al., 2013b). For detection and enumeration of bacteria in meat, there are many methods

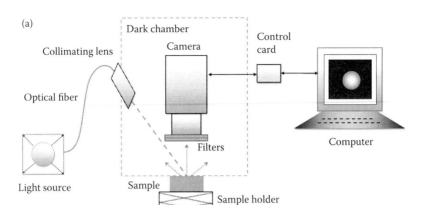

FIGURE 11.21 Portable meat freshness detection device: (a) schematic drawing and (b) device.

available, which include the plate-culturing method, ATP bioluminescence tests, and polymerase chain reaction. While these methods can give accurate results in the laboratory, they are time consuming and labor intensive (Ellis et al., 2002). Therefore, rapid and nondestructive methods that can be used by the meat industry for real-time detection of bacterial contamination in meat are urgently needed.

Although numerous optical techniques have shown the potential for detection and enumeration of bacteria in meat, few of them are currently being adopted for industrial applications. Light scattering is influenced by the structural and physical characteristics of meat, such as density, particle size, and muscle structures (Peng et al., 2011), and it is promising for predicting meat quality such as TVC. In Section 11.3.1, we have presented our research findings on using hyperspectral scattering imaging, combined with the Gompertz function, for rapid and nondestructive prediction of TVC in pork (Song et al., 2014; Tao and Peng, 2015).

A TVC detection device was developed in our laboratory (Song et al., 2014; Zhao et al., 2014). The hardware mainly consists of image collecting, height adjustment, and motion control systems, which is shown in Figure 11.22a.

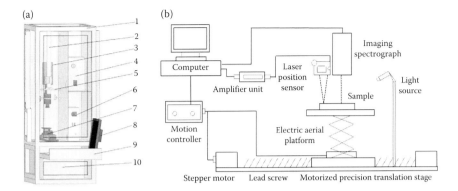

FIGURE 11.22 Meat TVC detection device: (a) schematic drawing (1—shield box, 2—stand column, 3—hyperspectral imaging unit, 4—light source, 5—sensor holder, 6—control interface, 7—displacement platform, 8—computer monitor, 9—monitor-mounting bracket, and 10—computer mainframe) and (b) height adjustment system.

The scattering image collection system mainly consists of a high-performance CCD camera (SensiCam QE, The Cooke Corp., now PCO-Tech Inc., Romulus, MI) and its control unit, an imaging spectrograph (ImSpector V10E, Spectral Imaging Ltd., Oulu, Finland) with the spectral range between 400 and 1100 nm, a specially assembled light unit with a quartz tungsten halogen lamp as the light source (Oriel Instruments, Irvine, CA), and a vertical-position sensor holder.

To reduce measurement errors, the position sensor is used to maintain a 25-cm distance between the sample and lens automatically, while the distance between the sample and light source is kept at 15 cm. A schematic drawing of the height adjustment system is shown in Figure 11.22b.

A motion control unit that consists of a motorized precision translation stage and an electric aerial platform is used to control the stepper motor to realize multipoint measurements. The motorized precision translation stage mounted with the electric aerial platform is used to adjust the height of the sample. The sample is moved up and down through the stepper motor with a fixed speed, and it halts during the signal acquisition.

Software was designed for the detection of meat TVC. Through the software, the spectrometer is triggered for data collection and saving as well as other operations. After the completion of sample spectral data acquisition, the four-parameter Gompertz distribution function is used to fit the scattering profiles of individual wavebands (400–1100 nm). The software automatically displays the TVC prediction result for the sample, using the MLR prediction model developed for the Gompertz parameters of optimal wavelengths (Song et al., 2014).

The software design was implemented by hybrid programming of Microsoft VS 2010 language and MATLAB® data processing platform, so that the program is transplantable. The system allows fast, real-time detection. The software operation is simple, versatile, and can be implemented in MATLAB (Zhao et al., 2014).

Experiments in commercial slaughtering houses demonstrated that this system is user friendly and has high detection accuracy (≥90%), low relative error (<4%), high speed (1–3 samples per second), and good reliability. The TVC detection device can play an important role in quality inspection and control of meat for the industry.

11.5 CONCLUSIONS

This chapter provides an overview of principles, methods, and techniques in spectral scattering imaging for the detection and inspection of meat quality and safety. Applications of hyperspectral and multispectral scattering techniques for multiple meat quality and safety attributes are presented. Several spectral scattering imaging-based inspection prototypes are also presented; their main features and potential for commercial applications are discussed.

Despite the fact that significant progress has been made, the spectral scattering technique still faces many technical difficulties for online analysis of meat quality and safety. First, the distance between the detection probe and the sample influences the prediction results, since the sample thickness varies. It is important to keep the sample at a constant distance during the imaging. Solutions to the problem could be realized through the hardware by automatically adjusting the position of the detection probe or sample, or through the software by means of correcting the original spectral signal. Second, sample surface features influence the prediction results. Owing to its irregular shape and uneven surface, original spectral signals need to be corrected. One way to solve this problem is through a symmetrical light supplement from different directions to decrease spectral signal distortion due to the irregular surface. In addition, the structural factors of meat, such as spatial variation in the muscle structure and type and direction of the muscle, also have adverse effects on the accuracy and reliability of prediction results. Finally, the optical characteristics of individual optical components (i.e., camera, spectrograph, light source, object lens, etc.) may vary from one system to another. This means that prediction models developed from a specific system are not transferrable to another system. By properly addressing these issues, the spectral scattering imaging technique should have a bright future for applications in the meat industry.

To meet commercial meat inspection demands, more effort should be given to enhance hardware and software for portable and movable devices for meat quality and safety evaluation. Optimal wavelengths should be first determined by the hyperspectral scattering imaging system for a particular application, and they can then be utilized in a multispectral scattering imaging system for fast image acquisition and processing. With further research and development, scattering imaging can become one of the most important techniques for real-time inspection of meat quality and safety.

ACKNOWLEDGMENTS

The authors gratefully acknowledge the Special Fund for Agro-scientific Research in the Public Interest (Project No. 201003008), and National Science and Technology Support Program (Project No. 2012BAH04B00), China for providing funding support for the research related to this chapter.

REFERENCES

Antoniewski, M. N., and S. A. Barringer. 2010. Meat shelf-life and extension using collagen/gelatin coatings: A review. *Critical Reviews in Food Science and Nutrition* 50(7): 644–653.

Barbin, D. F., G. ElMasry, D.-W. Sun, P. Allen, and N. Morsy. 2013. Non-destructive assessment of microbial contamination in porcine meat using NIR hyperspectral imaging. *Innovative Food Science and Emerging Technologies* 17: 180–191.

Barbin, D. F., C. M. Kaminishikawahara, A. L. Soares, I. Y. Mizubuti, M. Grespan, M. Shimokomaki, and E. Y. Hirooka. 2015. Prediction of chicken quality attributes by near infrared spectroscopy. *Food Chemistry* 168: 554–560.

Bhowmick, A. R., and S. Bhattacharya. 2014. A new growth curve model for biological growth: Some inferential studies on the growth of *Cirrhinus mrigala*. *Mathematical Biosciences* 254: 28–41.

Cen, H., R. Lu, and K. Dolan. 2010. Optimization of inverse algorithm for estimating the optical properties of biological materials using spatially resolved diffuse reflectance. *Inverse Problems in Science and Engineering* 18(6): 853–872.

Cen, H., R. Lu, F. Mendoza, and R. M. Beaudry. 2013. Relationship of the optical absorption and scattering properties with mechanical and structural properties of apple tissue. *Postharvest Biology and Technology* 85: 30–38.

Chao, K., C.-C. Yang, M. S. Kim, and D. E. Chan. 2008. High throughput spectral imaging system for wholesomeness inspection of chicken. *Applied Engineering in Agriculture* 24(4): 475–485.

Cheng, J. H., D.-W. Sun, X. A. Zeng, and H. B. Pu. 2014. Non-destructive and rapid determination of TVB-N content for freshness evaluation of grass carp (*Ctenopharyngodon idella*) by hyperspectral imaging. *Innovative Food Science Emerging Technologies* 21: 179–187.

Cluff, K., G. K. Naganathan, J. Subbiah, R. Lu, C. R. Calkins, and A. Samal. 2008. Optical scattering in beef steak to predict tenderness using hyperspectral imaging in the VIS-NIR region. *Sensory and Instrumentation for Food Quality* 2: 189–196.

Cluff, K., G. K. Naganathan, J. Subbiah, A. Samal, and C. R. Calkins. 2013. Optical scattering with hyperspectral imaging to classify longissimus dorsi muscle based on beef tenderness using multivariate modeling. *Meat Science* 95: 42–50.

Dale, L. M., A. Thewis, C. Boudry, I. Rotar, P. Dardenne, V. Baeten, and J. A. F. Pierna. 2013. Hyperspectral imaging applications in agriculture and agro-food product quality and safety control: A review. *Applied Spectroscopy Reviews* 48(1/4): 142–159.

Davis, C. C. 1996. *Lasers and Electro-Optics: Fundamentals and Engineering*. New York, NY: Cambridge University Press.

Dissing, B. S., O. S. Papadopoulou, C. Tassou, B. K. Ersbøll, J. M. Carstensen, E. Z. Panagou, and G.-J. Nychas. 2013. Using multispectral imaging for spoilage detection of pork meat. *Food and Bioprocess Technology* 6(9): 2268–2279.

Ellis, D. I., D. Broadhurst, D. B. Kell, J. J. Rowland, and R. Goodacre. 2002. Rapid and quantitative detection of the microbial spoilage of meat by Fourier transform infrared spectroscopy and machine learning. *Applied and Environmental Microbiology* 68(6): 2822–2828.

ElMasry, G., D. F. Barbin, D. W. Sun, and P. Allen. 2012a. Meat quality evaluation by hyperspectral imaging technique: An overview. *Food Science and Nutrition* 52(8): 689–711.

ElMasry, G., D. W. Sun, and P. Allen. 2011. Non-destructive determination of water holding capacity in fresh beef by using NIR hyperspectral imaging. *Food Research International* 44(9): 2624–2633.

ElMasry, G., D. W. Sun, and P. Allen. 2012b. Near-infrared hyperspectral imaging for predicting colour, pH and tenderness of fresh beef. *Journal of Food Engineering* 110: 127–140.

Farrell, T. J., M. S. Patterson, and B. Wilson. 1992. A diffusion-theory model of spatially resolved steady-state diffuse reflectance for the noninvasive determination of tissue optical properties *in vivo*. *Medical Physics* 19: 879–888.

Fekedulegn, D., M. P. Mac Siurtain, and J. J. Colbert. 1999. Parameter estimation of nonlinear growth models in forestry. *Silva Fennica* 33(4): 327–336.

Feng, Y.-Z., and D.-W. Sun. 2013. Determination of total visible count (TVC) in chicken breast fillets by near-infrared hyperspectral imaging and spectroscopic transforms. *Talanta* 105: 244–249.

Forrest, J. C., M. T. Morgan, C. Borggaard, A. J. Rasmussen, B. L. Jespersen, and J. R. Andersen. 2000. Development of technology for the early post mortem prediction of water holding capacity and drip loss in fresh pork. *Meat Science* 55(1): 115–122.

Grau, R., A. J. Sanchez, J. Giron, E. Iborra, A. Fuentes, and J. M. Barat. 2011. Nondestructive assessment of freshness in packaged sliced chicken breast using SW-NIR spectroscopy. *Food Research International* 44(1): 331–337.

Holmer, S. F., R. O. McKeith, D. D. Boler, A. C. Dilger, J. M. Eggert, D. B. Petry, F. K. McKeith, K. L. Jones, and J. Killefer. 2009. The effect of pH on shelf-life of pork during aging and simulated retail display. *Meat Science* 82(1): 86–93.

Iqbal, A., D.-W. Sun, and P. Allen. 2013. Prediction of moisture, color and pH in cooked, pre-sliced turkey hams by NIR hyperspectral imaging system. *Journal of Food Engineering* 117(1): 42–51.

Kamruzzaman, M., G. ElMasry, D. W. Sun, and P. Allen. 2012. Prediction of some quality attributes of lamb meat using near-infrared hyperspectral imaging and multivariate analysis. *Analytica Chimica Acta* 714: 57–67.

Khamis, A., Z. Ismail, K. Haron, and A. T. Mohammed. 2005. Nonlinear growth models for modeling oil palm yield growth. *Journal of Mathematics and Statistics* 1(3): 225–233.

Kim, M. S., K. Chao, D. E. Chan, W. Jun, A. M. Lefcourt, S. R. Delwiche, S. Kang, and K. Lee. 2011. Line-scan hyperspectral imaging platform for agro-food safety and quality evaluation: System enhancement and characterization. *Transactions of the ASABE* 54(2): 703–711.

Kreyenschmidt, J., A. Hübner, E. Beierle, L. Chonsch, A. Scherer, and B. Petersen. 2010. Determination of the shelf life of sliced cooked ham based on the growth of lactic acid bacteria in different steps of the chain. *Journal of Applied Microbiology* 108: 510–520.

Lawrence, K. C., W. R. Windham, B. Park, and R. J. Buhr. 2003. Hyperspectral imaging system for identification of faecal and ingesta contamination on poultry carcasses. *Journal of Near Infrared Spectroscopy* 11: 261–281.

Li, C., Y. Peng, X. Tang, and A. Sasao. 2013. A portable system for prediction of pork freshness parameters using multispectral imaging technology. *ASABE Annual International Meeting*, Paper No. 1587037, Kansas City, MO.

Li, C., Y. Peng, W. Wang, and X. Tang. 2012. Device for nondestructive detection system of pork freshness based on multispectral imaging technology. *Transactions of the Chinese Society for Agricultural Machinery* 43(S1): 202–206.

Li, Y., L. Zhang, Y. Peng, X. Tang, K. Chao, and S. Dhakal. 2011. Hyperspectral imaging technique for determination of pork freshness attributes. *SPIE/Defense, Security and Sensing, Sensing for Agriculture and Food Quality and Safety*, Paper No. 8027-16, Orlando, FL.

Liao, Y., Y. Fan, and F. Cheng. 2012. On-line prediction of pH values in pork using visible/near-infrared spectroscopy with wavelet de-noising and variables selection methods. *Journal of Food Engineering* 109(4): 668–675.

Lin, H., J. Zhao, Q. Chen, and Y. Zhang. 2013. Rapid detection of total viable count (TVC) in pork meat by hyperspectral imaging. *Food Research International* 54: 821–828.

McDonald, K., and D. W. Sun. 1999. Predictive food microbiology for the meat industry: A review. *International Journal of Food Microbiology* 52: 1–27.

Naganathan, G. K., L. M. Grimes, J. Subbiah, C. R. Calkins, A. Samal, and G. E. Meyer. 2008. Partial least squares analysis of near-infrared hyperspectral images for beef tenderness prediction. *Sensing and Instrumentation for Food Quality and Safety* 2(3): 178–188.

Panagou, E. Z., O. Papadopoulou, J. M. Carstensen, and G.-J. E. Nychas. 2014. Potential of multispectral imaging technology for rapid and non-destructive determination of the microbiological quality of beef filets during aerobic storage. *International Journal of Food Microbiology* 174: 1–11.

Papadopoulou, O., E. Z. Panagou, C. C. Tassou, and G.-J. E. Nychas. 2011. Contribution of Fourier transform infrared (FTIR) spectroscopy data on the quantitative determination of minced pork meat spoilage. *Food Research International* 44(10): 3264–3271.

Pedersen, D. K., S. Morel, H. J. Andersen, and S. B. Engelsen. 2003. Early prediction of water-holding capacity in meat by multivariate vibrational spectroscopy. *Meat Science* 65(1): 581–592.

Peng, Y., and R. Lu. 2005. Modeling multispectral scattering profiles for prediction of apple fruit firmness. *Transactions of the ASABE* 48(1): 235–242.

Peng, Y., and R. Lu. 2007. Prediction of apple fruit firmness and soluble solids content using characteristics of multispectral scattering images. *Journal of Food Engineering* 82(2): 142–152.

Peng, Y., and W. Wang. 2008. Prediction of pork meat total viable bacteria count using hyperspectral imaging system and support vector machines. *Food Processing Automation Conference*, Paper No. 701P0508, Reno, NV.

Peng, Y., J. Wu, and J. Chen. 2009a. Prediction of beef quality attributes using hyperspectral scattering imaging technique. *ASABE Annual International Meeting*, Paper No. 096424, Reno, NV.

Peng, Y., J. Zhang, W. Wang, Y. Li, J. Wu, H. Huang, X. Gao, and W. Jiang. 2011. Potential prediction of the microbial spoilage of beef using spatially resolved hyperspectral scattering profiles. *Journal of Food Engineering* 102(2): 163–169.

Peng, Y., J. Zhang, J. Wu, and H. Hang. 2009b. Hyperspectral scattering profiles for prediction of the microbial spoilage of beef. *SPIE/Defense, Security and Sensing*, Paper No. 7315-25, Orlando, FL.

Phongpa-Ngan, P., S. E. Aggrey, J. H. Mulligan, and L. Wicker. 2014. Raman spectroscopy to assess water holding capacity in muscle from fast and slow growing broilers. *LWT— Food Science and Technology* 57: 696–700.

Qiao, J., N. Wang, M. O. Ngadi, A. Gunenc, M. Monroy, C. Gariépyb, and S. O. Prasher. 2007. Prediction of drip-loss, pH, and color for pork using a hyperspectral imaging technique. *Meat Science* 76(1): 1–8.

Qin, J., and R. Lu. 2007. Measurement of the absorption and scattering properties of turbid liquid foods using hyperspectral imaging. *Applied Spectroscopy* 61(4): 388–396.

Qin, J., and R. Lu. 2008. Measurement of the optical properties of fruits and vegetables using spatially resolved hyperspectral diffuse reflectance imaging technique. *Postharvest Biology and Technology* 49: 355–365.

Qin, J. W., K. L. Chao, M. S. Kim, R. Lu, and T. F. Burks. 2013. Hyperspectral and multispectral imaging for evaluating food safety and quality. *Journal of Food Engineering* 118: 157–171.

Raab, V., S. Bruckner, E. Beierle, Y. Kampmann, B. Petersen, and J. Kreyenschmidt. 2008. Generic model for the prediction of remaining shelf life in support of cold chain management in pork and poultry supply chains. *Journal on Chain and Network Science* 8: 59–73.

Shackelford, S. D., T. L. Wheeler, D. A. King, and M. Koohmaraie. 2012. Field testing of a system for online classification of beef carcasses for longissimus tenderness using visible and near-infrared reflectance spectroscopy. *Journal of Animal Science* 90: 978–988.

Song, Y., Y. Peng, H. Guo, L. Zhang, and J. Zhao. 2014. A method for assessing the total viable count of fresh meat based on hyperspectral scattering technique. *Spectroscopy and Spectral Analysis* 34(3): 741–745.

Stanbridge, L. H., and A. R. Davies. 1998. The microbiology of meat and poultry. In Davies, A., and R. Board (eds.), *The Microbiology of Chill Stored Meat*, 174–219. London: Blackie Academic & Professional.

Tao, F. 2013. Study on the rapid and nondestructive detection methods and antimicrobial delivery for the bacterial contamination of pork. The PhD thesis of China Agricultural University, May, 2013, No. B10209185, Beijing, China.

Tao, F., and Y. Peng. 2014. A method for nondestructive prediction of pork meat quality and safety attributes by hyperspectral imaging technique. *Journal of Food Engineering* 126: 98–106.

Tao, F., and Y. Peng. 2015. A nondestructive method for prediction of total viable count in pork meat by hyperspectral scattering imaging. *Food Bioprocess Technology* 8: 17–30.

Tao, F., Y. Peng, and Y. Li. 2011. Detection of bacterial contamination of pork using hyperspectral scattering technique. *ASABE Annual International Meeting*, Paper No. 1110805, Louisville, KY.

Tao, F., Y. Peng, Y. Li, K. Chao, and S. Dhakal. 2012a. Simultaneous determination of tenderness and *Escherichia coli* contamination of pork using hyperspectral scattering technique. *Meat Science* 90: 851–857.

Tao, F., Y. Peng, Y. Song, H. Guo, and K. Chao. 2012b. Improving prediction of total viable counts in pork based on hyperspectral scattering technique. *SPIE/Defense, Security and Sensing, Sensing for Agriculture and Food Quality and Safety*, Paper No. 8369-9, Baltimore, MD.

Tao, F., X. Tang, Y. Peng, and S. Dhakal. 2012c. Classification of pork quality characteristics by hyperspectral scattering technique. *ASABE Annual International Meeting*, Paper No. 121341184, Dallas, TX.

Tao, F., W. Wang, L. Y. Li, Y. Peng, J. Wu, J. Shan, and L. Zhang. 2010. A rapid nondestructive measurement method for assessing the total plate count on chilled pork surface. *Spectroscopy and Spectral Analysis* 30(12): 3405–3409.

Tuchin, V. V. 2000. Tissue image contrasting using optical immersion technique. *Biomedical Photonics and Optoelectronic Imaging, Proceedings of SPIE* 4224: 351–365.

Wang, W., Y. Peng, H. Huang, and J. Wu. 2011. Application of hyperspectral imaging technique for the detection of total viable bacteria count in pork. *Sensor Letters* 9: 1–7.

Wang, W., Y. Peng, and J. Wu. 2010a. Prediction of pork water-holding capacity using hyperspectral scattering technique. *ASABE Annual International Meeting*, Paper No. 1008571, Pittsburgh, PA.

Wang, W., Y. Peng, and X. Zhang. 2010b. Study on modeling method of total viable count of fresh pork meat based on hyperspectral imaging system. *Spectroscopy and Spectral Analysis* 30(2): 411–415.

Wu, D., and D. W. Sun. 2013. Advanced applications of hyperspectral imaging technology for food quality and safety analysis and assessment: A review—Part II: Applications. *Innovative Food Science and Emerging Technologies* 19: 15–28.

Wu, J., Y. Peng, J. Chen, W. Wang, X. Gao, and H. Huang. 2010. Study of spatially resolved hyperspectral scattering images for assessing beef quality characteristics. *Spectroscopy and Spectral Analysis* 30(7): 1815–1819.

Wu, J., Y. Peng, Y. Li, W. Wang, J. Chen, and S. Dhakal. 2012. Prediction of beef quality attributes using VIS/NIR hyperspectral scattering imaging technique. *Journal of Food Engineering* 109(2): 267–273.

Xia, J. J., E. P. Berg, J. W. Lee, and G. Yao. 2007. Characterizing beef muscles with optical scattering and absorption coefficients in VIS-NIR region. *Meat Science* 75: 78–83.

Xiong, Z., D. W. Sun, X. A. Zeng, and A. Xie. 2014. Recent developments of hyperspectral imaging systems and their applications in detecting quality attributes of red meats: A review. *Journal of Food Engineering* 132: 1–13.

Zhang, H., Y. Peng, W. Wang, S. Zhao, and S. Dhakal. 2013. Nondestructive real-time detection system for assessing main quality parameters of fresh pork. *Transactions of the Chinese Society for Agricultural Machinery* 44(4): 147–151.

Zhang, L., Y. Li, Y. Peng, W. Wang, F. Tao, and J. San. 2012. Determination of pork freshness attributes by hyperspectral imaging technique. *Transactions of the CSAE* 28(7): 254–259.

Zhang, L., Y. Peng, S. Dhakal, Y. Song, J. Zhao, and S. Zhao. 2013a. Rapid non-destructive assessment of pork edible quality by using VIS/NIR spectroscopy technique. *SPIE/Defense, Security and Sensing, Sensing for Agriculture and Food Quality and Safety*, Paper No. 8721-6, Baltimore, MD.

Zhang, L., Y. Peng, S. Dhakal, F. Tao, Y. Song, and S. Zhao. 2013b. Spoilage detection of chilled meat during shelf life by using hyperspectral imaging technique. *ASABE Annual International Meeting*, Paper No. 131587037, Kansas City, MO.

Zhang, L., Y. Peng, Y. Liu, S. Dhakal, J. Zhao, and Y. Zhu. 2013c. Nondestructive evaluation of chilled meat TVC by comparison between reflection and scattering spectral profiles from hyperspectral images. *CIGR Section VII International Technical Symposium on Advanced Food Processing and Quality Management*, Paper No. P128, November 3–7, Guangzhou, China.

Zhao, J., Y. Peng, H. Guo, F. Tao, and L. Zhang. 2014. Control and analysis software design of hyperspectral imaging system for detection in agricultural food quality. *Transactions of the Chinese Society for Agricultural Machinery* 45(9): 210–215.

Zhao, J., J. Zhai, M. Liu, and J. Cai. 2006. The determination of beef tenderness using near-infrared spectroscopy. *Spectroscopy and Spectral Analysis* 264: 640–642.

12 Light Scattering Applications in Milk and Dairy Processing

Czarena Crofcheck

CONTENTS

12.1 Introduction 319
12.2 Light Scattering in Participating Media 320
12.3 Determining the Size Distribution 321
12.4 Monitoring Composition 323
12.5 Monitoring Milk Coagulation and Synersis during Cheesemaking 324
12.6 Food Quality and Safety 327
12.7 Concluding Remarks 327
References 327

12.1 INTRODUCTION

Milk is a complex biological fluid composed of water, fat, protein, lactose, citric acid, and inorganic compounds (Walstra and Jenness, 1984). Although every constituent plays at least a minor role in the scattering of light, the majority of scattering is due to milk fat globules and proteins. The most prevalent protein in milk is casein, which exists as a colloidal dispersion of particles known as casein micelles. The remaining proteins are referred to as whey proteins. In addition to scattering effects, fat globule size and concentration have been shown to have large effects on transmission measurements (Ben-Gera and Norris, 1968).

Light scattering by fat globules and casein micelles causes milk to appear turbid and opaque. These two components scatter light differently because of the differences in size, number, and optical properties (e.g., index of refraction) of the particles. The average particle diameter of casein micelles falls in the range of 0.130–0.160 µm (Ruettiman and Ladisch, 1987) and milk fat in the form of globules in the range of 0.1–10 µm for unhomogenized milk (Mulder and Walstra, 1974). Skim milk appears slightly blue because small casein micelles predominately scatter shorter wavelengths of visible light (blue). Whole milk appears white because larger fat globules multiple-scatter all wavelengths of incident light.

The surfaces of fat globules and casein micelles are different and, as a result, may scatter light differently. Surrounding the fat globule is a membrane, approximately 2% by weight of the entire globule. Therefore, the surface of the fat globule is smooth and "solid." Fat globules are not simple emulsion droplets in solution because of this

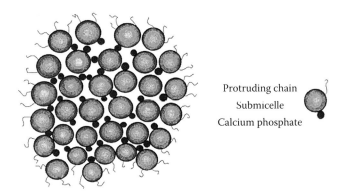

FIGURE 12.1 Structure of a casein micelle showing submicelles, calcium phosphate, and protruding chains that make up the "hairy" layer.

membrane. On the other hand, casein micelles are made up of smaller submicelles (Figure 12.1), and their composition is susceptible to the movement of serum proteins and water. The outer surfaces of the micelles are covered with hair-like macropeptides, so that the micelles appear to be covered with a "hairy" layer (Walstra and Jenness, 1984).

Casein makes up approximately 80% of the total proteins in milk. The volume fraction of casein micelles depends upon whether the "hairy" layer of macropeptides surrounding the micelle is taken into consideration. The volume fraction given by Walstra and Jenness (1984) is 0.06 without the "hairy" layer and 0.12 with the "hairy" layer. The effective thickness of the hairy layer is at least 5 nm. The volume fraction of fat globules is approximately 0.042 (Walstra and Jenness, 1984), but changes as fat is added during the milk standardization process. Approximate particle densities for casein micelles and fat globules are 10^{14} and 10^{10} per mL, respectively (Walstra and Jenness, 1984).

12.2 LIGHT SCATTERING IN PARTICIPATING MEDIA

In studying light scattering in turbid media, the important aspects of system geometry are the boundaries and placement of the light source and detector. Investigations with various boundary conditions have been done, including medium bound by two parallel planes (Bolt and ten Bosch, 1993), and semi-infinite (Haskell et al., 1994), and infinite (Haskell et al., 1994) medium. Actual scattering and absorption of light can vary with source–detector distance and position. The source and the detector can be positioned side by side (separated by a distance r) (Ishimaru, 1977) or on either side of the sample (separated by a depth h) (Ishimaru and Kuga, 1982). The latter orientation is associated with the medium bound by two parallel planes.

The behavior of the light can be studied with a point source (Bonner et al., 1987; Haskell et al., 1994) or a cylindrical beam (Ishimaru and Kuga, 1982; Bolt and ten Bosch, 1993). Typically, a point source can be created using an integrating sphere, which produces uniform scattering of light in all directions. The inside of the sphere is covered with a diffuse white reflective coating, such that it can create uniform scattering in every direction. In contrast, light is considered collimated when the beam is cylindrical, emitting straight into the sample and then scattering radially. Additional information about the properties of the milk materials can be determined using the full Mueller matrix of scattering data, which also includes the changes in the polarization of light (Crofcheck et al., 2005).

Imaging techniques can also be used to measure important parameters for milk characterization. Hyperspectral imaging, where spectral and spatial data are collected together, can be used to measure the absorption and scattering properties of turbid foods over the spectral range of 530–900 nm (Qin and Lu, 2007).

12.3 DETERMINING THE SIZE DISTRIBUTION

Many researchers investigated the size distribution of casein micelles (Holt et al., 1973; Slattery, 1977; Brooker and Holt, 1978; McMahon and Brown, 1984). The most popular methods have been light scattering techniques and electron microscopy (Schmidt et al., 1974b; Holt et al., 1975; Schmidt et al., 1977).

Light scattering can be categorized into elastic and inelastic scattering. Elastic light scattering shows no change in light wavelength and frequency during the scattering process; examples include Rayleigh scattering and Mie scattering. Inelastic light scattering shows changes in both light wavelength and frequency during the scattering process, and examples include Raman scattering, Brillouin scattering, and Thomson scattering. Rayleigh scattering is the scattering of light by tiny practices (smaller than the wavelength of incident light) such as molecules or atoms in all directions uniformly and is wavelength dependent. Mie scattering is the scattering of light by large particles or molecules (equal to or larger than the wavelength of incident light) in a nonuniform pattern. Mie scattering is wavelength independent and the scattering intensity is proportional to the size of particle. The larger the particle size, the more light is scattered in the forward direction than in the backward direction. Raman scattering shows alteration in photon energy as light scatters off a molecule. The shift in photon energy is dependent on the vibrational state of the molecule. If the scattered photon energy is lower than that of the incident light, the transfer of energy is referred to as Stokes Raman scattering. If the scattered photon energy is higher than that of the incident wavelength, the transfer of energy is referred to as anti-Stokes Raman scattering. Brillouin scattering is the interaction of light with thermally excited acoustic photons in a physical medium. Thomson scattering is the scattering of electromagnetic waves by free charged particles. Previous research on the particle size distribution of casein micelles were based on both elastic and inelastic scattering models (Line et al., 1971; Holt et al., 1975).

Instead of light wavelength, electron microscopy uses electrons to create a magnified image of the sample in order to achieve a much higher magnification and resolving power than conventional optical microscopy. Electrons are emitted from a high-energy electron beam, pass through electromagnetic lenses, and accelerate down the microscopic column toward the sample at an accelerating voltage range of 80–300 kV. The path of electron beam is focused and controlled by a series of electromagnetic lenses. The resolution of electron microscopy is proportional to the electron wavelength. Increasing the acceleration voltage will decrease the electron wavelength resulting in high-resolution images. On the other hand, excessively high resolution can hinder the image contrast as the scattering of electrons is inversely proportional to the velocity of the acceleration voltage. Another limitation to electron microscopy is the limited frequency a sample can be exposed to electron irradiation. A fraction of the radiation can break chemical bonds and destroy the sample rapidly. Thus, a low dosage of electrons is often used for high-resolution imaging.

The original form of electron microscopy is called transmission electron microscopy (TEM). Since then many types of electron microscopy have been developed. Among them, cryo-transmission electron microscopy (cryo-TEM) and scanning electron microscopy (SEM) are the two most popular microscopy types used for biological samples. Cryo-TEM operates similarly to a TEM with the capability of creating an image of a frozen hydrated sample. A thin film sample is spread onto an electron microscopy grid and held in a specialized holder that is cooled in either liquid nitrogen or helium. The cryogenic temperature reduces the amount of electron irradiation damage imposed on the sample thus allowing a higher-resolution image to be taken using higher voltage compared to TEM (Milne et al., 2013). The rapid freezing of the sample also enables imaging of a biological sample at its most innate state without staining or other sample treatment. However, most biological samples are sensitive to radiation and their images can be quite noisy. SEM uses a focused electron beam to scan the surface of a solid sample at a selected point location in a raster scan pattern. The electrons interact with the atoms of the sample producing various detectable signals as a measure of the sample's surface and crystalline texture, composition, orientation, etc. Detectable signals include secondary electrons, backscattered electrons, diffracted backscattered electrons, characteristic x-rays, and light (cathodoluminescene). The secondary electron detector is the most common detector found in SEM next to the backscattered electron detector and provides the sample's topography and morphology data. The backscattered detector shows the contrast in the composition of a multiphase sample. The characteristic x-ray detector is used for elementary analysis and continuum x-rays measurement. The cathodoluminescence detector is used to detect signals produced from a luminescent sample. Since SEM is a selected point scanning instrument, its resolution is dependent on the system that produces the electron beam and the electron wavelength. Resolution is also limited by the volume of sample interacting with the electron beam. There are a few operating requirements for SEM. The sample must be composed of a dried solid material that can withstand high vacuum and fit into the microscopic chamber (maximum vertical dimension of 40 mm and horizontal dimension of 10 cm) (Swapp, 2015). The sample's surface must be electrically conductive; otherwise, an electrically conductive coating such as gold, gold/palladium, platinum, or iridium is to be applied to sample. In addition, any sample that has a tendency of decrepitating at low pressure will not be suitable for SEM.

In other methods, the micelles are fractionated by size using either ultracentrifugation or chromatographic means (Schmidt et al., 1974a; McGann et al., 1979; Anderson et al., 1984; Zbikowski et al., 1992). The average particle diameter can be determined with sedimentation, diffusion, or light scattering techniques (McMahon and Brown, 1984). When multiple regression analysis is applied to sedimentation experiments, molecular weight distributions can be determined (Nakai and van De

Voort, 1979). Typically, light scattering experiments result in larger values for the average diameter than the values obtained using electron microscopy (Ruettimann and Ladisch, 1987).

Similarly, the size distribution of fat has been investigated using spectroturbidimetry, ordinary and fluorescence microscopy, photomicrography, Coulter counting, gravity separation, and field-flow fractionation (Walstra, 1969; Walstra et al., 1969). Size distribution as a function of homogenization pressure was studied by Walstra (1975). Fat globules in raw milk were fractionated by size gravity separation (Ma and Barbano, 2000) and the size distribution in whole and skim milk was measured using a Coulter particle size analyzer (Attaie and Ritcher, 2000). Recently, Konokhova et al. (2014) utilized scanning flow cytometry to determine the size, shape, and refractive index of milk fat globules.

12.4 MONITORING COMPOSITION

As the benefits of online process control are realized, the need increases for precise, online monitoring systems capable of providing tight control of critical parameters. One important parameter is the consistency and composition of food processing materials, which dictate the raw material selection and/or the processing method and conditions. Monitoring and/or controlling the consistency of food processing materials can have a substantial quality and economic impact. Measurement of fat and protein content is important because of the increase in quality requirements in the food industry, where monitoring light absorbance and scattering makes it possible to develop real-time, online sensors (Dufour, 2011).

Spectroscopic analysis has been studied widely for fat and protein composition characterization in raw and homogenized milk, where wavelengths include infrared (IR), near IR (NIR), and far-NIR (1100–2500 nm) (Dufour, 2011). Visible light backscattering, where the scatter is stronger than in NIR, has been used to a lesser extent (Crofcheck et al., 2000; Muniz et al., 2009).

Scattering typically complicates the spectroscopic analysis of opaque media. At the same time, the scattering measurement may be able to decipher additional quantitative values. The amount of scattering measured at different wavelengths will change based on the number and size of the particles in suspension, such that the relative composition can be determined with careful calibration. Bogomolov et al. (2012) utilized visible light scatter measurements (400 and 1000 nm) to quantitatively analyze fat and total protein in milk. In that study, multivariate data analysis was capable of quantifying the individual scatter spectra of both fat and protein, and the results were supported by Mie theory calculations. Dahm and Dahm (2013) assumed that the sample can be divided into plane parallel layers and calculated the absorption, remission, and transmission for each of these hypothetical layers, which allowed them to separate the effects of scattering and absorption.

Fourier transform (FT)-NIR was used to predict fat, protein, and casein in milk using wavelengths from 1000 to 2500 nm (Jankovska and Sustova, 2003), while FT-IR has been used to predict the casein content with wavelengths of 3000–10,000 nm (Luginbuhl, 2002). The success of these techniques facilitated the development of a portable short-wave NIR charged coupled devices (CCD) spectrometer to measure fat, casein, and whey proteins with root mean squared errors of predictions of 0.06–0.12 wt% and correlation coefficients of 0.82–0.89 (Kalinin et al., 2013).

Specific processing applications of this technology have been investigated for determining the interface and duration of transition time between two different fluids (skim milk or raw milk transitioned with water) through a pipe system with various flow velocities and pipe lengths utilizing an optical sensor at 880 nm (Danao and Payne, 2003).

12.5 MONITORING MILK COAGULATION AND SYNERESIS DURING CHEESEMAKING

One important application of light scattering technology is the use of optical sensors to monitor and control cheesemaking operations during coagulation, where milk is converted from a liquid to a gel and syneresis, and the whey is expelled from the curd. Selecting the right time to cut the curd (the gel resulting from the acid coagulation of skim milk) is the single most important step in the production of a high-quality cottage cheese (Perry and Carroad, 1980; Emmons and Beckett, 1984). Controlling the extent of syneresis is crucial in cheesemaking, because it is directly related to the dry matter content and the composition of the resulting curd (Fagan et al., 2007).

When light is directed into milk, the majority of the light is scattered or transmitted forward, while very little light is scattered backward (back toward the incoming light). As the milk hardens into a gel and the milk proteins crosslink, the amount of light scattered backward increases. In other words, as the particles grow in size, the likelihood that light will encounter them and thus be scattered backward increases. The sensor directs light into the milk sample via an optical fiber and the light scattered backward (referred to as light backscattering) is received by a second fiber, as shown in Figure 12.2. By measuring the increase in light backscattering, the extent of coagulation (or crosslinking) can be monitored.

Fiber optic sensors have been developed for monitoring the changes in backscatter during the enzymatic coagulation of milk (Payne et al., 1993; Payne, 1995) and the culture of cottage cheese (Payne et al., 1997; Crofcheck et al., 1999). These sensors work well for applications where a change in reflectance provides the needed information. However, a limitation of the reflectance sensors cited above is that their output is not a physical property but a relative measure of light energy in volts that depends on fiber size, amplifier gain, etc. In some cases, these reflectance sensors can be calibrated to a physical property (e.g., fat content).

Optical sensor technologies, based on either light backscatter or transmission, have been proven as a successful tool for monitoring milk coagulation. The fiber optic light backscatter sensor, CoAguLite™, is a well-established online sensor technology developed to monitor coagulation and milk syneresis that is used to predict both clotting and cutting times (Payne et al., 1993; Castillo et al., 2000).

In an effort to further improve the CoAguLite sensor, Fagan et al. (2007) tested an online optical sensor with a large field of view (LFV) relative to curd particle size, for detecting light backscatter to monitor coagulation and syneresis during

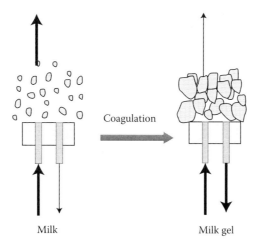

FIGURE 12.2 Orientation of a typical backscatter sensor, showing the relative amount of light transmitted and backscattered in milk before coagulation and in milk gel after coagulation.

cheesemaking. Results showed that the LFV sensor was sensitive to aggregation of casein micelles and development of curd firmness. The sensor response during syneresis was related to changes in curd moisture and whey fat content.

NIR light scattering technology was used to monitor milk coagulation and to predict the cutting time in milk mixtures (cow, sheep, and goat milk). Coagulation time was measured using an NIR fiber optic light backscatter sensor and small-amplitude oscillatory rheometry (SAOR). A prediction model on cutting time was successfully developed (Abdelgawad et al., 2014).

The International Dairy Federation (IDF) Standard 157: 2007/ISO 11815 measures milk-clotting activity by visually determining the Berridge clotting time (i.e., time of appearance of flocks of renneted standard milk substrate on the wall of a rotating test tube). The IDF Standard 157: 2007 method is somewhat subjective because it depends on an operator's skill to consistently identify milk flocculation. Using an optical method would reduce the subjectivity and improve the process. An optical method (NIR light backscatter at 880 nm) was investigated as an alternative method to detect milk-clotting time (Tabayehnejad et al., 2012). No significant differences were found between the clotting time found with the subjective Berridge clotting time and the clotting time found with the objective optical method.

NIR spectral data (2000–2500 nm) were used to develop an algorithm to determine an automatic cutting time (Lyndgaard et al., 2012). Data from the entire coagulation process were analyzed by principal component analysis as a function of time, and results indicated that milk coagulation could be separated into three stages: κ-casein proteolysis, micelle aggregation, and network formation. Two real-time models were created, one for the entire coagulation process, and one for the composite three-stage model. These were compared to the experimental NIR data and both models fitted very well ($R^2 > 0.99$) to the experimental data and served as a first step for a simulation model.

IR light backscatter and dynamic SAOR were used to monitor the gel formation of cottage cheese (Castillo et al., 2006). The NIR light backscatter sensor CoAguLite (Reflectronics, Inc., Lexington, Kentucky) was used to measure the backscattering light at 880 nm during coagulation. The sensor was comprised of two optical fibers to emit light into the milk and to transmit the backscattered light from milk to a photodetector. A data acquisition system and a customized algorithm was used to collect, store, and analyze the data to determine the optical time parameters based on the reflective ratio (R) and its first and second derivatives (t_{max} and t_{2max}), respectively. Second, dynamic SAOR was used to measure the viscous and elastic-like properties of gel as a function of temperature. The analysis was done by imposing a sinusoidally varying strain on the gel and measuring the resulting stress with a Couette measuring geometry in a ThermoHaakeRS1 rheometer (ThermoHaake GMBh, Karlsruhe, Germany). The changes in the elastic (G') and viscous (G'') modulus and the loss tangent (tan $\delta = G''/G'$) were monitored. The reference rheological parameter of clotting time was determined by the software Rheowin.

The light backscatter profile was explained by a complex combination of casein micelle aggregation, curd firming, and micelle demineralization. The second minimum of the backscatter second derivative and the rheological gelation time were highly correlated but not significantly different, suggesting that they both corresponded to the beginning of the gel.

Single wavelength (980 nm), broad spectrum, and color coordinates were examined for monitoring syneresis during cheesemaking (Mateo et al., 2010). Measurements were made via an online visible–NIR optical sensor. The developed model successfully predicted the curd moisture content based on single-wavelength NIR and visible spectrum (color coordinates) together with other parameters. The broad spectrum showed the potential to detect solids in whey.

An NIR reflectance sensor with an LFV and a fiber optic connection to a spectrometer measuring light backscattering at 980 nm was used to monitor syneresis online during cheesemaking (Mateo et al., 2009). The objective was to predict whey yield, curd moisture content, and fat loss to whey yield over various curb stirring speed and cutting programs. The most accurate model developed was for prediction of whey yield involving light backscatter, milk fat, and cutting intensity.

Photon density wave (PDW) spectroscopy was used to study the melting and crystallization of milk fat in homogenized fresh milk at different temperatures (Ruiz et al., 2012). PDW spectroscopy is an optical in-line process analytical technique that is based on the radiation transport theory. A PDW is generated by inserting an intense modulated light (i.e., laser diodes) into a turbid but low absorption material. The amplitude and phase of the wave are characterized by the absorption and scattering properties of the material. PDW is quantified by the coefficients of absorption (μ_a) and reducing scattering (μ'_s) representing the chemical composition and the physical properties of the material, respectively. As demonstrated in the work of Ruiz et al. (2012), the absorption coefficients reflect the changes in the milk fat content, while the reduced scattering coefficients reflect the physical changes in the fat droplets and casein micelles coinciding with the phase transitions of milk fat during the cooling and heating of milk. Continuous measurement of the coefficients showed that the crystallization of milk fat took several hours to occur under an isothermal condition.

12.6 FOOD QUALITY AND SAFETY

Beyond detecting the size, amount, and crosslinking of fat and proteins, light can also be used to detect contaminates and changes in the protein structure as a result of denaturation. Owing to its high nutritional value and widespread consumption, milk is one of the foodstuffs most targeted for adulteration–dilution with a less expensive component (Moore et al., 2012). Both deliberate adulteration and unintentional contamination effects would be minimized with fast, nondestructive detection, which light-based technologies are capable of.

Visible backscatter light sensors have been used to monitor β-lactoglobulin denaturation during heat treatment at 80°C (Lamb et al., 2013). Results showed that the whey protein denatured in a first-order response (P-value < 0.0001) with R^2 values similar (~0.80) to those found when typical biotechnology was used to quantify denaturation.

12.7 CONCLUDING REMARKS

Light scatter presents a very useful opportunity to characterize a complex biological fluid such as milk. The future use of light scatter will be propelled by two interacting developments. One is the continued miniaturization of optical technologies, which generally increases reliability, reduces costs, and makes it feasible for many different optical detection configurations. The second is the development of more sophisticated data analysis technologies that allow useful information to be obtained from ever-increasing quantities of data. The future will see fluorescence and Raman technologies gain increasing attention. From a processing standpoint, light scatter technologies will provide tighter control of processing operations and result in ever-increasing consistent products produced.

REFERENCES

Abdelgawad, A. R., B. Guamis, and M. Castillo. Using a fiber optic sensor for cutting time prediction in cheese manufacture from a mixture of cow, sheep and goat milk. *Journal of Food Engineering* 125, 2014: 157–168.

Anderson, M., M. C. A. Griffin, and C. Moore. Fixation of bovine casein micelles for chromatography on controlled pore glass. *Journal of Dairy Research* 51(4), 1984: 615–622.

Attaie, R. and R. L. Ritcher. Size distribution of fat globules in goat milk. *Journal of Dairy Science* 83(5), 2000: 940–944.

Ben-Gera, I. and K. H. Norris. Influence of fat concentration on the absorption spectrum of milk in the near-infrared region. *Israel Journal of Agricultural Research* 18, 1968: 117–124.

Bogomolov, A., S. Dietrich, B. Boldrini, and R. W. Kessler. Quantitative determination of fat and total protein in milk based on visible light scatter. *Food Chemistry* 134(1), 2012: 412–418.

Bolt, R. A. and J. J. ten Bosch. Method for measuring position-dependent volume reflection. *Applied Optics* 32(24), 1993: 4641–4645.

Bonner, R. F., R. Nossal, S. Havlin, and G. H. Weiss. Model for photon migration in turbid biological media. *Journal of the Optical Society of America A* 4(3), 1987: 423–432.

Brooker, B. E. and C. Holt. Natural variation in the average size of bovine casein micelles: III. Studies on colostrum by electron by microscopy and light scattering. *Journal of Dairy Research* 45(3), 1978: 355–362.

Castillo, M., J. A. Lucey, and F. A. Payne. The effect of temperature and inoculum concentration on rheological and light scatter properties of milk coagulated by a combination of bacterial fermentation and chymosin. Cottage cheese-type gels. *International Dairy Journal* 16(2), 2006: 131–146.

Castillo, M., F. A. Payne, C. L. Hicks, and M. B. Lopez. Predicting cutting and clotting time of coagulating goat's milk using diffuse reflectance: Effect of pH, temperature and enzyme concentration. *International Dairy Journal* 10(8), 2000: 551–562.

Crofcheck, C. L., F. A. Payne, C. L. Hicks, M. P. Mengüç, and S. E. Nokes. Fiber optic sensor response to low levels of fat in skim milk. *Journal of Food Process Engineering* 23, 2000: 163–175.

Crofcheck, C. L., F. A. Payne, and S. E. Nokes. Predicting the cutting time of cottage cheese using backscatter measurements. *Transactions of the ASAE* 42(4), 1999: 1039–1045.

Crofcheck, C. L., J. Wade, J. N. Swamy, M. M. Aslan, and M. P. Mengüç. Effect of fat and casein particles in milk on the scattering of elliptically-polarized light. *Transactions of the ASAE* 48(3), 2005: 1147–1155.

Dahm, K. D. and D. J. Dahm. Separating the effects of scatter and absorption using the representative layer. *Journal of Near Infrared Spectroscopy* 21(5), 2013: 351–357.

Danao, M. C. and F. A. Payne. Determining product transitions in a liquid piping system using a transmission sensor. *Transactions of the ASAE* 46(2), 2003: 415–421.

Dufour, E. Recent advances in the analysis of dairy product quality using methods based on the interactions of light with matter. *International Journal of Dairy Technology* 64(2), 2011: 153–165.

Emmons, D. B. and D. C. Beckett. Effect of pH at cutting and during cooking on cottage cheese. *Journal of Dairy Science* 67(10), 1984: 2200–2209.

Fagan, C. C., M. Castillo, F. A. Payne, C. P. O'Donnell, M. Leedy, and D. J. O'Callaghan. Novel online sensor technology for continuous monitoring of milk coagulation and whey separation in cheesemaking. *Journal of Agricultural and Food Chemistry* 55(22), 2007: 8836–8844.

Haskell, R. C., L. O. Svaasand, T. Tsay, T. Feng, M. S. McAdams, and B. J. Tromberg. Boundary conditions for the diffusion equation in radiative transfer. *Journal of the Optical Society of America A* 11(10), 1994: 2727–2741.

Holt, C., D. G. Dalgleish, and T. G. Parker. Particle-size distributions in skim milk. *Biochimica Biophysica Acta* 328(2), 1973: 428–432.

Holt, C., T. G. Parker, and D. G. Dalgleish. Measurement of particle sizes by elastic and quasi-elastic light scattering. *Biochimica Biophysica Acta* 400(2), 1975: 283–292.

Ishimaru, A. Theory and application of wave propagation and scattering in random media. *Proceedings of IEEE* 65(7), 1977: 1030–1061.

Ishimaru, A. and Y. Kuga. Attenuation constant of a coherent field in a dense distribution of particles. *Journal of the Optical Society of America* 72(10), 1982: 1317–1320.

Jankovska, R. and K. Sustova. Analysis of cow milk by near-infrared spectroscopy. *Czech Journal of Food Sciences* 21(4), 2003: 123–128.

Kalinin, A., V. Krasheninnikov, S. Sadovskiy, and E. Yurova. Determining the composition of proteins in milk using a portable near infrared spectrometer. *Journal of Near Infrared Spectroscopy* 21(5), 2013: 409–415.

Konokhova, A. I., A. A. Rodionov, K. V. Gilev, I. M. Mikhaelis, D. I. Strokotov, A. E. Moskalensky, M. A. Yurkin, A. V. Chernyshev, and V. P. Maltsev. Enhanced characterisation of milk fat globules by their size, shape and refractive index with scanning flow cytometry. *International Dairy Journal* 39(2), 2014: 316–323.

Lamb, A., F. A. Payne, Y. L. Xiong, and M. Castillo. Optical backscatter method for determining thermal denaturation of beta-lactoglobulin and other whey proteins in milk. *Journal of Dairy Science* 96(3), 2013: 1356–1365.

Lin, S. H. C, R. K. Dewan, V. A. Bloomfield, and C. V. Morr. Inelastic light-scattering study of size distribution of Bovine milk casein micelles. *Biochemistry* 10(25), 1971: 4788–4793.

Luginbuhl, W. Evaluation of designed calibration samples for casein calibration in Fourier transform infrared analysis of milk. *Lebensmittel-Wissenschaft Und-Technologie—Food Science and Technology* 35(6), 2002: 554–558.

Lyndgaard, C. B., S. B. Engelsen, and F. W. J. van den Berg. Real-time modeling of milk coagulation using in-line near infrared spectroscopy. *Journal of Food Engineering* 108(2), 2012: 345–352.

Ma, Y. and D. M. Barbano. Gravity separation of raw bovine milk: Fat globule size distribution and fat content of milk fractions. *Journal of Dairy Science* 83(8), 2000: 1719–1727.

Mateo, M. J., D. J. O'Callaghan, C. D. Everard, M. Castillo, F. A. Payne, and C. P. O'Donnell. Evaluation of on-line optical sensing techniques for monitoring curd moisture content and solids in whey during syneresis. *Food Research International* 43(1), 2010: 177–182.

Mateo, M. J., D. J. O'Callaghan, C. D. Everard, C. C. Fagan, M. Castillo, F. A. Payne, and C. P. O'Donnell. Influence of curd cutting programme and stirring speed on the prediction of syneresis indices in cheese-making using NIR light backscatter. *LWT—Food Science and Technology* 42(5), 2009: 950–955.

McGann, T. C. A., R. D. Kearney, and W. J. Donnelly. Developments in column chromatography for the separation and characterization of casein micelles. *Journal of Dairy Research* 46(2), 1979: 307–311.

McMahon, D. J. and R. J. Brown. Composition, structure, and integrity of casein micelles: A review. *Journal of Dairy Science* 67(3), 1984: 499–512.

Milne, J. L. S., M. J. Borgnia, A. Bartesaghi, E. E. H. Tran, L. A. Earl, D. M. Schauder, J. Lengyel, J. Pierson, A. Patwardhan, and S. Subramaniam. Cryo-electron microscopy: A primer for the non-microscopist. *FEBS Journal* 280(1), 2013: 28–45.

Moore, J. C., J. Spink, and M. Lipp. Development and application of a database of food ingredient fraud and economically motivated adulteration from 1980 to 2010. *Journal of Food Science* 77(4), 2012: R118–R126.

Mulder, H. and P. Walstra. *The Milk Fat Globule.* Wageningen: Pudoc, 1974.

Muniz, R., M. A. Perez, C. de la Torre, C. E. Carleos, N. Corral, and J. A. Baro. Comparison of principal component regression (PCR) and partial least square (PLS) methods in prediction of raw milk composition by VIS-NIR spectrometry. Application to development of on-line sensors for fat, protein and lactose contents. XIX IMEKO World Congress Fundamental and Applied Metrology, Lisbon, Portugal, 2009.

Nakai, S. and F. van De Voort. Application of multiple regression analysis to sedimentation equilibrium data of $\alpha s1$- and κ-casein interactions for calculation of molecular weight distributions. *Journal of Dairy Research* 46(2), 1979: 283–290.

Payne, F. A. Automatic control of coagulum cutting in cheese manufacturing. *Applied Engineering in Agriculture* 11(5), 1995: 691–697.

Payne, F. A., C. L. Hicks, and P.-S. Shen. Predicting optimal cutting time of coagulating milk using diffuse reflectance. *Journal of Dairy Science* 76, 1993: 48–61.

Payne, F. A., Y. Zhou, R. C. Sullivan, and S. E. Nokes. Radial backscatter profiles in milk in the wavelength range of 400 to 1000 nm. ASAE Paper No. 97–6097, ASAE, St. Joseph, Michigan, 1997.

Perry, C. A. and P. A. Carroad. Influence of acid related manufacturing practices on properties of cottage cheese curd. *Journal of Food Science* 45(4), 1980: 794–797.

Qin, J. and R. Lu. Measurement of the absorption and scattering properties of turbid liquid foods using hyperspectral imaging. *Applied Spectroscopy* 61(4), 2007: 388–396.

Ruettiman, K. W. and M. R. Ladisch. Casein micelles: Structure, properties and enzymatic coagulation. *Enzyme Microbiology Technology* 9(10), 1987: 578–589.

Ruiz, S. V., R. Hass, and O. Reich. Optical monitoring of milk fat phase transition within homogenized fresh milk by photon density wave spectroscopy. *International Dairy Journal* 26(2), 2012: 120–126.

Schmidt, D. G., P. Both, B. W. Van Markwijk, and W. Buchheim. The determination of size and molecular weight of casein micelles by means of light-scattering and electron microscopy. *Biochimica Biophysica Acta* 365(1), 1974a: 72–79.

Schmidt, D. G., J. Koops, and D. Westerbeek. Properties of artificial casein micelles. 1. Preparation, size distribution and composition. *Netherlands Milk Dairy Journal* 31, 1977: 328–341.

Schmidt, D. G., C. A. Van Der Spek, W. Buchheim, and A. Hinz. On the formation of artificial casein micelles. *Milchwissenschaft* 29, 1974b: 455–459.

Swapp, S. Scanning Electron Microscopy (SEM). http://serc.carleton.edu/research_education/geochemsheets/techniques/SEM.html. Accessed on May 8, 2015.

Slattery, C. W. Model calculations of casein micelle size distributions. *Biophysical Chemistry* 6(1), 1977: 59–64.

Tabayehnejad, N., M. Castillo, and F. A. Payne. Comparison of total milk-clotting activity measurement precision using the Berridge clotting time method and a proposed optical method. *Journal of Food Engineering* 108(4), 2012: 549–556.

Walstra, P. Studies on milk fat dispersion. II. The globule-size distribution of cow's milk. *Netherlands Milk Dairy Journal* 23, 1969: 99–110.

Walstra, P. Effect of homogenization of the fat globule size distribution in milk. *Netherlands Milk Dairy Journal* 29, 1975: 279–294.

Walstra, P. and R. Jenness. *Dairy Chemistry & Physics*. New York: Wiley, 1984.

Walstra, P., H. Oortwijn, and J. J. de Graaf. Studies on milk fat dispersion. I. Methods for determining globule-size distributions. *Netherlands Milk Dairy Journal* 23, 1969: 12–36.

Zbikowska, A., J. Dziuba, H. Jaworska, and A. Zaborniak. The influence of casein micelle size on selected functional properties of bulk milk proteins. *Polish Journal of Food and Nutrition Sciences* 1(1), 1992: 23–32.

13 Dynamic Light Scattering for Measuring Microstructure and Rheological Properties of Food

Fernando Mendoza and Renfu Lu

CONTENTS

13.1 Introduction 332
13.2 Static versus Dynamic Light Scattering 333
13.3 Principle of Measuring DLS 335
13.4 Instrument Setup 337
13.5 Autocorrelation Function and DLS Data Analysis 338
    13.5.1 Autocorrelation Function 338
    13.5.2 Hydrodynamic Size 340
    13.5.3 Polydisperse Samples 341
    13.5.4 Methods to Calculate Size Distribution 342
        13.5.4.1 Cumulant Method 342
        13.5.4.2 Inverse Laplace Transform or CONTIN Algorithm 343
13.6 Effects of Concentration 344
    13.6.1 Multiple Scattering 344
    13.6.2 Cross-Correlation 345
    13.6.3 Two-Color DLS 346
    13.6.4 3D Cross-Correlation DLS 346
13.7 Microstructure and Rheological Property Measurements in Foods 348
    13.7.1 Diffusing Wave Spectroscopy 348
        13.7.1.1 Principle of DWS 350
        13.7.1.2 Applications in Food Materials 351
    13.7.2 Microrheology 352
        13.7.2.1 Basic Concepts 352
        13.7.2.2 DLS Microrheology 354
13.8 Conclusions 355
Nomenclature 356
References 356

13.1 INTRODUCTION

Nowadays, microstructure and dynamical behavior of viscoelastic food systems have attracted considerable attention of food designers and developers, and attempts have been made to relate the microstructure of food systems to their bulk, or macroscopic, properties and stability (i.e., particles forces/interactions). Knowledge and characterization of particle size distribution, microstructure, and rheological properties and their changes in complex fluids, such as colloidal aggregates, surfactant solutions, and polymer blends, are of fundamental importance to better understand, create, and control structure formation in soft materials (Moschakis 2013). In food materials, the changes occurring during colloidal destabilization may have important implications on the structural characteristics of the final matrix, and thus impact the final texture, sensory perception, and consumer acceptance of the food (Alexander and Corredig 2013). A typical manifestation of formulation instability in a colloidal suspension, for example, is an increase in particle size due to the aggregation of individual compounds or particles. As the particle size increases, the efficacy of particle dispersion is diminished, primarily due to the decrease in the active surface area (Mattison et al. 2003).

Conventional rheology, which studies the deformation of materials with time in the presence of stress, provides an important tool to study the mechanical and flow properties of soft materials, which is especially significant for food colloids (Chen et al. 2010). However, bulk rheological measurements only describe the overall mechanical and flow response of a material on a macroscopic scale; that is, where structural and rheological heterogeneities are averaged over some macrodimensional scale. They do not provide any information on local variations in the microstructure and their contribution to the overall rheological behavior of a material. It is, therefore, essential to establish correlations between particle forces/interaction at the micro and nano level and bulk mechanical and flow properties in the design, development, and quality control of foods. Accordingly, it is necessary to probe the rheology of foods over shorter length scales (Moschakis 2013). Overall, the viscoelastic response of complex food colloids and fluids is length- and time-scale dependent, encoding information on intrinsic dynamic correlations and structure at molecular, mesoscopic, and macroscopic scales (Sonn-Segev et al. 2014).

Dynamic light scattering (DLS) is a noninvasive analytical technique that is used to determine the size distribution profile of small particles in suspension or polymers in solution (Berne and Pecora 2000), typically at the submicron scale. DLS measures the fluctuations of scattered light intensity resulting from Brownian motion of particles suspended in a fluid. Brownian motion refers to the random motion of particles suspended in a fluid or gas resulting from their collision with atoms or molecules in the fluid or gas. DLS can provide detailed information about the dynamics of the scattering medium, and the technique has been successfully extended to analyze opaque media that exhibit a very high degree of multiple scattering (Pine et al. 1990; Pusey 1999; Alexander and Corredig 2013). DLS is currently used to characterize the size of various food particles, including proteins, polymers, micelles, carbohydrates, and nanoparticles; it is also used to probe the rheological behavior of complex fluids such as concentrated polymer solutions and formulations over short length scales (Rao 2014). The DLS technique is often used in combination with other techniques to simultaneously obtain more complete information on viscoelastic behavior or structural information that cannot be assessed by DLS alone. An example is *optical microrheology*, which studies the deformation and flow of a soft material at a length scale well below the macroscopic (Scheffold et al. 2004). It typically involves the use of a colloidal probe particle with DLS schemes for measuring local rheology at the colloidal length scale.

DLS is now considered a standard experimental technique in scientific research and industrial application. DLS instruments have entered laboratories in the pharmaceutical, coating, earth material, powder (Sapsford et al. 2011), and food industries (Alexander and Dalgleish 2006; Calzolai et al. 2012; Malvern 2015), and even in the diagnosis of pathological conditions of body fluid (Maurer and Pittendreigh 2012). The popularity of DLS within the food industry and in food product development is attributed to its wide size and sample concentration ranges, as well as its low volume requirements for complete analysis. DLS is suitable for a broad concentration of approximately 10^8–10^{12} particles/mL and the particle size range of 10–1000 nm (Sapsford et al. 2011).

It should be mentioned that the term "dynamic light scattering" covers different techniques for measurement of particle size, which are based on measuring the dynamic changes of the scattered light intensity. The widely used name at present is *photon correlation spectroscopy*, which relates to the autocorrelation technique that is most frequently used in commercial instruments. In the past, *quasi-elastic light scattering* (QELS) was the frequently used name in relation to the type of interaction between particles and light. Diffusing wave spectroscopy (DWS), an extension of the conventional DLS scheme, is the name used for applications in concentrated dispersions, where multiple scattering leads to the diffusive transport of light (Merkus 2009). While DLS relies on the suppression of multiple scattering, DWS works in the limit of very strong multiple scattering, where a diffusion model can be used to describe the propagation of light across the sample (Mezzenga et al. 2005).

13.2 STATIC VERSUS DYNAMIC LIGHT SCATTERING

Light scattering is a consequence of the interaction of light with the electric field of a particle or small molecule. Such interaction induces a dipole in the particle electric field that oscillates with the same frequency as that of the incident light. Inherent to the oscillating dipole is the acceleration of charge, which leads to the release of energy in the form of scattered light (Bancarz et al. 2008). Therefore, when light passes through a solution containing particles or molecules, depending on the optical parameters of the system, part of the light will be scattered. This scattered light may be analyzed either in terms of its intensity, that is, *static scheme*, or in terms of its fluctuation, that is, *dynamic scheme*.

Static light scattering (SLS) measures scattering intensities due to light–particle interaction at various spatial locations (i.e., as a function of scattering angle at a fixed wavelength of light) (Chu 2008). SLS represents the time-averaged intensity of scattered light over a duration of approximately 1 s. Parameters derived using this technique are the size of particle suspensions (with diameters >10 nm) and polymer molecular weight (M_w, or average molar mass) with accuracy of approximately 5%. In addition, by measuring the scattering intensity at many angles (multiangle light scattering or MALS technique), the root mean squared radius, also named the radius of gyration (R_g), can be calculated. By measuring the scattering intensity for many samples of various concentrations, the second virial coefficient (B_{22}), a measure of the pairwise interaction potential between neighboring particles in a bioformulation, can also be calculated. However, in SLS, the angular dependence information is required for accurate measurements of both molecular weight and size for all macromolecules with radii greater than 1%–2% of the incident wavelength. In many biomolecules (<20 nm), the radius of gyration is not easily measurable since only a minor angular variation in the intensity is exhibited. This means that measurements may be performed at a single angle, provided that the concentration and the refractive index are known. Nonetheless, an average M_w may be determined from the average scattered intensity. An SLS experiment requires the preparation of several known concentrations of the sample, and samples must be relatively clean and dust-free.

DLS refers to the measurement and interpretation of the intensity fluctuation of light scattered on a microsecond time scale. Diffusion coefficient and particle size information can be obtained from the analysis of these fluctuations. The parameters derived are the *translation diffusion coefficient* (D_T) and the *hydrodynamic radius* (R_h or Stokes radius) with uncertainty of approximately 10% for a monodisperse sample. D_T is the property representing the velocity of a particle undergoing Brownian motion and R_h is the diameter of a sphere that has the same translational diffusion coefficient as the particle. Also, the Z-average hydrodynamic radius and polydispersity index are obtained approximately using the method of cumulant analysis or Laplace transformation at a given scattering angle (Sun 2011). DLS technique became a reality only after laser—a coherent, collimated, stable, and high-intensity light source—was invented in the 1960s. The measurements are largely done in batch mode, but can be also done in in-line mode combined with a fractionation step, such as size exclusion chromatography or flow field fractionation (Xu 2015). Among the optical techniques that can directly work with viscoelastic fluid samples, DLS technique and its improved schemes are the most widely used (Brar and Verma 2011).

In summary, SLS provides information on a sample's average structure, while DLS analyzes temporal fluctuations in the scattered light, providing information on a sample's dynamics, typically Brownian motions. Figure 13.1 depicts a schematic representation of SLS and DLS. For transparent or dilute samples, there is a direct and relatively simple relationship between the properties of the material and the intensity and temporal fluctuations of the scattered light (Pusey 1999). While both SLS and DLS techniques are easy to use, fast, and relatively low in cost, and are well suited for samples of nanoparticles of a single size (i.e., monodisperse), they can give misleading results when used to analyze samples containing particles of different sizes (Calzolai et al. 2012). SLS experiments can be also done on turbid suspensions if they are combined with novel cross-correlation DLS techniques (Urban and Schurtenberger 1998). The capabilities and limitations of both techniques are summarized in Table 13.1.

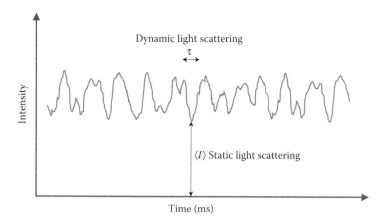

FIGURE 13.1 Schematic representation of light scattered from a solution of macromolecules showing the SLS represented by the average intensity I that reflects the molecular weight of the macromolecules, and the DLS represented by the fluctuations in the intensity with a characteristic fluctuation time τ that reflects the diffusion coefficient of the molecules.

13.3 PRINCIPLE OF MEASURING DLS

Particles and macromolecules in solution undergo Brownian motion, which arises from collisions between the particles and the solvent molecules. As a consequence of this particle motion, light scattered from the particle ensemble will fluctuate with time. In the DLS technique, the intensity of the scattered light by an ensemble of particles is measured at a given angle as a function of time (Mattison et al. 2003). The Brownian motion of the dispersed particles determines the rate of changes of the scattered light intensity. Larger particles travel more slowly, whereas smaller particles travel faster. Small particles move faster than large particles because the driving forces on them are the same (i.e., collision with the solvent molecules), but the large particles encounter a larger retarding force (friction) from the surrounding solvent. Since the particles can move a short distance very quickly, in order to capture the intensity fluctuations, it is often necessary to measure the scattered intensity at very short time intervals, typically 5,000,000 times per second (Øgendal 2013).

As the particles move through the suspension, they pass through the laser beam, which causes some of the light to scatter. The temporal intensity changes are then converted to a mean translational diffusion coefficient (or a set of diffusion coefficients, see Section 13.5.1). Fast intensity changes are related to a rapid decay of the correlation function and a larger diffusion coefficient. The diffusion coefficient is then converted to particle size by means of the Stokes–Einstein equation. According to Stokes' law, a perfect sphere traveling through a viscous liquid feels a drag force proportional to the frictional coefficient. The diffusion coefficient (D_T) of a spherical particle is proportional to its mobility; substituting the frictional coefficient of a perfect sphere from Stokes' law by the liquid's viscosity and sphere's radius, we have Stokes–Einstein equation.

For this conversion, therefore, the particles are assumed to be spherical and without interaction.

TABLE 13.1
Static versus Dynamic Light Scattering: Capabilities and Limitations

	SLS	DLS
Capabilities	Fast and accurate determination of macromolecular sizing (for diameters >10 nm), polymer molecular weight (M_w), radius of gyration (R_g), and second virial coefficient (B_{22}) and also shape. Single measurement should be sufficient to determine M_w. Can determine oligomeric state of modified polypeptide (glycosylated protein, conjugated with polyethylene glycol, protein–lipid–detergent complexes, and protein–nucleic acid complexes). Instruments are relatively cheap and small and measurements are fast.	Allow accurate, reliable, and repeatable particle size analysis in 1 or 2 min. Good for macromolecular sizing (sizes <1 nm and molecules with M_w <1000 Da). Allow measurements in the native environment of the material. Simple or no sample preparation is needed. Measurements can be done on high concentration, turbid samples. Low volume requirement (~2 µL). Good for detecting trace amounts of aggregate. In bath mode, very fast determination of particle size and detection of aggregates, and evaluation of polydispersity of sample with great dynamic range. Well suited to study kinetics of aggregation. Capable for high-volume analysis by using a plate reader setup.
Limitations	Needs to know in advance: weight concentration, C, and the refractive index increment dn/dC. Measures of average M_w need fractionation to resolve different oligomeric states or fitting data to an association model. Size range is rather limited (~10–1000 nm). Possible loss of sample during filtration and fractionation. Limitation on solvent choices. Needs extra hardware modification for samples that absorb laser light (633 nm). Sample preparation and cuvette cleaning is extremely critical to avoid dust.	Requires knowledge of the viscosity (η) of the solvent at the measurement temperature (T) for mean size estimation. Autocorrelation function cannot be described by single exponential (cumulant fit). Measures of hydrodynamic radius (R_h) are affected by shape and aggregation. Cannot discriminate between shape effects and changes in oligomeric states, that is, nonspherical shape mimics effects from oligomerization. Needs fractionation to resolve oligomers when present in a mixture.

Colloidal particles in a liquid dispersion, however, contain an attached layer of ions and molecules from the dispersion medium that moves with the particles. Therefore, their hydrodynamic particle size is somewhat larger than the size of the particle (Merkus 2009).

The DLS technique works without exact knowledge of the sample concentration, and it has been used with success in structural biology, chemistry, and physics (Bergfords 1999; Sun 2011). The only requirement is that enough light must be scattered to achieve sufficient statistical accuracy of the correlation function.

Dynamic Light Scattering for Measuring

The scattering vector (**q**, or momentum transfer) is the most important parameter in DLS. The wave vector determines the length scale over which molecular motions are detected (Xu et al. 2015). This is defined as the difference between vectors of the incident and scattered waves, and is given by

$$|\mathbf{q}| = \frac{4\pi n}{\lambda} \sin\left(\frac{\theta}{2}\right) \tag{13.1}$$

where θ is the scattering angle, n is the refractive index of the medium (solution), and λ is the wavelength of the beam.

13.4 INSTRUMENT SETUP

The typical setup for a DLS instrument is given in Figure 13.2. The essential elements of the apparatus to observe the DLS process are light source, cuvette sample holder, digital correlator, and a detector. In most DLS instruments, a monochromatic, vertically polarized, coherent helium neon (HeNe) laser with a fixed wavelength of 633 nm is used as the light source and an optical lens is often applied to focus the light beam into the measurement zone. A laser power output of 2–5 mW is often used; but higher power (up to about 25 mW) is recommended for measurement of very small particles in view of their poor scattering behavior (Merkus 2009). The well-collimated beam enables the direction of scattering to be established with precision; stability in the light source is always important when comparing samples under different conditions. Emission over a narrow line-width or wavelength range is another important attribute of laser light sources allowing a wave vector to be defined precisely (Horne 2011).

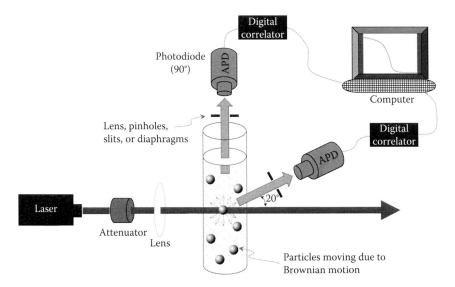

FIGURE 13.2 Setup of a DLS instrument showing the typical configuration with a detector fixed to 20° and 90° of the beam (APD, avalanche photodiode detector).

The sample is located in a round or square measurement cuvette of about 1 cm path length, conditioned in a temperature-controlled environment (±0.3°C). In some instruments, the cuvette cell is surrounded by an index-matching liquid, to compensate for refractive index differences between the cell wall and its surroundings. The light source should be focused on the center of this cuvette. The scatterer should be of a size that the DLS scheme is applicable. In a conventional DLS scheme, the volume fraction of the scatterer should be low enough to ensure that only a single scattering occurs, that is, any photon received at the detector is scattered only once in its passage through the cuvette to avoid multiple scattering (Urban and Schurtenberger 1998).

Finally, the light detector that is placed behind the cuvette is made up of a lens and a set of pinholes, slits, or diaphragms at some angle with respect to the transmitted beam. The detectors convert the scattered light intensity (number of photons per time) into electrical signals. Typical detectors used in a DLS instrument are the avalanche photodiode detector (APD) or a photomultiplier tube. While light is scattered by the particles at all angles, a DLS instrument normally only detects the scattered light at one angle, conventionally 90°; in some cases, the detection angle may be variable in the range of 20–160° through application of a goniometer or optical fibers (Merkus 2009). In DLS instrumentation, the correlation summations are performed using an integrated digital correlator, which is a logic board comprising operational amplifiers that continually add and multiply short time scale fluctuations in the measured scattering intensity to generate the correlation curve for the sample.

13.5 AUTOCORRELATION FUNCTION AND DLS DATA ANALYSIS

After data acquisition, a computer is used to analyze the spectrum based on a correlation function, and then to convert the characteristic decay of this correlation function into a diffusion coefficient distribution and a particle size distribution. This analytical procedure is explained in the following subsections.

13.5.1 AUTOCORRELATION FUNCTION

The *autocorrelation function* (for evenly spaced data) is a second-order statistic measuring the degree of nonrandomness in an apparently random data set. When a focused laser beam travels through the sample cell, the suspended particles scatter the incident laser beam in all directions. The intensities of the scattered lights fluctuate with time due to the Brownian motion of particles.

In DLS, a digital correlator continually adds and multiplies these short time scale fluctuations in the measured scattering intensity to generate the autocorrelation function. The typical quantity calculated from these intensity fluctuations in a DLS experiment is the intensity autocorrelation function given by

$$g'_2(\tau) = \langle I(t) \cdot I(t+\tau) \rangle = \int_0^\infty I(t) \cdot I(t+\tau) \cdot dt \qquad (13.2)$$

Dynamic Light Scattering for Measuring

where $I(t)$ is the time-varying measured scattered intensity, τ is the variable lag time, the angle brackets denote an ensemble average (equivalent to a time average for describing dynamical systems), and the dependence on the scattering vector **q** has been suppressed (Berne and Pecora 2000). The autocorrelation procedure is illustrated in Figure 13.3a, where the fluctuating light scattering intensity $I(t)$ is plotted as a function of t (j being the number of delay channels, N is the number available in the correlator, and τ is the time span of each channel assuming a linear spacing of channels). Hence, the autocorrelation function is calculated as a sum of products of intensities measured with a time separation of τ as $I(0).I(t=\tau)$, $I(0).I(t=2\tau)$, $I(0).I(t=3\tau)$,..., $I(0).I(t=N\tau)$, repeating this procedure over a period of time (minutes to hours) and accumulating the autocorrelation function $g'_2(\tau)$. Hence, after one cycle has been completed, the time grid shifts up one channel (1 becomes 0, 2 becomes 1, etc.) and the cycle is repeated. Normally, $g'_2(\tau)$ is calculated for approximately 300 different values of τ (with ~5,000,000 intensity measurements per second). This has to be done simultaneously with the intensity measurements, meaning that approximately 1,500,000,000 multiplications and 1,500,000,000 additions have to be carried out every second (Øgendal 2013). $g'_2(\tau)$ is a monotonically decaying function that decays from an initial value equal to the time average of the intensity squared, $\langle I^2 \rangle$, to the square of the average intensity, $\langle I \rangle^2$, being the amplitude of the correlation function the variance of these intensity fluctuations, $\langle I^2 \rangle - \langle I \rangle^2$ (Horne 2011).

The correlation coefficients are then normalized and the decay rate of the autocorrelation function is related to the particle's diffusion coefficient D_T. Using the Siegert relation (Siegert 1943) to fit the autocorrelation function of monodisperse particles, the normalized autocorrelation function is written as

$$g_2(\tau) = \frac{\langle I(t) \cdot I(t+\tau) \rangle}{\langle I(t) \rangle^2} = B + \beta[g_1(\tau)]^2 \qquad (13.3)$$

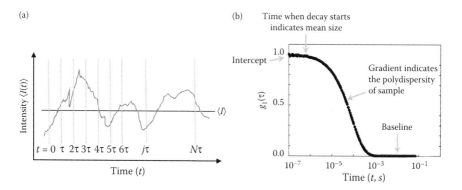

FIGURE 13.3 Schematic representation of the autocorrelation function. (a) Fluctuating light scattering intensity as a function of time and the derivation of the correlation function. After one cycle has been completed, the time grid shifts up one channel (1 becomes 0, 2 becomes 1, etc.) and the cycle is repeated. (b) The shape of a typical correlation function $g_1(\tau)$ for scattered light, as measured by DLS. These functions are generated directly from the fluctuations of the detected intensity of the (singly) scattered photons.

where B is the baseline and β denotes the coherence factor in the instrumentation setup. Thus, the field of autocorrelation function or dynamic structure function describes the temporal decay of a particular orientation of the sample within the scattering volume. At short times, the system is nearly stationary and so the value of the correlation function is approximately equal to unity; that is, the particles have insufficient time to move very far from their initial positions, and hence the intensity signals are very similar. Owing to the random nature of the forces applied to the system by Brownian motion, the correlation of the system at time τ with the initial state approaches zero at longer times (see Figure 13.3b).

For monodisperse particle solution, the decay rate of the correlation function can be approximated as a single exponential as

$$g_1(\tau) = e^{-\Gamma\tau} \tag{13.4}$$

The exponential decay constant is given by

$$\Gamma = \mathbf{q}^2 D_T \tag{13.5}$$

where D_T is the diffusion coefficient and $\mathbf{q}$ is the known magnitude of the scattering wave vector defined in Equation 13.1.

Thus, all of the information regarding the motion or diffusion of the particles in the solution is embodied within the measured correlation curve (Mattison et al. 2003). These intensity fluctuations provide information on the sizes and size distributions of the suspended particles (see further details in the following subsections), or information on the viscosity of the sample if the particle sizes are known. An example of a typical DLS autocorrelation function is shown in Figure 13.3b.

13.5.2 Hydrodynamic Size

The hydrodynamic size (also referred to as hydrodynamic radius [R_h] or hydrodynamic diameter [D_h]) is defined as the radius of a hard sphere that diffuses at the same rate as the particle under examination. For monodisperse samples, consisting of a single particle size group, the correlation curve can fit to a single exponential form as given in Equation 13.4. Then, for spherical particles undergoing Brownian motion with no interaction with others, the Stokes–Einstein equation relates the particles' translational diffusion coefficient D_T to the particle's hydrodynamic radius R_h as follows (Miller 1924):

$$R_h = \frac{k_b T}{6\pi \eta D_T} \tag{13.6}$$

where $k_b = 1.38 \cdot 10^{-23}$ J/mol is Boltzmann's constant, T is the absolute temperature, and η is the dynamic viscosity of the suspending medium. Using this simple model of particle motion in solution, the exponential decay can be related directly to the hydrodynamic radii of the scattering particles if the viscosity of the suspending medium is known.

Dynamic Light Scattering for Measuring 341

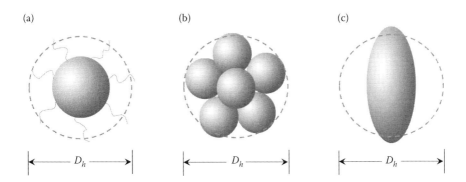

FIGURE 13.4 Hydrodynamic size for particles with the same translational diffusion coefficient, D_T, due to (a) ionized particles, (b) particle's interaction showing aggregation, and (c) nonspherical.

It should be noted that the hydrodynamic radius can differ from the radius measured by other means such as transmission electron microscopy since surfactants or double layers that form around the particle in the test solution will influence the diffusive motion of the particle. If the particles are nonspherical, the equivalent or apparent radius of a sphere with the same D_T is calculated. Because the parameter obtained by DLS is the collective diffusion coefficient of the scatterers, involving the movement of the particles within their dispersing medium, there will be a layer of solvent molecules moving with the particle, especially in cases where the surface is rough or has protruding parts, or is partially draining. Hence, DLS measures the apparent hydrodynamic radius of the scatterers (Alexander and Dalgleish 2006). Figure 13.4 depicts examples of apparent hydrodynamic radius.

13.5.3 Polydisperse Samples

In most cases, samples of interest are not monodisperse but encompass a range of sizes. Samples such as milk, yogurt, juice, and other colloidal emulsions contain scatterers with a finite range of sizes. In a DLS experiment, this polydispersity appears itself as a range of decay times in the correlation function where each size contributes according to its intensity-weighted mass fraction in the system and produces a nonlinear decay in the intensity correlation curve. Smaller particles introduce faster decay and larger particles slower decay (Horne 2011). Therefore, the correlation function defined as a single exponential in Equation 13.4 cannot be used for polydisperse particles, and the system must be modeled as an integral of exponential decays. In such a case, the exponential Equation 13.4 becomes an intensity-weighted summation over all sizes present and, for a continuous distribution, the summation can be replaced by an integral of exponential decays represented as

$$g_1(\tau) = \int_0^\infty G(\Gamma) e^{-\Gamma \tau} \cdot d\Gamma \tag{13.7}$$

where the integral of $G(\Gamma)$ is the intensity-weighted distribution of decay constants often normalized to 1, which must be determined to obtain the underlying particle size distribution. However, the inversion of this integral, to solve for the distribution of the particle sizes, is ill-conditioned, meaning that it poses particular problems where the coefficients or constants are estimated from experimental results; and hence, the mathematical solution is very unstable. Consequently, various methods and boundary conditions have been proposed to extract size distribution information from an autocorrelation function, which are described in the following subsection.

Overall polydispersity in a sample is defined in terms of the particle size distribution width. If the DLS data analysis assumes a Gaussian distribution of particles, given as a single peak, polydispersity of the sample is the peak width at half its height. Thus, percent polydispersity can be calculated by dividing polydispersity by mean hydrodynamic radius (R_h) and expressing it as a percentage. A protein sample with less than 20% polydispersity is regarded as monodisperse (Khurshid et al. 2014).

13.5.4 Methods to Calculate Size Distribution

Size distribution is obtained from the correlation function by using various mathematical algorithms. Much effort has been devoted to the development of accurate methods for the analysis of an autocorrelation function obtained from DLS methods (Provencher 1979, 1982), and especially in the problem of deconvoluting or inverting the integral equation presented in Equation 13.7. In fact, most manufacturers of DLS equipment provide specific software applications for this purpose. Moreover, many of the currently proposed methods for particle size analysis are in general only applicable to the specific type of samples and/or they do not provide complete information about the size distribution. Hence, their results need to be interpreted and used with caution (Horne 2011). The characteristics of the most used methods are discussed below.

13.5.4.1 Cumulant Method

For an apparent monodisperse particle size distribution system (such as a protein sample with <20% polydispersity), the most robust and straightforward fitting strategy is to apply the method of cumulants (Koppel 1972). The method of cumulants is the most common for extracting the distribution of decay constants from Equation 13.4 and hence particle sizes (Frisken 2001). This single exponential or cumulant fit of the correlation curve is the fitting procedure recommended by the International Standards Organization. The size distribution obtained is a plot of the relative intensity of light scattered by particles in various size classes and is therefore known as an intensity size distribution. The method uses a moment-generating function that models the mean exponential decay (first moment), the variance (second moment), the skew (third moment), etc., of a distribution by expressing the function as a Taylor series expansion. The logarithm of a moment-generating function, and so a cumulant fit is given by

$$\ln[g_2(\tau)-1] = \ln\left(\frac{\beta}{2}\right) - \bar{\Gamma}\tau + \frac{\kappa_2}{2!}\tau^2 - \frac{\kappa_3}{3!}\tau^3 \quad (13.8)$$

where $\bar{\Gamma}$ represents the mean decay, κ_2 is the second cumulant (equal to the second moment), and κ_3 is the third cumulant (equal to the third moment). The method of cumulants is simple and widely used for characterizing a reasonably narrow distribution without forcing $G(\Gamma)$ to fit a fixed functional form. However, when $G(\Gamma)$ is broad or bimodal (e.g., when "dust" or other contaminants are present), difficulties can arise because of the slow convergence or actual divergence of the cumulant expansion (Hassan and Kulshreshtha 2006). Moreover, inclusion of moments beyond κ_3 in a Taylor series may cause overfitting of the data, making $\bar{\Gamma}$, the variance and skew parameters less precise (Chu 2008).

The hydrodynamic size extracted using this method is an average value, weighted by the particle scattering intensity. Because of the intensity weighting, the cumulant size is defined as the average hydrodynamic radius (R_h) or *Z-average* diameter. Thus, the method gives an average diffusion coefficient $(\bar{D}_T)$ and a *polydispersity index* defined by

$$\text{Polydispersity} = \frac{k_2}{\bar{\Gamma}^2} \tag{13.9}$$

Polydispersity in the area of light scattering is used to describe the width of the particle size distribution. The measurement depends on the size of the particle core, the size of surface structures, particle concentration, and the type of ions in the medium.

13.5.4.2 Inverse Laplace Transform or CONTIN Algorithm

For multimodal or polydisperse particle solutions, the size distribution from DLS is derived from a deconvolution of the measured intensity autocorrelation function of the sample. Generally, this deconvolution is accomplished using a common algorithm such as inverse Laplace algorithm known as CONTIN or using a nonnegatively constrained least squares (NNLS) fitting algorithm (Mattison et al. 2003). Overall, more complex size distributions are most accurately modeled with the CONTIN algorithm developed by Provencher (1982). However, greater care is required in the application of this method due to its use of an ill-defined numerical Laplace transform and assumptions regarding the physical and optical properties of the particle population. The problem of deconvoluting or inverting the integral equation presented in Equation 13.7 is a challenge. These inversion problems are ill-conditioned; the solutions are not necessarily unique, and as many independent pieces of prior knowledge as available should be employed in justifying any particular selection (Provencher 1979; Horne 2011). Polydisperse size distributions can be accurately determined with this method, whereas this is impossible using a cumulant analysis. The CONTIN program resolves the ill-posed Laplace inversion to obtain the diffusion coefficient distribution from the measured correlation function using the regularized NNLS method.

Knowledge of particle sizes and size distribution of a colloidal suspension, powder system, and soft material is a prerequisite for most production and processing operations. Particle size and size distribution have a significant effect on the mechanical strength, density, electrical, and thermal properties of the finished food product.

13.6 EFFECTS OF CONCENTRATION

Despite the mathematical considerations presented above, DLS is now widely used as a convenient and nondestructive method for particle sizing in biomaterials. The technique is suitable for the characterization of colloidal particles and formulations over a wide range of sizes from a few nanometers to several micrometers (i.e., ranging from 1 nm to 1 µm). However, its application in investigations of the dynamic behavior of concentrated and/or strongly interacting colloidal suspensions, and also fluid food products has often been considered to be complicated due to the very strong multiple scattering effect (Nicolai et al. 2001). Hence, in addition to particle–particle interactions, multiple scattering becomes the main concern invalidating common DLS experiments.

13.6.1 MULTIPLE SCATTERING

As the sample concentration is increased, the probability that a scattered photon will interact with another particle or molecule and be scattered again significantly increases; and especially for fluids with larger particles and those with high refractive index contrast. This rescattering effect is called *multiple scattering*. Multiple scattering is a process in which light is being scattered by a sequence of particles for many times; it thus changes the light intensity data being collected. Measuring with DLS in high concentrated solutions may spoil the measured correlation function. The multiple scattering process and the effect on the plot of scattering intensity as a function of time are graphically represented in Figure 13.5.

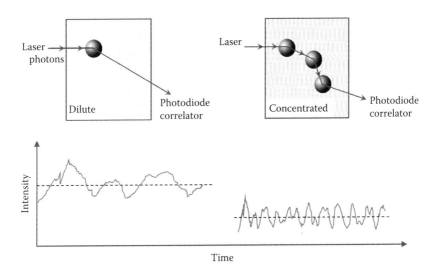

FIGURE 13.5 Schematic representation of singly light scattering in a dilute solution and multiply light scattering wherein photons are scattered multiple times by the particles before being detected.

When the sample concentration increases, three typical changes in the intensity fluctuations can be identified from the correlation curve: (i) the scattering intensity decreases, due to the reduction in the light scattered in the direction of the detector with each rescattering event; (ii) the intercept of the correlation curve decreases, due to the reduction in measured scattering intensity; and (iii) the correlation curve initially decays faster, but often with smaller slope, due to the randomness of the multiple scattering events (Malvern 2015). Consequently, sizing measurement estimations are significantly affected for samples exhibiting multiple scattering since an increase in sample concentration will lead to an apparent decrease in the mean size and often an apparent increase in the polydispersity or distribution width.

The concentration at which multiple scattering occurs is also dependent on the size of the particles being measured as well as the amount of excessive scattered light produced by the particles or molecules studied. However, there is no clear delineation between samples exhibiting only single scattering and those exhibiting multiple scattering, but instead a region where the magnitude of multiple scattering becomes more important to influence scattering measurements and size calculations. The diffusive limit of multiple scattering is governed by the photon mean free path, or the distance that an average photon travels in a sample before encountering a scattering particle. Using Mie theory, it is possible to calculate the mean free path when the particle diameter, refractive index of particles and dispersant, laser wavelength, and sample concentration are all known (Jillavenkatesa et al. 2001). The mean free path measurement then defines the concentration- and size-dependent minimum particle separation distance for single-scattering samples (Pine et al. 1990).

In biological materials, the traditionally used method to overcome the problem of multiple scattering is the dilution of turbid samples or contrast variation, which aim to avoid the multiple scattering of light. These methods, however, cause a change of the sample composition. Since for many systems it is not clear how the sample properties depend on the composition and the concentration of the sample, there is always a risk in measuring artifacts. Other interesting approaches aim at sufficiently suppressing contributions from multiple scattering from the measured photon correlation data using a cross-correlation technique (Schätzel 1991; Urban and Schurtenberger 1999), the two-color DLS technique (Segrè et al. 1995), and also the three-dimensional (3D) cross-correlation technique where the general idea is to isolate singly scattered light and suppress undesired contributions from multiple scattering in a DLS experiment. Other recent approaches to this problem are the *fiber optic QELS* and *DWS*, but they are limited in obtaining useful information and generally cannot provide important features such as the polydispersity of the particles.

13.6.2 CROSS-CORRELATION

In cross-correlation DLS, two simultaneously scattering experiments are performed on the same scattering volume and with identical scattering wave vectors(q), but with the detectors being placed at different spatial positions, which are separated by the characteristic length scale of a single speckle (Zakharov et al. 2006). It has been shown theoretically (Schätzel 1991) and experimentally (Urban and Schurtenberger 1998) that when correlating the intensities, only singly scattered photons contribute to the correlation function. The effect of uncorrelated multiply scattered photons is to reduce the intercept of the correlation function in proportion to their relative intensity (Nicolai et al. 2001). Therefore, the intensity of singly scattered photons can be reduced from the reduction of the intercept. Suppression is achieved because multiple scattering occurs over a larger volume in space, and therefore gives rise to speckles in the far field smaller than the detector separation distance. The main disadvantage of this method is that it produces a cross-correlation signal whose amplitude is significantly below that of the ideal case, and the signal degrades rapidly in moderately turbid systems.

Nicolai et al. (2001), in studying the structure factor of turbid protein gels of β-lactoglobulin with light scattering, showed that the effect of multiple scattering on the scattered light intensity can be effectively corrected using cross-correlation DLS even if the transmission or fraction of singly scattered light is only 1%. Although the method was developed to suppress multiple scattering for DLS, this method has also been shown to be effective for SLS measurements of moderate concentrate turbid samples (Block and Scheffold 2010).

13.6.3 Two-Color DLS

Two-color DLS is another approach to suppress the influence of multiple scattering in the optical signal. This technique employs a more standard goniometer-based measurement setup that provides a large range of measurable scattering vectors. In this case, the technique uses two lasers operating at different wavelengths and two detectors having distinct bandpass filters to capture scattering information from each single laser (Segrè et al. 1995). While the two-color technique has proven to be effective at extracting single-scattering information from highly turbid systems, it is technically challenging to obtain and maintain precise alignment of the illumination and detection optics, especially as the scattering vector is varied, making implementation and operation tedious and difficult (Pusey 1999).

13.6.4 3D Cross-Correlation DLS

One particularly powerful technique for suppressing multiple scattering is the *3D cross-correlation* experiment (3D-DLS). This technique relies on the cross-correlation of two measurements to extract single-scattering information from the same scattering volume and the same nominal scattering vector. Several research groups (Schätzel 1991; Pusey 1999) have demonstrated the feasibility of such an experiment and clearly shown that DLS needs not be restricted to dilute suspensions of small particles, but that it can be used successfully to characterize extremely turbid soft matter such as undiluted skim milk or turbid heat-set protein gels (Urban and Schurtenberger 1999; Nicolai et al. 2001).

In this scheme, the single-scattering information is common to both measurements, while multiple scattering information is uncorrelated, thereby leading to its effective suppression (Urban and Schurtenberger 1998). However, one important drawback of the 3D technique is that one photon detector measures the scattered light intensity at the desired scattering vector, but also receives a contribution at a second undesired scattering vector given by the relative geometry to the second illumination beam operating at the same wavelength.

A significant improvement to this method is achieved by modulating the two incident laser beams temporally and gating the detector outputs at frequencies exceeding the time scale of the system dynamics. This robust modulation scheme eliminates cross-talk between the two beam-detector pairs and leads to a four-fold improvement in the 3D cross-correlation intercept (Block and Scheffold 2010). The illumination beams are alternately activated with high-speed intensity modulators and the detectors are gated in unison. A schematic of the modulated 3D hardware and methodology is shown in Figure 13.6.

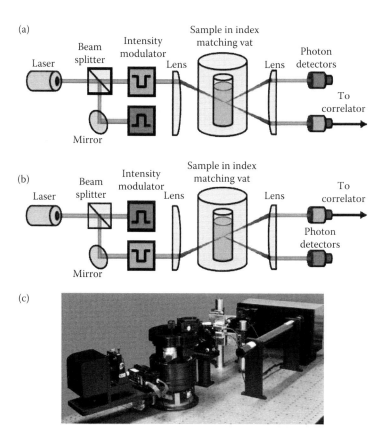

FIGURE 13.6 Schematic representation of modulated 3D cross-correlation light instrument showing (a) and (b) the two states wherein one of the modulators is active and one detector is gated; (c) photo of a commercial 3D-DLS system. (Reprinted from Block, I.D. and Scheffold, F. *Rev. Sci. Instrum.* 81, 123107, 2010.)

13.7 MICROSTRUCTURE AND RHEOLOGICAL PROPERTY MEASUREMENTS IN FOODS

Food materials such as gels and concentrated polymer solutions can have complex structures with characteristic or specific length and time scales. The response to shear strain is an important way of characterizing and understanding their structure and rheological behavior (Popescu et al. 2002). Many food materials, due to their inherent microstructural features, have the ability to partially store and partially dissipate energy when exposed to shear. These materials are known as *viscoelastic materials*. The viscoelastic materials can be characterized by the stress relaxation modulus or complex shear modulus $G^*(t)$, which describes the magnitudes and time scales of the relaxation of the stress in the bulk material to a fixed strain after a step shear strain. The Fourier transform of these moduli is the frequency-dependent complex shear modulus $G^*(\omega)$. The real part of the complex modulus $G'(\omega)$ measures the in-phase response of the medium to an oscillatory strain, thus giving information on the elastic (storage) property of the material. The out-of-phase response is given by the imaginary part $G''(\omega)$, which is related to the viscous behavior (loss) of the material (Dasgupta et al. 2002).

The viscoelastic properties of food are traditionally measured by means of a mechanical rheometer. However, the computed stress relaxation properties are the result of bulk measurements of a composite system, whose structural and rheological heterogeneities are averaged over some length scale. Rheological properties of viscoelastic foods arise from the composition and physical characteristics of various microscopic and molecular components and, more importantly, from the structural arrangement or architecture of these constituents. Hence, analyses at the micro or nano scales are needed for more accurate and realistic characterization of viscoelastic materials. Nowadays, experimental procedures relating both the macro- and microrheological parameters $G'(\omega)$ and $G''(\omega)$ with food microstructure in colloid aggregation, stability, and macroscopic mechanical behavior are essential in food research and product development.

DLS is one of the most popular experimental techniques in the investigation of colloidal suspensions, and it has played an important role in recent developments on measuring the dynamic properties of colloidal systems (Scheffold and Schurtenberger 2003). A significant limitation of DLS, however, has been the lack of a general scheme for applying it to systems that exhibit strong multiple scattering. Through a series of experiments (Urban and Schurtenberger 1998; Dasgupta et al. 2002; Lederer and Schöpe 2012), DLS has been extended to study fluid materials that exhibit a high degree of multiple scattering. Two relatively recent applications of conventional DLS technique have demonstrated the capability of determining the complex shear modulus $G^*(\omega)$ in complex soft materials; they are DWS (Alexander and Dalgleish 2006; Nik et al. 2011; Alexander and Corredig 2013) and the DLS or DWS-based *microrheology*, both of which have begun to play a major role in the characterization of complex food systems (Amin et al. 2012; Malvern 2015).

13.7.1 DIFFUSING WAVE SPECTROSCOPY

While in conventional DLS experiments the sample has to be almost transparent and hence often highly diluted, DWS extends conventional DLS to media with strong multiple scattering, treating the transport of light as a diffusion process (Scheffold et al. 2004). The DWS exploits the diffusive nature of light in strongly scattering media and relates the temporal fluctuations of multiply scattered light to the motion of the scatterers (Pine et al. 1990; Moschakis 2013). In other words, DWS measures the properties of multiply scattered light in suspensions, in which all detected photons are known to have been multiply scattered.

Since its inception in 1987, significant progress has been made in the theory and methodology of DWS technique for its application to various complex soft materials, including foods. Two different experimental setups are commonly used for DWS: transmission and backscattering. In transmission DWS, the laser impinges on the sample on one side and the scattered light is collected on the other side. In backscattering DWS, the scattered light is collected on the same side as the incoming laser light. A graphical representation of these two setups is shown in Figure 13.7. Depending on the experimental conditions, both techniques can be adopted to perform simultaneous DWS and rheology experiments; however, a clear understanding of the theory of light scattering under shear flow is required in order to avoid

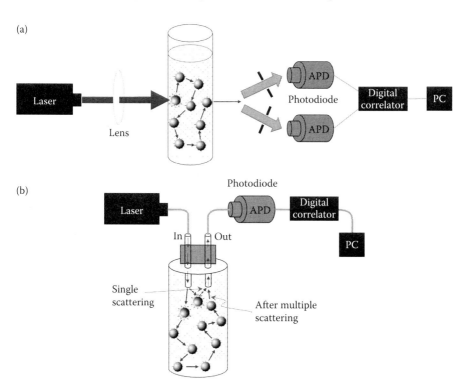

FIGURE 13.7 Schematic representation of two DWS setups: (a) forward-scattering experiment where the light is split and passed to the two detectors and then cross-correlated to remove noise and (b) backscattering experiment where the light is guided and collected by optic fibers and then passed to the detector and correlator. With setups (a) and (b), it is possible to detect both single and multiple scattering light. (Reproduced from Alexander, M. and Dalgleish, D.G. *Food Biophys.* 1, 2–13, 2006.)

obtaining erroneous or misleading measurements and interpretations (Bicout and Maynard 1993).

13.7.1.1 Principle of DWS

DWS technique is based on diffusion approximation for the description of the propagation of light in a multiple scattering medium. With this approach, the phase correlations of the scattered waves within the medium are ignored, and only the scattered intensities are considered (Pine et al. 1990). Thus, the path followed by an individual photon can be described as a *random walk*.

In strongly scattering media, the propagation of light can be treated as a diffusion process, that is, by assuming that each detected photon has undergone a random walk through the suspension. Because all such paths are sampled, the overall correlation function obtained is markedly nonexponential. Therefore, to obtain the correlation function, the contributions from all paths are summed, each weighted by the probability that the light follows that path or undergoes a random walk. This is where diffusion approximation comes into play because it allows the problem of summation (i.e., integration), to be reduced to the solution of a diffusion equation for the light subjected to boundary conditions set by the geometry of the experiment (Horne 2011). The size, shape, and positioning of sample, source, and detector(s) are important factors in the experimental setup.

The fluctuations in the intensity of backscattered light can be correlated and described by Equation 13.3. For a laminar flow of a viscous fluid sample of infinite thickness and in the range of shear rates where the Brownian motion is not affected by the laminar shear, the relative motion contributed by particle diffusion and convective shear will be decoupled. The backscattering autocorrelation function has the following expression (Alexander and Corredig 2013):

$$g_1(\tau) = \frac{1}{1-\gamma l^*/L} \frac{\sinh\left(L/l^* \left\{6\left[\tau/\tau_B + (\tau/\tau_S)^2\right]\right\}^{1/2} (1-\gamma l^*/L)\right)}{\sinh\left(L/l^* \left\{6\left[\tau/\tau_B + (\tau/\tau_S)^2\right]\right\}^{1/2}\right)} \quad (13.10)$$

where L is the width of the sample cell, l^* is the photon transport mean free path, and γ is the minimum distance a photon must travel to be completely decorrelated (Wu et al. 1990). Equation 13.10 consists of two totally independent decay times, τ_B, corresponding to Brownian motion, and τ_S, corresponding to the convective shear flow. In the case where $\tau_S \to 0$, the correlation function reverts to the typical decay function characteristic of stochastic motion. If instead $\tau_S \gg \tau_B$ then Brownian motion can be neglected and the correlation function has a decay characteristic of convective motion.

Similarly, fluctuations in the intensity of scattered light in transmission mode can be characterized by Equation 13.3 as for DLS, and, in samples with the thickness $L \gg l^*$ (i.e., $L/l^* \gg 10$) (where l^* is the transport mean free path) and the time $t \ll \tau$ (correlation time of the sample), the correlation function for the transmission can be described by (Weitz and Pine, 1993):

$$g_1(\tau) \approx \frac{\left((L/l^*) + (4/3)\right)\sqrt{6t/\tau}}{(1+(8t/3\tau))\sinh\left[L/l^* \sqrt{6t/\tau}\right] + (4/3)\sqrt{6t/\tau}\cosh\left[L/l^* \sqrt{6t/\tau}\right]} \quad (13.11)$$

where all the variables (*t*, time and τ, decay time) have been defined above. Assuming that a completely diffusing particle motion occurs, this correlation function is nearly exponential in time, with a characteristic decay time of

$$\tau = \tau_0 \left(\frac{l^*}{L}\right)^2 \quad (13.12)$$

Although DWS can provide information on the average size of particles in a sample, it cannot provide information on size distributions. Because of the multiple scattering and random walk nature of the photon path, all particles within a scattering sample potentially contribute to the intensity of each scattered path. This means that, without specific information at any given **q**, only average knowledge of the particles giving rise to the scattered light is possible. For this reason, size distribution within a sample is not measurable (Alexander and Dalgleish 2006; Horne 2011).

13.7.1.2 Applications in Food Materials

DWS was recently used to study the effect of soy protein composition (specific glycinin and β-conglycinin subunits) on the rheological properties of soymilk during acidification. Nik et al. (2011) showed that transmission DWS and conventional rheology in combination were able to describe the beginning stages of aggregation of soymilk, and also monitor the acid gelation, by measuring two parameters: the inverse photon transport mean free path or turbidity parameter ($1/l^*$) which provides structural information of the early stages of sol–gel transition (Alexander and Dalgleish 2004), and the diffusion coefficient from which the hydrodynamic size or apparent radius (R_h) can be derived. It was concluded that by using DWS on different soybean lines, the soy glycinin and β-conglycinin subunits had a significant effect on the formation of gel structure, although no large differences in gelation pH were observed (Nigro et al. 2015).

More recently, Alexander and Corredig (2013) designed a novel setup to investigate the dynamics of gelation of model dairy-based colloids, using a rheometer coupled with DWS in backscattering mode. The Rheo-DWS setup (Figure 13.8) is attached to a commercial controlled-stress rheometer. In the experiment, a HeNe laser (wavelength 632.5 nm, power 150 mW) impinges on the side of the concentric cylinder geometry (bob-and-cup cuvette) of the rheometer, encased in a stainless-steel cylindrical jacket donned with an opening to allow the laser beam to pass through. The stainless jacket is hollow and connected to an external water bath control to keep the sample at 30°C. The backscattered light collection setup uses a single fiber optic bifurcated and fed into two matched photomultipliers to analyze the data. In that study, the light scattering and rheological parameters were measured as a function of time for skim milk and concentrated (4×) skim milk with the addition of rennet.

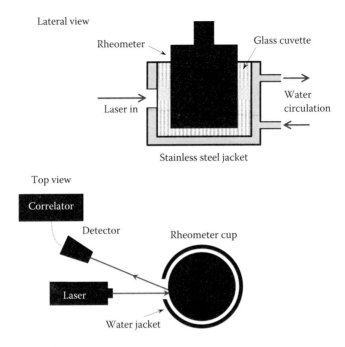

FIGURE 13.8 Lateral and top views of the Rheo-DWS setup developed by Alexander and Corredig (2013). Diagram of the cuvette with its temperature jacket and window to allow passage of the laser light. The sample (8 mL) was incubated at 30°C in a specially built concentric cylinder geometry (1 mm fixed gap). A time sweep was run at 0.01 controlled strain, 1.0 Hz frequency. The sol–gel transition was defined as the time at which the values of G' (storage modulus) and G'' (loss modulus) were equal.

Results showed that the gelation points as measured by transmission DWS and Rheo-DWS were equal for both experimental methods and in good agreement with conventional rheological measurements. The results were also in agreement with other reported studies (Sandra et al. 2011). The study of Alexander and Corredig (2013) demonstrated the potential of the proposed setup to follow the dynamics of colloidal destabilization and structure formation *in situ* while subjecting them to simultaneous rheological measurements. Moreover, the results suggested the need for continued development and application of this and other novel DWS setups for the investigation of dynamically evolving food structures. Compared to classical rheology, DWS makes it possible to significantly increase the range of accessible frequencies, thereby opening up new possibilities for the study of many complex materials.

13.7.2 Microrheology

13.7.2.1 Basic Concepts

A particularly interesting application of DLS or DWS in fundamental soft material research as well as in food science has been *optical microrheology*. Microrheology techniques are usually classified into two main categories: active techniques that require particle manipulation by external forces and passive techniques that are based on thermal fluctuations of the embedded particles (Waigh 2005). In DLS or DWS-based optical microrheology, there is no external driving force applied to the tracer microparticles and the local particle motion is driven solely by Brownian forces generated by the thermal energy k_bT; that is, the particles receive a random displacement by collision with other molecules with no external force (Moschakis 2013). The thermally driven motion of colloidal particles suspended in a complex fluid, such as a protein solution, is intimately linked to the rheological properties of the suspending fluid (Amin et al. 2012). Thus, the underlying idea is to study the thermal response of small (colloidal) particles embedded in the system under study. By analyzing the thermal motion of the particle, it is possible to obtain quantitative information about the loss and storage moduli, $G'(\omega)$ and $G''(\omega)$, over an extended range of frequencies (Gisler and Weitz 1998). Since DWS allows access to a broad range of time scales, this results in an unprecedented large frequency range covered by DWS-based optical microrheology.

Particles embedded/dispersed in a medium undergo Brownian motion and are thus displaced from their original positions over time (Sartor 2015). For a particle moving freely due to thermal fluctuations in a purely viscous (Newtonian) fluid, the mean square displacement (MSD) increases linearly with time. The slope of this increase is directly related to the diffusion coefficient, D_T, of the particle. For purely diffusive motion, the d-dimensional MSD is given by

$$\langle \Delta r^2(\tau) \rangle = 2dD_T\tau = 2d\left(\frac{k_BT}{6\pi\eta R}\right)\tau \qquad (13.13)$$

where d represents the measured d-dimensional MSD and R is the particle radius. From this the fluid viscosity is given by

$$\eta = \frac{2dk_BT}{6\pi R}\frac{\tau}{\langle \Delta r^2(\tau) \rangle} \qquad (13.14)$$

When the MSD for a viscoelastic material is plotted as a function of time on a double logarithmic plot, it is given by

$$\langle \Delta r^2(\tau) \rangle \propto \tau^p \qquad (13.15)$$

where the slope p is equal to

$$p = \frac{d\log\langle \Delta r^2(\tau) \rangle}{d\log\tau} \qquad (13.16)$$

In viscoelastic materials, the slope p falls between the viscous limit ($p = 1$) and the elastic limit ($p = 0$). For $0 < p < 1$, at very short lag times, the embedded particles exhibit Newtonian response and the MSD rises diffusively. However, at intermediate lag times, the MSD reaches a plateau, indicating that the particles are becoming constrained and the elasticity prevails. The lower the plateau, the stronger the elasticity. At even longer lag times, the particles may in some cases escape (diffuse out) from the confined space (cage) as the network relaxes the stress and the MSD grows linearly with time reflecting a diffusive behavior. The viscosity estimated at long displacements can be related to the macroscopic (bulk) viscosity (Moschakis 2013).

13.7.2.2 DLS Microrheology

In optical microrheology, the average movement of an ensemble of dispersed probe particles is tracked using DLS or its extension DWS. The technique has significant potential for measuring low-viscosity, weakly structured samples, such as polymer and protein solutions, because it allows access to the very high frequencies needed to measure critical short time scale dynamics of these systems. An additional benefit of optical microrheology is the ability to work with much smaller sample volumes than are typically needed for a mechanical rotational rheometer.

The MSD of the particles is determined by the viscoelasticity of the medium. Consequently, measuring the MSD offers information about the viscoelasticity of the sample to be extracted, so long as the size of the dispersed particles is known. In a DLS experiment, the autocorrelation function ($g_1(\tau)$) of the scattered light can be written as a function of the MSD of the scatterers $_D r^2$ with time, where q is the scattering vector:

$$g_1(\tau) \propto e^{-q^2 \langle \Delta r^2(\tau) \rangle} \tag{13.17}$$

or

$$g_1(\tau) = g_1(0) \cdot e^{-(1/6)q^2 \langle \Delta r^2(\tau) \rangle} \tag{13.18}$$

where $g_1(0)$ is the value of the correlation at zero time, or intercept, which is expected to be lower than 1 due to optical effects (Berne and Pecora 2000). Then, the complex shear modulus of the medium as a function of angular frequency $G^*(\omega)$ can be obtained by performing a DLS experiment, and through a unilateral Laplace transform of the MSD using the generalized Stokes–Einstein relationship (Mason and Weitz 1995):

$$G^*(\omega) = \frac{k_B T}{\pi R i \omega \langle \Delta r^2(i\omega) \rangle} \tag{13.19}$$

in which the elastic modulus $G'(\omega)$ and viscous modulus $G''(\omega)$ are derived using the Euler relationships (Mason 2000).

Finally, the relative viscosities (η_r) for action can be obtained using the following equation:

$$\eta_r = \frac{\eta}{\eta_s} \tag{13.20}$$

where η_s is the solvent viscosity and η is the solution viscosity.

Recently, Amin et al. (2012) illustrated the applicability of DLS-based optical microrheology for determining the rheological response of dilute protein solutions as they started to form insoluble aggregates under the influence of thermal stress. The authors also showed the capabilities of the technique for quick measurement of the viscosity in protein solutions. Comparisons between the relative viscosities computed by DLS microrheology and those obtained from a commercial multicapillary viscometer (Viscotek DVS from Malvern Instruments Ltd., Worcestershire, UK) for bovine serum albumin (BSA) solutions with concentrations between 10 and 100 mg/mL, showed an excellent agreement. Hence, DLS microrheology can be used to make quick and accurate measurements on protein solutions.

Amin et al. (2012) also evaluated the evolution of the elastic modulus $G'(\omega)$ with temperature for a 10-mg/mL BSA solution for a range of temperatures. A weak elastic response at 70°C was observed, but the response at 80°C increased by more than an order of magnitude. In addition, the frequency dependence of $G'(\omega)$ became weaker with increasing temperature. The power law (p) showed a relatively poor fitting for $G'(\omega)$ at 70°C as the data were noisy due to the presence of very weak elasticity at this temperature, but improved at 80°C as the elasticity increased. The study showed that the MSD obtained from DLS-based optical microrheology is very sensitive to the evolution of weak viscoelasticity in the aggregating dilute solutions. It was found that the BSA concentration impacted strongly on the viscoelastic response, with the 2-mg/mL sample hardly exhibiting a detectable elastic response, while the 10-mg/mL sample exhibited a strong evolution of the elasticity with temperature as the aggregation process progressed. The technique could therefore be effectively employed to explore how formulation conditions (concentration, electrolyte, pH, etc.) impact on the evolution of the viscoelastic response.

13.8 CONCLUSIONS

In recent years, significant progress has been made in the development of modern DLS techniques. DLS has proven to be a powerful tool for the study and testing of dynamical process in simple and complex fluids. Combining DLS technology with complementary light scattering technologies and other characterization technologies such as microrheology, we can obtain more complete information about complex soft materials, which, otherwise, cannot be assessed by DLS alone. Over the years, the DLS technique has been extensively studied, improved, and extended for different applications in complex soft materials and food matrices in the lab research and by industry. DLS is rapid and noninvasive for determination of protein size. However, like any other particle sizing technique, DLS has its disadvantages. It is particularly important to use the DLS technique with prior knowledge of the composition and physical interaction of the studied material as available, if meaningful results are to be obtained. Today, the potential of the DLS technique to follow the dynamics of colloidal destabilization and structure formation *in situ* while subjecting them to simultaneous rheological measurements has been widely demonstrated in various food materials.

NOMENCLATURE

Γ	decay constant
D_T	diffusion coefficient
$g(\tau)$	correlation function
$G^*(t)$	stress relaxation modulus
$G'(\omega)$	real part of the complex modulus that gives a measure of the elasticity of the material
$G''(\omega)$	imaginary part of the complex modulus which is related to the viscosity of the material
k_b	Boltzmann's constant
n	refractive index
$\mathbf{q}$	scattering vector
η	viscosity
R_h	hydrodynamic radius
λ	wavelength
θ	scattering angle

REFERENCES

Alexander, M. and Corredig, M. On line diffusing wave spectroscopy during rheological measurements: A new instrumental setup to measure colloidal instability and structure formation *in situ*. *Food Research International* 54, 2013: 367–72.

Alexander, M. and Dalgleish, D.G. Application of transmission diffusing wave spectroscopy to the study of gelation of milk by acidification and rennet. *Colloids and Surface B: Biointerfaces* 38, 2004: 83–90.

Alexander, M. and Dalgleish, D.G. Dynamic light scattering techniques and their applications in food science. *Food Biophysics* 1, 2006: 2–13.

Amin, S., Rega, C. and Jankevics, H. Detection of viscoelasticity in aggregating dilute protein solutions through dynamic light scattering-based optical microrheology. *Rheologica Acta* 51, 2012: 1–14.

Bancarz, D., Huck, D., Kaszuba, M., Pugh, D. and Ward-Smith, S. Particle sizing in the food and beverage industry. In Irudayaraj, J. and Reh, C. (eds.), *Nondestructive Testing of Food Quality*, 165–96. Oxford, UK: Blackwell Publishing Ltd, 2008.

Bergfords, T. Dynamic light scattering. In Bergfords, T. (ed.), *Protein Crystallization: Strategies, Techniques and Tips*, 29–38. La Jolla, California: International University Line, 1999.

Berne, B. and Pecora, R. *Dynamic Light Scattering: With Applications to Chemistry, Biology, and Physics*. New York: Dover Publications, Inc., 2000.

Bicout, D. and Maynard, R. Diffusing wave spectroscopy in inhomogeneous flows. *Physica A* 199, 1993: 387–411.

Block, I.D. and Scheffold, F. Modulated 3D cross-correlation light scattering: Improving turbid sample characterization. *Review of Scientific Instruments* 81, 2010: 123107.

Brar, S.K. and Verma M. Measurement of nanoparticles by light-scattering techniques. *Trends in Analytical Chemistry* 30, 2011: 4–17.

Calzolai, L., Gilliland, D. and Rossi, F. Measuring nanoparticles size distribution in food and consumer products: A review. *Food Additives and Contaminants* 29, 2012: 1183–93.

Chen, D.T.N., Wen, Q., Janmey, P.A., Crocker, J.C. and Yodh, A.G. Rheology of soft materials. *Annual Review of Condensed Matter Physics* 2, 2010: 301–22.

Chu B. Dynamic light scattering. In Borsali, R. and Pecora, R. (eds.), *Soft Matter Characterization*, 335–369. New York, NY, USA: Springer, 2008.

Dasgupta, B.R., Tee, S.-Y., Crocker, J.C., Frisken, B.J. and Weitz, D.A. Microrheology of polyethylene oxide using diffusing wave spectroscopy and single scattering. *Physical Review E* 65, 2002: 051505-1-10.

Frisken, B.J. Revisiting the method of cumulants for the analysis of dynamic light scattering data. *Applied Optics* 40, 2001: 4087–91.

Gisler, T. and Weitz, D.A. Tracer microrheology in complex fluids. *Current Opinion in Colloid & Interface Science* 3, 1998: 586–92.

Hassan, P.A. and Kulshreshtha, S.K. Modification to the cumulant analysis of polydispersity in quasielastic light scattering data. *Journal of Colloid and Interface Science* 300, 2006: 744–8.

Horne, D.S. Analytical methods light scattering techniques. In Fuquay, J.W., Fox, P.F., and McSweeney, P.L.H. (eds.), *Encyclopedia of Dairy Sciences*, 2nd Edn, 133–40, Amsterdam, Netherlands: Academic Press, 2011.

Jillavenkatesa, A., Dapkunas, S.J. and Lum, L.-S.H. Particle size characterization, NIST Recommended Practice Guide. Special Publication 960-1, 164 pages. Washington, DC: US Government Printing Office, 2001.

Koppel, D.E. Analysis of macromolecular polydispersity in intensity correlation spectroscopy: The method of cumulants. *The Journal of Chemical Physics* 57, 1972: 4814–21.

Khurshid, S., Saridakis, E., Govada, L. and Chayen, N.E. Porous nucleating agents for protein crystallization. *Nature Protocols* 9, 2014: 1621–33.

Lederer, A. and Schöpe, H.J. 2012. Easy-use and low-cost fiber-based two-color dynamic light-scattering apparatus. *Physical Review E* 85, 2012: 031401-1–8.

Malvern Instruments Limited. *Application of Dynamic Light Scattering (DLS) to Protein Therapeutic Formulations: Principles, Measurements and Analysis—2. Concentration Effects and Particle Interactions.* Malvern Instruments Limited, Worcestershire, UK. http://www.copybook.com/pharmaceutical/malvern-instruments-ltd/articles/application-of-dls-to-proteintherapeutic-formulations-2). Accessed Jan 22, 2016.

Mason, T.G. Estimating the viscoelastic moduli of complex fluids using the generalized Stokes–Einstein equation. *Rheologica Acta* 39, 2000: 371–8.

Mason, T.G. and Weitz, D.A. Optical measurements of frequency dependent linear viscoelastic moduli of complex fluids. *Physical Review Letters* 74, 1995: 1250–3.

Mattison, K., Morfesis, A. and Kaszuba, M. A primer on particle sizing using dynamic light scattering. *American Biotechnology Laboratory* 21, 2003: 20–2.

Maurer, E. and Pittendreigh, C. Dynamic light scattering for *in vitro* testing of body fluids. U.S. Patent 8323922 B2, 2012.

Merkus, H.G. *Particle Size Measurements*. The Netherlands: Springer Science + Business Media B.V., 2009.

Mezzenga, R., Schurtenberger, P., Burbidge, A. and Michel, M. Understanding foods as soft materials. *Nature Materials* 4, 2005: 729–40.

Miller, C.C. The Stokes–Einstein law for diffusion in solution. *Proceedings of the Royal Society of London. Series A, Containing Papers of a Mathematical and Physical Character* 106, 1924: 724–49.

Moschakis, T. Microrheology and particle tracking in food gels and emulsions. *Current Opinion in Colloid & Interface Science* 19, 2013: 311–23.

Nicolai, T., Urban, C. and Schurtenberger, P. Light scattering study of turbid heat-set globular protein gels using cross-correlation dynamic light scattering. *Journal of Colloid and Interface Science* 240, 2001: 419–24.

Nigro, V., Angelini, R., Bertoldo, M., Castelvetro, V., Ruocco, G. and Ruzicka, B. Dynamic light scattering study of temperature and pH sensitive colloidal microgels. *Journal of Non-Crystalline Solids* 407, 2015: 361–6.

Nik, A.M., Alexander, M., Poysa, V., Woodrow, L. and Corredig, M. Effect of soy protein subunit composition on the rheological properties of soymilk during acidification. *Food Biophysics* 6, 2011: 26–36.

Nobbmann, U., Connah, M., Fish, B., Varley, P., Gee, C., Mulot, S., Chen, J. et al. Dynamic light scattering as relative tool for assessing the molecular integrity and stability of monoclonal antibodies. *Biotechnology and Genetic Engineering Reviews* 24(1), 2007: 117–28.

Øgendal, L. *Light Scattering—A Brief Introduction*. University of Copenhagen, September 16, 2013. (http://igm.fys.ku.dk/~lho/personal/lho/LS_brief_intro.pdf.). Accessed May 10, 2015.

Pine, D.J., Weitz, D.A., Zhu, J.X. and Herbolzheimer, E. Diffusing-wave spectroscopy: Dynamic light scattering in the multiple scattering limit. *Journal of Physics France* 51, 1990: 2101–27.

Popescu, G., Dogariu, A. and Rajagopalan, R. Spatially resolved microrheology using localized coherence volumes. *Physical Review E* 65, 2002: 041504-1–8.

Provencher, S.W. Inverse problems in polymer characterization: Direct analysis of polydispersity with photon correlation spectroscopy. *Macromolecular Chemistry and Physics* 180, 1979: 201–9.

Provencher, S.W. CONTIN: A general purpose constrained regularization program for inverting noisy linear algebraic and integral equations. *Computer Physics Communications* 72, 1982: 229–42.

Pusey, P. Suppression of multiple scattering by photon cross-correlation techniques. *Current Opinion Colloid Interface Science* 4, 1999: 177–85.

Rao, M.A. *Rheology of Fluid, Semisolid, and Solid Foods*, 3rd edn. New York: Springer, 2014.

Sandra, S., Cooper, C., Alexander, M. and Corredig, M. Coagulation properties of ultrafiltered milk retentates measured using rheology and diffusing wave spectroscopy. *Food Research International* 44, 2011: 951–6.

Sapsford, K.E., Tyner, K.M., Dair, B.J., Deschamps, J.R. and Medintz, I.L. Analyzing nanomaterial bioconjugates: A review of current and emerging purification and characterization techniques. *Analytical Chemistry* 83, 2011: 4453–88.

Sartor, M. *Dynamic Light Scattering to Determine the Radius of Small Beads in Brownian Motion in a Solution*. San Diego, California: UCSD, University of California. (http://physics.ucsd.edu/neurophysics/courses/physics_173_273/dynamic_light_scattering_03.pdf.). Accessed May 1, 2015.

Schätzel, K. Suppression of multiple-scattering by photon cross-correlation techniques. *Journal of Modern Optics* 38, 1991: 1849–65.

Scheffold, F., Romer, S., Cardinaux, F., Bissig, H., Stradner, A., Rojas-Ochoa, L.F., Trappe, V. et al. New trends in optical microrheology of complex fluids and gels. *Progress in Colloid and Polymer Science* 123, 2004: 141–6.

Scheffold, F. and Schurtenberger, P. Light scattering probes of viscoelastic fluids and solids. *Soft Materials* 1, 2003: 139–65.

Segrè, P.N., Van Megen, W., Pusey, P.N., Schatzel, K. and Peters, W.J. Two-colour dynamic light scattering. *Journal of Modern Optics* 42, 1995: 1929–52.

Siegert, A.J.F. On the fluctuations in signals returned by many independently moving scatters. Report 465, Massachusetts Institute of Technology Radiation Laboratory, 1943.

Sonn-Segev, A., Bernheim-Groswasser, A., Diamant, H. and Roichman, Y. Viscoelastic response of a complex fluids at intermediate distances. *Physical Review Letters* 112, 2014: 088301-1-5.

Sun, Y. Investigating static and dynamic light scattering. Cornell University Library, eprint arXiv: 1110.1703v1 [physics.chem-ph], 2011. (http://arxiv.org/pdf/1110.1703.pdf). Accessed Jan 22, 2016.

Urban, C. and Schurtenberger, P. Characterization of turbid colloidal suspensions using light scattering techniques combined with cross-correlation methods. *Journal of Colloid and Interface Science* 207, 1998: 150–8.

Urban, C. and Schurtenberger, P. Application of a new light scattering technique to avoid the influence of dilution in light scattering experiments. *Physical Chemistry Chemical Physics* 1, 1999: 3911–5.

Waigh, T.A. Microrheology of complex fluids. *Reports on Progress Physics* 68, 2005: 685–742.

Weitz, D.A. and Pine, D.J. Diffusing-wave spectroscopy. In Brown, W. (ed.), *Dynamic Light Scattering: The Method and Some Applications*, 652–720. Oxford: Oxford University Press, 1993.

Wu, X.L., Pine, D.J., Chaikin, P.M., Huang, J.S. and Weitz, D.A. Diffusing wave spectroscopy in a shear flow. *Journal of Optical Society of America B* 7, 1990: 15–20.

Xu, C., Cai, X., Zhang, J. and Liu, L. Fast nanoparticle sizing by image dynamic light scattering. *Particuology* 19, 2015: 82–5.

Xu, R. Light scattering: A review of particle characterization applications. *Particuology* 18, 2015: 11–21.

Zakharov, P., Bhat, S., Schurtenberger, P. and Scheffold, F. Multiple-scattering suppression in dynamic light scattering based on a digital camera detection scheme. *Applied Optics* 45, 2006: 1756–64.

14 Biospeckle Technique for Assessing Quality of Fruits and Vegetables

Artur Zdunek, Piotr Mariusz Pieczywek, and Andrzej Kurenda

CONTENTS

14.1 Biospeckle Phenomena 361
14.2 Biological Sources of Biospeckle Activity 363
14.3 Image Analysis for Assessment of Biospeckle Activity 366
    14.3.1 Global Activity Measures 366
        14.3.1.1 Speckle Contrast 366
        14.3.1.2 Space–Time Speckle/Time History Speckle Pattern-Based Measures 367
        14.3.1.3 Spatial–Temporal Correlation Technique 369
    14.3.2 Analysis of Biospeckle Spatial Activity 369
        14.3.2.1 Fujii Method 369
        14.3.2.2 Generalized Differences 370
        14.3.2.3 Laser Speckle Contrast Analysis 371
        14.3.2.4 Modified Laser Speckle Imaging 372
        14.3.2.5 Temporal Difference Method 373
        14.3.2.6 Motion History Image 373
        14.3.2.7 Empirical Mode Decomposition 374
        14.3.2.8 Normal Vector Space Statistics 374
        14.3.2.9 Analysis in Spectral Domain 375
        14.3.2.10 Multivariate Speckle Pattern Analysis 375
14.4 Evaluation of Fruit and Vegetable Quality 376
14.5 Conclusions 381
References 382

14.1 BIOSPECKLE PHENOMENA

When laser light illuminates a rough object, due to the optical interference of multi-backscattered electromagnetic waves, a speckle pattern is formed on an observation plane. The speckle pattern is composed of dark and light spots due to superposition of the waves, which is static for nonliving matter. When the laser light illuminates a biological sample, the speckle patterns are no longer stable due to the physical and chemical processes occurring inside the sample. This phenomenon is called the dynamic laser speckle, or the biospeckle for biological samples. The movement of speckles is analyzed to characterize biospeckle activity. Laser light can penetrate biological tissue, particularly in the range of the red light. Therefore, biospeckle activity carries some information about the sample from a depth related to penetration of a tissue by laser light.

In the case of botanical materials such as fruits and vegetables, the sources of apparent biospeckle activity are the Brownian motions, which are the random motions of particles suspended in a fluid (a liquid or a gas) resulting from their collision with the quick-moving atoms or molecules in the gas or liquid, and biological processes such as cyclosis, growth, transport, etc., and this activity is affected by pigments, pathogen infections, or other mechanical and physiological defects. Several different applications have been shown to monitor aging, maturation, and organ development in fruits and vegetables. Biospeckle activity is also affected by light absorption by tissue pigments and by the presence of disorders in tissues such as pathogens, surface diseases, etc. Therefore, the biospeckle method can provide information about a wide range of living processes in fruits and vegetables. The biospeckle method, due to the low power laser light used, is nondestructive for plants. Moreover, typical systems for biospeckle measurements are relatively simple and inexpensive (Figure 14.1). These features make this method attractive to many applications, which require fast and nondestructive sampling.

Most systems for biospeckle evaluation are made of a red He–Ne or a diode laser, camera with frame grabber, some optic components (for beam expansion, setting area of interest or polarization of the light), and software for the biospeckle activity evaluation (Figure 14.1). An experimental procedure to record the biospeckle usually consists of three steps. (1) Illumination of the sample by laser light, which may be extended or collimated to a chosen region of interest. (2) Recording of video or stack of images of biospeckles. Most often, two approaches are taken here—collecting

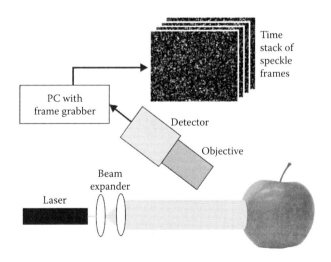

FIGURE 14.1 Typical system for biospeckle evaluation.

whole frames with known lag time or collecting a single line with known lag time to construct the time history matrix. (3) Evaluation of biospeckle activity from the time history speckle frames.

Since the technique is not complicated from the point of view of hardware, at present, major effort is placed on decomposition of the signal to follow or quantify processes of interest. This requires two indispensable steps: interpretation of the biological and physical origins of biospeckle phenomena and development of image analysis methods for assessment of biological activity. This chapter will review recent developments in these subjects.

14.2 BIOLOGICAL SOURCES OF BIOSPECKLE ACTIVITY

Dynamic speckle patterns are observed in materials characterized by spatiotemporal instability. Biological materials are a special group of interference pattern–generating objects due to the presence of a number of physiological processes, which in most cases are the major sources of dynamic light scattering. Biospeckle activity can be recorded in every organism; however, the diversity of organisms, variety of physiological processes, or differences in shape and structure cause a large variety of possible activity sources. Because the physiological state of organisms is constantly changing, biospeckle dynamics are also dependent on, for example, stage of development, health, instantaneous content of scattering centers or light-attenuating pigments, and are therefore variable in time. Thus, on the one hand, the interpretation of the observed changes of biospeckle activity is difficult because it requires specific knowledge about the object, and on the other hand, it creates great potential for the application of this optical technique for evaluating a wide range of physiological parameters in all kinds of organisms.

In living organisms, in addition to physiological activity, physical processes such as Brownian motion and diffusion take place, and therefore biospeckle activity contains components originating from these physiological and physical processes. Consequently, the speckle "signature" contains information about the state of the organism, which is encoded in the form of frequency fluctuations of the intensity of the speckles in the interference image (Oulamara et al., 1989).

When physiological processes are related to the movement, they become the major sources of biospeckle. One can divide all physiological movement–generating processes of organisms or inside organisms related to biospeckle into four groups: independent active movements of microorganisms (I), changes of the shape of organs or tissues (II), secondary transport of the cells, particles, and fluids in the scale of organism (III), and transport of the organelles and particles at the scale of a single cell (IV).

Active movement of microorganisms as a source of biospeckle activity has been used to detect bacteria in liquid culture media (Zheng et al., 1994; Cole and Tinker, 1996), viability of gametes (Carvalho et al., 2009), fungi in bean seeds (Braga et al., 2005), motility of nematode parasites exposed to different anthelmintic drugs (Pomarico et al., 2004), and motility of sea crustaceans at different temperatures (Ebersberger et al., 1986).

Variations in biospeckle activity induced by surface changes of organs and tissues have been used mainly in medicine and plant physiology. The rate of blood flow

(Fujii et al., 1985) and functional state of the human brachial biceps (Tanin et al., 1993) were studied by analysis of biospeckle activity registered on human skin and muscular tissue, respectively. In plant research, movement (Aizu and Asakura, 1996) and rate of growth (Briers, 1977) of organs were tracked with the use of dynamic speckles as a measure of organ displacement and local change of the shape of the organ surface as a result of elongation.

Dynamic light scattering caused by long-distance transport of cells, particles, or fluids was used in blood flow velocity measurements by analysis of biospeckle patterns obtained directly from blood vessels of different organs (Aizu and Asakura, 1991; Briers, 2001; Murari et al., 2007; Zhu et al., 2007). In the case of plants, a process analogous to the circulation of the bloodstream is the transport of water and nutrients in vascular bundles. Velocity of water transport in the xylem is measured in millimeters per second (Taiz and Zeiger, 2006); however, under favorable conditions, in some plant species, this may be on a scale of meters per second (Nobel, 1991). Such velocities of water transport are sufficient to induce vibration in the conductive elements and surrounding tissues as a result of cavitation (Milburn and Johnson, 1966; Davis et al., 1999), which in turn could be followed by biospeckle activity analysis. Thus, biospeckle can presumably be a noncontact method of analysis of the velocity of water transport in plant organs.

Cellular metabolic processes, including active transport of organelles and particles, are the most common group of factors generating biospeckle dynamics (Figure 14.2). A direct relationship between biospeckle activity and rate of plant metabolism has been proven in experiments showing their dependence on temperature (Kurenda et al., 2012). A decrease in the temperature of apple tissue from 30°C to 5°C, which decelerates apple metabolism, resulted in a decrease in biospeckle activity by about 50%. A similar effect was obtained in apples subjected to high hydrostatic pressure (Kurenda et al., 2014). The inhibition of metabolic processes by the destruction of

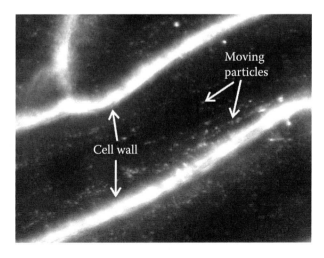

FIGURE 14.2 Intracellular cyclosis in onion. Organelles and particles move around the cell, which causes a dynamic laser speckle pattern (biospeckles).

cells after application of hydrostatic pressure results in the transient reduction of biospeckle activity. A subsequent increase in biospeckle activity in pressure-treated and stored apples is probably caused by an increase in particle motility as a result of an increased rate of autolytic enzymatic reactions, leading to tissue browning and decomposition.

Major metabolic processes, which are sources of biospeckle activity, have been identified using inhibitors of selected physiological processes (Kurenda et al., 2013). Inhibitors of actin cytoskeleton reconstruction and ion channels decrease biospeckle activity by about 20% and 40%, respectively. This result suggests that groups of processes of active intracellular movement, which are dependent on actin motor proteins and transmembrane ion transport, are among the most important generators of dynamic light scattering at the cellular scale.

The above information confirms the assumptions presented in the pioneering work of Briers (1975a, b). He suggested that dynamic light scattering arises mainly as a result of chloroplast movements, because according to scattering theory the size and color of these organelles optimally correspond to the laser wavelengths used. In fact, the movements of the chloroplasts, other organelles, macromolecules, cytoplasmic streaming, organization of cell structure, and cell wall formation are dependent on the function of motor proteins.

Cytoplasmic streaming is a directional flow of cytosol and organelles that can reach up to 100 µm/s (Shimmen and Yokota, 2004). Small organelles, as well as dissolved particles moving with the stream, act as scattering centers, and can dynamically scatter laser light. However, other actin-based processes such as a vesicle transport, cytoskeleton spatial reorganization, excretion of enzymes, and placement of cell wall components into extracellular space also contribute to the increase in the overall biospeckle activity.

Another group of processes indirectly affecting light scattering is connected with transmembrane ion transport. Ion channels and proton pumps regulate the intracellular content of inorganic ions, which allows for the regulation of cell turgor, and density of intracellular compartments, and creates ion gradients used for secondary transmembrane transport of specific macromolecules. Movement of molecules, changes of cell hydration, density of compartments, and tension of cell membrane and cell walls further enhance the dynamics of light scattering.

A change in the dynamics of light scattering can be obtained not only by a change in the velocity of intracellular components but also by their number and physical properties, and in particular their ability for light absorption. An increase in both biospeckle activity and starch granule content observed in apples suggests that growth in the number of moving scattering centers in the cells can increase dynamic light scattering (Zdunek and Cybulska, 2011). An extracellular process that can also potentially increase the number of scattering centers in the plant tissue is enzymatic cell wall degradation during aging or damage. As has been shown, physiological degradation of the carrot cell wall increases the number of pectic and hemicellulose oligosaccharides (Cybulska et al., 2015), which can move passively in extracellular space. In turn, a decrease in biospeckle activity with an increase in chlorophyll content in apples (Zdunek and Herppich, 2012) indicates that red laser light absorption presumably reduces the amount of backscattered light, shortens photon paths in apple tissue, and decreases the number of internal reflections on moving cellular elements, finally decreasing dynamic scattering.

The above information suggests that, in tissues with a high rate of metabolism, dynamic light scattering is relatively high, while in organs with a low rate of metabolism or high pigment content, dynamic light scattering is relatively low. Spatial analysis of biospeckle activity confirmed this suggestion. Spatial analysis for apples shows a significant local decrease of biospeckle activity in the place where metabolism inhibitors are applied (Kurenda et al., 2013), while a local increase in biospeckle activity is registered in the part of the root subjected to thigmostimuli (Ribeiro et al., 2014).

A multitude of metabolic processes in living organisms that can potentially affect biospeckle activity suggest a possible application of this technique for the assessment of a wide range of physiological parameters in different types of organisms. However, at present, existing methods of image analysis allow only for the evaluation of the overall physiological conditions of the organism. This problem can be solved by the development of new methods of biospeckle image analysis and new models on the relationship between specific components of biospeckle activity and corresponding physiological processes.

14.3 IMAGE ANALYSIS FOR ASSESSMENT OF BIOSPECKLE ACTIVITY

14.3.1 GLOBAL ACTIVITY MEASURES

14.3.1.1 Speckle Contrast

One of the first measures of speckle pattern activity was proposed by Briers (1975a, b, 1978). He observed that when photography of speckle pattern was implemented with an exposure time that was in the order of the fluctuation period, the contrast of the speckle pattern was closely related to speckle intensity fluctuations. In single-exposure laser photography, areas of high fluctuations of speckle pattern were blurred and thus the contrast reduced, whereas in areas with stationary patterns, contrast remained high. He assumed that for biological media, the speckle pattern is composed of two separate components: one produced by stationary scatterers and another by moving scatterers. Based on this so-called speckle contrast, the ratio of the mean intensity of the light from the moving scatterers to the total intensity of the scattered light was defined. Two measures of speckle contrast were proposed. The first approach involved the measurement of the first-order temporal statistics of the fluctuations followed by spatial averaging. The spatial contrast is expressed by the following equation:

$$\rho = 1 - \frac{\sigma}{\langle I \rangle} \quad (14.1)$$

where σ is the spatial variance or standard deviation of the pixel intensities and $\langle I \rangle$ is the mean intensity of the speckle pattern. Later, an alternative method was developed, in which the time average of the fluctuating speckle pattern was followed by the measurement of first-order spatial statistics. This measure, called temporal contrast, is expressed as

$$\rho = 1 - \left[1 - \frac{\langle \sigma_t^2(x,y) \rangle_{x,y}}{\langle I \rangle^2}\right]^{1/2} \quad (14.2)$$

where $\sigma_t^2(x,y)$ is the temporal variance of the intensity fluctuations at point (x,y) of the speckle pattern and $\langle ... \rangle_{x,y}$ indicates spatial averaging. $\langle I \rangle$ is the mean intensity of the speckle pattern in the temporal and spatial domain. Both temporal statistics of fluctuating speckle patterns provide information about the relative contributions of stationary and moving scatterers to the speckle pattern. Even though both measures are velocity dependent, they do not provide direct information on the velocity of moving speckles nor the frequency components of intensity fluctuations. Instead, they can be treated as relative estimates of speckle fluctuations.

14.3.1.2 Space–Time Speckle/Time History Speckle Pattern-Based Measures

Another approach to analyze the time evolution of speckle patterns is based on the examination of properties of space–time speckle (STS) patterns (Oulamara et al., 1989), also known as time history speckle pattern (THSP) (Arizaga et al., 1999). The THSP is a matrix composed of a number of successive images of dynamic speckle phenomena. The THSP is created using a selected line extracted from each image and placed side by side as column vectors. Lines are arranged sequentially in chronological order. The horizontal direction of the THSP matrix corresponds to time-scale speckle intensity fluctuations, while the vertical direction indicates the spatial position of the sampling point along a selected line.

Arizaga et al. (1999) suggested an approach to characterize speckle time evolution based on the inertia moment (IM) of the co-occurrence matrix calculated from the THSP. Co-occurrence matrix is an N by N matrix, where N is the number of possible gray levels of the THSP image. Each entry ij of this matrix indicates the number of occurrences of gray level i followed by a gray level j along the horizontal, temporal direction of THSP. Low activity of speckle fluctuations is indicated by a high concentration of values around the diagonal, while for high activity the values of the co-occurrence matrix spread outside the diagonal, forming a cloud-like shape. The IM of the co-occurrence matrix is defined as the sum of the matrix values multiplied by the squared distance from the principal diagonal (expressed in terms of row and column indexes).

Studies on numerical simulations of speckle patterns showed that IM is sensitive to the mean length of speckle grains (Arizaga et al., 1999). The IM value decreases with the increasing length of speckles in the THSP image, as the regions of the co-occurrence matrix far from diagonal become less populated. Furthermore, when the ratio of grain length to the width of the time window (total length of the THSP in horizontal direction) is greater than or equal to one-third, the IM reaches the saturation state. It has been shown that IM is sensitive to the sudden jumps of intensity, which result in an increase in IM values. Therefore, for comparison purposes, the illumination of the specimen should be kept constant during the measurements. Squaring operations carried out during the calculation of IM induce nonlinear weighting on the co-occurrence matrix values, in such a way that entries located far from the diagonal are much more emphasized. This implies that the IM is more efficient in analyzing high frequencies.

To overcome this limitation, an absolute value of differences (AVD) has been introduced as an alternative to the routine IM method (Braga et al., 2011). This approach is based on calculation of the absolute value of the distance of each entry of the co-occurrence matrix from the principal diagonal. The AVD showed better sensitivity to the whole spectral composition than the standard IM method and presented better results with regard to monitoring the biospeckle process.

The spectral content of THSP lines was also analyzed by means of the coherence function, derived from the cross-spectrum (Braga et al., 2008; Nobre et al., 2009). This technique was used to obtain spectral information in distinct frequency ranges encoded between two lines of the same THSP, as well as between pair of THSPs. The results obtained with simulated and experimental data showed that cross-spectrum analysis is more sensitive to the lower frequencies of speckle intensity fluctuations. This analysis also put into question the reliability of using a fewer number of lines in a THSP image, since different lines within the same THSP showed a distinct spectrum content. Cross-spectrum analysis was considered to be a complementary measure to the IM method.

Passoni et al. (2005) proposed wavelet entropy (WE) as a measure of the activity of dynamic speckle phenomena. Wavelet-based entropy enables the analysis of the degree of order/disorder of a complex signal. The basic concept of analyzing speckle time evolution using this approach is to apply wavelet transform to each row of the THSP, divided into temporal windows of equal lengths. For each temporal window, the relative wavelet energy and then the Shannon entropy are calculated. In case of stationary process analysis, the WE is calculated for each row of the image and the mean value from all image rows is obtained in order to obtain a single image descriptor. For the nonstationary process, an entropy mean value corresponding to each time segment is calculated by considering the whole row set. It was reported that this technique permitted both qualitative and quantitative estimations within a shorter time, while requiring less data compared to other methods of THSP analysis. However, compared to IM, the wavelet transform demands complex computational operations, which are susceptible to some arbitrary choices such as that of a mother wavelet (Ribeiro et al., 2014).

The histogram difference method was developed to analyze the texture of a single THSP image for seed specimen classification purposes (Fernández et al., 2003). The difference histogram (DH) is computed from the auxiliary image, which is obtained by subtracting the original THSP image of its shifted duplicate. The calculated histogram contains information about the second-order statistics of the original image that characterizes its texture. The displacement value is chosen arbitrarily to achieve the best possible results. The classification of DH is carried out by means of Bayesian decision rule and the assumption of the multinomial distribution of the histogram values. The histogram difference method is considered to be appropriate for the classification of seeds, and in general any other type of specimen whose mean speckle life time can be a significant measure of sample biological activity.

14.3.1.3 Spatial–Temporal Correlation Technique

When two images of dynamic speckle pattern are taken in a short time interval, the corresponding local regions of these images show a high degree of correlation. When the interval separating two images increases, the corresponding local regions decorrelate, that is, over time the speckle pattern degenerates from the initial state. For a homogenous biospeckle pattern, the fluctuations of speckle intensity are equal in each region, and therefore the correlation peaks in all locations show similar degrees of biospeckle pattern degradation in time. In this case, a set of local correlation peaks can be replaced by one averaged value equal to the value of correlation calculated between two images. This observation has been exploited in the spatial–temporal correlation technique (Zdunek et al., 2007). This technique is based on the correlation analysis of two or more speckle patterns, where one is considered as an image of a reference state. With images captured in the given temporal order, these coefficients can be expressed as functions of the biospeckle pattern movement speed. Each such dependency is equivalent to the temporal degradation of a correlation peak.

Biospeckle activity evolution can be evaluated by calculating temporal changes in the correlation coefficient $C^{k\tau}$, where k is the frame number and τ is the lag time between frames. In practice, $C^{k\tau}$ is calculated as the correlation coefficient of the data matrix, consisting of intensities of pixels of the first frame with the data matrices of the following frames of biospeckles. Typically, the lag time is limited by detector capabilities and is most often around 1/15 s. A more pronounced decrease in $C^{k\tau}$ (larger decorrelation) reflects higher biospeckle activity, and the value $BA = 1 - C^{k\tau}$ is therefore used for a more meaningful representation. To simplify the analysis in practical application, it is enough to take just two snapshots of biospeckles with a lag of several seconds and calculate the correlation coefficient between them (Zdunek and Cybulska, 2011; Adamiak et al., 2012; Kurenda et al., 2012; Szymanska-Chargot et al., 2012; Zdunek and Herppich, 2012). This "longer" lag time is suitable for certain biological samples and may be used as a compromise between the time of measurement and the biospeckle decorrelation ratio.

14.3.2 Analysis of Biospeckle Spatial Activity

14.3.2.1 Fujii Method

Global measures of speckle activity are suitable for monitoring spatially homogenous processes that show similar activity levels in different regions of the same object. However, in biological samples, there may be a variation in activity in different regions of the same object. When a large enough area is illuminated, the speckle pattern depicts the spatial variability of speckle fluctuations that corresponds to underlying dynamic processes. This speckle activity can be presented by means of spatial activity maps, which are locally calculated measures assigned to specific locations.

One of the first methods for evaluation of speckle spatial activity is Fujii's index, which was applied for observations of blood flow (Fujii et al., 1985, 1987). The Fujii method is based on the calculation of weighted sums of the absolute differences of gray-level intensity for each pixel of the time series of dynamic speckle patterns. The value of Fujii's index at a specified location is defined as

$$F(x,y) = \sum_k \frac{|I_k(x,y) - I_{k+1}(x,y)|}{I_k(x,y) + I_{k+1}(x,y)} \quad (14.3)$$

where k is the image index from the time series $k = 1...N$, and I_k and I_{k+1} are the intensity values of a pixel at a location given by coordinates x and y.

The presence of the weighting factor in the denominator of Fujii's index results in a nonlinear response that emphasizes both the large differences as well as the small ones that involve values from the limits of the detector's dynamic range. This leads to one of the major drawbacks of this method. Owing to the weighting term, the same value in the denominator can be dealt in a completely different way. For instance, the transition of gray level from 0 to 1 gives the highest value of Fujii's index, while the same absolute transition from 254 to 255 gives the lowest possible nonzero value. In practice, this results in false speckle activity generated by detector noise in darker, less illuminated regions of the speckle pattern.

One of the possible solutions to this problem is to use an alternative method, the so-called parameterized Fujii method (Minz et al., 2014a, b). The proposed method is given by

$$F_p(x,y,g_r) = \sum_k [(|I_k(x,y) - I_{k+1}(x,y)|)(255 - |g_r - I_k(x,y)| + 255 - |g_r - I_{k+1}(x,y)|)] \quad (14.4)$$

where k is the image index from the time series $k = 1...N$, I_k and I_{k+1} are the intensity values of a pixel at location given by coordinates x and y, and g_r is the reference gray level. A new weighting term that introduces the reference gray level in the calculations allows the adjustment of the range of the emphasized speckle intensity values. When the reference gray level rises, higher values are more pronounced, while the influence of lower values is limited. For low values of the reference gray level, only transitions from relatively low to low pixel intensity counts.

Another possible approach for overcoming the drawbacks of the original Fujii method is to use the frequency decomposition carried out by the wavelet transform, to filter out all redundant frequencies from the original signal (Braga et al., 2009, 2012; Cardoso et al., 2011).

14.3.2.2 Generalized Differences

A simplified variation of the Fujii's algorithm was proposed called the generalized differences (GDs) method (Arizaga et al., 2002). The main difference between these two approaches lies in the elimination of the weighting process and temporal order of successive frames from the latter. The GD is therefore defined as the total sum of absolute differences between the pixel intensities among all possible frame combinations, and can be written as

$$GD(x,y) = \sum_{k=1}^{N-1} \sum_{l=k+1}^{N} |I_k(x,y) - I_l(x,y)| \qquad (14.5)$$

where k and l are the frame indexes and I_k is the intensity value of a pixel with x and y coordinates. Lack of the temporal order implies that the algorithm includes the differences between pixel values from nonconsecutive frames, so this method does not preserve any information about the frequency of signal.

For a static speckle pattern, where values of pixels do not change over the temporal window, the GD is equal to zero. For a signal composed with half of the pixel intensity values equal to the upper bound of dynamic range and the second half equal to the lower bound value, the GD reaches its maximum value. For this case, owing to the lack of temporal order, the maximum value of GD will be reached regardless of the frequency of occurrence of these values (their temporal order).

The resulting activity maps show fewer visible contours of examined objects when compared to the Fujii method. This algorithm is also more time consuming. In the variation of the GD algorithm proposed by Saúde et al. (2010), the modulus of the difference of pixel intensity is replaced by the squared value of difference. This measure, denoted by GD*, is defined as follows:

$$GD^*(x,y) = \sum_{k=1}^{N-1} \sum_{l=k+1}^{N} (I_k(x,y) - I_l(x,y))^2 \qquad (14.6)$$

Images of biospeckle activity generated by the GD* algorithm are characterized by a higher contrast compared to the standard GD. Despite having qualitative similarities, these two approaches are not directly comparable. Nevertheless, both methods include only the spread of values, showing a qualitative relationship to variance. This limitation was partially eliminated in the weighted generalized difference (WGD) algorithm by introducing an additional weighting term, the value of which varied along each summation depending on the time scale (Arizaga et al., 2002). This measure is defined as

$$WGD(x,y) = \sum_{k=1}^{N-1} \sum_{l=k+1}^{N} |I_k(x,y) - I_l(x,y)| w_p \qquad (14.7)$$

where $p = l - k$ and indicates the temporal distance from reference frame k. The weight values vary with temporal distance p, emphasizing fast or slow variations in speckle intensity.

14.3.2.3 Laser Speckle Contrast Analysis

Laser speckle contrast analysis (LASCA) is the one of the most popular online routine methods for speckle pattern analysis. Originally, it was developed as a local version of the Briers contrast method (Briers 1975a, b, 1978, 2007; Briers and Webster, 1996), and used to monitor capillary blood flow. Up to now, a few variations of the original algorithm have been developed (Draijer et al., 2009). In the most basic form, LASCA is calculated as the ratio of the standard deviation to the mean of the intensities captured for each pixel in the local square window of the single image, and it can be expressed as

$$C \equiv \frac{\sigma}{\langle I \rangle} = \frac{\sqrt{\langle I^2 \rangle - \langle I \rangle^2}}{\langle I \rangle} \tag{14.8}$$

where $\langle I \rangle$ and $\langle I \rangle^2$ are the average value of pixel intensities and the average value of squared pixel intensities, respectively, calculated in the local square window. This version of the algorithm is also called spatially derived contrast. The most commonly used window sizes are 3×3, 5×5, and 7×7 pixels. Contiguous and overlapping windows are used for biospeckle pattern mapping. Both types of windows reduce spatial resolution by a factor equal to the length of the window edge. Window size is adjusted according to the size of the speckles, in order to provide the highest possible statistical validity while maintaining a reasonable spatial resolution.

Dunn et al. (2001) improved standard spatially derived contrast using temporal frame averaging (sLASCA). The novelty is that the modified algorithm operates on a predetermined number of raw speckle images. Each image is processed using basic LASCA and then local contrast values are further averaged out over all captured speckle temporal frames.

The time-integrated variation of the first-order temporal statistics of the dynamic speckle pattern (tLASCA) is called temporally derived contrast and is similarly calculated using multiple images of speckle phenomena. The definition of contrast is the same as in the case of the standard LASCA, except that the standard deviation and mean intensity are calculated for each pixel individually from all collected frames. Therefore, the effective spatial resolution of the original image is maintained. To achieve a better visual effect, smoothing of the final map is filtered using averaging or Gauss kernel. The higher spatial resolution in this method is paid with lower temporal resolution.

As in the case of the Fujii method and other intensity-based methods, the overall performance and ability to distinguish different activity regions suffers from non-uniform sample illumination and ambient light variations. The processing time of this method is relatively fast, which allows for real-time observations, although it reduces image quality.

14.3.2.4 Modified Laser Speckle Imaging

A technique in which the velocity information coded in the speckle pattern is obtained through the first-order temporal statistics of a time-averaged speckle image has been proposed (Cheng et al., 2003). The activity map obtained by means of the modified laser speckle imaging (mLSI) technique can be computed as

$$\mathrm{mLSI}(x,y) = \frac{\langle I^2_{x,y,t} \rangle_N - \langle I_{x,y,t} \rangle^2_N}{\langle I_{x,y,t} \rangle^2_N} \tag{14.9}$$

where $I_{x,y,t}$ and $I^2_{x,y,t}$ are the instantaneous intensity and instantaneous square intensity of the pixel at the tth frame, with x and y coordinates. $\langle I_{x,y,t}\rangle_N$ and $\langle I^2_{x,y,t}\rangle_N$ are the mean intensity and mean of squared intensities of the x, y pixel over N consecutive frames, respectively. The mLSI value is said to be inversely proportional to the velocity of the scattering particles. Compared to LASCA, the blood flow map obtained by the mLSI possessed higher spatial resolution and provided additional information about changes in blood perfusion in small blood vessels. This method also showed lower susceptibility to artifacts arising from stationary speckle. A fundamental disadvantage of mLSI is that it does not take into account the camera exposure time. Furthermore, averaging over a number of consecutive frames reduces its temporal resolution.

Nevertheless, the experimental results suggested that this is a suitable method for imaging the full field of blood flow without scanning, with the ability to provide a relatively high spatial resolution compared to other imaging methods (Cheng et al., 2008).

14.3.2.5 Temporal Difference Method

The Fujii and GD method have one drawback in common in that the resulting matrix describes the speckle activity as a whole during the observation interval, but misses the evolution of the activity (temporal resolution) (Martí-López et al., 2010). In order to describe speckle activity in time, a sequence of matrices separated by time interval as a statistical descriptor is proposed. In the temporal difference method, two speckle images of an object, separated by a time interval, are subtracted from one another to detect whether or not the speckle structure has changed. This approach offers easy interpretation of the measurement results, with low to modest computational costs. The initial test showed promising results and the possibility of transforming this method into a standard laboratory method.

14.3.2.6 Motion History Image

The motion history image (MHI) approach is a temporal template method that represents movements based on timestamps of pixels from a set of static pictures of motion sequence. This technique is widely used in various applications such as human motion recognition, object tracking, motion analysis, and other related applications and fields. Since MHI is related to the temporal characteristic of captured phenomena, it is justified to assume that this method is capable of monitoring biospeckle activity. Applications of silhouette-based MHI variant for online biospeckle assessment were proposed (Godinho et al., 2012). Silhouette images are generated by subtracting two sequenced images from a buffer of historical images, depicting the most recent motion. For each silhouette, the binary threshold is applied. The final MHI is created using the procedure of weighting the binary silhouettes stored in buffer, with respect to the "lifetime" of each image. The MHI method was tested on biological and nonbiological samples and showed better results when compared to the alternative online method—LASCA. The MHI produced similar results to the known offline methods such as Fujii and GD.

14.3.2.7 Empirical Mode Decomposition

The empirical mode decomposition (EMD) is a method defined by the sifting optimization algorithm, and results in a decomposition of the data into n-empirical modules called the intrinsic mode functions (IMFs) and a residue (Federico and Kaufmann, 2006). This method allows the separation of different activity levels of the speckle patterns and characterizes them in the time–frequency domain. Using EMD, any given temporal signal $I(t)$ can be expressed as

$$I(t) = \sum_{j=1}^{n} a_j(t)\cos\phi_j(t) + r(t) \tag{14.10}$$

where $r(t)$ is the residue value, while $a_j(t)$ and $\phi_j(t)$ are the instantaneous amplitude and the phase defined by using the Hilbert transform on the j-IMF component, respectively. Number of modes n is empirically determined. The EMD method is applied to the intensity signal along the time axis for each pixel of the sequence of dynamic speckle pattern. The average instantaneous energy and average instantaneous frequency calculated over a window containing N pixels were used as speckle spatial activity descriptors. The application of this method was demonstrated only in a limited way on the detection of bruised regions of fruits.

14.3.2.8 Normal Vector Space Statistics

Most methods used for monitoring speckle pattern activity are based on measurements of pixel intensity variations. A major drawback of these methods is that they are heavily influenced by varying light conditions. A normal vector-based approach as a new descriptor of time-varying speckle pattern to address this issue was proposed (Zhong et al., 2013).

In this approach, the speckle pattern is treated as a topographic surface, where high-intensity values correspond to hills and low values to valleys. Based on this, the dynamics of the speckle pattern at specific locations are measured by means of the temporal statistics of the directions of local normal vectors. The local normal vectors of the digital speckle image are calculated using the normal vector voting method (Page et al., 2003). The speckle pattern is considered as a triangulated surface and a local normal vector is calculated from particular pixel intensity and its neighbors. The geodesic distance between two pixels is used to select a neighboring set of pixels. Before the calculation of normal vectors, surface smoothing is applied to the speckle pattern image to reduce the impact of noise. The statistics of the rotation of the normal vector are evaluated via three algorithms, adopted from intensity-based dynamic speckle pattern analysis—GD of angle of rotation, modified approximated entropy, and modified sample entropy. The proposed method was able to provide consistent results under various external light conditions and performed better in comparison to most of the intensity-based methods. The simulated experiments showed that results provided by this method are not affected by nonuniform illumination and reflectivity, time-varying ambient light, or sample-to-sample varying ambient light. Experiments carried out for surface activity detection of attached leaves showed the capabilities of this method to monitor variation in water status in biological samples

(Zhong et al., 2014). Robustness and reliability enables this method to be applied to a broad range of applications, especially where measurements are carried out under harsh external light conditions.

14.3.2.9 Analysis in Spectral Domain

Many studies showed that biospeckle activity is related to biological processes in living organisms. However, there are a number of processes taking place at the same time and their effects may coincide with natural phenomena such as water evaporation, diffusion, and mechanical deformation of the structure. The above-described pixel intensity-based methods are not able to independently distinguish different activity levels caused by various processes and biological phenomena. Owing to the complexity of biological materials, alternative techniques of biospeckle activity analysis related to spectra domains have been developed. Typically, frequency analysis involves signal decomposition by means of discrete Fourier or wavelet transform and reconstruction of the signal by means of inverse transform with selected frequencies only. Other applications employ direct analysis of frequency bands by means of statistical descriptors such as mean amplitude value of harmonic frequencies or signal entropy, etc. (Braga et al., 2007, 2009; Sendra et al., 2007).

Research showed a possibility of using biospeckle analysis to separate artificially fungi-contaminated bean seeds from healthy ones (Rabelo et al., 2011). The seeds were examined using the IM of THSP/STS and frequency values of the THSP signals. Frequency analysis was carried out by means of fast Fourier transform of the convoluted signal. Both methods of analyzing dynamic biospeckle patterns were capable of identifying the presence of microorganisms in the seeds. However, the methods used were unable to identify the type of pathogenic agent. It was concluded that frequency analysis may be used to complement the information provided by other techniques such as IM, for improving the overall effectiveness of biospeckle laser imaging of biological materials. Wavelet transform was used in conjunction with traditional biospeckle laser methods—Fujii, GDs, and THSPs, in order to identify frequency bands of water activity in biological materials, particularly maize and bean seeds (Cardoso et al., 2011). The frequency analysis allowed the mapping of activities that only occur at certain frequencies in the seeds associated with particular areas in which they operate, as in the case of activities present in the embryo as well as those present in the endosperm. The ability of segmenting different biological activity zones by means of the WE calculated for THSP of each individual pixel was also proven for the detection of bruised apple zones and for the corn seed viability test (Passoni et al., 2005).

14.3.2.10 Multivariate Speckle Pattern Analysis

Both fast Fourier and wavelet transforms are data filtering methods that operate in the frequency domain, allowing for signal decomposition and analysis of selected bands only. In a similar sense, principal component analysis (PCA), a multivariate statistical tool, has been proposed as an alternative to the spectral analysis of the dynamic speckle signal (Ribeiro et al., 2014). This technique consists of applying the PCA as a preprocessing tool for biospeckle signals. During preprocessing, the sequence of images of speckle patterns is reorganized into a data matrix, in which each image is treated as a single variable and each pixel of this image as a separate observation. By means of PCA, this data matrix is transformed to a set of statistically uncorrelated coordinates, expressed in the PCA score domain. At this point, the contribution of each component is studied and some of the principal components may be eliminated. Then, inverse PCA transform is applied and the reconstructed data are analyzed. Similarly to the frequency analysis methods, PCA is used in combination with existing methods such as Fujii and GD. The PCA-based method was tested on real data examples, specifically the biological activity images of the endosperm and embryo of the maize seed. The results showed significant improvements in the visual quality of activity images. The PCA approach allowed the association of specific principal components to biological phenomena, thus enabling the definition of markers of samples' bioactivity. Applied as a filtering tool, it provided a nonparametric and adaptive method of the decomposition of biospeckle activity regions with fast computational processing (Ribeiro et al., 2014).

14.4 EVALUATION OF FRUIT AND VEGETABLE QUALITY

The first work of biospeckles on fruits was reported in 1989 when differences in the temporal speckle activity of apples, oranges, and tomatoes were found (Oulamara et al., 1989). The biospeckle activities of different commodities are shown in Figure 14.3. Temporal fluctuations of speckles decrease with aging, and therefore, biospeckle could be used for monitoring the shelf life of fruits and vegetables (Xu et al., 1995; Zhao et al., 1997). Several examples have been presented in the scientific literature. A shelf life experiment for apples showed a decrease in biospeckle activity with days of storage (Figure 14.4), and this activity correlated very well with the firmness of fruits (Zdunek et al., 2007, 2008). It was shown that biospeckles followed the biological variations of oranges due to senescence at postharvest time (Rabelo et al., 2005). Moreover, the activity obtained for the central calomel base (apex) allowed differentiation between oranges according to fruit freshness. Other fruits and vegetables such as potatoes, radish, tomatoes, and soybean also showed a decrease in biospeckle activity with aging (Xu et al., 1995; Zhao et al., 1997).

The biochemical processes in fruits and vegetables during maturation such as starch and pigment transformations may also change biospeckle activity. This is due to a change in scattering particles and light propagation through tissue. Therefore, biospeckle activity could be used for monitoring metabolism-related changes in fruits and vegetables, although the number of studies on this topic is still very small. It was shown that, in apples, starch granule degradation caused a significant decrease in biospeckle activity, probably due to a lower number of scattering centers (Zdunek and Cybulska, 2011). Studies on chlorophyll content in apples showed that biospeckle activity increased when chlorophyll was degraded due to red light absorption (Zdunek and Herppich, 2012). The absorption may have limited the amount of light that could penetrate the tissue. Similar results on the relationship between chlorophyll and biospeckle activity were obtained for tomatoes during ripening (Romero et al., 2009).

The temperature of the investigated material is the key factor influencing measured biospeckle activity. It was shown that storage temperature affected the measured biospeckle fluctuation for apples (Kurenda et al., 2012). The biospeckle

Biospeckle Technique for Assessing Quality of Fruits and Vegetables

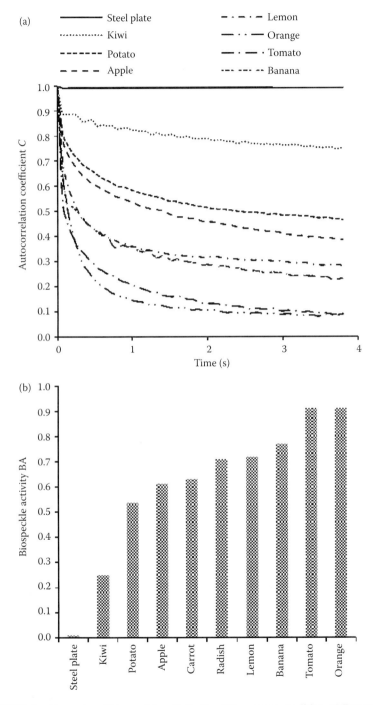

FIGURE 14.3 Examples of biospeckle activity for different commodities. (a) Decorrelation of dynamic speckle pattern. (b) Estimated biospeckle activity: BA = 1 − $C(t = 4s)$, where C is the autocorrelation coefficient.

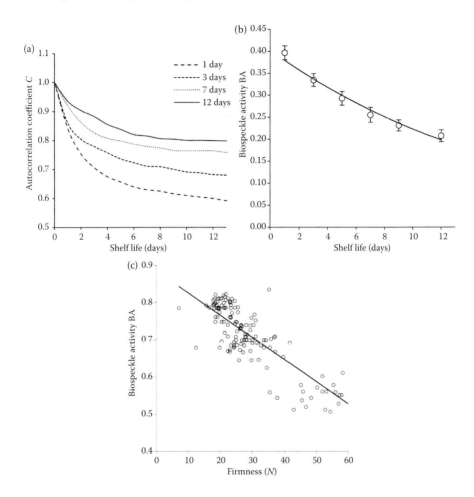

FIGURE 14.4 Shelf life effect on biospeckle activity of apples. (a) Decorrelation of dynamic speckle pattern during 12 days of shelf life. (b) Biospeckle activity decrease during shelf life. (c) Relationship of firmness to biospeckle activity for apples during 12 days of shelf life. ((a) and (b) Reprinted with permission of the Institute of Agrophysics, publisher of International Agrophysics, Zdunek, A., Muravsky, L.I., Frankevych, L., Konstankiewicz, K. 2007. *International Agrophysics*, 21, 305–310; (c) Reprinted with permission of the Institute of Agrophysics, publisher of Acta Agrophysica, Zdunek, A., Frankevych, L., Konstankiewicz, K., Ranachowski, Z. 2008. *Acta Agrophysica*, 11, 303–315.)

activity dropped with temperature as presented in Figure 14.5. Interpretation in this case is straightforward; with decreasing temperature the metabolism of tissue and Brownian motions slow down, resulting in lower biospeckle activity.

Defects under the surface alter the propagation of light and can affect observed biospeckle activity. Therefore, the biospeckle method is useful for the detection of damage and disease in fruits and vegetables (Figure 14.6). For example, for apples, several dynamic speckle techniques (biospeckle methods) were used for internal damage detection (Pajuelo et al., 2003). The moment of inertia of the co-occurrence

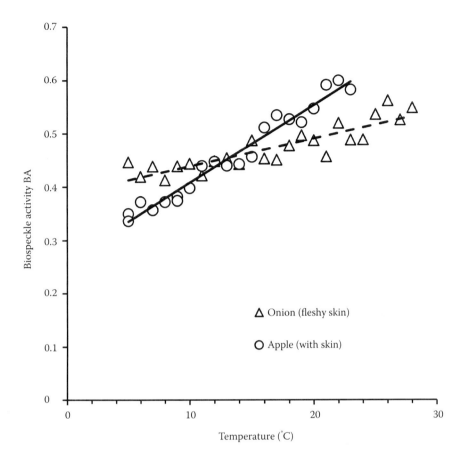

FIGURE 14.5 Relationship of biospeckle activity to temperature for apples and onions.

matrix and the statistical cumulants calculation used for biospeckle activity evaluation in bruised apples showed that, after a mechanical injury, biospeckle variations were lowered and further fluctuations decreased with time. Moreover, the biospeckle method allowed visualization of the bruising area by such methods as the WGD, the LASCA, the Konishi method (Pajuelo et al., 2003), or by the entropy wavelet methods (Passoni et al., 2005). All of these techniques showed that the bruised region could be recognized from the undamaged part of the fruit (Figure 14.6).

Fungal infection changes the structure and biological conditions of fruit tissue, which are visible in the late stage of infection. Fungi spores attack cells, causing their senescence. Since spores are physical objects and the infection affects the metabolism of the tissue, one may expect that the biospeckle method is sensitive for this type of disease. This hypothesis was confirmed for apples for bull's eye rot development (Adamiak et al., 2012). Three stages of biospeckle activity were observed (Figure 14.7): at first biospeckle activity decreased, then biospeckle activity increased significantly, and the first symptoms of bull's eye rot infection were noted, and finally biospeckle activity suddenly dropped, which was associated

FIGURE 14.6 Real view of a raw biospeckle image (a) and spatial biospeckle activity of the apple after local mechanical bruising treatment near the center of the apple (b), visualization by Fujii (c) and LASCA (d) method. This example shows that biospeckle imaging allows the distinction of intact and damaged areas (dark spot near the center of the apple) even when fruit damage is externally still invisible.

with extensive rotting of the tissue. Interestingly, the increase in biospeckle activity was also observed on the apparently healthy parts of the infected fruits. The same effect was observed for infected apples by injection of spores under the skin; in this case, biospeckle activity also increased in uninfected places (Adamiak et al., 2012). Moreover, in the experiment, the increase in biospeckle activity was observed before symptoms were visible on the skin. Both observations drew the conclusion that biospeckles are a very promising method for early detection of infected fruits. Biospeckles were also implemented for the detection of fungal contamination in seeds (Braga Jr et al., 2005; Rabelo et al., 2011). The GD and IM methods, as well as Fujii's algorithm, were used for the evaluation of biospeckle activity in bean seeds. Bean seeds inoculated with fungi exhibited greater biospeckle activity compared to the control group, and thus the effect of contamination is similar to that previously described for apples. In addition, the biospeckle method is able to distinguish fungi species. The GD and Fujii's image techniques also displayed the presence of fungi.

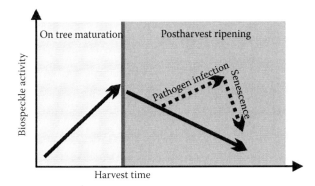

FIGURE 14.7 Schematic changes of biospeckle activity during fruit maturation. Dashed lines present a possible scenario when fruit is infected by a pathogen.

Relatively new applications of the biospeckle method include the monitoring of plant development. Since the method is sensitive to growth and motion, the changes in processes occurring in cells during development may be monitored (Braga et al., 2009). For example, biospeckle activity increases during preharvest apple development (Szymanska-Chargot et al., 2012). A significant correlation is usually observed between biospeckle activity and soluble solids content, starch content, and firmness, showing that this method has the potential to be used for nondestructive evaluation of these properties in the preharvest period.

14.5 CONCLUSIONS

The biospeckle method is considered a promising method due to its nondestructive character and unique sensitivity to a mixture of physical and biological processes related to the movement of particles in a studied material, although no commercial system is available in the market yet. Instead, a setup is relatively easy and inexpensive, and many mathematical methods are available to analyze biospeckle activity that could be tailored to the studied material and problem of interest. Further development in this area is possible with the use of frequency analysis that will allow distinction of processes from the biospeckle temporal pattern. Such an approach could be used for either local or spatial evaluation of biological material. From the point of hardware, an important obstacle of the method is its sensitivity to vibration. This hampers the use of present systems close to noisy or vibrating machines; therefore, this problem must be technically solved for successful application in industry conditions. For further development of this method, the use of multiple lasers is particularly interesting. One may say that it would be a kind of biospeckle spectroscopy since, similar to other optical methods, the propagation of light (absorption and scattering) in fruits and vegetables for instance largely depends on the laser wavelength. However, it needs in-depth interpretation of the relation of laser light with the medium at the microscopic scale, therefore involving basic science as in the case of other spectroscopic methods. Moreover, the evaluation of intact fruits and vegetables needs scaling up of such interpretation and consideration of biological variability and physiological spatiotemporal changes related to maturation.

REFERENCES

Adamiak, A., Zdunek, A., Kurenda, A., Rutkowski, K. 2012. Application of the biospeckle method for monitoring bull's eye rot development and quality changes of apples subjected to various storage methods—Preliminary studies. *Sensors*, 12, 3215–3227.

Aizu, Y., Asakura, T. 1991. Bio-speckle phenomena and their application to the evaluation of blood flow. *Optics and Laser Technology*, 23, 205–219.

Aizu, Y., Asakura, T. 1996. Bio-speckle. In: A. Consortini (Ed.), *Trends in Optics. Research, Development and Applications*, Academic Press, San Diego, California, pp. 27–49.

Arizaga, R., Cap, N., Rabal, H., Trivi, M. 2002. Display of the local activity using dynamical speckle patterns. *Optics Engineering*, 41, 287–294.

Arizaga, R., Trivi, M., Rabal, H. 1999. Speckle time evolution characterization by the co-occurrence matrix analysis. *Optics and Laser Technology*, 31, 163–169.

Braga, R.A., Dupuy, L., Pasqual, M., Cardoso, R.R. 2009. Live biospeckle laser imaging of root tissues. *European Biophysics Journal*, 38, 679–686.

Braga, R.A., Nobre, C.M.B., Costa, A.G., Sáfadi, T., da Costa, F.M. 2011. Evaluation of activity through dynamic laser spackle using the absolute value of the differences. *Optics Communications*, 284, 646–650.

Braga, R.A., Silva, W.S., Sáfadi, T., Nobre, C.M.B. 2008. Time history speckle pattern under statistical view. *Optics Communications*, 281, 2443–2448.

Braga, R.A. Jr, Cardoso, R.R., Bezerra, P.S., Wouters, F., Sampaio, G.R., Varaschin, M.S. 2012. Biospeckle numerical values over spectral image maps of activity. *Optics Communications*, 285, 553–561.

Braga, R.A. Jr, Horgan, G.W., Enes, A.M., Miron, D., Rabelo, G.F., Barreto Filho, J.B. 2007. Biological feature isolation in biospeckle laser images. *Computers and Electronics in Agriculture*, 58, 123–132.

Braga, R.A. Jr, Rabelo, G.F., Granato, L.R., Santos, E.F., Machado, J.C., Arizaga, R., Rabal, H.J., Trivi, M. 2005. Detection of fungi in beans by the laser biospeckle technique. *Biosystems Engineering*, 91, 465–469.

Briers, J.D. 1975a. Wavelength dependence of intensity fluctuations in laser speckle patterns from biological specimens. *Optics Communications*, 13, 324–326.

Briers, J.D. 1975b. A note on the statistics of laser speckle patterns added to coherent and incoherent uniform background fields, and a possible application for the case of incoherent addition. *Optical and Quantum Electronics*, 7, 422–424.

Briers, J.D. 1977. The measurement of plant elongation rates by means of holographic interferometry: Possibilities and limitations. *Journal of Experimental Botany*, 28, 493–506.

Briers, J.D. 1978. The statistics of fluctuating speckle patterns produced by a mixture of moving and stationary scatterers. *Optical and Quantum Electronics*, 10, 364–366.

Briers, J.D. 2001. Laser Doppler, speckle and related techniques for blood perfusion mapping and imaging. *Physiological Measurement*, 22, 35–66.

Briers, J.D. 2007. Laser speckle contrast imaging for measuring blood flow. *Optica Applicata*, 37, 139–152.

Briers, J.D., Webster, S. 1996. Laser speckle contrast analysis (LASCA): A non-scanning, full-field technique for monitoring capillary blood flow. *Journal of Biomedical Optics*, 1, 174–179.

Cardoso, R.R., Costa, A.G., Nobre, C.M.B., Braga, R.A. 2011. Frequency signature of water activity by biospeckle laser. *Optics Communications*, 284, 2131–2136.

Carvalho, P.H.A., Barreto, J.B., Braga, R.A. Jr, Rabelo, G.F. 2009. Motility parameters assessment of bovine frozen semen by biospeckle laser (BSL) system. *Biosystems Engineering*, 102, 31–35.

Cheng, H., Luo, Q., Zeng, S., Chen, S., Cen, J., Gong, H. 2003. Modified laser speckle imaging method with improved spatial resolution. *Journal of Biomedical Optics*, 8, 559–564.

Cheng, H., Yan, Y., Duong, T.Q. 2008. Temporal statistics analysis of laser speckle images and its application to retinal blood-flow imaging. *Optics Express*, 16, 10214–10219.

Cole, J.A., Tinker, M.H. 1996. Laser speckle spectroscopy—A new method for using small swimming organisms as biomonitors. *Bioimaging*, 4, 243–253.

Cybulska, J., Zdunek, A., Kozioł, A. 2015. The self-assembled network and physiological degradation of pectins in carrot cell walls. *Food Hydrocolloids*, 43, 41–50.

Davis, S.D., Sperry, J.S., Hacke, U.G. 1999. The relationship between xylem conduit diameter and cavitation caused by freezing. *American Journal of Botany*, 86, 1367–1372.

Draijer, M., Hondebrink, E., van Leeuwen, T., Steenbergen, W. 2009. Review of laser speckle contrast techniques for visualizing tissue perfusion. *Laser and Medical Science*, 24, 639–651.

Dunn, A.K., Bolay, H., Moskowitz, M.A., Boas, D.A. 2001. Dynamic imaging of cerebral blood flow using laser speckle. *Journal of Cerebral Blood Flow & Metabolism—Nature*, 21, 195–201.

Ebersberger, J., Weigelt, G., Li, Y. 1986. Coherent motility measurements of biological objects in a large volume. *Optics Communications*, 58, 89–91.

Federico, A., Kaufmann, G.H. 2006. Evaluation of dynamic speckle activity using the empirical mode decomposition method. *Optics Communications*, 267, 287–294.

Fernández, M., Mavilio, A., Rabal, H., Trivi, M. 2003. Characterization of viability of seeds by using dynamic speckles and difference histograms. *Progress in Pattern Recognition, Speech and Image Analysis. Lecture Notes in Computer Science*, 2905, 329–333.

Fujii, H., Asakura, T., Nohira, K., Shintomi, Y., Ohura, T. 1985. Blood flow observed by time-varying laser speckle. *Optics Letters*, 10, 104–106.

Fujii, H., Nohira, K., Yamamoto, Y., Ikawa, H., Ohura, T. 1987. Evaluation of blood flow by laser speckle image sensing. *Applied Optics*, 26, 5321–5325.

Godinho, R.P., Silva, M.M., Nozela, J.R., Braga, R.A. 2012. Online biospeckle assessment without loss of definition and resolution by motion history image. *Optics and Laser Engineering*, 50, 366–372.

Kurenda, A., Adamiak, A., Zdunek, A. 2012. Temperature effect on apple biospeckle activity evaluated with different indices. *Postharvest Biology and Technology*, 67, 118–123.

Kurenda, A., Pieczywek, P.M., Adamiak, A., Zdunek, A. 2013. Effect of cytochalasin b, lantrunculin b, colchicine, cycloheximid, dimethyl sulfoxide and ion channel inhibitors on biospeckle activity in apple tissue. *Food Biophysics*, 8, 290–296.

Kurenda, A., Zdunek, A., Schlüter, O., Herppich, W.B. 2014. VIS/NIR spectroscopy, chlorophyll fluorescence, biospeckle and backscattering to evaluate changes in apples subjected to hydrostatic pressures. *Postharvest Biology and Technology*, 96, 88–98.

Martí-López, L., Cabrera, H., Martínez-Celorio, R.A., González-Peña, R. 2010. Temporal difference method for processing dynamic speckle patterns. *Optics Communication*, 283, 4972–4977.

Milburn, J.A., Johnson, R.P.C. 1966. The conduction of sap. II. Detection of vibrations produced by sap cavitation in *Ricinus* xylem. *Planta*, 69, 43–52.

Minz, P.D., Nirala, A.K. 2014a. Bio-activity assessment of fruits using generalized difference and parameterized Fujii method. *Optik*, 125, 314–317.

Minz, P.D., Nirala, A.K. 2014b. Intensity based algorithms for biospeckle analysis. *Optik*, 125, 3633–3636.

Murari, K., Li, N., Rege, A., Jia, X., All, A., Thakor, N. 2007. Contrast-enhanced imaging of cerebral vasculature with laser speckle. *Applied Optics*, 46, 5340–5346.

Nobel, P.S. 1991. *Physicochemical and Environmental Plant Physiology*, Academic Press, San Diego, California, 483–496.

Nobre, C.M.B., Braga, R.A. Jr, Costa, A.G., Cardoso, R.R., da Silva, W.S., Sáfadi, T. 2009. Biospeckle laser spectral analysis under inertia moment, entropy and cross-spectrum methods. *Optics Communication*, 282, 2236–2242.

Oulamara, A., Tribillon, G., Duobernoy, J. 1989. Biological activity measurements on botanical specimen surfaces using a temporal decorrelation effect of laser speckle. *Journal of Modern Optics*, 36, 165–179.

Page, D.L., Sun, Y., Koschan, A.F., Paik, J., Abidi, M.A. 2003. Normal vector voting: Crease detection and curvature estimation on large, noisy meshes. *Graphical Models*, 64, 199–229.

Pajuelo, M., Baldwin, G., Rabal, H., Cap, N., Arizaga, R., Trivi, M. 2003. Bio-speckle assessment of bruising in fruits. *Optics and Laser Engineering*, 40, 13–24.

Passoni, I., Dai Pra, A., Rabal, H., Trivi, M., Arizaga, R. 2005. Dynamic speckle processing using wavelets based entropy. *Optics Communication*, 246, 219–228.

Pomarico, J.A., Di Rocco, H.O., Alvarez, L., Lanusse, C., Mottier, L., Saumell, C., Arizaga, R., Rabal, H., Trivi, M. 2004. Speckle interferometry applied to pharmacodynamic studies: Evaluation of parasite motility. *European Biophysics Journal*, 33, 694–699.

Rabelo, G.F., Braga, R.A. Jr, Fabbro, I.M.D., Arizaga, R., Rabal, H.J., Trivi, M.R. 2005. Laser speckle techniques in quality evaluation of orange fruits. *Revista Brasileira de Engenharia Agrícola e Ambiental*, 9, 570–575.

Rabelo, G.F., Enes, A.M., Braga, R.A. Jr, Dal Fabro, I.M. 2011. Frequency response of biospeckle laser images of bean seeds contaminated by fungi. *Biosystems Engineering*, 110, 297–301.

Ribeiro, K.M., Braga, R.A. Jr, Horgan, G.W., Ferreira, D.D., Sáfadi, T. 2014. Principal component analysis in the spectral analysis of the dynamic laser speckle patterns. *Journal of the European Optical Society—Rapid Publications*, 9, 14009.

Romero, G.G., Martinez, C.C., Alanis, E.E., Salazar, G.A., Broglia, V.G., Alvarez, L. 2009. Bio-speckle activity applied to the assessment of tomato fruit ripening. *Biosystems Engineering*, 103, 116–119.

Saúde, A.V., de Menezes, F.S., Freitas, P.L.S., Rabelo, G.F., Braga, R.A. 2010. On generalized differences for biospeckle image analysis. *Proceedings of the 23rd Conference on Graphics, Patterns and Images (SIBGRAPI)*, August 30th–September 3rd, Gramado, Brazil, 209–215.

Sendra, H., Murialdo, S., Passoni, L. 2007. Dynamic laser speckle to detect motile bacterial response of *Pseudomonas aeruginosa*. *Journal of Physics: Conference Series*, 90, 012064.

Shimmen, T., Yokota, E. 2004. Cytoplasmic streaming in plants. *Current Opinion in Cell Biology*, 16, 68–72.

Szymanska-Chargot, M., Adamiak, A., Zdunek, A. 2012. Pre-harvest monitoring of apple fruits' development with the use of the biospeckle method. *Scientia Horticulturae*, 145, 23–28.

Taiz, L., Zeiger, E. 2006. *Plant Physiology*, Chapter 4, 4 edn, Sinauer Associates, Sunderland.

Tanin, L.T., Rubanov, A.S., Markhvida, I.V., Dick, S.C., Rachkovsky, L.I. 1993. In: G. von Bally, S. Khanna (Eds), *Optics in Medicine, Biology and Environmental Research: Proceedings of the International Conference on Optics within Life Sciences (OWLS I) (149)*. Amsterdam, The Netherlands: Elsevier Science Publishers B.V.

Xu, Z., Joenathan, C., Khorana, B.M. 1995. Temporal and spatial properties of the time-varying speckles of botanical specimens. *Optical Engineering*, 34, 1487–1502.

Zdunek, A., Cybulska, J. 2011. Relation of biospeckle activity with quality attributes of apples. *Sensors*, 11, 6317–6327.

Zdunek, A., Frankevych, L., Konstankiewicz, K., Ranachowski, Z. 2008. Comparison of puncture test, acoustic emission and spatial–temporal speckle correlation technique as methods for apple quality evaluation. *Acta Agrophysica*, 11, 303–315.

Zdunek, A., Herppich, W.B. 2012. Relation of biospeckle activity with chlorophyll content in apples. *Postharvest Biology and Technology*, 64, 58–63.

Zdunek, A., Muravsky, L.I., Frankevych, L., Konstankiewicz, K. 2007. New non-destructive method based on spatial–temporal speckle correlation technique for evaluation of apples quality during shelf-life. *International Agrophysics*, 21, 305–310.

Zhao, Y., Wang, J., Wu, X., Williams, F.W., Schmidt, R.J. 1997. Point-wise and whole-field laser speckle intensity fluctuation measurements applied to botanical specimens. *Optics and Laser Engineering*, 28, 443–456.

Zheng, B., Pleass, C., Ih, C. 1994. Feature information extraction from dynamic biospeckle. *Applied Optics*, 33, 231–237.

Zhong, X., Wang, X., Cooley, N., Farrell, P., Foletta, S., Moran, B. 2014. Normal vector based dynamic laser speckle analysis for plant water status monitoring. *Optics Communications*, 313, 256–262.

Zhong, X., Wang, X., Cooley, N., Farrell, P., Moran, B. 2013. Dynamic laser speckle analysis via normal vector space statistics. *Optics Communications*, 305, 27–35.

Zhu, D., Lu, W., Weng, Y., Cui, H., Luo, Q. 2007. Monitoring thermal-induced changes in tumor blood flow and microvessels with laser speckle contrast imaging. *Applied Optics*, 46, 1911–1917.

15 Raman Scattering for Food Quality and Safety Assessment

Jianwei Qin, Kuanglin Chao, and Moon S. Kim

CONTENTS

15.1 Introduction 387
15.2 Principles of Raman Scattering 388
15.3 Raman Measurement Techniques 390
    15.3.1 Backscattering Raman Spectroscopy 390
    15.3.2 Transmission Raman Spectroscopy 390
    15.3.3 Spatially Offset Raman Spectroscopy 391
    15.3.4 Surface-Enhanced Raman Spectroscopy 394
    15.3.5 Raman Chemical Imaging 394
15.4 Raman Instruments...397
    15.4.1 Major Components of Raman Systems 397
        15.4.1.1 Excitation Sources 397
        15.4.1.2 Wavelength Separation Devices...398
        15.4.1.3 Detectors 400
    15.4.2 Raman Measurement Systems 403
15.5 Raman Data Analysis Techniques 405
    15.5.1 Data Preprocessing 405
    15.5.2 Qualitative Analysis...409
    15.5.3 Quantitative Analysis... 414
15.6 Applications for Food Quality and Safety 416
15.7 Conclusions..422
References..423

15.1 INTRODUCTION

As food producers need to comply with more strict rules from regulatory agencies and satisfy customers' demands for safer and higher-quality foods, inspection of food quality and safety is becoming more crucial in modern food production systems. Conventional approaches of evaluating only end products have been gradually replaced by systematic methods that require food materials and ingredients to be examined at every step along the production chain. These advances have brought new challenges and opportunities for sensing technology development. Novel techniques that are able to carry out inspections effectively and efficiently have great potential to deal with food quality and safety problems in the real world. Depending on how molecules in samples interact with electromagnetic radiations, a variety of nondestructive optical sensing techniques, such as x-ray, ultraviolet (UV), visible, fluorescence, Raman, infrared, and terahertz, have been investigated for various food quality and safety applications with different advantages and disadvantages. The physical principles of these techniques are well established, and the progress in analyzing different food and agricultural products mostly relies on technological advances rather than fundamental discoveries.

Raman scattering effect was first experimentally observed in 1928 by Indian physicists C. V. Raman and K. S. Krishnan. Since its discovery, Raman scattering has come a long way to become one of today's advanced measurement techniques. The obstacles that prevent Raman scattering technique from being widely and routinely used (e.g., weak signal, fluorescence interference, low detection efficiency, and slow data processing) have been overcome by a series of technological advances, such as introductions of small diode lasers, long optical fibers, Fourier transform (FT) Raman spectrometers, charge-coupled devices (CCDs), efficient laser rejection filters, and compact, powerful personal computers (McCreery, 2000). Propelled by increasing interests from both academia and industry, the Raman scattering technique has been developed rapidly during past decades to satisfy the needs of various applications. This chapter presents Raman scattering technology for assessing food quality and safety. Emphasis is put on introducing and demonstrating Raman spectroscopy and imaging techniques for practical uses in food analysis. The main topics include principles of Raman scattering, Raman measurement techniques (e.g., backscattering Raman spectroscopy, transmission Raman spectroscopy, spatially offset Raman spectroscopy [SORS], surface-enhanced Raman spectroscopy [SERS], and Raman chemical imaging [RCI]), Raman instruments (e.g., excitation sources, wavelength separation devices, detectors, measurement systems, and calibration methods), and Raman data analysis techniques (e.g., data preprocessing, qualitative analysis, and quantitative analysis). Finally, Raman spectroscopy and imaging applications for food quality and safety evaluation are also reviewed.

15.2 PRINCIPLES OF RAMAN SCATTERING

Raman scattering is a physical phenomenon based on the interaction of light radiation with molecular vibrations. When a sample is exposed to a monochromatic light beam with high energy, such as a laser, the incident light is absorbed and scattered after photons interact with the molecules. The scattered light consists of both elastic and inelastic scattering, as illustrated in Figure 15.1. The elastically scattered light is called Rayleigh scattering, which is the predominant form of scattering with the same frequency (or wavelength) of the incident radiation. The inelastically scattered light occurs due to the energy transfer between the photons and the molecules. The photons either lose energy to the molecules when exciting them from the ground state to the excited state (Stokes scattering) or gain energy from the molecules in an opposite process (anti-Stokes scattering). The Stokes scattering and the anti-Stokes scattering are collectively called Raman scattering (Smith and Dent, 2005). The frequency of Raman scattering is shifted from that of the incident light by the vibrational energy

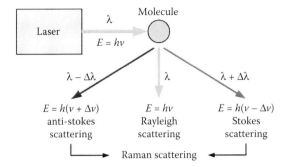

FIGURE 15.1 Principle of Raman scattering. λ is the excitation wavelength, $\Delta\lambda$ is the wavelength change, ν is the excitation frequency, $\Delta\nu$ is the frequency change, h is Planck constant, and E is the energy of a single photon.

that is decreased or increased from the photon–molecule interactions. The molecular information thus can be obtained through analysis for the frequency shift of the Raman scattering light. Generally, there are much more molecules in the ground state than those in the excited state. The intensity of the Stokes scattering, which is proportional to the number of the molecules excited from the ground state to the excited state, is thus much stronger than that of the anti-Stokes scattering. Typical Raman measurements only record the lower-frequency (or longer-wavelength) Stokes scattering information. The anti-Stokes scattering is only used in some special applications (e.g., coherent anti-Stokes Raman scattering). Raman scattering is intrinsically very weak since the probability that a Raman photon appears is normally in the order of one out of 10^6–10^8 scattered photons (Smith and Dent, 2005). The intensity of Raman scattering is proportional to the intensity of the incident laser and the reciprocal of the fourth power of the excitation wavelength (McCreery, 2000).

A Raman spectrum is generally presented by plotting the intensities of the inelastically scattered portion of the incident light (i.e., number of Raman photons) versus the shifts of the frequency from that of the excitation source (i.e., Raman shift). Unlike broad peaks commonly observed in visible and infrared spectra, a Raman spectrum is typically featured by a series of narrow and sharp peaks. The position of each peak is related to a particular molecular vibration at a fixed frequency, and it can be used to analyze the composition of a sample. The intensity of the Raman peak is linearly proportional to the concentration of molecules (Pelletier, 2003), which can be used for quantitative analysis of the analyte. The Raman shift is essentially a relative unit with respect to the excitation frequency, making it easy to compare Raman spectra regardless of the laser wavelengths. The spectral dimension of a Raman spectrum is traditionally expressed as wavenumber (i.e., the number of waves per unit length) in cm^{-1} instead of Δ cm^{-1}. Wavelength and wavenumber can be converted to each other. The wavelength of the Raman spectrum can be calculated using the wavenumber of the Raman shift by the following equation:

$$\lambda_R = \left(\frac{1}{\lambda_L} - \frac{\tilde{\nu}_R}{10^7} \right)^{-1} \tag{15.1}$$

where λ_R is the wavelength of the Raman spectrum in nm, λ_L is the wavelength of the laser in nm, and $\tilde{v}_R$ is the wavenumber of the Raman shift in cm^{-1}. For example, a Raman peak at the wavenumber of 673.0 cm^{-1} that is excited by a 785.0-nm laser will be observed at the wavelength of 828.8 nm. If a 1064.0-nm laser is used for excitation, the same peak will shift to 1146.1 nm. It is also easy to derive from Equation 15.1 that the wavelength of zero Raman shift is always the same with the laser wavelength.

15.3 RAMAN MEASUREMENT TECHNIQUES

15.3.1 Backscattering Raman Spectroscopy

Backscattering geometry is the most widely used mode for Raman scattering collection owing to its simplicity and the convenience in carrying out the experiment. The geometry is similar to the reflectance mode commonly used in visible and near-infrared (NIR) spectroscopy measurement. In this mode, the laser and the detector are arranged on the same side of the sample. The detector acquires backscattered Raman signals from the laser incident point. Since the laser light is generally converged to a small spot, it is critical to align the axis of the detector to the incident laser point because misalignment will deteriorate Raman signal collection. The laser can be projected on the sample surface normally or obliquely. The incident angle of the laser is also important for efficiently acquiring Raman signals. The oblique incidence of the laser will usually make the measurement system susceptible to the height of the samples. As illustrated in Figure 15.2a, the detector axis may coincide with the laser spot obliquely projected on a sample surface. However, samples with higher or lower surfaces will change the position of the laser spot. As a result, the detector will mismatch the Raman signals generated from the excitation spot. Such a problem can be overcome by using an optical layout based on a 45° dichroic beamsplitter, which provides normal laser incidence on the sample surface (Figure 15.2b). The beamsplitter reflects the laser wavelength and passes the longer Raman-shifted wavelengths. The detector axis is thus always aligned with the laser spot (or the Raman axis) regardless of the sample height. Consequently, the Raman signals from the samples with different heights can be efficiently collected by the detector. The configuration based on the 45° dichroic beamsplitter or its variants are commonly adopted in commercial Raman systems (e.g., Raman microscopes and fiber-optic Raman probes). The backscattering geometry is predominantly used in various Raman applications for food quality and safety evaluation.

15.3.2 Transmission Raman Spectroscopy

Despite its popularity for Raman scattering measurement, the backscattering method has a limitation of retrieving internal sample information. Backscattering Raman measurement gives a strong weight to the surface layers of the sample, which generally cannot be used to quantify the overall bulk content of the heterogeneous samples. Transmission Raman spectroscopy is capable of determining the bulk composition of a sample, especially for small individual items with a diffusely

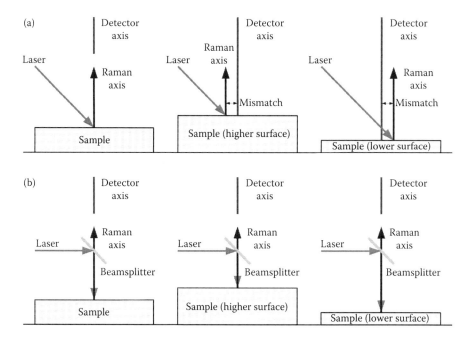

FIGURE 15.2 Optical configurations of backscattering Raman spectroscopy: (a) oblique and (b) normal laser incidence on the sample surface with different heights.

scattering and weakly absorbing internal condition (Matousek and Parker, 2006). In this mode, the laser and the detector are arranged on different sides of the sample. The detector acquires the forward-scattered Raman signals that pass through the sample. Although the geometry is similar to the transmission experiment frequently performed in visible and near-infrared spectroscopy, their measurement principles are fundamentally different in that the sample information obtained by transmission Raman spectroscopy is based on the forward-scattered Raman photons (with wavelength different from the laser) rather than the absorption by the molecules (with the same wavelength as that of the laser). The practical usage of transmission Raman spectroscopy was recently improved for pharmaceutical applications (Matousek and Parker, 2006). The technique greatly suppresses the Raman and fluorescence signals originated from the surface of the samples (e.g., tablet coatings and capsule shells), making it suitable for bulk composition analysis of diffusing and translucent materials. It has also been used for evaluating granular agricultural products, such as composition analysis of single soybeans (Schulmerich et al., 2012) and corn kernels (Shin et al., 2012) and differentiation of the geographical origins of rice (Hwang et al., 2012).

15.3.3 SPATIALLY OFFSET RAMAN SPECTROSCOPY

Transmission Raman spectroscopy provides overall content information of the sample without the capability of separating the information from individual layers.

SORS is a relatively new Raman measurement technique developed for noninvasive retrieval of layered internal information from diffusely scattering media (Matousek et al., 2005). The SORS technique aims to acquire subsurface information by collecting Raman scattering signals from a series of surface positions laterally offset from the excitation laser (Figure 15.3). The offset spectra exhibit different sensitivities to the Raman signals from surface and subsurface layers. As the source–detector distance increases, the contribution of the Raman signals from the deep layers gradually outweighs that from the top layer. Pure Raman spectra of the individual layers can be extracted by applying a spectral mixture analysis technique to an array of SORS spectra. The chemical information of the subsurface layers thus can be obtained based on the decomposed Raman spectra. Early applications of the technique were primarily in biomedical and pharmaceutical fields, such as noninvasive evaluation of human bone *in vivo* (Matousek et al., 2006) and authentication of pharmaceutical products through packaging (Eliasson and Matousek, 2007). Recently, the SORS technique has been used for assessing food and agricultural products, such as nondestructive evaluation of internal maturity of tomatoes (Qin et al., 2012) and quality analysis of salmon through the skin (Afseth et al., 2014).

An example of SORS measurement for retrieving chemical information under the tomato outer pericarp is demonstrated in Figure 15.4. A 10-mm-thick red tomato pericarp (Figure 15.4a) was placed on a Teflon slab, which was used as a subsurface reference material known to exhibit identifiable Raman peaks. Raman spectra were acquired in an offset range of 0–5 mm with a step size of 0.2 mm using a 785-nm laser (Figure 15.4b). Fluorescence-corrected Raman spectra at four offset positions were selected to demonstrate the general pattern of the SORS data (Figure 15.4c). Raman peaks of the Teflon slab under the tomato pericarp were observed at all the offset positions between 0 and 5 mm. Three Raman peaks attributed to the lycopene in the red pericarp (i.e., 1001, 1151, and 1513 cm^{-1}) gradually diminished with increasing offset distance. The SORS data were analyzed using self-modeling mixture analysis (SMA) to extract the pure component spectrum of each layer. The Raman spectra of the red pericarp and the Teflon slab were successfully separated (Figure 15.4d). The decomposed spectra of the red pericarp and the Teflon slab are similar to the reference spectra of lycopene and Teflon, respectively. Similar results were obtained for a 10-mm-thick green tomato pericarp placed on the Teflon slab.

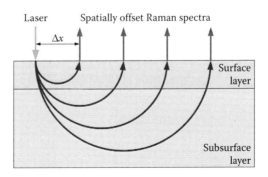

FIGURE 15.3 SORS technique.

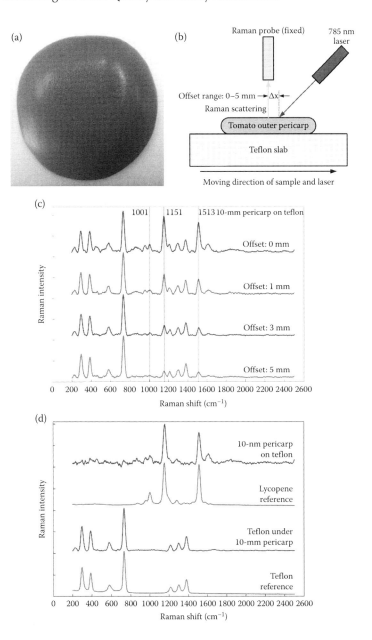

FIGURE 15.4 SORS for subsurface detection through tomato outer pericarp: (a) a 10-mm-thick red tomato outer pericarp, (b) experimental setup for SORS measurement of the tomato outer pericarp placed on a Teflon slab, (c) fluorescence-corrected Raman spectra at four selected offset positions, and (d) pure component spectra of the pericarp and the Teflon slab extracted from SMA and reference spectra of lycopene and Teflon. (Reprinted from *Postharvest Biology and Technology*, 71, Qin, Jianwei, Kuanglin Chao, and Moon S. Kim, Nondestructive evaluation of internal maturity of tomatoes using spatially offset Raman spectroscopy, 21–31, Copyright 2012, with permission from Elsevier.)

These results suggested that it is possible to obtain subsurface chemical information through tomato outer pericarps by SORS coupled with SMA, which forms a basis to develop an SORS-based nondestructive method for evaluating the internal maturity of tomatoes (Qin et al., 2012).

15.3.4 SURFACE-ENHANCED RAMAN SPECTROSCOPY

Intensity of normal Raman scattering is inherently weak due to the fact that only one out of 10^6–10^8 scattered photons originates from Raman scattering. SERS is a technique that can amplify conventional Raman signals by several orders of magnitude when molecules are attached on or in close proximity to particles or surfaces of noble metals (e.g., gold and silver). Electromagnetic enhancement and chemical enhancement are two mechanisms attributed to the SERS enhancement, and the electromagnetic effect is generally considered larger than the chemical effect for the signal amplification. Lasers, spectrometers, and detectors used in normal Raman measurements can be indistinguishably used in SERS measurements. The key difference between the normal and the enhanced Raman is the enhancing media uniquely used in the SERS test. Based on the state of the material, the SERS enhancing media can be broadly divided into two categories: colloidal suspensions of nanoparticles and micro-textured solid substrates (Schlücker, 2011). In practice, the target analyte is usually dissolved in an aqueous solution. The solution then can be either blended with a nanoparticle suspension or deposited on a solid substrate. The enhanced Raman signals are collected from the analyte adsorbed on the media. The SERS technique can enhance the intensity of the normal Raman signals by a factor up to 10^{12} (Le Ru et al., 2007), enabling it to be used for trace detection or even single molecule detection. Such high sensitivity has realized many analytical applications and drawn considerable research interest, especially in the new methods and materials for fabricating novel SERS substrates (Fan et al., 2011). Applications of SERS technique for food and agricultural products are mainly in the area of food safety inspection (Craig et al., 2013). Example applications include detections of foodborne pathogenic bacteria (Chu et al., 2008), melamine in human foods (Lin et al., 2008), and pesticide residues on fruit peels (Liu et al., 2012).

15.3.5 RAMAN CHEMICAL IMAGING

Owing to the natural size of the laser spot, Raman spectral data is conventionally collected from a point on the sample surface, which cannot cover a large surface area due to the size limitation of the point measurement. Spatial information, which is important for food quality and safety inspection, thus cannot be obtained by the traditional Raman spectroscopy method. RCI is a technique that equips Raman spectroscopy with the capability of spatial information acquisition. Sample composition, spatial distribution, and morphological features of interesting targets can be visualized in chemical images created using both Raman and spatial information. The RCI technique has been developed as a useful tool with many applications in different fields, such as biomedicine, pharmaceuticals, agriculture, archeology, forensics, mineralogy, and threat detection (Stewart et al., 2012). Currently, most

RCI research and applications use commercial Raman imaging instruments, which utilize global (wide-field), point, and line lasers as excitation sources. Global excitations are used in most Raman microscopes, in which a relatively large sample area is illuminated by a defocused laser spot. Raman spectra over the entire excited area are collected using filters (e.g., liquid crystal tunable filters [LCTFs]) for wavelength selection (Morris et al., 1996). The point lasers are usually used in systems that combine FT-Raman spectrometers and XY positioning stages for point-scan imaging (Schulz et al., 2005). The line lasers, which can be formed by either spreading a laser spot using a scanning mirror (Markwort et al., 1995) or expanding a laser beam using cylindrical or similar optics (Christensen and Morris, 1998), have been used in line-scan Raman microscopes. Regardless of the configurations, the current commercial RCI systems generally conduct imaging measurements at subcentimeter scales. Typical size of Raman microscopic images is measured at a few hundred micrometers (Liu et al., 2009a). Such small spatial coverage is the main restriction for food evaluation since they cannot be used for inspecting samples with large surface areas.

Efforts have been made to remedy the lack of tools for macro-scale RCI for food quality and safety research. For example, a bench-top point-scan Raman imaging system was developed for this purpose (Qin et al., 2010). The large spatial range (i.e., on the order of centimeters) and the high spatial resolution (e.g., 0.1 mm) of a two-axis motorized positioning table enabled the system to be used for inspecting large food and agricultural products, such as detecting lycopene changes from tomato cross-sections during ripening (Qin et al., 2011). Figure 15.5 shows an example of using this system for simultaneous detection of four types of adulterants (i.e., ammonium sulfate, dicyandiamide, melamine, and urea) mixed in milk powder (Qin et al., 2014c). Raman images were acquired from a 25 × 25 mm^2 area of each milk–adulterant mixture using a 785-nm laser (Figure 15.5a). Unique Raman peaks for each adulterant were identified from their reference spectra (Figure 15.5b). Fluorescence-corrected Raman images of a 5.0% milk–adulterant mixture at four identified Raman peaks of the adulterants are shown in Figure 15.5c. Raman chemical images were created based on the single-band images in Figure 15.5c to visualize identification and distribution of the four adulterant particles in the milk powder (Figure 15.5d). One limitation of this system is the long sampling time due to the point-scan image acquisition in two spatial dimensions. Generally, the scan time is measured by hours, which is the bottleneck that prevents it from conducting fast inspection tasks. Recently, high-throughput macro-scale RCI has been realized on a newly developed line-scan hyperspectral system that utilizes a 24-cm-long 785-nm line laser as the excitation source (Qin et al., 2014a). The system is able to image a large sample area with a short sampling time (typically in minutes), making it suitable for rapid evaluation of food quality and safety.

The measurement techniques presented in this section are representative rather than comprehensive. There are many other Raman techniques that have been developed for different purposes, such as stand-off Raman spectroscopy for remote detection of minerals at a distance up to 66 m using a telescope (Sharma et al., 2003), coherent anti-Stokes Raman scattering for tissue imaging of a live mouse at video rate (Evans et al., 2005), and shifted excitation Raman difference spectroscopy for

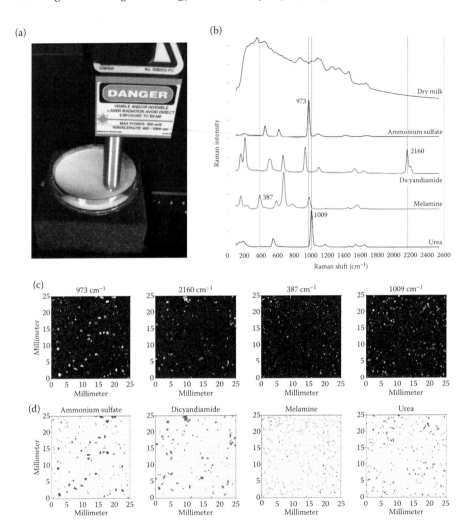

FIGURE 15.5 Macro-scale RCI for simultaneous detection of multiple adulterants in dry milk: (a) image acquisition using a 785-nm laser, (b) Raman spectra of skim milk powder and four chemical adulterants, (c) fluorescence-corrected Raman images of a 5.0% milk–adulterant mixture at selected Raman peaks of the adulterants, and (d) chemical images of individual adulterants in the milk powder. (With permission from Springer, *Journal of Food Measurement and Characterization*, Development of a Raman chemical imaging detection method for authenticating skim milk powder, 8, no. 2, 2014c, 122–31, Qin, Jianwei, Kuanglin Chao, Moon S. Kim, Hoyoung Lee, and Yankun Peng.)

rejection of fluorescence from meat samples (Sowoidnich and Kronfeldt, 2012). Also, novel Raman techniques are continuously emerging to create new detection possibilities that cannot be achieved by existing methods. For instance, SORS has been combined with SERS and stand-off Raman spectroscopy to generate two new techniques: surface-enhanced SORS (Stone et al., 2011) and stand-off SORS (Zachhuber et al., 2011), respectively. The subsurface probing capability of the SORS technique is thus extended to the enhanced detection of the SERS technique and the remote detection of the stand-off Raman spectroscopy. Such combinations make it possible to conduct SERS measurements in deep tissue or through bone (Sharma et al., 2013) and detect chemicals concealed in distant opaque containers (Zachhuber et al., 2011). Adoption of existing and newly developed Raman techniques for food quality and safety applications is likely to expand in the near future.

15.4 RAMAN INSTRUMENTS

15.4.1 MAJOR COMPONENTS OF RAMAN SYSTEMS

Key components of a Raman system generally include an excitation source, a wavelength separation device, and a detector. These components are presented in the following sections.

15.4.1.1 Excitation Sources

Lasers are powerful light sources that are widely used for Raman excitations owing to their highly concentrated energy, perfect directionality, and true monochromatic emission. Light from lasers is generated through stimulated emission, which usually occurs inside a resonant optical cavity filled with a gain medium (e.g., gas, dye solution, semiconductor, and crystal). They can operate in continuous-wave mode or pulse mode in terms of temporal continuity of the output. Since Raman scattering intensity is proportional to the laser intensity and the reciprocal of the fourth power of the laser wavelength (i.e., $1/\lambda^4$) (McCreery, 2000), selection of the laser is a critical part in assembling an efficient Raman measurement system. Raman signals can be enhanced by boosting the laser intensity or by lowering the excitation wavelength. However, high laser intensity and short excitation wavelength are commonly associated with sample degradation/burning and strong fluorescence. Hence, in practice, laser selection is usually a compromise among the factors of maximizing the intensity of the Raman signals, minimizing the risk of the sample degradation, diminishing the interference of fluorescence, and optimizing the sensitivity of the detector. Food and biological materials usually generate strong fluorescence signals when excited by visible lasers (e.g., 488, 532, and 633 nm). Diode lasers operating at 785 and 830 nm (typically with power of tens to hundreds of milliwatts) are usually used to weaken the fluorescence. Nd:YAG lasers operating at 1064 nm, which are widely used in FT-Raman systems, can minimize the fluorescence interference. However, the Raman intensity is largely reduced at the same time. The laser power usually needs to be adjusted to above 1 W to compensate for the loss of the Raman intensity. UV lasers (Efremov et al., 2008) and ultrashort pulse lasers (Matousek et al., 1999) can also remove the fluorescence interference by introducing spectral and time gaps between Raman and fluorescence signals, respectively. But high risk of sample damage by the UV excitation and high cost and complexity of the instruments for temporal filtering have limited their widespread applicability.

Since the linewidth of an observed Raman band is a result of convoluting the linewidth of a laser with the inherent linewidth of a vibrational band (Matousek and

Morris, 2010), it is critical to use lasers with narrow linewidths for excitation purposes. Raman spectra generated using wide linewidth excitation sources will exhibit broad, poorly resolved peaks. Generally, a laser linewidth (defined as full width at half maximum [FWHM]) better than 1 cm^{-1} is adequate to be used for most Raman excitations (McCreery, 2000). At 785 nm, the 1 cm^{-1} linewidth can be translated to a wavelength of approximately 0.06 nm. In practice, lasers with the FWHM linewidth less than 0.1 nm are commonly used in various Raman applications. Besides lasers, narrowband light-emitting diodes have begun to be used for Raman excitations (Adami and Kiefer, 2013), although at present they cannot compete with lasers as routine Raman excitation sources because of their lower intensities and broader linewidths (e.g., a few nanometers). Meanwhile, the output light other than the center wavelength of the laser will create Rayleigh scattering, which will interfere with the weak Raman scattering signals. Therefore, effective removal of the extraneous emissions from the laser is also important to obtain high-quality Raman signals. Interference bandpass filters are usually used to clean up the laser output by blocking the off-line wavelengths. Such bandpass filters are commercially available with optical densities up to six (i.e., light at wavelengths outside the center wavelength of the filter is attenuated by a factor of 10^6), which is generally sufficient for most Raman applications.

15.4.1.2 Wavelength Separation Devices

Wavelength separation devices disperse Raman scattering light into different wavelengths and project the dispersed light to the detectors. Such devices can be divided into three categories: dispersive spectrometers, FT spectrometers, and electronically tunable filters.

Dispersive Raman spectrometers are constructed based on diffraction gratings, which spatially separate the incoming light into different wavelengths. The degree of dispersion is related to groove spacing on the grating surface. The narrower the groove spacing, the higher the dispersion (or resolution) for the incident light. There are two main types of diffraction gratings: transmission gratings and reflection gratings. A reflection-grating-based dispersive Raman spectrometer is illustrated in Figure 15.6a. The basic structure includes a pair of spherical mirrors coupled with a convex reflection grating. The lower mirror guides light from the entrance slit to the reflection grating, where the beam is dispersed into different wavelengths. The upper mirror then reflects the dispersed light to the detector, where a continuous Raman spectrum is formed. The spectrometers based on transmission gratings operate in a similar manner, except that light is dispersed after passing through the gratings. Owing to the advantages such as low cost, no moving parts, small size, and easy configuration, the dispersive spectrometers are widely used in single-point Raman spectroscopy measurements. Dispersive imaging spectrographs with the capacity of acquiring spatial information have also been developed for Raman imaging applications (Qin et al., 2014a). The Raman spectrographs work in much narrower wavelength ranges (e.g., 770–980 nm) than visible and near-infrared spectrographs (e.g., 400–1000 and 900–1700 nm). Given the same number of CCD pixels, the spectral resolution of the Raman spectrographs is much higher than that of the visible and near-infrared spectrographs. Such high resolution is necessary since Raman

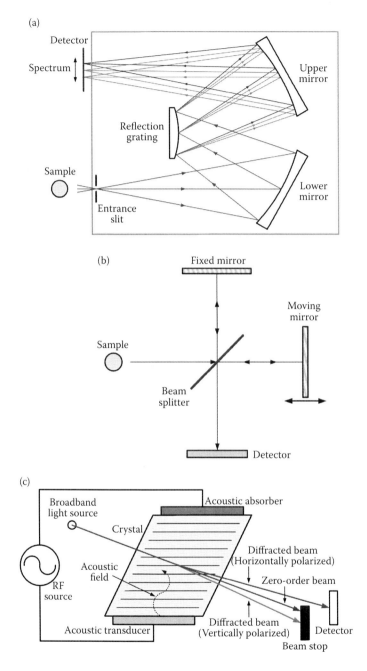

FIGURE 15.6 Wavelength separation devices for Raman measurement systems: (a) dispersive spectrometer based on a reflection grating, (b) FT spectrometer based on a Michelson interferometer, and (c) AOTF.

spectrum is generally featured as sharp peaks, which are uncommon in relatively broad visible/near-infrared and fluorescence spectra of food and biological materials.

Other than dispersing light into different wavelengths, FT-Raman spectrometers acquire light signals all together in a form of interferogram, which carries spectral information of the broadband light. Figure 15.6b shows an FT-Raman spectrometer based on a Michelson interferometer. It consists of a beamsplitter and two flat mirrors (fixed and moving mirrors) that are perpendicular to each other. Light from the sample is divided into two beams at the beamsplitter. The light is partially reflected to the fixed mirror, and the remaining is transmitted through the beamsplitter to the moving mirror, which moves in a direction parallel with the incident light. The beams reflected back from the two mirrors are recombined by the beamsplitter. The moving mirror introduces an optical path difference between the two beams. An interferogram is then generated and collected by the detector. An inverse FT to the interferogram can reveal the wavelengths of the light accurately in a broad spectral region. The spectral resolution of the FT spectrometers is determined by the distance traveled by the moving mirror, and generally it is higher than the dispersive spectrometers. The FT-Raman spectrometers use 1064 nm lasers as common excitation sources, which can significantly reduce fluorescence signals, especially for food and biological samples. However, this advantage has been gradually diminished as new compact 1064-nm lasers and near-infrared sensitive detectors were recently introduced for dispersive Raman systems.

Beside dispersive and FT spectrometers, electronically tunable filters can also be used to separate wavelengths. There are two major types of tunable filters: acousto-optic tunable filters (AOTFs) and LCTFs. An AOTF isolates a single wavelength from broadband light based on light–sound interactions in a crystal (Figure 15.6c). An acoustic transducer generates high-frequency acoustic waves that change the refractive index of the crystal, by which light is diffracted into two first-order beams. The zero-order beam and the undesired diffracted beams are blocked by the beam stop. The AOTF diffracts light at one particular wavelength at a time, and the passing wavelength can be controlled by varying the frequency of the radio frequency (RF) source. An LCTF utilizes electronically controlled liquid crystal cells to transmit light at a specific wavelength. The LCTF is constructed by a series of optical stacks, each consisting of a retarder and a liquid crystal layer between two polarizers. Each stage transmits light as a sinusoidal function of the wavelength, and all the stages function together to transmit a single wavelength. The controller can shift the narrow bandpass region by applying an electric field to each liquid crystal layer. Advantages of the electronically tunable filters include small size, larger aperture, accessibility of random wavelength, and flexible controllability. However, their spectral resolution is generally lower than that of the dispersive and FT spectrometers. AOTFs and LCTFs have been used in building different Raman systems, especially for portable Raman spectroscopy (Yan and Vo-Dinh, 2007; Sakamoto et al., 2012) and microscope systems (Morris et al., 1994).

15.4.1.3 Detectors

Given the weakness of Raman scattering, it is crucial to use a sensitive and low-noise detector to collect scattering signals. Currently, CCDs are the mainstream detectors used in Raman measurement systems. Since they were first adopted for Raman applications in the mid-1980s, the CCDs have replaced almost all other detectors, such as single-channel detectors (e.g., photomultiplier tubes) and early multichannel detectors (e.g., intensified photodiode arrays) (McCreery, 2000). The CCD sensor is composed of many small photodiodes (called pixels) that are made of light-sensitive materials. Each photodiode converts incident photons to electrons, generating an electrical signal proportional to total light exposure. The rectangular CCD sensor is positioned with one dimension parallel to the direction of wavelength dispersion, and the other parallel to the entrance slit of most dispersive spectrometers, which can be used in various readout modes (e.g., full vertical binning, single-track, multiple-track, and imaging). The CCDs used in the Raman systems generally require high quantum efficiency (QE) and low dark noise to maximize the quality of the Raman signals.

QE is used to quantify the spectral response of the CCDs, and it is primarily governed by the substrate materials used to make the photodiodes. Silicon is intensively used as sensor material for making CCDs that work in visible and short-wavelength NIR regions (e.g., 400–1000 nm). A typical QE curve of the silicon CCDs is a bell-shaped curve with QE values declining toward both UV and NIR regions. The silicon CCDs are generally used for visible lasers (e.g., 488, 532, and 633 nm). Deep depletion CCDs, which enhance the spectral response toward the red end of the spectrum using controlled doping of the silicon, can be used for lasers of longer wavelengths (e.g., 785 and 830 nm). For the NIR region, indium gallium arsenide (InGaAs), an alloy of indium arsenide (InAs) and gallium arsenide (GaAs), is the common substrate material of the CCD sensors. Standard InGaAs CCDs have fairly flat and high QE in the spectral region of 900–1700 nm. An extended wavelength range (e.g., 1100–2600 nm) can be achieved by changing the percentages of InAs and GaAs for making the sensors. The InGaAs CCDs are generally used to collect Raman signals excited by NIR lasers (e.g., 1064 nm).

In addition to the high QE, the dark noise of the CCDs needs to be minimized to ensure the best possible signal-to-noise ratio (SNR) for Raman signals. The dark noise arising from the photodiodes on the sensor surface can be reduced by lowering the temperature of the CCDs. Generally, longer-wavelength detection results in higher dark noise. Hence, the operating temperature must be low to prevent the weak Raman scattering signals from being buried in the dark noise, especially for the NIR region. Typical temperature of the CCDs by air cooling is in the range of −70 to −20°C. Liquid cooling using water or coolants can further reduce the temperature to −100°C. The SNR of the Raman signals can also be improved by changing the readout mode of the CCDs. Figure 15.7 demonstrates an example of using a single-track mode for Raman spectral acquisition. With this method, a rectangular area is defined on the CCD sensor to only include pixels that are illuminated by the incoming light (Figure 15.7a, 27 × 1024 on a CCD with 256 × 1024 pixels in this case). All the rows within this specified area are vertically binned together. As a result, a single Raman spectrum is obtained for each measurement. The highest intensity (CCD count) of the 27 Raman spectra within the defined area is less than 1500 (Figure 15.7b). After vertical binning, the intensity of the final Raman spectrum (Figure 15.7c) is at least one order higher than that of the individual spectra. The contribution

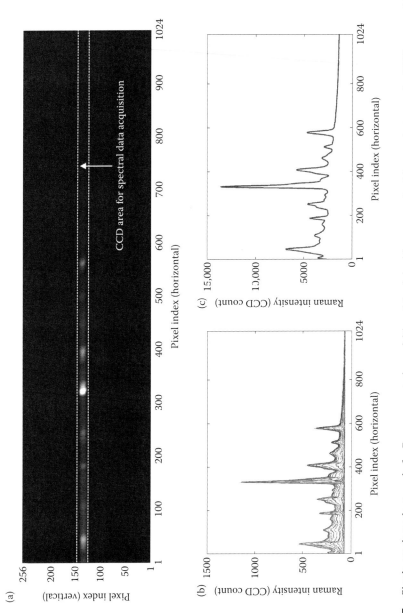

FIGURE 15.7 Single-track readout mode for Raman spectral acquisition: (a) original image acquired from polystyrene by CCD sensor, (b) individual Raman spectra extracted in a defined area (27 × 1024) on the CCD (between two dashed lines in (a), and (c) final Raman spectrum after vertical binning). (From Qin, J., K. Chao, and M. S. Kim, *Trans. ASABE*, 53(6), 1873–82, 2010. With permission from ASABE.)

of the dark noise is minimized by excluding the pixels that are not in the illuminated area. Further enhancement for Raman scattering detection can be achieved using high-performance CCDs, such as electron-multiplying CCDs and intensified CCDs.

15.4.2 Raman Measurement Systems

Driven by growing interests of both academia and industry, Raman instruments and measurement systems have been developed rapidly to meet requirements from various applications. A broad variety of integrated systems and modular components are now commercially available for different usages, such as research-grade Raman systems used in laboratories, portable Raman systems used in fields, and process monitoring and control systems used in different industries (Mukhopadhyay, 2007). Owing to the increasing demands from existing and new disciplines and the decreasing costs for manufacturing major components in Raman systems (e.g., lasers, filters, spectrometers, and CCDs), the overall costs of commercial Raman systems have gradually decreased from hundreds of thousands of dollars toward tens of thousands of dollars. Such price drops have helped extend their applicability in academic and industrial areas. The integrated bench-top Raman systems (e.g., FT-Raman spectrometers and Raman microscopes), which provide solutions for well-defined applications, are routinely used in many research laboratories. New instruments, such as miniature Raman spectrometers and battery-powered handheld instruments operated in a point-and-shoot manner (Carron and Cox, 2010), are continuously appearing in the market. On the other hand, custom-designed systems, which use modular components such as lasers, filters, spectrometers, CCDs, and sample handling units, generally are able to provide more flexibility and versatility than the integrated commercial systems, since many aspects of the modular systems (e.g., excitation wavelength, filter bandwidth, spectrometer resolution, CCD sensitivity, and sample control) can be customized and optimized for particular applications. Developments and uses of novel custom-designed systems usually will lead to new research possibilities and opportunities.

A custom-designed Raman system is demonstrated in Figure 15.8a (Qin et al., 2010). The system uses a 785-nm laser with a linewidth of 0.1 nm and maximum power of 350 mW as the excitation source. A fiber-optic Raman probe is used to focus the laser on the sample surface and acquire Raman scattering signals. A bifurcated optical fiber bundle is used to deliver the laser light to the probe and transfer the collected Raman signals to the detection module. A 16-bit CCD camera with 1024×256 pixels and QE greater than 90% at 800 nm is used to acquire Raman signals. The CCD is thermoelectrically cooled to $-70°C$ during spectral acquisition to minimize the dark noise. A reflection-grating-based Raman imaging spectrometer is mounted to the camera. The spectrometer accepts light through a $5 \text{ mm} \times 100 \text{ μm}$ input slit, and detects a Raman shift range of -98 to 3998 cm^{-1} (or a wavelength range of 799–1144 nm) with a spectral resolution of 3.7 cm^{-1}. A two-axis motorized positioning table is used to move the samples in two perpendicular directions, with a displacement resolution of 6.35 μm across a square area of $127 \times 127 \text{ mm}^2$. The Raman probe, the positioning table, and the sample materials are placed in a closed black box to avoid the influence of ambient light.

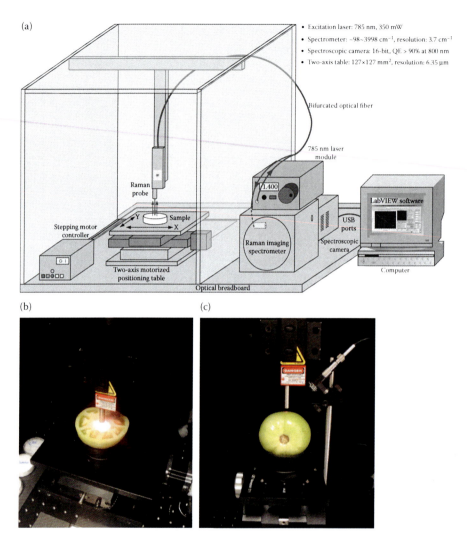

FIGURE 15.8 Custom-designed Raman system for macro-scale chemical imaging and flexible spectroscopy measurement: (a) schematic diagram of the system, (b) image acquisition from cross-section of a cut-open tomato for lycopene detection, and (c) SORS measurement of an intact tomato for internal maturity evaluation.

System software was developed using LabVIEW (National Instruments, Austin, Texas) to fulfil functions such as camera control, data acquisition, sample movement, and synchronization. The Raman data are saved in the format of band interleaved by pixel, which can be analyzed by commercial software packages such as ENVI (Exelis Visual Information Solutions, Boulder, Colorado and MATLAB® [MathWorks, Natick, Massachusetts]).

The system was developed primarily for macro-scale RCI of different food and agricultural products, such as detecting lycopene changes from tomato cross-sections during ripening (Figure 15.8b) (Qin et al., 2011) and screening multiple adulterants mixed into milk powder (Qin et al., 2013). In addition to imaging, the system can also be configured to perform versatile spectroscopic measurements, such as SORS for nondestructive evaluation of internal maturity of tomatoes (Figure 15.8c) (Qin et al., 2012) and temperature-dependent Raman spectroscopy for investigation of isomerization mechanism for endosulfan (Schmidt et al., 2014). Such flexibilities for both imaging and spectroscopic measurements are usually not available from integrated commercial Raman systems.

The custom-designed systems generally need proper spectral and spatial calibrations before obtaining meaningful data. Spectral calibrations aim to define the wavelengths (or wavenumbers) for the pixels along the spectral dimension of the acquired data, and they are generally required for both spectroscopic and imaging systems. The calibration results can be used to determine the range and the interval of the spectral data. For Raman systems, it is more useful to use wavenumbers as references instead of wavelengths commonly used in absolute spectral calibrations. Usually, a narrow-linewidth single-wavelength laser and chemicals with known relative wavenumber shifts are used to calibrate Raman systems. A guide of Raman shift standards has been established by the American Society for Testing and Materials (ASTM) International (ASTM Standards, 2007), which provides Raman shift wavenumbers of eight standard chemicals that cover a wide wavenumber range (i.e., 85–3327 cm^{-1}). On the other hand, spatial calibrations are performed to determine the range and the resolution of the spatial information, and usually they are only required for the imaging systems. The calibration results are useful for adjusting the field of view and estimating the spatial detection limit.

Figure 15.9 shows an example of spectral and spatial calibrations for the custom-designed Raman system demonstrated in Figure 15.8a. The spectral calibration was performed using a 785-nm laser and two Raman shift standards (polystyrene and naphthalene) (Qin et al., 2010). After identifying the pixel positions and the corresponding wavenumbers of 12 selected Raman peaks from the two chemicals (Figure 15.9a), a quadratic regression model was established to determine all the wavenumbers along the spectral dimension. The system was found to cover a wavenumber range of 102–2538 cm^{-1}. The spatial calibration was conducted by imaging a standard test chart using a point-scan method with a step size of 0.1 mm for both scan directions (Figure 15.9b). The diameter for the smallest dots in the central area is 0.25 mm, and the distance between these adjacent dots is 0.50 mm. The outermost large dots are positioned within a 50-mm square. No image distortions were observed and the 0.25-mm dots can be clearly resolved owing to the small step sizes used to scan the chart.

15.5 RAMAN DATA ANALYSIS TECHNIQUES

15.5.1 Data Preprocessing

Data preprocessing aims to remove noises, artifacts, and useless signals due to the test environments and the imperfections of the components in the measurement systems. In Raman scattering measurements, fluorescence signals are usually generated

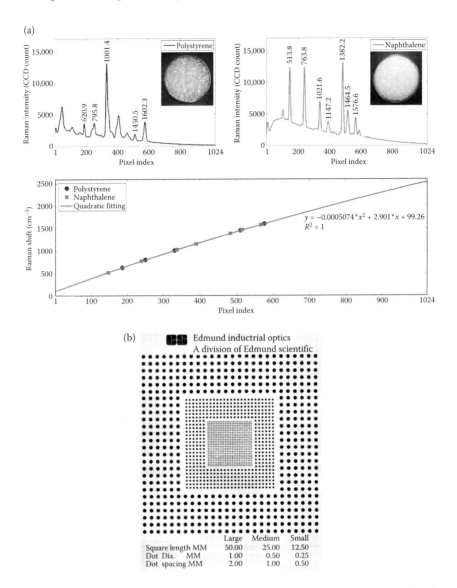

FIGURE 15.9 Calibrations for a custom-designed Raman system: (a) spectral calibration using a 785-nm laser and two Raman shift standards (polystyrene and naphthalene) and (b) spatial calibration by imaging a resolution test chart with a step size of 0.1 mm for both scan directions. (From Qin, J., K. Chao, and M. S. Kim, *Trans.* ASABE, 53(6), 1873–82, 2010. With permission from ASABE.)

during laser–sample interactions, especially for food and biological materials. The fluorescence intensity is typically several orders more intense than that of the Raman scattering, which will easily overwhelm the weak signals that are valuable to sample evaluation. Therefore, removal of the fluorescence is a major preprocessing procedure for practical Raman applications for food quality and safety inspection. Many methods have been developed for removing the underlying fluorescence baseline in the Raman data, which can be separated into two general categories: hardware and software correction methods. The hardware correction methods utilize special instruments or components to suppress the fluorescence based on its physical nature. For example, fluorescence is generated in a short time measured by nanoseconds after the instantaneous production of Raman scattering. This delay can be used to separate the Raman and fluorescence signals in time domain using time gating methods with ultrashort pulse lasers (Matousek et al., 1999). Other instrument-based correction methods include UV excitation methods (Efremov et al., 2008), shifted excitation methods (Cooper et al., 2013), methods using special optical components and configurations (Cormack et al., 2007; Le Ru et al., 2012), etc. Extra instruments or components are generally needed for the hardware correction methods, which will raise the overall cost of the measurement systems. On the other hand, the software correction methods use various algorithms to eliminate the fluorescence based on its mathematical nature. For example, polynomial curve fitting provides a simple and effective correction method based on the fact that most fluorescence baselines can be modeled by polynomial functions of different degrees. Other algorithm-based correction methods include least squares, wavelet transformation, Fourier transformation, derivatives, etc. (Schulze et al., 2005; Zhang et al., 2010).

Figure 15.10 demonstrates an example of fluorescence background correction for Raman data acquired from cross-sections of cut-open tomatoes using the imaging system illustrated in Figure 15.8b. A modified polynomial curve-fitting method, which uses an iterative comparison approach to identify and prevent Raman peaks from being used in the curve-fitting process (Lieber and Mahadevan-Jansen, 2003), was used for the correction. This method is effective for eliminating fluorescence background and retaining Raman features at the same time. To minimize high-frequency noise in the raw data, a first-order Savitzky–Golay filter was applied first to the spectra. After smoothing, the curve-fitting method with an eighth-order polynomial was applied to all Raman spectra. The fitted baseline at each hyperspectral pixel was then subtracted from the original spectrum to generate the Raman spectrum with a near-flat background. As shown in Figure 15.10a, the fitting method gave a good fit for the fluorescence background at both high and low intensities. Removal of the fluorescence baseline enhanced the Raman peaks, as can be seen in the corrected spectra of the locular tissues from "Red" and "Breaker" tomatoes. For comparison, the corrected seed spectra plotted in Figure 15.10b are almost entirely flat without notable Raman features. Corrected results for the images are demonstrated in Figure 15.10c using single-band images at four wavenumbers (corresponding to four Raman peaks of lycopene) of four tomatoes at selected ripeness stages. The original Raman images were dominated by the strong fluorescence from tomato seeds and adjacent areas. After the correction, the influence of the seed fluorescence was largely diminished. The areas of the locular tissues and the outer pericarps became major features. The brightness patterns of the tomato cross-sections in the corrected images generally revealed the changes in lycopene content that occur during the postharvest ripening process. The fluorescence correction formed a basis to detect lycopene in the tomatoes using a spectral matching method with pure lycopene as reference (Qin et al., 2011).

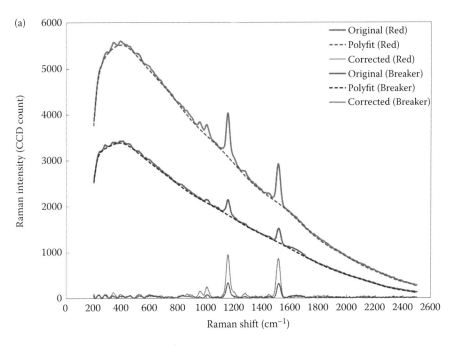

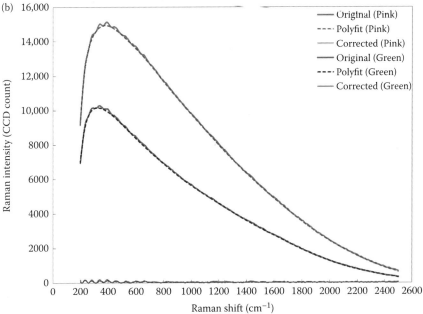

FIGURE 15.10 Fluorescence background correction for Raman spectra and images from cross-sections of cut-open tomatoes: original and corrected spectra of (a) locular tissues of red and breaker tomatoes and (b) seeds of pink and green tomatoes. *(Continued)*

Raman Scattering for Food Quality and Safety Assessment

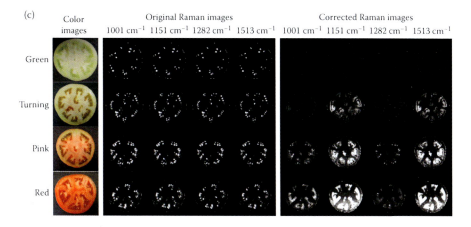

FIGURE 15.10 (*Continued*) Fluorescence background correction for Raman spectra and images from cross-sections of cut-open tomatoes: (c) original and corrected single-band images of tomatoes at selected ripeness stages. (Reprinted from *Journal of Food Engineering*, 107, no. 3–4, Qin, Jianwei, Kuanglin Chao, and Moon S. Kim., Investigation of Raman chemical imaging for detection of lycopene changes in tomatoes during postharvest ripening, 277–88, Copyright 2011, with permission from Elsevier.)

Besides the fluorescence baseline correction discussed above, many other preprocessing methods can be applied to the Raman spectra and images to make the data independent of the measurement systems and the test conditions, such as dark current subtraction, spectral smoothing, spike removal, spectral normalization, image masking, and spatial filtering. These preprocessing methods can be used in spectral or spatial domains to form a base for the raw Raman data to be suitable for qualitative/quantitative analysis.

15.5.2 Qualitative Analysis

Qualitative analysis intends to identify substances on the basis of their Raman characteristics, and it usually involves extracting Raman signatures of pure compositions and classifying the compositions by comparing the extracted spectra with the reference spectra. SMA is an effective method to extract pure Raman signals from different compositions in a mixture. SMA uses an alternating least squares approach with added constraints to decompose a data matrix into the outer product of pure component spectra (or factors) and contributions (or scores). It is a useful tool to resolve a mixture of compounds without knowing the prior spectral information of the individual components (Windig and Guilment, 1991). Performing SMA requires the expected number of pure components to be predefined. For a mixture consisting of an unknown number of compositions, it is desirable to overestimate the number, and then inspect the resolved spectra to determine the appropriate number of pure components. SMA is usually conducted to a set of spectra with different spectral contributions from different compositions. For hyperspectral image data, the hypercube needs to be unfolded in the spatial domain so that each single-band image becomes a vector. The three-dimensional image data is consequently transformed to a two-dimensional (2D) matrix, on which SMA can be performed in the same manner as that used for regular spectral data. After SMA, each score vector for the selected pure components is folded back to form a 2D contribution image with same dimensions of the single-band image.

After pure component spectra are extracted from the mixture analysis, spectral matching algorithms are usually used to perform statistical comparisons between the resolved spectra and the reference spectra previously saved in a spectral library. Various spectral similarity metrics have been developed for target detection and spectral classification, such as spectral angle mapper (SAM), spectral correlation mapper (SCM), Euclidean distance (ED), and spectral information divergence (SID) (Chang, 2000). SAM, SCM, ED, and SID calculate angle, correlation, distance, and divergence between two spectra, respectively. The smaller the values of these metrics, the smaller the differences between two spectra. Whole spectra are usually used to calculate the similarity metrics mentioned above. Since Raman spectra are typically featured by a series of sharp peaks, unique Raman peaks of potential compositions can also be used for the purpose of identification at selected Raman shift positions (Qin et al., 2014c). The identification methods based on the whole spectra are usually used for off-line data analysis, and the results are more accurate than those from using several selected Raman peaks. However, the classification algorithms based on a few unique Raman peaks are generally faster and simpler than those developed using whole spectra, which makes them suitable for fast screening and real-time applications. Caution should be given in selecting the unique peaks since close peaks among the potential targets can cause misclassifications

Figure 15.11 shows an example of an SMA for a mixture of four chemical adulterants (i.e., ammonium sulfate, dicyandiamide, melamine, and urea, 1/4 for each by weight) previously found in milk powder (Qin et al., 2013). The imaging system shown in Figure 15.5a scanned a 25×25 mm^2 area of the mixture placed in a Petri dish with a diameter of 47 mm, resulting in a $100 \times 100 \times 1024$ hypercube (1024 bands). All 10,000 spectra are plotted in Figure 15.11a, in which Raman peaks of individual chemicals can be observed. The first four pure component spectra extracted from SMA using eight components are shown in Figure 15.11b. Identification of each resolved spectrum was based on its SID values with respect to the reference Raman spectra of the chemicals. For example, the SID values between the first extracted spectrum and the reference spectra of ammonium sulfate, dicyandiamide, melamine, and urea were 0.17, 0.99, 1.13, and 1.27, respectively. Thus, this spectrum was identified as ammonium sulfate since it had the least spectral difference with the reference of ammonium sulfate. When compared to the reference spectra, the extracted spectra retrieved almost all the spectral features (e.g., Raman peak positions and intensities) for each chemical. Besides eight components, three to seven components were also tried in SMA for the same data. The first four resolved spectra from four to seven components were similar to those from the eight components. However, when only three components were used, the spectra of the individual chemicals started mixing with each other, suggesting that a sufficiently large number of pure components are necessary for SMA to effectively identify all possible constituents in a mixture. The contribution maps of each identified chemical are shown in Figure 15.11c. The score values in the contribution images are proportional to the concentrations of each chemical. Thus, the pixels with high

Raman Scattering for Food Quality and Safety Assessment

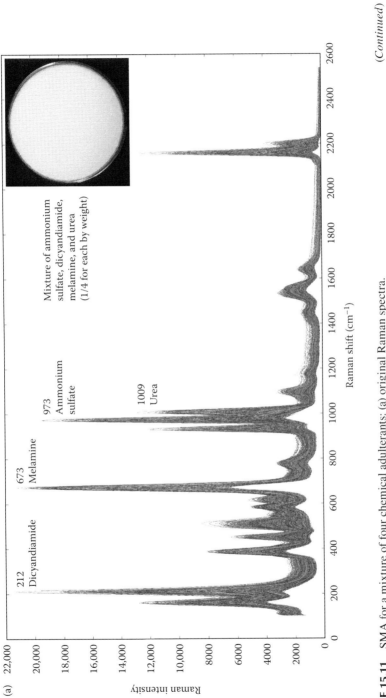

FIGURE 15.11 SMA for a mixture of four chemical adulterants: (a) original Raman spectra.

(Continued)

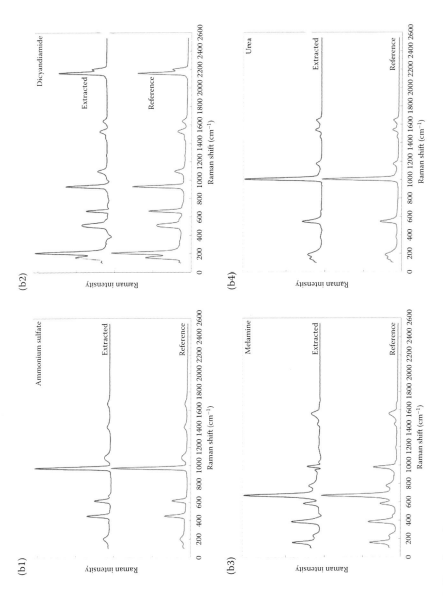

FIGURE 15.11 (*Continued*) SMA for a mixture of four chemical adulterants: (b) extracted pure component spectra. (*Continued*)

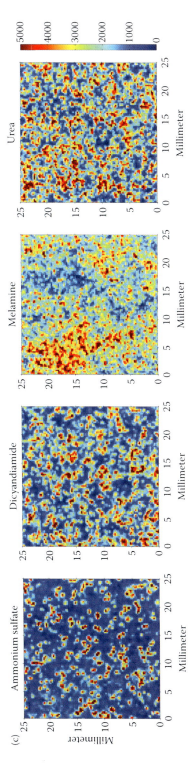

FIGURE 15.11 (*Continued*) SMA for a mixture of four chemical adulterants: (c) contribution maps. (Reprinted from *Food Chemistry*, 138, no. 2–3, Qin, Jianwei, Kuanglin Chao, and Moon S. Kim., Simultaneous detection of multiple adulterants in dry milk using macro-scale Raman chemical imaging, 998–1007, Copyright 2013, with permission from Elsevier.)

intensities likely represent the chemical particles in each map. The above results demonstrated that SMA is capable of identifying and locating the individual compositions in a mixture based on their unique Raman characteristics.

15.5.3 QUANTITATIVE ANALYSIS

Quantitative analysis aims to determine the amount or concentration of an analyte based on Raman spectral information. Raman scattering intensity is proportional to the number of molecules being sampled, and it can be related to the analyte concentration using an equation similar to the Lambert–Beer law for light absorption measurement (Sun, 2008). This proportional relationship between the scattering intensity and the analyte concentration forms a base for quantitative Raman analysis. For the analytes dissolved in a liquid, the principle of linear superposition can be used for quantitative analysis since analytes are evenly distributed in the solution (Pelletier, 2003). The Raman spectrum of the solution is the weighted sum of Raman spectra of all compositions in the liquid mixture. The weights of individual compositions are associated with their concentrations in the sample. Univariate or multivariate regression models (e.g., multivariate linear regression and partial least squares) can be established to quantify the concentrations of the analytes. On the other hand, linear superposition is usually not applicable for a mixture of solid or powder samples because the analytes are not evenly distributed in the mixture. It is not practical for the point measurement of conventional Raman spectroscopic technique to cover the spatial variation of the analytes. RCI is generally needed for quantitative analysis of the compositions in the solid or powder mixtures. Chemical images can be used to visualize sample composition, spatial distribution, and morphological features for targets of interest. The number of pixels for the targets of interest can be used to estimate the concentrations of the analytes.

An example of using Raman chemical images for quantitative analysis of powder mixtures is demonstrated in Figure 15.12. Ammonium sulfate, dicyandiamide, melamine, and urea were mixed into skim milk powder as chemical adulterants in six concentrations ranging from 0.1% to 5.0%. Raman images were acquired from a 25×25 mm^2 area of each milk-plus-four-adulterant mixture using the imaging system illustrated in Figure 15.5a. Chemical images in Figure 15.12a were created by combining the binary images of individual adulterants (see Figure 15.5d) for three equal parts of each concentration (Qin et al., 2014c). Identification and distribution of the multiple adulterants are clearly displayed in the background of the milk powder. The number of the chemical pixels generally increases with the increasing concentration of the adulterants. To investigate the relationship between the pixel number and the adulterant concentration, a linear regression analysis was applied to the total number of the adulterant pixels at each concentration and the concentration of each adulterant. Highly linear relationships were observed in the results shown in Figure 15.12b. The correlation coefficients (r) between the adulterant concentrations and the pixel numbers of ammonium sulfate, dicyandiamide, melamine, and urea were 0.994, 0.995, 0.994, and 0.996, respectively. The high correlations suggest great potential for using chemical images for quantitative assessment of adulterant concentrations in milk powder.

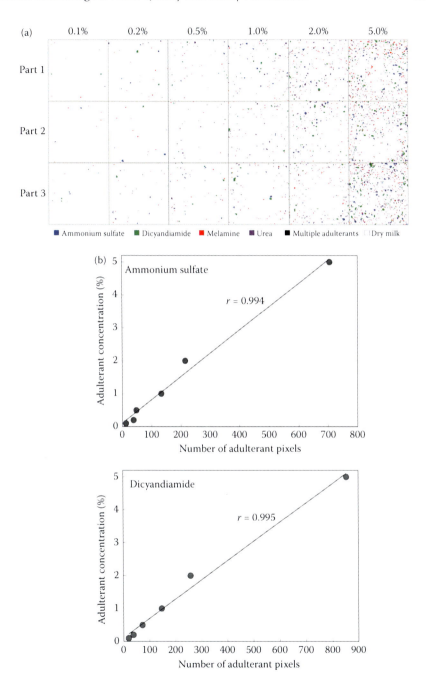

FIGURE 15.12 Raman chemical images for quantitative analysis of powder mixtures: (a) chemical images of four adulterants mixed in skim milk powder at six concentrations, and (b) correlations between number of adulterant pixels and concentration of each adulterant in the milk powder. (*Continued*)

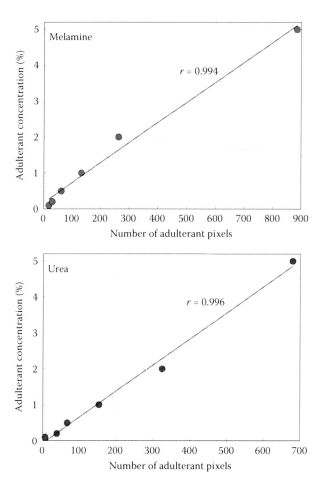

FIGURE 15.12 (*Continued*) Raman chemical images for quantitative analysis of powder mixtures: (b) correlations between number of adulterant pixels and concentration of each adulterant in the milk powder. (With permission from Springer, *Journal of Food Measurement and Characterization*, Development of a Raman chemical imaging detection method for authenticating skim milk powder, 8, no. 2, 2014c, 122–31, Qin, Jianwei, Kuanglin Chao, Moon S. Kim, Hoyoung Lee, and Yankun Peng.)

15.6 APPLICATIONS FOR FOOD QUALITY AND SAFETY

The Raman scattering technique is suitable for measuring solid, aqueous, and gaseous samples, and it can detect subtle chemical and biological changes based on peak positions and scattering intensities in Raman spectral and image data. The Raman technique has many advantages for analyzing chemical and biological materials, such as high specificity, nondestructive measurement, insensitivity to water, little sample preparation, detection capability through glass or polymer packages, and complement to infrared spectroscopy technique. Raman spectroscopy and imaging techniques have been used in food quality and safety evaluation (Yang and Ying,

2011). Food and agricultural products are complex systems, which can be considered as mixtures of different types of molecules from a chemistry point of view. Raman fingerprint information can be used to investigate molecules of interest in a complex food matrix. Analytes that are suitable for Raman analysis include the intrinsic major (e.g., proteins, fats, and carbohydrates) and minor components (e.g., carotenoids, fatty acid, and inorganics) as well as extrinsic components (e.g., bacteria and adulterants) of food (Li-Chan, 1996). During the past decades, the Raman scattering technique has been widely investigated for analyzing physical, chemical, and biological properties of a broad range of food and agricultural products. Various applications have been reported for analyzing solid and liquid foods, such as fruits and vegetables, meat, grains, food powders, beverages, oils, and foodborne pathogenic bacteria.

Table 15.1 summarizes representative Raman spectroscopy and imaging applications in quality and safety evaluation of food and agricultural products. As shown in the table, backscattering Raman spectroscopy is the principal measurement mode used in various applications for inspecting the external features of different types of products, such as the quality evaluation of olives, differentiation of closely related meat (e.g., beef and horse meat, and chicken and turkey), detection of deoxynivalenol in barley, nutritional analysis of milk powder, quantification of glucose in sports drinks, authentication of edible oils, and discrimination of *Campylobacter* species. When it comes to the evaluation of internal attributes, transmission Raman spectroscopy and SORS are usually used to provide overall content information and layered internal information, respectively. Example applications include composition analysis of single soybeans and corn kernels using transmission Raman spectroscopy, and nondestructive evaluation of internal maturity of tomatoes and quality analysis of salmon through the skin using SORS.

Figure 15.13 shows an example of evaluating the internal maturity of tomatoes using SORS. The Raman system illustrated in Figure 15.8c was used to collect spatially offset spectra from 160 tomatoes at seven ripeness stages (Figure 15.13a) using a source–detector distance ranging from 0 to 5 mm with a step size of 0.2 mm (Qin et al., 2012). Representative pure component spectra from self-modeling mixture analyses for the SORS data are plotted in Figure 15.13b. Three Raman peaks due to carotenoids inside the tomatoes started showing at the mature green stage. Two peaks appeared consistently at 1001 and 1151 cm^{-1}, and the third peak was gradually shifted from 1525 cm^{-1} (lutein at mature green stage) to 1513 cm^{-1} (lycopene at red stage) owing to the loss of lutein and β-carotene and the accumulation of lycopene during tomato ripening. Raman peak changes were quantified by SID with pure lycopene as the reference. The SID values steadily decreased from the immature green tomatoes to the red tomatoes (Figure 15.13c), demonstrating the declining trend of the spectral differences with respect to the pure lycopene. The distribution of the SID values can be used to evaluate the internal maturity of the tomatoes. For example, it is not easy to differentiate the maturity status of the tomatoes at immature green, mature green, and breaker stages using the exterior appearance because their surface color is generally all green. Appropriate threshold values can be applied to separate these three groups based on their SID values, and such information can be used to determine the proper harvest time for tomatoes.

TABLE 15.1
Representative Raman Scattering Applications for Food Quality and Safety

Class	Product	Application	Measurement Technique	Laser (nm)	References
Fruits and vegetables	Apiaceae vegetables	Polyacetylenes analysis	Raman chemical imaging	1064	Roman et al. (2011)
	Apple	Chlorpyrifos detection	Backscattering Raman spectroscopy	785	Dhakal et al. (2014)
	Apricot	Amygdalin distribution in seeds	Raman chemical imaging	785	Krafft et al. (2012)
	Banana	Thiabendazole detection	Surface-enhanced Raman spectroscopy	532, 1064	Müller et al. (2014)
	Citrus	Fruit classification	Backscattering Raman spectroscopy	514	Feng et al. (2013)
	Carrot	Carotenoid and polyacetylene evaluation	Backscattering Raman spectroscopy	785, 830, 1064	Killeen et al. (2013)
	Fruits	Pesticide residues detection	Surface-enhanced Raman spectroscopy	532	Liu et al. (2012)
	Mango	Epicuticular wax characterization	Backscattering Raman spectroscopy	514	Prinsloo et al. (2004)
	Olive	Quality evaluation	Backscattering Raman spectroscopy	1064	Muik et al. (2004)
	Tomato	Lycopene and β-carotene evaluation	Backscattering Raman spectroscopy	1064	Baranska et al. (2006)
	Tomato	Quality evaluation	Backscattering Raman spectroscopy	785	Nikbakht et al. (2011)
	Tomato	Lycopene generation pattern	Raman chemical imaging	785	Qin et al. (2011)
	Tomato	Internal maturity evaluation	Spatially offset Raman spectroscopy	785	Qin et al. (2012)
	Vegetables	Methamidophos detection	Surface-enhanced Raman spectroscopy	785	Xie et al. (2012)
Meats	Beef	Sensory quality evaluation	Backscattering Raman spectroscopy	785	Beattie et al. (2004)
	Chicken	Enrofloxacin detection	Surface-enhanced Raman spectroscopy	785	Xu et al. (2014)
	Fish	Quality evaluation	Backscattering Raman spectroscopy	785	Marquardt and Wold (2004)
	Horse meat	Beef and horse meat differentiation	Backscattering Raman spectroscopy	671	Ebrahim et al. (2013)
	Lamb	Shear force and cooking loss evaluation	Backscattering Raman spectroscopy	671	Schmidt et al. (2013)
	Meat	Meat species differentiation	Shifted excitation Raman difference spectroscopy	671, 783	Sowoidnich and Kronfeldt (2012)

(Continued)

TABLE 15.1 (Continued)
Representative Raman Scattering Applications for Food Quality and Safety

Class	Product	Application	Measurement Technique	Laser (nm)	References
	Pork	Sensory quality evaluation	Backscattering Raman spectroscopy	780	Wang et al. (2012)
	Salmon	Fatty acid and carotenoid evaluation	Spatially offset Raman spectroscopy	830	Afseth et al. (2014)
	Shrimp	White spot analysis	Backscattering Raman spectroscopy	1064	Careche et al. (2002)
	Turkey	Chicken and turkey differentiation	Backscattering Raman spectroscopy	785	Ellis et al. (2005)
Grains	Barley	Deoxynivalenol detection	Backscattering Raman spectroscopy	1064	Liu et al. (2009b)
	Corn	Protein evaluation	Transmission Raman spectroscopy	785	Shin et al. (2012)
	Oat	Globulin conformation	Backscattering Raman spectroscopy	1064	Ma et al. (2000)
	Rice	Geographical origin differentiation	Transmission and backscattering Raman spectroscopy	785	Hwang et al. (2012)
	Soybean	Protein and oil evaluation	Transmission Raman spectroscopy	785	Schulmerich et al. (2012)
	Wheat	Protein distribution	Raman chemical imaging	633	Piot et al. (2000)
Powders	Chili powder	Sudan dye detection	Backscattering Raman spectroscopy	1064	Haughey et al. (2015)
	Milk powder	Adulterants detection	Backscattering Raman spectroscopy	785	Qin et al. (2013)
	Milk powder	Melamine detection	Backscattering Raman spectroscopy	785	Okazaki et al. (2009)
	Milk powder	Nutritional analysis	Backscattering Raman spectroscopy	1064	Moros et al. (2007)
	Starch	Amylose quantification	Backscattering Raman spectroscopy	1064	Almeida et al. (2010)
	Wheat flour	Azodicarbonamide detection	Raman chemical imaging	785	Qin et al. (2014b)
	Wheat gluten	Melamine detection	Surface-enhanced Raman spectroscopy	785	Lin et al. (2008)
Beverages	Apple juice	Yeast detection	Backscattering Raman spectroscopy	785	Mizrach et al. (2007)
	Coffee	Coffee species differentiation	Backscattering Raman spectroscopy	1064	Rubayiza and Meurens (2005)
	Liquor	Ethyl carbamate detection	Surface-enhanced Raman spectroscopy	633	Yang et al. (2013)
	Sport drink	Glucose quantification	Backscattering Raman spectroscopy	633	Delfino et al. (2011)
	Tea	Methidathion detection	Surface-enhanced Raman spectroscopy	785	Yao et al. (2013)
	Tomato juice	Quality evaluation	Surface-enhanced Raman spectroscopy	532	Malekfar et al. (2010)

(Continued)

TABLE 15.1 (Continued)
Representative Raman Scattering Applications for Food Quality and Safety

Class	Product	Application	Measurement Technique	Laser (nm)	References
Oils	Citrus oil	Composition analysis	Backscattering Raman spectroscopy	1064	Schulz et al. (2002)
	Edible oil	Oil species differentiation	Backscattering Raman spectroscopy	633	Yang et al. (2005)
	Edible oil	Unsaturation degree evaluation	Coherent anti-Stokes Raman spectroscopy	800	Kim et al. (2014)
	Hazelnut oil	Hazelnut and olive oil differentiation	Backscattering Raman spectroscopy	780	López-Díez et al. (2003)
	Olive oil	Adulterants detection	Backscattering Raman spectroscopy	1064	Baeten et al. (1996)
	Vegetable oil	Lipid oxidation monitoring	Backscattering Raman spectroscopy	1064	Muik et al. (2005)
Bacteria	Bacteria	Foodborne bacteria detection	Backscattering Raman spectroscopy	1064	Yang and Irudayaraj (2003)
	Bacteria	Foodborne bacteria detection	Surface-enhanced Raman spectroscopy	785	Chu et al. (2008)
	Campylobacter	*Campylobacter* detection	Backscattering Raman spectroscopy	514, 532	Lu et al. (2012)
	Escherichia coli	*Escherichia coli* detection	Surface-enhanced Raman spectroscopy	785	Temur et al. (2010)
	Listeria	*Listeria* detection	Backscattering Raman spectroscopy	785	Oust et al. (2006)
	Salmonella	*Salmonella* detection	Surface-enhanced Raman spectroscopy	785	Assaf et al. (2014)
Miscellaneous	Butter	Margarine and butter differentiation	Backscattering Raman spectroscopy	785	Uysal et al. (2013)
	Chocolate	Composition analysis	Raman chemical imaging	532, 785	Larmour et al. (2010)
	Fat	Fat differentiation	Backscattering Raman spectroscopy	1064	Abbas et al. (2009)
	Honey	Floral origin differentiation	Backscattering Raman spectroscopy	780	Goodacre et al. (2002)
	Syrup	Adulterants detection	Backscattering Raman spectroscopy	1064	Paradkar et al. (2002)

Raman Scattering for Food Quality and Safety Assessment

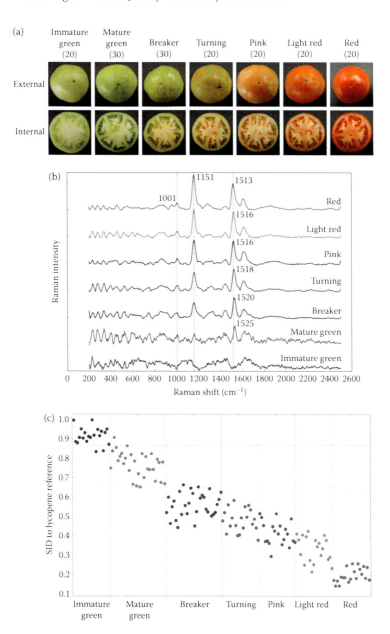

FIGURE 15.13 SORS for nondestructive evaluation of internal maturity of tomatoes: (a) tomato samples at different ripeness stages (sample numbers tested at each stage are marked in parentheses), (b) representative pure component spectra from self-modeling mixture analyses for the SORS data acquired in a source–detector range of 0–5 mm, and (c) SID values between pure component spectra of all 160 tested tomatoes and reference Raman spectrum of lycopene. (Reprinted from *Postharvest Biology and Technology*, 71, Qin, Jianwei, Kuanglin Chao, and Moon S. Kim., Nondestructive evaluation of internal maturity of tomatoes using spatially offset Raman spectroscopy, 21–31, Copyright 2012, with permission from Elsevier.)

SERS is primarily used in the area of food safety inspection. The high sensitivity of the SERS technique has enabled it to be used for trace detection of analytes in a variety of food and agricultural products, such as detections of pesticide residues on fruit peels, enrofloxacin in chicken muscles, melamine in wheat gluten, and ethyl carbamate in alcoholic beverages. The SERS technique has also been used for detecting and differentiating various foodborne pathogenic bacteria, such as *Campylobacter*, *Escherichia coli*, *Listeria*, and *Salmonella*. When spatial information is important for food quality and safety applications, RCI is generally used to visualize sample composition, spatial distribution, and morphological features of targets of interest using both Raman and spatial information. The RCI technique eliminates the size limitation of point measurements by the traditional Raman spectroscopy technique, and has begun to find applications for evaluating food and agricultural products, such as analyzing polyacetylenes in Apiaceae vegetables, detecting lycopene changes from tomato cross-sections during ripening, investigating protein distribution in wheat kernels, screening multiple adulterants mixed into milk powder, and evaluating compositions in chocolate.

It can also be found from Table 15.1 that near-infrared lasers with typical output wavelengths of 785 and 1064 nm are the main sources used to excite food samples to minimize fluorescence interference. Both integrated commercial Raman systems and custom-designed Raman systems are used to meet different measurement requirements. Dispersive Raman spectrometers with 785-nm diode lasers and FT-Raman spectrometers with 1064-nm Nd:YAG lasers are two major tools for Raman applications in the field of food quality and safety evaluation, while visible lasers (e.g., 514, 532, 633, and 671 nm) are also utilized when fluorescence signals are relatively weak in some particular circumstances (e.g., liquid food samples and SERS applications).

15.7 CONCLUSIONS

This chapter has presented Raman scattering technologies in the quality and safety evaluation of food and agricultural products. The chapter put an emphasis on the introduction and demonstration of Raman spectroscopy and imaging techniques for practical uses in food analysis, including principles of Raman scattering, Raman measurement techniques, Raman instruments, and Raman data analysis techniques. The common practices and applications for food quality and safety evaluation were also reviewed to demonstrate the current status of Raman scattering techniques. Driven by growing interests from both academia and industry, Raman scattering technologies have advanced rapidly during the past decades. Novel Raman measurement techniques are continuously emerging to create new detection possibilities that cannot be achieved by existing methods. Improved and new hardware components are constantly being introduced to build high-performance Raman spectroscopy and imaging systems. The fast-growing computing capacity of computers will meet the challenges of handling large datasets and processing Raman spectra and images for rapid and online applications. The advances in measurement techniques, Raman instruments, and data analysis techniques will boost the further development of Raman scattering technologies and broaden their applications for evaluating food quality and safety in the future.

REFERENCES

Abbas, Ouissam, Juan A. F. Pierna, Rafael Codony, Christoph von Holst, and Vincent Baeten. Assessment of the discrimination of animal fat by FT-Raman spectroscopy. *Journal of Molecular Structure* 924–926, 2009: 294–300.

Adami, Renata, and Johannes Kiefer. Light-emitting diode based shifted-excitation Raman difference spectroscopy (LED-SERDS). *Analyst* 138, no. 21, 2013: 6258–61.

Afseth, Nils Kristian, Matthew Bloomfield, Jens Petter Wold, and Pavel Matousek. A novel approach for subsurface through-skin analysis of salmon using spatially offset Raman spectroscopy (SORS). *Applied Spectroscopy* 68, no. 2, 2014: 255–62.

Almeida, Mariana R., Rafael S. Alves, Laura B. L. R. Nascimbem, Rodrigo Stephani, Ronei J. Poppi, and Luiz Fernando C. de Oliveira. Determination of amylose content in starch using Raman spectroscopy and multivariate calibration analysis. *Analytical and Bioanalytical Chemistry* 397, no. 7, 2010: 2693–701.

Assaf, Ali, Christophe B. Y. Cordella, and Gerald Thouand. Raman spectroscopy applied to the horizontal methods ISO 6579:2002 to identify *Salmonella* spp. in the food industry. *Analytical and Bioanalytical Chemistry* 406, no. 20, 2014: 4899–910.

ASTM Standards. *E1840-96: Standard Guide for Raman Shift Standards for Spectrometer Calibration.* West Conshohocken, Pennsylvania: ASTM, 2007.

Baeten, Vincent, Marc Meurens, Maria T. Morales, and Ramon Aparicio. Detection of virgin olive oil adulteration by Fourier transform Raman spectroscopy. *Journal of Agricultural and Food Chemistry* 44, no. 8, 1996: 2225–30.

Baranska, Malgorzata, Wolfgang Schütz, and Hartwig Schulz. Determination of lycopene and beta-carotene content in tomato fruits and related products: Comparison of FT-Raman, ATR-IR, and NIR spectroscopy. *Analytical Chemistry* 78, no. 24, 2006: 8456–61.

Beattie, Rene J., Steven J. Bell, Linda J. Farmer, Bruce W. Moss, and Desmond Patterson. Preliminary investigation of the application of Raman spectroscopy to the prediction of the sensory quality of beef silverside. *Meat Science* 66, no. 4, 2004: 903–13.

Careche, Mercedes, Ana Herrero, and Pedro Carmona. Raman analysis of white spots appearing in the shell of argentine red shrimp (*Pleoticus muelleri*) during frozen storage. *Journal of Food Science* 67, no. 8, 2002: 2892–5.

Carron, Keith, and Rick Cox. Qualitative analysis and the answer box: A perspective on portable Raman spectroscopy. *Analytical Chemistry* 82, no. 9, 2010: 3419–25.

Chang, Chein-I. An information theoretic-based approach to spectral variability, similarity and discriminability for hyperspectral image analysis. *IEEE Transactions on Information Theory* 46, no. 5, 2000: 1927–32.

Christensen, Kenneth A., and Michael D. Morris. Hyperspectral Raman microscopic imaging using Powell lens line illumination. *Applied Spectroscopy* 52, no. 9, 1998: 1145–7.

Chu, Hsiaoyun, Yaowen Huang, and Yiping Zhao. Silver nanorod arrays as a surface-enhanced Raman scattering substrate for foodborne pathogenic bacteria detection. *Applied Spectroscopy* 62, no. 8, 2008: 922–31.

Cooper, John B., Mohamed Abdelkader, and Kent L. Wise. Sequentially shifted excitation Raman spectroscopy: Novel algorithm and instrumentation for fluorescence-free Raman spectroscopy in spectral space. *Applied Spectroscopy* 67, no. 8, 2013: 973–84.

Cormack, Iain G., Michael Mazilu, Kishan Dholakia, and C. Simon Herrington. Fluorescence suppression within Raman spectroscopy using annular beam excitation. *Applied Physics Letters* 91, no. 2, 2007: 023903.

Craig, Ana Paula, Adriana S. Franca, and Joseph Irudayaraj. Surface-enhanced Raman spectroscopy applied to food safety. *Annual Review of Food Science and Technology* 4, 2013: 369–80.

Delfino, Ines, Carlo Camerlingo, Marianna Portaccio, Bartolomeo Della Ventura, Luigi Mita, Damiano G. Mita, and Maria Lepore. Visible micro-Raman spectroscopy for determining glucose content in beverage industry. *Food Chemistry* 127, no. 2, 2011: 735–42.

Dhakal, Sagar, Yongyu Li, Yankun Peng, Kuanglin Chao, Jianwei Qin, and Langhua Guo. Prototype instrument development for non-destructive detection of pesticide residue in apple surface using Raman technology. *Journal of Food Engineering* 123, 2014: 94–103.

Ebrahim, Halah A., Kay Sowoidnich, and Heinz-Detlef Kronfeldt. Raman spectroscopic differentiation of beef and horse meat using a 671 nm microsystem diode laser. *Applied Physics B: Lasers and Optics* 113, no. 2, 2013: 159–63.

Efremov, Evtim V., Freek Ariese, and Cees Gooijer. Achievements in resonance Raman spectroscopy review of a technique with a distinct analytical chemistry potential. *Analytica Chimica Acta* 606, no. 2, 2008: 119–34.

Eliasson, Charlotte, and Pavel Matousek. Noninvasive authentication of pharmaceutical products through packaging using spatially offset Raman spectroscopy. *Analytical Chemistry* 79, no. 4, 2007: 1696–701.

Ellis, David I., David Broadhurst, Sarah J. Clarke, and Royston Goodacre. Rapid identification of closely related muscle foods by vibrational spectroscopy and machine learning. *Analyst* 130, no. 12, 2005: 1648–54.

Evans, Conor L., Eric O. Potma, Mehron Puoris'haag, Daniel Côté, Charles P. Lin, and X. Sunney Xie. Chemical imaging of tissue *in vivo* with video-rate coherent anti-Stokes Raman scattering microscopy. *PNAS* 102, no. 46, 2005: 16807–12.

Fan, Meikun, Gustavo F. S. Andrade, and Alexandre G. Brolo. A review on the fabrication of substrates for surface enhanced Raman spectroscopy and their applications in analytical chemistry. *Analytica Chimica Acta* 693, no. 1–2, 2011: 7–25.

Feng, Xinwei, Qinghua Zhang, and Zhongliang Zhu. Rapid classification of citrus fruits based on Raman spectroscopy and pattern recognition techniques. *Food Science and Technology Research* 19, no. 6, 2013: 1077–84.

Goodacre, Royston, Branka S. Radovic, and Elke Anklam. Progress toward the rapid non-destructive assessment of the floral origin of European honey using dispersive Raman spectroscopy. *Applied Spectroscopy* 56, no. 4, 2002: 521–7.

Haughey, Simon A., Pamela Galvin-King, Yen-Cheng Ho, Steven E. J. Bell, and Christopher T. Elliott. The feasibility of using near infrared and Raman spectroscopic techniques to detect fraudulent adulteration of chili powders with Sudan dye. *Food Control* 48, 2015: 75–83.

Hwang, Jinyoung, Sukwon Kang, Kangjin Lee, and Hoeil Chung. Enhanced Raman spectroscopic discrimination of the geographical origins of rice samples via transmission spectral collection through packed grains. *Talanta* 101, 2012: 488–94.

Killeen, Daniel P., Catherine E. Sansom, Ross E. Lill, Jocelyn R. Eason, Keith C. Gordon, and Nigel B. Perry. Quantitative Raman spectroscopy for the analysis of carrot bioactives. *Journal of Agricultural and Food Chemistry* 61, no. 11, 2013: 2701–8.

Kim, Jinsun, Jang H. Lee, and Do-Kyeong Ko. Determination of degree of unsaturation in edible oils using coherent anti-Stokes Raman scattering spectroscopy. *Journal of Raman Spectroscopy* 45, no. 7, 2014: 591–5.

Krafft, Christoph, Claudia Cervellati, Christian Paetz, Bernd Schneider, and Jűrgen Popp. Distribution of amygdalin in apricot (*Prunus armeniaca*) seeds studied by Raman microscopic imaging. *Applied Spectroscopy* 66, no. 6, 2012: 644–9.

Larmour, Iain A., Karen Faulds, and Dunca Graham. Rapid Raman mapping for chocolate analysis. *Analytical Methods* 2, no. 9, 2010: 1230–2.

Le Ru, Eric C., Evan Blackie, Matthias Meyer, and Pablo G. Etchegoin. Surface enhanced Raman scattering enhancement factors: A comprehensive study. *Journal of Physical Chemistry C* 111, no. 37, 2007: 13794–803.

Le Ru, Eric C., Lina C. Schroeter, and Pablo G. Etchegoin. Direct measurement of resonance Raman spectra and cross sections by a polarization difference technique. *Analytical Chemistry* 84, no. 11, 2012: 5074–9.

Li-Chan, Eunice C. Y. The applications of Raman spectroscopy in food science. *Trends in Food Science & Technology* 7, no. 11, 1996: 361–70.

Lieber, Chad A. and Anita Mahadevan-Jansen. Automated method for subtraction of fluorescence from biological Raman spectra. *Applied Spectroscopy* 57, no. 11, 2003: 1363–7.

Lin, Mengshi, Lili He, Joseph Awika et al. Detection of melamine in gluten, chicken feed, and processed foods using surface enhanced Raman spectroscopy and HPLC. *Journal of Food Science* 73, no. 8, 2008: T129–34.

Liu, Bianhua, Guangmei Han, Zhongping Zhang et al. Shell thickness-dependent Raman enhancement for rapid identification and detection of pesticide residues at fruit peels. *Analytical Chemistry* 84, no. 1, 2012: 255–61.

Liu, Yongliang, Kuanglin Chao, Moon S. Kim, David Tuschel, Oksana Olkhovyk, and Ryan J. Priore. Potential of Raman spectroscopy and imaging methods for rapid and routine screening of the presence of melamine in animal feed and foods. *Applied Spectroscopy* 63, no. 4, 2009a: 477–80.

Liu, Yongliang, Stephen R. Delwiche, and Yanhong Dong. Feasibility of FT-Raman spectroscopy for rapid screening for DON toxin in ground wheat and barley. *Food Additives and Contaminants* 26, no. 10, 2009b: 1396–401.

López-Díez, E. Consuelo, Giorgio Bianchi, and Royston Goodacre. Rapid quantitative assessment of the adulteration of virgin olive oils with hazelnut oils using Raman spectroscopy and chemometrics. *Journal of Agricultural and Food Chemistry* 51, no. 21, 2003: 6145–50.

Lu, Xiaonan, Qian Huang, William G. Miller et al. Comprehensive detection and discrimination of *Campylobacter* species by use of confocal micro-Raman spectroscopy and multilocus sequence typing. *Journal of Clinical Microbiology* 50, no. 9, 2012: 2932–46.

Ma, Ching-Yung, Manoj K. Rout, Wing-Man Chan, and David L. Phillips. Raman spectroscopic study of oat globulin conformation. *Journal of Agricultural and Food Chemistry* 48, no. 5, 2000: 1542–7.

Malekfar, Rasoul, Ali M. Nikbakht, Sara Abbasian, Fatemeh Sadeghi, and M. Mozaffari. Evaluation of tomato juice quality using surface enhanced Raman spectroscopy. *Acta Physica Polonica A* 117, no. 6, 2010: 971–3.

Markwort, Lars, Bert Kip, Edouard Da Silva, and Bernord Roussel. Raman imaging of heterogeneous polymers: A comparison of global versus point illumination. *Applied Spectroscopy* 49, no. 10, 1995: 1411–30.

Marquardt, Brian J. and Jens Petter Wold. Raman analysis of fish: A potential method for rapid quality screening. *Lebensmittel-Wissenschaft & Technologie* 37, no. 1, 2004: 1–8.

Matousek, Pavel, Edward R. C. Draper, Allen E. Goodship, Ian P. Clark, Kate L. Ronayne, and Anthony W. Parker. Noninvasive Raman spectroscopy of human tissue *in vivo*. *Applied Spectroscopy* 60, no. 7, 2006: 758–63.

Matousek, Pavel, Ian P. Clark, Edward R. C. Draper et al. Subsurface probing in diffusely scattering media using spatially offset Raman spectroscopy. *Applied Spectroscopy* 59, no. 4, 2005: 393–400.

Matousek, Pavel, and Michael D. Morris. *Emerging Raman Applications and Techniques in Biomedical and Pharmaceutical Fields*. New York: Springer, 2010.

Matousek, Pavel, and Anthony W. Parker. Bulk Raman analysis of pharmaceutical tablets. *Applied Spectroscopy* 60, no. 12, 2006: 1353–7.

Matousek, Pavel, Michael Towrie, A. Stanley, and Anthony W. Parker. Efficient rejection of fluorescence from Raman spectra using picosecond Kerr gating. *Applied Spectroscopy* 53, no. 12, 1999: 1485–9.

McCreery, Richard L. *Raman Spectroscopy for Chemical Analysis*. New York: John Wiley and Sons, 2000.

Mizrach, Amos, Ze'ev Schmilovitch, Raya Korotic, Joseph Irudayaraj, and Roni Shapira. Yeast detection in apple juice using Raman spectroscopy and chemometric methods. *Transactions of the ASABE* 50, no. 6, 2007: 2143–9.

Moros, Javier, Salvador Garrigues, and Miguel de la Guardia. Evaluation of nutritional parameters in infant formulas and powdered milk by Raman spectroscopy. *Analytica Chimica Acta* 593, no. 1, 2007: 30–8.

Morris, Hannah R., Clifford C. Hoyt, and Patrick J. Treado. Imaging spectrometers for fluorescence and Raman microscopy: Acousto-optic and liquid-crystal tunable filters. *Applied Spectroscopy* 48, no. 7, 1994: 857–66.

Morris, Hannah R., Clifford C. Hoyt, and Patrick J. Treado. Liquid crystal tunable filter Raman chemical imaging. *Applied Spectroscopy* 50, no. 6, 1996: 805–11.

Muik, Barbara, Bernhard Lendl, Antonio Molina-Diaz, and Maria J. Ayora-Canada. Direct monitoring of lipid oxidation in edible oils by Fourier transform Raman spectroscopy. *Chemistry and Physics of Lipids* 134, no. 2, 2005: 173–82.

Muik, Barbara, Bernhard Lendl, Antonio Molina-Diaz, Domingo Ortega-Calderon, and Maria J. Ayora-Canada. Discrimination of olives according to fruit quality using Fourier transform Raman spectroscopy and pattern recognition techniques. *Journal of Agricultural and Food Chemistry* 52, no. 20, 2004: 6055–60.

Mukhopadhyay, Rajendrani. Raman flexes its muscles. *Analytical Chemistry* 79, no. 9, 2007: 3265–70.

Müller, Csilla, Leontin David, Vasile Chis, and Simona C. Pinzaru. Detection of thiabendazole applied on citrus fruits and bananas using surface enhanced Raman scattering. *Food Chemistry* 145, 2014: 814–20.

Nikbakht, Ali M., Teymour T. Hashjin, Rasoul Malekfar, and Barat Gobadian. Nondestructive determination of tomato fruit quality parameters using Raman spectroscopy. *Journal of Agricultural Science and Technology* 13, no. 4, 2011: 517–26.

Okazaki, Shigetoshi, Mitsuo Hiramatsu, Kunio Gonmori, Osamu Suzuki, and Anthony T. Tu. Rapid nondestructive screening for melamine in dried milk by Raman spectroscopy. *Forensic Toxicology* 27, no. 2, 2009: 94–7.

Oust, Astrid, Trond Moretro, Kristine Naterstad et al. Fourier transform infrared and Raman spectroscopy for characterization of *Listeria* monocytogenes strains. *Applied and Environmental Microbiology* 72, no. 1, 2006: 228–32.

Paradkar, Manish M., Joseph Irudayaraj, and Sivakesava Sakhamuri. Discrimination and classification of beet and cane sugars and their inverts in maple syrup by FT-Raman. *Applied Engineering in Agriculture* 18, no. 3, 2002: 379–83.

Pelletier, Michael J. Quantitative analysis using Raman spectrometry. *Applied Spectroscopy* 57, no. 1, 2003: 20A–42A.

Piot, Olivier, Jean C. Autran, and Michel Manfait. Spatial distribution of protein and phenolic constituents in wheat grain as probed by confocal Raman microspectroscopy. *Journal of Cereal Science* 32, no. 1, 2000: 57–71.

Prinsloo, Linda C., Wilma du Plooy, and Chris van der Merwe. Raman spectroscopic study of the epicuticular wax layer of mature mango (*Mangifera indica*) fruit. *Journal of Raman Spectroscopy* 35, no. 7, 2004: 561–7.

Qin, Jianwei, Kuanglin Chao, and Moon S. Kim. Raman chemical imaging system for food safety and quality inspection. *Transactions of the ASABE* 53, no. 6, 2010: 1873–82.

Qin, Jianwei, Kuanglin Chao, and Moon S. Kim. Investigation of Raman chemical imaging for detection of lycopene changes in tomatoes during postharvest ripening. *Journal of Food Engineering* 107, no. 3–4, 2011: 277–88.

Qin, Jianwei, Kuanglin Chao, and Moon S. Kim. Nondestructive evaluation of internal maturity of tomatoes using spatially offset Raman spectroscopy. *Postharvest Biology and Technology* 71, 2012: 21–31.

Qin, Jianwei, Kuanglin Chao, and Moon S. Kim. Simultaneous detection of multiple adulterants in dry milk using macro-scale Raman chemical imaging. *Food Chemistry* 138, no. 2–3, 2013: 998–1007.

Qin, Jianwei, Kuanglin Chao, and Moon S. Kim. A line-scan hyperspectral system for high-throughput Raman chemical imaging. *Applied Spectroscopy* 68, no. 6, 2014a: 692–5.

Qin, Jianwei, Kuanglin Chao, Byound-Kwan Cho, Yankun Peng, and Moon S. Kim. High-throughput Raman chemical imaging for rapid evaluation of food safety and quality. *Transactions of the ASABE* 57, no. 6, 2014b: 1783–92.

Qin, Jianwei, Kuanglin Chao, Moon S. Kim, Hoyoung Lee, and Yankun Peng. Development of a Raman chemical imaging detection method for authenticating skim milk powder. *Journal of Food Measurement and Characterization* 8, no. 2, 2014c: 122–31.

Roman, Maciej, Rafal Baranski, and Malgorzata Baranska. Nondestructive Raman analysis of polyacetylenes in *Apiaceae* vegetables. *Journal of Agricultural and Food Chemistry* 59, no. 14, 2011: 7647–53.

Rubayiza, Aloys B., and Marc Meurens. Chemical discrimination of arabica and robusta coffees by Fourier transform Raman spectroscopy. *Journal of Agricultural and Food Chemistry* 53, no. 12, 2005: 4654–9.

Sakamoto, Akira, Shukichi Ochiai, Hisamitsu Higashiyama et al. Raman studies of Japanese art objects by a portable Raman spectrometer using liquid crystal tunable filters. *Journal of Raman Spectroscopy* 43, no. 6, 2012: 787–91.

Schlücker, Sebastian. *Surface Enhanced Raman Spectroscopy: Analytical, Biophysical and Life Science Applications.* Weinheim, Germany: Wiley-VCH, John Wiley and Sons, 2011.

Schmidt, Heinar, Rico Scheier, and David L. Hopkins. Preliminary investigation on the relationship of Raman spectra of sheep meat with shear force and cooking loss. *Meat Science* 93, no. 1, 2013: 138–43.

Schmidt, Walter F., Cathleen J. Hapeman, Laura L. McConnell et al. Temperature-dependent Raman spectroscopic evidence of and molecular mechanism for irreversible isomerization of β-endosulfan to α-endosulfan. *Journal of Agricultural and Food Chemistry* 62, no. 9, 2014: 2023–30.

Schulmerich, Matthew V., Michael J. Walsh, Matthew K. Gelber et al. Protein and oil composition predictions of single soybeans by transmission Raman spectroscopy. *Journal of Agricultural and Food Chemistry* 60, no. 33, 2012: 8097–102.

Schulz, Hartwig, Malgorzata Baranska, and Rafal Baranski. Potential of NIR-FT-Raman spectroscopy in natural carotenoid analysis. *Biopolymers* 77, no. 4, 2005: 212–21.

Schulz, Hartwig, Bernhard Schrader, Rolf Quilitzsch, and Boris Steuer. Quantitative analysis of various citrus oils by ATR/FT-IR and NIR-FT Raman spectroscopy. *Applied Spectroscopy* 56, no. 1, 2002: 117–24.

Schulze, Georg, Andrew Jirasek, Marcia M. L. Yu, Arnel Lim, Robin F. B. Turner, and Michael W. Blades. Investigation of selected baseline removal techniques as candidates for automated implementation. *Applied Spectroscopy* 59, no. 5, 2005: 545–74.

Sharma, Shiv K., Paul G. Lucey, Manash Ghosh, Hugh W. Hubble, and Keith A. Horton. Stand-off Raman spectroscopic detection of minerals on planetary surfaces. *Spectrochimica Acta Part A: Molecular and Biomolecular Spectroscopy* 59, no. 10, 2003: 2391–407.

Sharma, Bhavya, Ke Ma, Mattew R. Glucksberg, and Richard P. Van Duyne. Seeing through bone with surface-enhanced spatially offset Raman spectroscopy. *Journal of the American Chemical Society* 135, no. 46, 2013: 17290–3.

Shin, Kayeong, Hoeil Chung, and Chul-won Kwak. Transmission Raman measurement directly through packed corn kernels to improve sample representation and accuracy of compositional analysis. *Analyst* 137, no. 16, 2012: 3690–6.

Smith, Ewen, and Geoffrey Dent. *Modern Raman Spectroscopy—A Practical Approach.* Chichester, UK: John Wiley and Sons, 2005.

Sowoidnich, Kay, and Heinz-Detlef Kronfeldt. Shifted excitation Raman difference spectroscopy at multiple wavelengths for *in-situ* meat species differentiation. *Applied Physics B: Lasers and Optics* 108, no. 4, 2012: 975–82.

Stewart, Shona, Ryan J. Priore, Matthew P. Nelson, and Patrick J. Treado. Raman imaging. *Annual Review of Analytical Chemistry* 5, 2012: 337–60.

Stone, Nicholas, Marleen Kerssens, Garvin R. Lloyd, Karen Faulds, Duncan Graham, and Pavel Matousek. Surface enhanced spatially offset Raman spectroscopic (SESORS) imaging—The next dimension. *Chemical Science* 2, no. 4, 2011: 776–80.

Sun, Da-Wen. *Modern Techniques for Food Authentication*. San Diego, California: Academic Press, Elsevier, 2008.

Temur, Erhan, Ismail H. Boyaci, Ugur Tamer, Hande Unsal, and Nihal Aydogan. A highly sensitive detection platform based on surface-enhanced Raman scattering for *Escherichia coli* enumeration. *Analytical and Bioanalytical Chemistry* 397, no. 4, 2010: 1595–604.

Uysal, Reyhan S., Ismail H. Boyaci, Huseyin E. Genis, and Ugur Tamer. Determination of butter adulteration with margarine using Raman spectroscopy. *Food Chemistry* 141, no. 4, 2013: 4397–403.

Wang, Qi, Steven M. Lonergan, and Chenxu Yu. Rapid determination of pork sensory quality using Raman spectroscopy. *Meat Science* 91, no. 3, 2012: 232–9.

Windig, Willem, and Jean Guilment. Interactive self-modeling mixture analysis. *Analytical Chemistry* 63, no. 14, 1991: 1425–32.

Xie, Yunfei, Godelieve Mukamurezi, Yingying Sun, Heya Wang, He Qian, and Weirong Yao. Establishment of rapid detection method of methamidophos in vegetables by surface enhanced Raman spectroscopy. *European Food Research and Technology* 234, no. 6, 2012: 1091–8.

Xu, Ying, Yiping Du, Qingqing Li et al. Ultrasensitive detection of enrofloxacin in chicken muscles by surface-enhanced Raman spectroscopy using amino-modified glycidyl methacrylate–ethylene dimethacrylate (GMA–EDMA) powdered porous material. *Food Analytical Methods* 7, no. 6, 2014: 1219–28.

Yan, Fei, and Tuan Vo-Dinh. Surface-enhanced Raman scattering detection of chemical and biological agents using a portable Raman integrated tunable sensor. *Sensors and Actuators B* 121, no. 1, 2007: 61–6.

Yang, Danting, and Yibin Ying. Applications of Raman spectroscopy in agricultural products and food analysis: A review. *Applied Spectroscopy Reviews* 46, no. 7, 2011: 539–60.

Yang, Danting, Haibo Zhou, Yibin Ying, Reinhard Niessner, and Christoph Haisch. Surface-enhanced Raman scattering for quantitative detection of ethyl carbamate in alcoholic beverages. *Analytical and Bioanalytical Chemistry* 405, no. 29, 2013: 9419–25.

Yang, Hong, and Joseph Irudayaraj. Rapid detection of foodborne microorganisms on food surface using Fourier transform Raman spectroscopy. *Journal of Molecular Structure* 646, no. 1–3, 2003: 35–43.

Yang, Hong, Joseph Irudayaraj, and Manish M. Paradkar. Discriminant analysis of edible oils and fats by FTIR, FT-NIR and FT-Raman spectroscopy. *Food Chemistry* 93, no. 1, 2005: 25–32.

Yao, Chaoping, Fansheng Cheng, Cong Wang et al. Separation, identification and fast determination of organophosphate pesticide methidathion in tea leaves by thin layer chromatography-surface-enhanced Raman scattering. *Analytical Methods* 5, no. 20, 2013: 5560–4.

Zachhuber, Bernhard, Christoph Gasser, Engelene T. H. Chrysostom, and Bernhard Lendl. Stand-off spatial offset Raman spectroscopy for the detection of concealed content in distant objects. *Analytical Chemistry* 83, no. 24, 2011: 9438–42.

Zhang, Zhimin, Shan Chen, and Yizeng Liang. Baseline correction using adaptive iteratively reweighted penalized least squares. *Analyst* 135, no. 5, 2010: 1138–46.

16 Light Scattering–Based Detection of Food Pathogens

Pei-Shih Liang*, Tu San Park*, and Jeong-Yeol Yoon

CONTENTS

16.1 Introduction 429
16.2 Nephelometric and Turbidimetric Detection of Antibody–Antigen Complex 430
16.3 Particle-Enhanced Light Scattering Immunoassays.....................................431
16.4 Food Samples Tested with Light Scattering Immunoassays........................436
16.5 Pathogen Identification Using Light Scattering Refraction Pattern Imaging....438
16.6 Direct Mie Scattering Detection from Food Sample...................................440
16.7 Summary 442
References..443

16.1 INTRODUCTION

The current methods for detecting foodborne pathogens are mostly destructive (i.e., samples need to be pretreated), and require time, personnel, and laboratories for analyses. During the past decade, regulatory standards in food safety have been strengthened, requiring quality assurance at every step in food production (Craig et al., 2013). The increased frequency of food inspections emphasizes the need for field-deployable and portable detection devices, toward the ultimate goal of real-time, online monitoring methods. Obviously, current methods, including plate culturing and colony counting, immunoassay, and polymerase chain reaction (PCR), cannot meet these demands. Delayed identification of food pathogens may lead to larger-scale outbreaks and recalls of food products and therefore greater economic loss.

Optical methods including light scattering-based techniques have gained a lot of attention recently due to their rapid and nondestructive nature. While (surface-enhanced) Raman scattering is one of the light scattering-based techniques for food pathogens detection (Chapter 15), in this chapter, we are going to describe the methods based on Mie scattering. Specifically, latex particle immunoagglutination assay and subsequent Mie scattering detection (i.e., particle-enhanced light scattering immunoassay) for detecting foodborne pathogens will be presented as a more sensitive and portable method than direct scattering detections. In addition to that, refraction

---

* These two authors contributed equally.

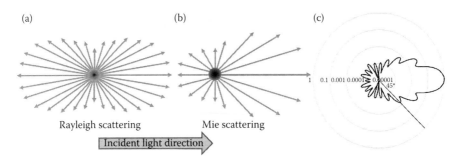

FIGURE 16.1 Schematic patterns of (a) dispersed Rayleigh scattering and (b) angle-dependent Mie scattering, and (c) result of Mie scattering calculation on a polar-logarithmic plot. (Part (c) image taken from Angus, S. V., Kwon, H.-J., Yoon, J.-Y. Field-deployable and near-real-time optical microfluidic biosensors for single-oocyst-level detection of *Cryptosporidium parvum* from field water samples. *Journal of Environmental Monitoring* 14, 2012: 3295–3304. Reproduced by permission of The Royal Society of Chemistry.)

pattern imaging and light scattering pattern analysis will also be reviewed as label-free methods to detect, quantify, and/or identify pathogens. Assays developed in portable or handheld format (especially smartphone based) will be introduced as well.

Mie scattering theory refers to the Mie solution to the scattering problem on a spherical object, and describes how much light is scattered and how the scatter intensities are changed according to the scattering angles (van de Hulst, 1983). Elastic scattering of light by molecules and particles can be categorized into two types, Rayleigh and Mie scattering, based on the ratio of the wavelength of incident light (λ) to the particle diameter (d). When $d \ll \lambda/10$, it is considered Rayleigh scattering, where the ratio of scattered light and incident light is a function of wavelength, scattering angle, refractive index of the particle, and the particle diameter (Figure 16.1a). As d increases ($d \approx$ or $> \lambda/10$), the scattered light intensity can then be predicted using the Mie scattering model, where the intensity is less dependent on wavelength, and more so on the other three parameters (i.e., particle size, refractive index, and scatter angle; Figure 16.1b).

Mie scattering intensities can be calculated using several free software programs (MiePlot by Philip Laven—http://www.philiplaven.com/mieplot.htm.; Mie Scattering Calculator by Scott Prahl—http://www.omlc.org/calc/mie_calc.html.). The size of particles, the wavelength of incident light, and the refractive indices of both particles and medium (1.0 if vacuum) need to be entered, and a plot of Mie scattering intensities can be generated as a function of the scattering angle (Figure 16.1c).

16.2 NEPHELOMETRIC AND TURBIDIMETRIC DETECTION OF ANTIBODY–ANTIGEN COMPLEX

Light scattering immunoassay is an antibody-utilized assay that exploits the change in scattered light (nephelometry) or blocked light (turbidimetry) from the immunological precipitation reaction (Figure 16.2a), to measure the amount of antigens (in this case, food pathogens) (Bangs Laboratories, 2008). In the past, antibodies were

Light Scattering–Based Detection of Food Pathogens

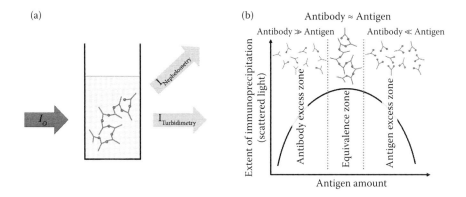

FIGURE 16.2 (a) Measurement of nephelometry and turbidimetry from light scattering immunoassay. (b) Extent of immunoprecipitation as measured by light scattering: Heidelberger–Kendall curve.

mixed with a specimen that might contain antigens; antibody–antigen complex would form and start to precipitate, and the precipitation was quantified by light scattering (nephelometry or turbidimetry). The ratio of antigen to antibody amount determines the degree of precipitation. As demonstrated by the Heidelberger–Kendall curve, the extent of precipitation is maximized when the antigen and antibody amounts are near equivalent to each other, whereas a ratio that is too high or too low causes the degree of precipitation to decrease (Figure 16.2b).

Specifically, turbidimetry is the measurement of the decrease in intensity of the incident beam after it passes through a solution, while nephelometry is the measurement of the light collected at an angle (often 30° or 90°) from the incident beam after being scattered by the solution. Although their detection principles are somewhat different, both turbidimetric and nephelometric detections show similar trends in quantifying antibody–antigen binding (thus immunoprecipitation), and target concentrations in samples can be determined. These assays have long been used for various types of antigen detection (and subsequently antibody detection as well), especially for clinical applications (Gosling, 1990; Spencer and Prince, 1980; van Munster et al., 1977).

Unfortunately, the sensitivity of this "direct" quantification was not satisfactory (detection limit is ~10^7 cells/mL, and may be as low as 10^6 cells/mL) (Stevens, 2010). The use of micrometer polymer particles (latex particles) has been suggested to amplify the light scattering intensities, known as particle-enhanced light scattering immunoassay (Blume and Greenberg, 1975).

16.3 PARTICLE-ENHANCED LIGHT SCATTERING IMMUNOASSAYS

One way to amplify the scattering signals from the antibody–antigen complex is to use micron-sized polymer particles with high refractive index (typically carboxylated polystyrene particles or sometimes referred to as carboxylated latex particles). Antibodies are chemically attached to these polymer particles (utilizing the carboxyl

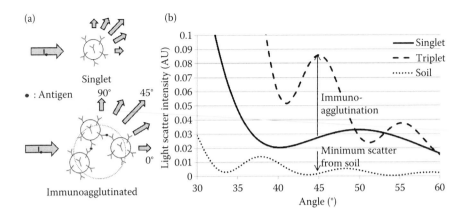

FIGURE 16.3 (a) Polystyrene particles are conjugated with antibodies (Y-shaped) to a target pathogen; they are agglutinated together via antibody–antigen binding, which alters the particle morphology and effective diameter, leading to the change (in this case an increase measured at 45°) in Mie scattering intensity. (b) Results of Mie scattering simulations for singlet polystyrene particles (920 nm, nonimmunoagglutinated), triplet polystyrene particles (immunoagglutinated), and silica particles (5 μm; soil particle contaminants in the sample matrix), showing at 45° the biggest change in Mie scattering intensity and the minimum background scattering from silica particles. (Part (b) image taken from Angus, S. V., Kwon, H.-J., Yoon, J.-Y. Field-deployable and near-real-time optical microfluidic biosensors for single-oocyst-level detection of *Cryptosporidium parvum* from field water samples. *Journal of Environmental Monitoring* 14, 2012: 3295–3304. Reproduced by permission of The Royal Society of Chemistry.)

groups on the particles). Antibody–antigen binding ("immuno") makes the particles to become "glued" to each other, or immunoagglutinated (Figure 16.3a). This type of assay is referred to as particle-enhanced light scattering immunoassay or latex particle immunoagglutination assay. Since these particles are much bigger (1–10 μm) than antibody or antigen molecules (a few tens of nanometers), the resulting light scattering should be substantially stronger. In addition, since the refractive index of the particles is higher than those of the antibodies and antigen, a bigger change in scattered or block light is expected. Recently, submicron polystyrene particles have shown further improved performance (with a detection limit of 10 cells/mL) in detecting pathogens (Fronczek et al., 2013; You et al., 2011). This particle-enhanced light scattering immunoassay follows the Mie scattering model, since the particle sizes ($d = 0.1–1$ μm in the case of submicron particles) are comparable to the wavelengths of visible light (400–750 nm). As seen in Figure 16.3b, the scattering intensity changes nonmonotonously against the angle and generates many different local maxima of scattering intensity.

There are several parameters of particles that contribute to the performance of the particle-enhanced light scattering assay: material choice, size, type, and density of surface functional groups, and storage conditions. The material choice of particles associates with its refractive index, which leads to different light scattering behaviors and the different optimum angle of Mie scattering detection. Polystyrene particles

Light Scattering–Based Detection of Food Pathogens

are popularly used due to their higher refractive index ($n = 1.59$ under typical visible wavelength) compared to that of water ($n = 1.33$) and sample matrix (typically proteins, with $n = 1.38 \sim 1.40$) (You et al., 2011). Other particles with much higher refractive index may be considered, such as titania particles ($n = 2.37$ under red light). However, the density of the particles has to be considered, since the particles must be suspended in medium (water) and should not precipitate without target presence or during the antibody conjugation procedure. The density of polystyrene is 1.05 g/cm^3 (close to that of water), while that of titania is approximately 4 g/cm^3. The size of the particle is also important in Mie scattering characteristics. Heinze and Yoon (2011) investigated the effect of particle size in a particle-enhanced light scattering assay of *Escherichia coli* (Figure 16.4). Three different sizes of particles were used: 10, 100, and 920 nm in diameter. Particle assays of 10 nm displayed the best linearity but the smallest change in scattering intensity (Figure 16.4a), whereas 100 nm particle assays showed better change in scattering intensity and lowest detection limit

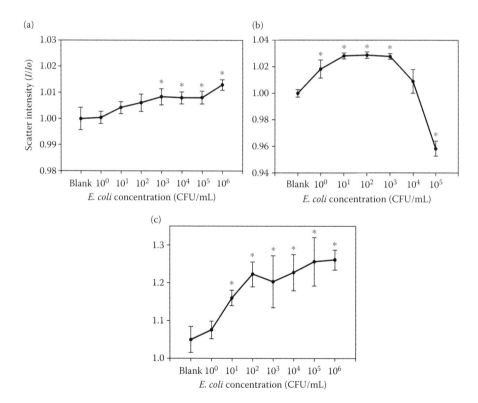

FIGURE 16.4 Standard curves produced through light scattering intensity measurements for the *E. coli* and 10 nm (a), 100 nm (b), and 920 nm (c) anti-*E. coli* conjugated particles, taken at 45° from incident light. *indicates significant difference from blank ($p < 0.05$). (Reprinted from *Colloids and Surfaces B: Biointerfaces*, 85, Heinze, B. C., Yoon, J.-Y., Nanoparticle immunoagglutination Rayleigh scatter assay to complement microparticle immunoagglutination Mie scatter assay in a microfluidic device, 168–173, Copyright 2011, with permission from Elsevier.)

(1 CFU/mL) but not linear (bell shape) (Figure 16.4b). However, 920 nm particle assays yielded a greater change in scattering intensity but with a nonmonotonic relationship (Figure 16.4c).

Considering the refractive index and size of the particle, polystyrene particles with the diameter of 700–1000 nm have commonly been used. Centrifuging/washing and microscopic observation (necessary for antibody conjugation) was also convenient with this size of polystyrene particles. Carboxylated polystyrene particles are commonly used to allow covalent coupling with the amino groups of antibodies. This covalent coupling also allows the long-term stability of the antibody-conjugated particles, up to 4–8 weeks in liquid suspension at 4°C or as lyophilized powder at room temperature (Fronczek et al., 2013; Kwon et al., 2010).

The particle-enhanced light scattering assay is essentially a one-step process and relatively easy to be automated, but with some potential danger of generating false-positive readings, caused by the nonspecific aggregation of less stable particles. This assay has then been developed in many different platforms, including two- and multiwell slides, polydimethylsiloxane (PDMS)-based microfluidics, and paper microfluidics. Light scattering detection can be made with a spectrometer setup (with optical fibers or optical waveguide channels on microfluidics) or a smartphone setup (plastic attachment and software application).

Earlier studies used two- or multiwell slide as a platform for these assays, where the sample solution and the liquid suspension of antibody-conjugated particles were either premixed and loaded on each well, or loaded directly into the well while mixing occurred purely by diffusion. In both platforms, the incident light comes in from one side of the slide (often the bottom side) and a detector (often an optical fiber and a spectrometer) oriented at an angle (determined based on the parameters mentioned before) is set on the other side. The two wells are used side by side to compare samples and the negative control whereas, the multiwell slide is later used to (1) reduce the reagent amount (smaller wells than in two-well slide) and (2) process multiple samples at once, especially helpful when generating a standard curve of serially diluted target concentrations (Figure 16.5a). Although these platforms are easy to build and operate, it is still difficult to achieve necessary reproducibility; typically, the error bars in a standard curve are quite substantial due to the different extent of diffusional mixing and antibody–antigen binding. More recent studies have incorporated microfluidic device (also known as lab-on-a-chip) as an improved platform in order to further improve the assay's reproducibility, sensitivity, and detection limit (Figure 16.5b).

Microfluidic devices are a network of channels and wells etched onto polymer, silicon, or glass substrates, and designed to conduct chemical and biological experiments on a miniature scale. Sample and reagent solutions are driven by pressure or electrokinetic forces in a highly controlled manner (Yoon and Kim, 2012). This enhanced microfluidic control provides added benefits to particle-enhanced light scattering assay, as the extents of diffusional mixing and antibody–antigen binding can be precisely controlled under a strict laminar flow condition, leading to a high level of reproducibility, higher signal-to-noise ratio, and lower detection limit. The reduced volume also helps lowering the cost of reagents as well as the reaction time (faster mixing) (Figure 16.6).

Light Scattering–Based Detection of Food Pathogens 435

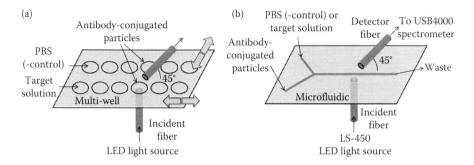

FIGURE 16.5 Forward 45° light scattering detection of particle immunoagglutination assays in a multiwell slide (a) and a y-channel microfluidic device (b) with "proximity" fiber optic setup. (Reprinted from *Journal of Virological Methods*, 178, Song, J.-Y. et al., Sensitive Mie scattering immunoagglutination assay of porcine reproductive and respiratory syndrome virus (PRRSV) from lung tissue samples in a microfluidic chip, 31–38, Copyright 2011, with permission from Elsevier.)

Another popular platform of microfluidic devices is paper microfluidics, on which the particle-enhanced light scattering assay was also demonstrated. Paper microfluidics have recently gained significant popularity due to its simplified and potentially low-cost fabrication (Martinez et al., 2008). Paper (cellulose) fibers can also function as a filter for various sample matrices (e.g., soil particles, plant and animal tissue debris, bacterial colonies, fecal matter, food materials, etc.). However, careful optimization is necessary to selectively detect Mie scattering from the antibody-conjugated particles while minimizing background scattering from the paper (cellulose) fibers (Park and Yoon, 2015). Park et al. (2013) have developed a multichannel

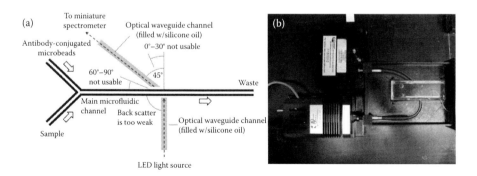

FIGURE 16.6 (a) Schematic of microfluidic chip with optical waveguide channel. (b) The experimental apparatus showing the light source (lower left), microfluidic device and tray (right), and the miniature spectrometer (top left). (Angus, S. V., Kwon, H.-J., Yoon, J.-Y. Field-deployable and near-real-time optical microfluidic biosensors for single-oocyst-level detection of *Cryptosporidium parvum* from field water samples. *Journal of Environmental Monitoring* 14, 2012: 3295–3304. Reproduced by permission of The Royal Society of Chemistry.)

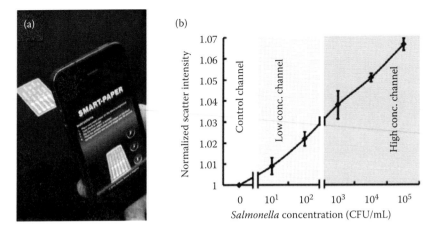

FIGURE 16.7 (a) Smartphone application for *Salmonella* detection from a multichannel microfluidic device. (b) A standard curve was constructed from the readout of three separate channels using the smartphone application under ambient light. (Park, T. S. et al. Smartphone quantifies *Salmonella* from paper microfluidics. *Lab on a Chip* 13, 2013: 4832–4840. Reproduced by permission of The Royal Society of Chemistry.)

paper microfluidic chip to quantify *Salmonella* utilizing a smartphone as an optical detector. The smartphone application prompted the user to position the smartphone at an optimized angle and distance from the multichannel paper microfluidics, such that optimum scattering detection can be made. The detection of the limit was comparable to the previous studies utilizing a spectrometer and PDMS microfluidics (Figure 16.7).

16.4 FOOD SAMPLES TESTED WITH LIGHT SCATTERING IMMUNOASSAYS

In the past decade, a substantial number of works have been published in detecting pathogens from various food samples, especially with particle-enhanced light scattering immunoassay in microfluidic platform. In these microfluidic platforms, the sample needs to be prepared in liquid form. Figure 16.8 shows some examples of how such liquid form samples can be prepared from fresh produce samples (potentially contaminated with pathogens such as *E. coli*). In Figure 16.8a, mortar and pestle are used to grind the iceberg lettuce sample and an appropriate solution (deionized and/or distilled water or phosphate buffered saline [PBS]) is added. Bigger food debris will be settled passively and thus removed, or a filter can additionally be used, resulting in relatively homogeneous solutions (Figure 16.8b). Alternatively, iceberg lettuce can be placed within a tube together with washing buffer, and hand shaken to isolate pathogens from the surface of the iceberg lettuce (Figure 16.8c). Using both methods, You et al. (2011) demonstrated 10 cells/mL detection of *E. coli* from iceberg lettuce with particle-enhanced light scattering immunoassay on a PDMS microfluidic platform.

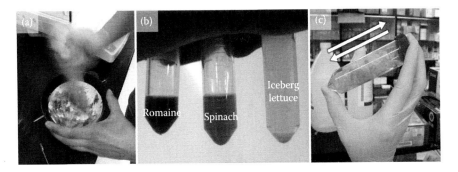

FIGURE 16.8 (a) Iceberg lettuce being ground using mortar and pestle. (b) Ground fresh produce samples added to PBS, which are ready to be used for light scattering immunoassays without further dilution. (c) Iceberg lettuce being washed with PBS in a tube.

Fecal samples and field water samples can also be tested with light scattering immunoassays. Figure 16.9a shows a 1% (w/v) suspension of chicken feces in PBS, where the same particle-enhanced light scattering immunoassays were performed on a microfluidic device with detection limit of 1 pg/mL avian influenza antigens (Heinze et al., 2010). Figure 16.9b shows several different field water samples (undiluted), where the same assays were performed for *E. coli* with a detection limit of 10 cells/mL (Park and Yoon, 2015).

Meat samples can also be tested with light scattering immunoassays. Although grinding or washing methods (Figure 16.8) can also be used for meat samples, there is another way to obtain liquid samples from meat. Packaged meat products typically generate some liquid over time, and this naturally occurring liquid can be used (typically diluted to 10% before the assays) in light scattering immunoassays (Figure 16.10). Again, using the same method mentioned above, 10 cells/mL detection limit was demonstrated for *Salmonella* from poultry packages (Fronczek et al., 2013).

FIGURE 16.9 (a) 1% w/v chicken feces in PBS. (b) Various field water samples containing algae, dust, and soil particles. (Part A image taken with kind permission from Springer Science + Business Media: *Analytical and Bioanalytical Chemistry*, Microfluidic immunosensor with integrated liquid core waveguides for sensitive Mie scattering detection of avian influenza antigens in a real biological matrix, 398, 2010, 2693–2700, Heinze, B. C. et al.)

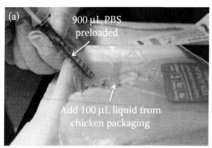
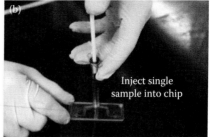

FIGURE 16.10 (a) A disposable syringe, prefilled with 900 mL PBS, taking 100 mL of liquid from a poultry package, thus diluting it to 10%. (b) The resulting dilution is applied directly, without any preprocessing, into the shared inlet of a microfluidic chip. (Reprinted from *Biosensors and Bioelectronics*, 40, Fronczek, C. F., You, D. J., Yoon, J.-Y., Single-pipetting microfluidic assay device for rapid detection of *Salmonella* from poultry package, 342–349, Copyright 2013, with permission from Elsevier.)

To use the above food samples for light scattering immunoassays, the optical parameters must be optimized prior to the assays in a way that the absorption and scattering of food samples would not undermine the light scattering signals from the antibody–antigen reaction. Since most food samples exhibit some coloration, wavelength(s) can be optimized to minimize absorption and scattering (collectively termed extinction) from food samples and maximize the light scattering signals from the antibody–antigen complex and/or antibody-conjugated particles. In addition, the angle of scattering detection can also be optimized to minimize the scattering from food samples and maximize the scattering from the antibody–antigen complex and/or antibody-conjugated particles. These optimizations have been successfully demonstrated in Fronczek et al. (2013) and You et al. (2011).

16.5 PATHOGEN IDENTIFICATION USING LIGHT SCATTERING REFRACTION PATTERN IMAGING

The traditionally approved gold standard in bacteria detection is culturing the sample on a culture plate. This conventional culture method requires dilution of the sample followed by distribution of the sample to an appropriate agar plate, where bacterial cells grow to form a colony. However, further tests such as metabolic or genetic fingerprinting, immunoassays, or PCR assays are necessary in order to identify the bacteria species. As an alternative, a label-free detection and identification of bacterial species has been demonstrated utilizing light scattering refraction images of bacterial colonies on a culture plate. This method is based on recognizing the refraction patterns from different bacteria colonies grown on a culture plate. Each bacterium produces colonies of different size, shape, and composition depending on its metabolic and genetic makeup. The micro- and mesoscopic properties of the colonies contribute to producing unique, characteristic forward

Light Scattering–Based Detection of Food Pathogens

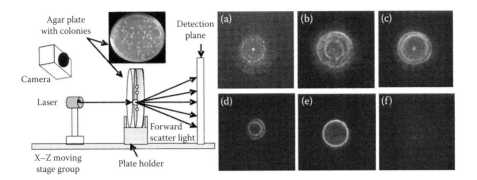

FIGURE 16.11 Left: Schematic representation of the laser scatterometer. Right: Scattering images of representative *Listeria* species. (a) *Listeria monocytogenes* ATCC19113. (b) *Listeria ivanovii* ATCC19119. (c) *Listeria innocua* F4248. (d) *Listeria seeligeri* LA-15. (e) *Listeria welshimeri* ATCC35897. (f) *Listeria grayi* LM37. (Reprinted from *Biosensors and Bioelectronics*, 22, Banada, P. P. et al., Optical forward-scattering for detection of *Listeria monocytogenes* and other *Listeria* species, 1664–1671, Copyright 2007, with permission from Elsevier.)

scattering patterns. In addition, the extracellular materials produced by different bacterial culture, along with cellular arrangements, may provide unique signatures that can differentiate the light scattering patterns (Figure 16.11) (Banada et al., 2007; Hirleman et al., 2008).

Banada et al. (2009) built a scattering image database for various bacterial culture plates. All bacteria samples were collected from real food materials (spinach, ground beef, hotdog, chicken, etc.) and were plated on brain heart infusion (BHI) or other agar plates. They demonstrated very high specificity of up to 100% and detection limit of 1 cell per 25 g sample. The only downside of this approach is that it takes 12–30 h, depending on the bacteria species, to have certain size of colonies (thus not a rapid method) (Figure 16.12). In order to reduce the culture time, Marcoux et al. (2014) developed a microscopic light scattering refraction image acquisition system to collect and analyze images of bacteria grown within microcolony size (cultured 6 h in *E. coli* case). However, this is still not fast and cannot be made portable.

There is another recent demonstration of combining microfluidic devices with refraction pattern imaging to detect and quantify *E. coli* (Yu et al., 2014). The study utilized microdroplets (diameter = 60 µm), a 632.8 nm laser as incident light, and two charge-coupled device (CCD) cameras for imaging. It was shown in this study that the degree of multiple reflection and refraction in the droplet was higher as *E. coli* concentration increased and the analysis of scattering patterns could detect *E. coli* at the single cell level, although due to the small size of the droplet, the equivalent detection limit remained rather high (10^7 cells/mL). This study showed a different execution of the refraction pattern imaging technique and may be further investigated toward sensitive detection of food pathogens.

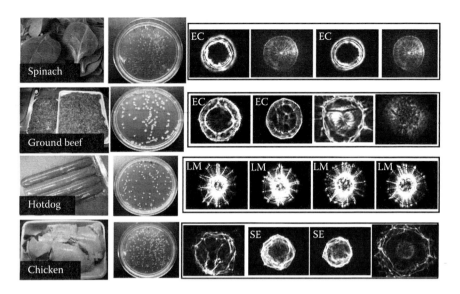

FIGURE 16.12 Bacterial rapid detection using optical scattering technology analysis of bacterial colonies from food specimens. Each food sample was spiked before analysis; spinach and ground beef with *E. coli* O157:H7 (EC) EDL933, hotdog with *L. monocytogenes* (LM) F4244, and chicken with *Salmonella enteritidis* (SE) PT1, enriched in the respective selective enrichment broth, and surface plated on BHI or selective media agar plates. Scattering images of representative colonies are presented; images marked with initials are identified from the database, while remainders are considered unknown background microflora. (Reprinted from *Biosensors and Bioelectronics*, 24, Banada, P. P. et al., Label-free detection of multiple bacterial pathogens using light-scattering sensor, 1685–1692, Copyright 2009, with permission from Elsevier.)

16.6 DIRECT MIE SCATTERING DETECTION FROM FOOD SAMPLE

While many of the Mie scattering detection of pathogens have been conducted using antibody-conjugated particles, it is still possible to directly detect Mie scattering caused by pathogens. For example, *E. coli* has a refractive index of 1.388 that is higher than that of water (1.33); therefore when in higher concentration (10^5 cells/mL and more), it is possible to measure the Mie scattering intensity caused from *E. coli* colonies.

However, this direct measurement is difficult at a lower concentration due to the lack of morphological features and the small difference in refractive indices, therefore it is not as useful in terms of rapid detection of food pathogens (i.e., detection limits are often high). The scattering patterns from the pathogen colonies, on the other hand, are found effective in identifying the strains of the pathogens (Hirleman et al., 2008), although colonies must be grown fully on culture plates (therefore not rapid). This aspect of pathogen detection where we can differentiate the strains of food pathogens without labeling has been discussed in detail earlier in Section 16.4.

There is one other attempt in detecting pathogens from the surface of food samples: Liang et al. (2014) demonstrated a detection method that utilizes the hydrophobicity of *E. coli* toward animal fat in ground beef. Many foodborne pathogens, including *E. coli*, are hydrophobic on their membrane surfaces; thus they tend to attach preferentially to fat surfaces (also hydrophobic). The cell fragments and proteins from *E. coli* interact with and aggregate around the fats within the ground beef sample, to form pseudo-colonies.

The morphology of *E. coli*—fat complex (pseudo-colonies) changes as the *E. coli* concentration increases: the pseudo-colony size increases, and eventually there are intact cells and real colonies of *E. coli* at high concentrations. The pseudo-colonies formed here have a slightly higher refractive index (around 1.40 from animal fat cells) than water and with more complex morphology. Since the sizes and ratios of the pseudo- and real colonies change with the concentrations, the optimum angle for the Mie scattering changes as well. In such demonstration, the detection was made by scanning at four different angles (15°, 30°, 45°, and 60°) from the incident light from an 880 nm light-emitting diode (LED) (Figure 16.13a). By finding the optimum angle among the four (one with the highest scattering intensity), one can then determine the range of *E. coli* concentration on a ground beef sample (10–10^2, 10^3–10^6, or 10^8 cells/mL) (Figure 16.13b). The detection was also integrated into a smartphone application. Through simple interface design, the user can take pictures of a ground beef sample from the four angles at a fixed distance, the images will be processed and the range of the *E. coli* concentration can be determined (Figure 16.14).

This method eliminates the use of antibody, and thus sacrifices the specificity (i.e., not able to differentiate between strains). Also, there is no standard curve for determining the exact concentration of the pathogen; however, the method is done

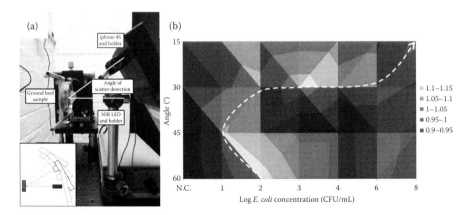

FIGURE 16.13 (a) The benchtop system consists of an iPhone 4S and its holder, a near infrared light-emitting diode (NIR LED) and its holder, and a ground beef sample and its holder. The angle of scatter detection refers to the angle between the iPhone camera and the NIR LED light source. (b) Surface plots that combine the normalized light intensities, the angles of scattering detection, and the log concentrations of *E. coli*. (From Liang, P.-S., Park, T.S., Yoon, J.-Y., *Sci. Rep.*, 4, 5953, 2014. With permission.)

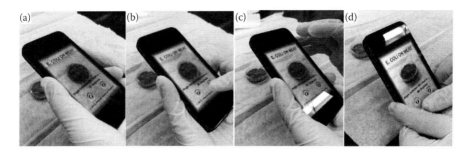

FIGURE 16.14 Photographs showing the operation of the smartphone application at the four specific angles of scattering detection: (a) 15°, (b) 30°, (c) 45°, and (d) 60°. (From Liang, P.-S., Park, T. S., Yoon, J.-Y., *Sci. Rep.*, 4, 5953, 2014. With permission.)

completely handheld and rapidly especially with the smartphone application, and can serve as a great preliminary screening tool for monitoring the contamination of meat products and help protect public food safety.

16.7 SUMMARY

Particle-enhanced light scattering immunoassay, which evolved from turbidimetric detection of antibody–antigen complex, has developed into various sensitive, rapid, and portable detection methods for food pathogens. With the help of microfluidic devices, both sample volumes and reaction time were reduced. With smartphone technologies, the assay no longer needs a spectrometer. Several examples showed minimum to no sample pretreatment and very low limits of detection. However, there are some elements that require careful control, including a fair amount of preparation of stable antibody-conjugated particles, the determination of the optimum angle for measuring scattered light from both the particles and the sample matrix. In addition to these, simultaneous multiple pathogen detection (i.e., multiplex detection) can complicate the method and future work needs to be done in order to better detect multiple foodborne pathogens in a more efficient matter.

Refraction image analysis of a bacteria colony has been proven to identify species and strains of bacteria grown on culture media plates with high specificity regardless of sample matrix. The time required for the bacteria colony to grow to a certain extent to obtain proper images has been somewhat long (typically overnight culture), but is getting shorter as the imaging systems improved. With recent advances in computer programs and accumulating database of image pattern recognition, this method can become a very promising alternative to identify unknown pathogens.

The direct Mie scattering detection method provided a new era of bacteria detection on food samples, which is nondestructive, needs no sample pretreatment, and is quick. Simply taking images of a sample from various angles provided the degree of bacterial contamination. This reagentless method sacrifices specificity for portability and rapidness and thus is much more suitable as a preliminary screening tool.

REFERENCES

Angus, S. V., Kwon, H.-J., Yoon, J.-Y. Field-deployable and near-real-time optical microfluidic biosensors for single-oocyst-level detection of *Cryptosporidium parvum* from field water samples. *Journal of Environmental Monitoring* 14, 2012: 3295–3304.

Banada, P. P., Guo, S., Bayraktar, B., Bae, E., Rajwa, B., Robinson, J. P., Hirleman, E. D., Bhunia, A. K. Optical forward-scattering for detection of *Listeria monocytogenes* and other *Listeria* species. *Biosensors and Bioelectronics* 22, 2007: 1664–1671.

Banada, P. P., Huff, K., Bae, E., Rajwa, B., Aroonnual, A., Bayraktar, B., Adil, A., Robinson, J. P., Hirleman, E. D., Bhunia, A. K. Label-free detection of multiple bacterial pathogens using light-scattering sensor. *Biosensors and Bioelectronics* 24, 2009: 1685–1692.

Bangs Laboratories. Tech Note #304: Light Scattering Assays. Fishers, Indiana: Bangs Laboratories, 2008. Available at http://www.bangslabs.com/technotes/304.pdf. (accessed December 16, 2014).

Blume, P., Greenberg, L. J. Application of differential light scattering to the latex agglutination assay for rheumatoid factor. *Clinical Chemistry* 21, 1975: 1234–1237.

Craig, A. P., Franca, A. S., Irudayaraj, J. Surface-enhanced Raman spectroscopy applied to food safety. *Annual Review of Food Science and Technology* 4, 2013: 369–380.

Fronczek, C. F., You, D. J., Yoon, J.-Y. Single-pipetting microfluidic assay device for rapid detection of *Salmonella* from poultry package. *Biosensors and Bioelectronics* 40, 2013: 342–349.

Gosling, J. P. A decade of development in immunoassay methodology. *Clinical Chemistry* 36, 1990: 1408–1427.

Heinze, B. C., Gamboa, J. R., Kim, K., Song, J.-Y., Yoon, J.-Y. Microfluidic immunosensor with integrated liquid core waveguides for sensitive Mie scattering detection of avian influenza antigens in a real biological matrix. *Analytical and Bioanalytical Chemistry* 398, 2010: 2693–2700.

Heinze, B. C., Yoon, J.-Y. Nanoparticle immunoagglutination Rayleigh scatter assay to complement microparticle immunoagglutination Mie scatter assay in a microfluidic device. *Colloids and Surfaces B: Biointerfaces* 85, 2011: 168–173.

Hirleman, E. D., Guo, S., Bhunia, A. K., Bae, E. System and method for rapid detection and characterization of bacterial colonies using forward light scattering. US Patent No. 7465560 B2, 2008.

Kwon, H.-J., Dean, Z. S., Angus, S. V., Yoon, J.-Y. Lab-on-a-chip for field *Escherichia coli* assays: Long-term stability of reagents and automatic sampling system. *JALA—Journal of Laboratory Automation* 15, 2010: 216–223.

Liang, P.-S., Park, T. S., Yoon, J.-Y. Rapid and reagentless detection of microbial contamination within meat utilizing a smartphone-based biosensor. *Scientific Reports* 4, 2014: 5953.

Marcoux, P. R., Dupoy, M., Cuer, A., Kodja, J.-L., Lefebvre, A., Licari, F., Louvet, R., Narassiguin, A., Mallard, F. Optical forward-scattering for identification of bacteria within microcolonies. *Applied Microbiology and Biotechnology* 98, 2014: 2243–2254.

Martinez, A. W., Phillips, S. T., Wiley, B. J., Gupta, M., Whitesides, G. M. FLASH: A rapid method for prototyping paper-based microfluidic devices. *Lab on a Chip* 8, 2008: 2146–2150.

Park, T. S., Li, W., McCracken, K. E., Yoon, J.-Y. Smartphone quantifies *Salmonella* from paper microfluidics. *Lab on a Chip* 13, 2013: 4832–4840.

Park, T. S., Yoon, J.-Y. Smartphone detection of *Escherichia coli* from field water samples on paper microfluidics. *IEEE Sensors Journal* 15, 2015: 1902–1907.

Song, J.-Y., Lee, C.-H., Choi, E.-J., Kim, K., Yoon, J.-Y. Sensitive Mie scattering immunoagglutination assay of porcine reproductive and respiratory syndrome virus (PRRSV) from lung tissue samples in a microfluidic chip. *Journal of Virological Methods* 178, 2011: 31–38.

Spencer, K., Prince, C. P. Kinetic immunoturbidimetric measurement of thyroxine binding globulin. *Clinical Chemistry* 26, 1980: 1531–1536.

Stevens, C. D. *Clinical Immunology and Serology: A Laboratory Perspective*, 3rd edn, Philadelphia, Pennsylvania: F. A. Davis Company, 2010.

Van de Hulst, H. C. Rigorous scattering theory for spheres of arbitrary size. In *Light Scattering by Small Particle*. Chapter 9, Mineola, New York: John Wiley and Sons, 1983.

Van Munster, P. J. J., Hoelen, G. E. J. M., Samwel-Mantingh, M., Holtman-van Meurs, M. A turbidimetric immune assay (TIA) with automated individual blank compensation. *Clinica Chimica Acta* 76, 1977: 377–388.

Yoon, J.-Y., Kim, B. Lab-on-a-chip pathogen sensor for food safety. *Sensors* 12, 2012: 10713–10741.

You, D. J., Geshell, K. J., Yoon, J.-Y. Direct and sensitive detection of foodborne pathogens within fresh produce samples using a field-deployable handheld device. *Biosensors and Bioelectronics* 28, 2011: 399–406.

Yu, J. Q., Huang, W., Chin, L. K., Lei, L., Lin, Z. P., Ser, W., Chen, H. et al. Droplet optofluidic imaging for λ-bacteriophage detection via co-culture with host cell *Escherichia coli*. *Lab on a Chip* 14, 2014: 3519–3524.

Index

A

Absolute value of differences (AVD), 368
Absorbance (Abs), 33
Absorption, 15, 29, 133, 227–233
    bands, 30
    coefficient, 31–33
    coefficient, 80, 82, 141, 191, 255
    and emission of photons, 29–31
    forms of photoluminescence, 31
    length, 32
    spectra, 232
    spectrum of absorption coefficient, 30
Acidity, 236
Acousto-optic tunable filters (AOTFs), 287, 400
Actin, 261
    actin-based processes, 365
Adding-doubling model (AD model), 81, 144
Additive errors, 118–119
Adenosine triphosphate (ATP), 275, 301
Agricultural Research Service (ARS), 16
Algorithm-based correction methods, 407
American Society for Testing and Materials (ASTM), 138, 405
Analytes, 417
Analytical process, 100
Analytical solution, 121–122
Angle of incidence, 93
Angle of transmission, 93
Angular dependence of scattering, 37
Animal tissues, 20
Anisotropic
    diffusion equation, 257
    diffusion theory, 256–258
    samples, 257
    scattering, 36
Anisotropy coefficient, *see* Anisotropy factor
Anisotropy factor, 35–39, 54, 91, 143, 233–234
Anti-Stokes scattering, 31, 388–389
    Raman scattering, 321
Antibodies, 430–431
Antibody–antigen
    binding, 432
    complex nephelometric and turbidimetric detection, 430–431
AOTFs, *see* Acousto-optic tunable filters (AOTFs)
Apple fruit, relationship between optical properties and structural properties of, 179–180
Area detectors, 288

ARS, *see* Agricultural Research Service (ARS)
ASTM, *see* American Society for Testing and Materials (ASTM)
Asymptotic confidence intervals for nonlinear models, 118
ATP, *see* Adenosine triphosphate (ATP)
Autocorrelation function, 338, 339
    autocorrelation procedure, 339
    correlation coefficients, 339–340
    in DLS, 338
    monodisperse particle solution, 340
Avalanche photodiode detector (APD), 338
AVD, *see* Absolute value of differences (AVD)

B

Backscattering
    autocorrelation function, 350
    backscattered detector, 322
    geometry, 390
    Raman spectroscopy, 390, 417
Bacteriological enumeration, 306–307
Ballistic scattering, 35
Bayesian decision rule, 368
Beam splitters, 287
Beef tenderness optical characterization, 271
    1-d scattering intensity, 276
    reduced scattering coefficient, 274, 276
    during rigor mortis, 275
    sarcomere length, 273–274
    Semimembranosus meat sample, 272
    theory of oblique-incidence reflectometry, 271–272
Beer–Lambert law, 33, 35, 134, 135
    for absorption measurement, 134–135
Beer's law, *see* Beer–Lambert law
Bench-top point-scan Raman imaging system, 395
β-lactoglobulin, 346
BHI, *see* Brain heart infusion (BHI)
Bidirectional reflectance distribution function (BRDF), 138
Biological materials, 345
Biological shift factor for firmness, 201
Biological sources of biospeckle activity, 363
    dynamic light scattering, 364, 366
    dynamic speckle patterns, 363
    intracellular cyclosis in onion, 364
    metabolic processes, 365
    in plant research, 364
Biological tissues, 163, 226

445

Biospeckle, 17
  activity, 362
  for biological samples, 362
Biospeckle spatial activity analysis, 369
  analysis in spectral domain, 375
  EMD method, 374
  Fujii method, 369–370
  GDs method, 370–371
  LASCA, 371–372
  MHI approach, 373
  mLSI technique, 372–373
  multivariate speckle pattern analysis, 375–376
  normal vector space statistics, 374–375
  temporal difference method, 373
Biospeckle technique, 361
  biological sources, 363–366
  botanical materials, 362
  dynamic laser speckle, 362
  fruit and vegetable quality evaluation, 376–381
  image analysis for assessment, 366–376
  system for biospeckle evaluation, 362
Birefringence, 13
Black body, 27
Body diffuse reflection, 26
Boltzmann equation, 51, 293–294
Botanical materials, 362
Botrytis aclada-infected onions, 232
Boundary conditions, 54
  EBC, 56–57
  Fresnel reflection at interface, 55
  PCBC, 55–56
  ZBC, 55
Bovine serum albumin (BSA), 355
"Braeburn" apples, 196, 197
Brain heart infusion (BHI), 439
BRDF, *see* Bidirectional reflectance distribution function (BRDF)
Brillouin scattering, 321
Brix-to-acid ratios, 236
°Brix, 236
Brownian motion, 17, 332, 335, 350, 362
Bruises, 244
BSA, *see* Bovine serum albumin (BSA)
Burkholderia cepacia-infected onions, 232

C

Carboxylated latex particles, 431
Carboxylated polystyrene particles, 434
Cartesian coordinate system, 84
Casein micelles, 319, 320
Cell(s), 20
  membrane, 20
Cellular metabolic processes, 364–365

Cellulose, 435
Central-processing unit (CPU), 97
Charge-coupled device (CCD), 165, 238, 285, 324, 388
  camera, 260, 439
  detectors, 288
  image acquisition, 271
Cheesemaking, milk coagulation and syneresis during, 324–326
Chemical enhancement, 394
Chemical images, 414
Chemiluminescence, 15
Classical electromagnetic theory, 2
Classic MC algorithms, 103
Classification models, 212–213, 217–218
CLSM, *see* Confocal laser-scanning microscopy (CLSM)
CMOS, *see* Complementary metal–oxide–semiconductor (CMOS)
Co-occurrence matrix, 367
Coagulation, 325
CoAguLite sensor, 326
Coefficient of variation (CV), 172
Coherence, 6
Coherent interference, *see* Constructive interference
Collagen content, 252–253
Collimated transmittance, 135
Combination bands, 14
Complementary metal–oxide–semiconductor (CMOS), 288
  detectors, 288
Computational methods, 82
Computational speed improvement, 93
  convolution, 94–95
  hybrid MC methods, 96
  parallel-computed MC algorithms, 96–97
  perturbation MC, 96
  photon packets, 93–94
  scaling MC, 95
  variance reduction techniques, 97
Concentration effects, 344
  3D cross-correlation DLS, 346–347
  cross-correlation, 345–346
  multiple scattering, 344–345
  two-color DLS, 346
Confidence intervals of parameters, 118
Confocal laser-scanning microscopy (CLSM), 179, 180
Consiglio per la Ricerca e la Sperimentazione in Agricoltura, Unità di ricerca per i processi dell'industria agroalimentare (CRA-IAA), 230
Constructive interference, 10
CONTIN algorithm, *see* Inverse Laplace transform

Index

Continuous-wave (cw), 48
  light illumination, 57–59
Conventional Henyey–Greenstein function, 258
Conventional rheology, 332
Conventional visible/near-infrared spectroscopy, 29
Convolution, 94–95
Correlation function, 341
Correlations, 209–210, 216–217
Counter-propagating diffuse flux, 144
"Coupled wave" equations, 262
CPU, see Central-processing unit (CPU)
"Cripps Pink" apples, 195
Critical angle, 8
Cross-correlation, 345–346
Cryo-transmission electron microscopy (Cryo-TEM), 322
CUDA toolkit, 97
Cumulant method, 342–343
Cumulant size, 343
Cumulative distribution function of free path, 86
Curl, 12
Current density, see Energy flux
Custom-designed Raman system, 403–404
CV, see Coefficient of variation (CV)
cw, see continuous-wave (cw)
Cylindrical symmetry, 87
Cytoplasm, 20
Cytoplasmic streaming, 365

D

Data collection, 111–112
Data preprocessing, 405
  fluorescence background correction for Raman spectra, 408–409
  fluorescence removal, 406–407
  hardware correction methods, 407
Data transformation, 122–123
Defects, 243–245
Deflection angle, 38
Demodulation method, 65
Deoxyribonucleic acid (DNA), 3
Derivatives, 76–77
Destructive interference, 10
Detectors, 400
  CCD sensor, 401
  high-performance CCDs, 403
  single-track readout mode for Raman spectral acquisition, 402
Diagnosis window, 4
Dichroic, 13
Difference histogram (DH), 368
Diffraction, 6, 10, 11
Diffuse/diffusion
  center, 272
  coefficient, 53, 335
  flux, 140
  model, 161
  reflectance, 27, 140
  theory, 90
  transmittance, 141
Diffuse reflection, 25
  from Lambertian surface, 27
  and transmission, 24–29
  types of medium, 26
Diffusing wave spectroscopy (DWS), 333, 348
  applications in food materials, 351–352
  principle, 350–351
  in transmission, 349
Diffusion approximation, 161
  equation, 52, 255
Diffusion approximation theory, 52–54; see also Radiative transfer theory
  model, 293–294
Diffusion equation, see Diffusion approximation equation
Dilute samples, 334
Dirac–Delta impulse response, 94
Direct measurement of optically thin (single scattering) tissue, 141
  light propagation models, 144
  measuring optical properties, 142
  refractive index, 143
  scattering phase function, 143
  scattering coefficient, 142
  total attenuation coefficient, 141
Direct methods, 160, 226
Direct Mie scattering detection, 440–442
Discrete ordinates method, 81
Divergence, 12, 76, 77
DNA, see Deoxyribonucleic acid (DNA)
Double-integrating spheres, 151
DWS, see Diffusing wave spectroscopy (DWS)
Dynamic laser speckle, 362
Dynamic light scattering (DLS), 17, 332, 333, 364, 366; see also Raman scattering
  autocorrelation function, 338–340
  concentration effects, 344–347
  DLS data analysis, 338
  hydrodynamic size, 340–341
  instrument setup, 337–338
  light scattered from solution of macromolecules, 335
  methods to calculating size distribution, 342–343
  microstructure and rheological property measurements in foods, 348–355
  optical microrheology, 333
  polydisperse samples, 341–342
  principle of measuring, 335–337
  SLS vs., 333–334, 336
Dynamic speckle patterns, 363

E

EBC, *see* Extrapolated boundary condition (EBC)
E. coli contamination, 304–306
ED, *see* Euclidean distance (ED)
Effective attenuation coefficient, 58, 271–272
Effegi penetrometer, 236
Elastic scattering, 15, 34
Electroluminescence, 15
Electromagnetic
 enhancement, 394
 radiation, 2
 spectrum, 2
Electromagnetic theory, 5, 6, 10; *see also* Ray theory
 dichroic, 13
 divergence, 12
 Maxwell's equations, 11, 12
 polarization, 10–11, 13
 propagation of electromagnetic radiation, 11
Electronic transitions, 29–30
Electron microscopy, 321–322
Emission of photons, 29–31
Empirical function, 178
Empirical mode decomposition method (EMD method), 374
Energy
 flux, 49
 gain, 50–51
 loss, 50
 of photon, 2
"Equi-intensity" profile, 257, 258
Euclidean distance (ED), 410
Excitation
 sources, 397–398
 state, 30
Exponential decay constant, 340
Extinction coefficient, 50, 86, 141
Extrapolated boundary condition (EBC), 55–57
Extrusion process, 254
Ex vivo methods, 134

F

Far-NIR, 323
Fast perturbation MC algorithm on GPU, 97
Fat globules, 319–320
Fecal samples, 437
Feed-forward neural network model, 299
FEM, *see* Finite element method (FEM)
Fiber-array probe SRS, 166–168
 measurement, 168–169
Fiber-optic Raman probe, 403
Fiber formation, 253
 CCD image acquisition, 271
 laser scanning system, 269–270
 in meat analog, 266
 "photon migration"-based method, 268–269
 reflectance profiles, 267
 single-and twin-screw extruders, 254
Fiber optic sensors, 324
Fibrous tissue, Monte Carlo simulation of light propagation in, 258–260
Fick's law, 52, 53, 61
Field water samples, 437
Filter wheel, 287
Finite element(s), 66
 application examples, 67–69
 finite element formulation, 66–67
 finite element mesh, 68
 formulation, 66–67
 MC simulations, 69
 modeling of light propagation in scattering media, 65
 numerical methods, 66
Finite element method (FEM), 44
Firmness, 236
Flesh tissues of Spanish yellow onions, 153
Flexible phase function, 103
Fluence rate, 48–49, 52, 56, 57, 70, 88
Fluorescence, 30
Flux, 48–49
Foodborne pathogens, 429, 441
Food materials, 332, 348, 387
 applications in, 351–352
Food products, MC simulation on, 97–103
Food quality and safety, 327
 applications for, 416
 food and agricultural products, 417
 Raman scattering applications for, 418–420
 SERS technique, 422
 SORS for nondestructive evaluation of tomatoes internal maturity, 421
Foods, 19
Food samples tested with light scattering immunoassays, 436–438
Food systems, 332
Forward problem, 112–113, 120
Fourier transform-infrared spectroscopy (FT-IR spectroscopy), 297
Fourier transform (FT), 61, 323, 375–376, 388
Fraction
 of parallel light component, 22
 of ports to whole surface area of sphere, 137
Frequency
 analysis, 375
 frequency-domain technique, 63
 frequency-modulated light illumination, 61–63
Fresh
 food products, 24
 fruit, 205–207

Freshness, 302–303
    detection device, 309
Fresnel equations
    at interface, 55
    for reflection and transmission, 21–24
Fresnel reflection, *see* Surface reflection
Fringe, 10
Fruit maturity, 198
    absorption and maturity, 204–208
    estimates of parameters, 202
    fresh fruit, 205–207
    nonlinear mixed effects regression analysis results, 199
    prediction of firmness decay during shelf life, 203
    processed fruit, 207–208
    rate constant, 198
    scattering and maturity, 208–209
    softening modeling, 200–204
    "Spring Bright" nectarines, 199, 200
Fruits and vegetables, 225
    biochemical processes in, 376
    biospeckle activity during fruit maturation, 381
    biospeckle activity for commodities, 377
    biospeckle method, 378–379
    fungal infection changes, 379–380
    GD and IM methods, 380–381
    optical properties of, 226–234
    quality assessment of, 234–246
    quality evaluation, 376
    shelf life effect, 378
FT-IR spectroscopy, *see* Fourier transform-infrared spectroscopy (FT-IR spectroscopy)
FT-Raman spectrometers, 400
FT, *see* Fourier transform (FT)
Fujii method, 369–370
Fujii's index, 369–370
Full width at half maximum (FWHM), 291, 398

G

Galerkin method, 66, 67
Gallium arsenide (GaAs), 401
Gamma rays, 2–3
GD, *see* "Golden Delicious" (GD)
Gegenbauer kernel phase function, 100
Generalized differences method (GDs method), 370–371
Geometrical theory, *see* Ray theory
Geometry-splitting MC algorithms, 97
Geometry tracker, *see* Ray tracer
Global positioning system (GPS), 5
Gloss-measuring techniques, 24, 25
Gloss, 24
"Golden Delicious" (GD), 176

Gompertz function, 291–292, 307
Goniophotometer, 36
Gradient, 76
"Granny Smith" apples (GS apples), 176
Graphics-processing units (GPUs), 97
Green's function, 57, 60, 94, 95

H

H-zone, 261
Halogen lamps, 285–286
Halogen lights, 285–286
Hardware correction methods, 407
Helium neon (HeNe), 337
Henyey–Greenstein function, 37, 38
Henyey–Greenstein phase function (HG phase function), 80, 90, 99, 100, 259, 264
HG phase function, *see* Henyey–Greenstein phase function (HG phase function)
HISRS technique, *see* Hyperspectral imaging-based spatially-resolved spectroscopic technique (HISRS technique)
Histogram difference method, 368
Hybrid MC methods, 96
Hydrodynamic diameter, *see* Hydrodynamic size
Hydrodynamic radius, *see* Hydrodynamic size
Hydrodynamic size, 340–341
Hyperspectral imaging-based spatially-resolved spectroscopic technique (HISRS technique), 170
Hyperspectral scattering
    imaging, 288–289
    technique, 304
Hyperspectral imaging, 170–173; *see also* Monochromatic imaging
    configuration, 241–242
    techniques, 284–285, 295–296

I

IAD method, *see* Inverse adding-doubling method (IAD method)
IDF Standard, *see* International Dairy Federation Standard (IDF Standard)
IDF Standard 157: 2007 method, 325
Illumination unit, 285–286
IM, *see* Inertia moment (IM)
Image processing algorithms, 304
Image analysis for biospeckle activity assessment, 366
    biospeckle spatial activity analysis, 369–376
    global activity measures, 366–369
Imaging
    spectrographs, 287
    techniques, 254, 321
IMFs, *see* Intrinsic mode functions (IMFs)
Impact bruising, 244

Incident diffuse flux, 144
Indirect methods, 134, 160, 226
Indium arsenide (InAs), 401
Indium gallium arsenide (InGaAs), 401
Inelastic scattering, 15, 34
Inertia moment (IM), 367
Information generator, 96
Infrared (IR), 323
  light, 4
Instrument-based correction methods, 407
Instrumental response function (IRF), 189
Integral-transformed diffusion model (ITDM), 122
Integrating sphere (IS), 134, 135, 227
  diffuse reflectance, 140
  diffuse transmittance, 141
  ideal and three-ports, 136
  optical measurements using, 138–141
  theory, 136–138
  total reflectance of incident light, 139
  total transmittance of sample, 140
Intensity, *see* Irradiance
Intensity size distribution, *see* Size distribution
Interactance, 28
Interference, 6, 10
  principle of, 10
Interferogram, 400
Internal conversion, *see* Nonradiative relaxation
Internal disorders, 213
  absorption and reduced scattering coefficients and, 214–216
  absorption and scattering spectra and, 213–214
  classification models, 217–218
  correlations and threshold values, 216–217
  mealiness, 213
International Dairy Federation Standard (IDF Standard), 325
Internet-based parallel computing, 97
Intrinsic mode functions (IMFs), 374
Inverse adding-doubling method (IAD method), 134, 149, 231; *see also* Kubelka–Munk method (K–M method)
  average values and standard deviations, 155
  double-integrating spheres, 151
  measuring optical properties of food, 152–155
  parallel slab geometry and reflectance and transmittance on slab geometry, 149
  single-integrating-sphere-based system, 152
  theory, 149–152
Inverse Laplace transform, 343
Inverse light transport problems, 112
Inverse problem, 113–114, 120
  importance, 114–115
Inverse radiation transport problems, *see* Inverse problem
In vitro experiments, 98

In vivo measurements, 17
In vivo methods, 134
In vivo optical property measurement techniques, 44, 57
Ion channels, 365
Ionizing radiation, 2–3
IR, *see* Infrared (IR)
IRF, *see* Instrumental response function (IRF)
Irradiance, 46–48
IS, *see* Integrating sphere (IS)
Isotropic
  diffusion equation, 257
  fluence rate, 162
  scattering, 36
ITDM, *see* Integral-transformed diffusion model (ITDM)

K

Kronecker delta function, 66
K/S ratio, 148
Kubelka–Munk method (K–M method), 134, 144; *see also* Inverse adding-doubling method (IAD method)
  applications in food and agriculture, 147–149
  K–M theory, 238
  K–M units, 301
  measurement techniques, 146–147
  model, 144–146
  reflectance and transmittance measurement, 146
  scattering coefficient, 226

L

Lab-on-a-chip device, 434
Lambertian radiator, 27
Lambertian reflection, 136
Lambertian surface, 26, 136
LAN, *see* Local area network (LAN)
Laplace operator, 77
Laplacian operator, *see* Laplace operator
Large field of view (LFV), 324
Laser(s), 286, 397
  backscattering imaging, 245
  light, 362
  scanning system, 269–270
Laser scattering
  experimental arrangement, 237
  image acquisition system, 238
Laser speckle contrast analysis (LASCA), 371–372
Latex particle immunoagglutination assay, *see* Particle-enhanced light scattering immunoassays
Law of refraction, 7
LCTF, *see* Liquid crystal-tunable filter (LCTF)

Index

LD function, *see* Lorentzian distribution function (LD function)
LD muscle, *see* M. *longissimus dorsi* muscle (LD muscle)
Least-squares support vector machines (LS-SVMs), 302
LEDs, *see* Light-emitting diodes (LEDs)
Legendre polynomials, 52
LeM, *see* Less mature (LeM)
Less mature (LeM), 199
LFV, *see* Large field of view (LFV)
Light, 1
   absorption, 159, 227
   backscattering, 324
   gamma rays, 2–3
   microwaves, 4–5
   radio waves, 5
   scattering and applications in food and agriculture, 15–18
   sources, 191, 285–286
   spectrum of electromagnetic radiation, 3
   spectrum of visible light, 4
   speed of, 2
   transfer in multilayered media, 163
   transport theory, 255
   UV light, 3
   visible and infrared light, 4
   x-rays, 2–3
Light-emitting diodes (LEDs), 285, 286, 441
Light interaction; *see also* Light transfer theory
   absorption, 29–33
   with food and biological materials, 19, 20
   with meat and meat analog, 254–256
   reflection and transmission, 21–29
   scattering, 33–40
Light propagation
   anisotropic diffusion theory, 256–258
   equi-intensity contours, 265
   in meat analogs, 256
   models, 97, 144
   Monte Carlo simulation, 258–261
   optical diffraction by single muscle fiber, 262–263
   optical reflectance in whole muscle, 264–266
   sarcomere structure, 266
Light scattering-based detection of food pathogens
   direct Mie scattering detection from food sample, 440–442
   food samples tested with light scattering immunoassays, 436–438
   Mie scattering theory, 430
   nephelometric and turbidimetric detection of antibody–antigen complex, 430–431
   particle-enhanced light scattering immunoassays, 431–436
Light scattering, 227, 295–296, 319, 333
   pathogen identification using light scattering refraction pattern imaging, 438–440
   techniques, 29
Light scattering immunoassay, 430–431
   food samples tested with, 436–438
Light transfer theory, 44; *see also* Light interaction
   analytical solutions to diffusion equation for semi-infinite scattering media, 57–65
   boundary conditions, 54–57
   diffusion approximation theory, 52–54
   finite element modeling of light propagation in scattering media, 65–69
   radiative transfer theory, 49–52
   radiometric quantities, 44–49
Line-scan
   method, 290
   system, 288–289
Line lasers, 395
Liquid crystal-tunable filter (LCTF), 173, 240, 287, 395
Local area network (LAN), 5
Logarithm-transformed diffusion model (LTDM), 122
Logistic model, 201
Lorentzian distribution function (LD function), 240, 290
Lorentzian function, 291, 297
LS-SVMs, *see* Least-squares support vector machines (LS-SVMs)
LTDM, *see* Logarithm-transformed diffusion model (LTDM)
Luminescence, 15
Luminescent light, 15

M

Magness–Taylor firmness tester (MT firmness tester), 236
MALS technique, *see* Multiangle light scattering technique (MALS technique)
Matte surface, *see* Lambertian surface
Maximum *a posteriori* estimation (MAP), 119
Maxwell's equations, 11, 12, 22, 39
MC simulation, *see* Monte Carlo simulation (MC simulation)
Mealiness, 213, 244
Mean free path, 86
   for absorption, 32
   path for scattering, 35
Mean square displacement (MSD), 353
Meat, 251
   collagen content, 252–253
   light interaction with, 254–256
   quality, 284

Meat (*Continued*)
  samples, 437
  sarcomere length, 252
  spectral scattering in, 284
Meat analog(s), 253
  light interaction with, 254–256
  light propagation in, 256–261
  optical characterization of beef tenderness, 271–276
  optical characterization of fiber formation in, 266–271
  unique vegetable protein-based food product, 254
Meat quality attributes assessment, 294
  color, 294–295
  detection of multi quality attributes, 299–300
  pH value, 298–299
  quality assurance and control, 294
  tenderness, 295–296
  water-holding capacity, 297–298
Meat safety attributes assessment, 300
  E. coli contamination, 304–306
  shelf-life estimation, 306–308
  TVB-N, 302–304
  TVC, 300–302
Meat tenderness, 251
  in beef industry, 253
  collagen content, 252–253
  tenderness, 252
Meshed MC with flexible phase function (MMC-fpf), 103
Metabolic processes, 365
MHI approach, *see* Motion history image approach (MHI approach)
Microfluidic devices, 434, 435
  with optical waveguide channel, 435
Microrheology, 352
  DLS, 354–355
  MSD, 353
  optical microrheology, 352–353
Microstructure and rheological property measurements in foods, 348
  DWS, 348–352
  microrheology, 352–355
Microwaves, 4–5
Mid-infrared spectroscopy, 14
Mie phase function, 99
Mie scattering, 35, 321; *see also* Raman scattering
  detection, 429–430
  Mie theory model, 39
  scattering coefficient, 35
  scattering event, 36
  scattering phase function and anisotropy factor, 35–39
  theory, 430

Mie theory, 101, 193, 345
  model, 39
Milk, 319
  coagulation and syneresis during cheesemaking, 324–326
Milk and dairy processing
  determining size distribution, 321–323
  food quality and safety, 327
  light scattering, 319
  monitoring composition, 323–324
  monitoring milk coagulation and syneresis during cheesemaking, 324–326
  in participating media, 320–321
MISRS technique, *see* Monochromatic imaging-based spatially-resolved spectroscopic technique (MISRS technique)
MLD, *see* Modified LD (MLD)
M. longissimus dorsi muscle (LD muscle), 265, 276
MLR models, *see* Multilinear regression models (MLR models)
mLSI technique, *see* modified laser speckle imaging technique (mLSI technique)
MMC-fpf, *see* Meshed MC with flexible phase function (MMC-fpf)
Modified Gompertz
  functions, 291–293
  model, 113
Modified HG phase function, 99, 100
modified laser speckle imaging technique (mLSI technique), 372–373
Modified LD (MLD), 240
Modified Lorentzian function, 291
Modified polynomial curve-fitting method, 407
MoM, *see* More mature (MoM)
Monochromatic imaging-based spatially-resolved spectroscopic technique (MISRS technique), 169–170
Monochromatic imaging, 169–170
Monte Carlo simulation (MC simulation), 59, 66, 69, 82, 112, 161, 234
  computational speed improvement, 93–97
  on food products, 97–103
  light propagation in fibrous tissue, 258–260
  MC techniques, 255
  model, 262
  nonparametric phase functions, 100–102
  parametric phase functions, 99–100
  from plane parallel to realistic geometries, 103
  principles of, 82–83
  propagation of photons, 82
  tracing photons through tissue, 83–93
More mature (MoM), 199
Motion history image approach (MHI approach), 373
Moving spherical coordinate system, 85

Index

M. psoas major muscle (PM muscle), 265
MSD, *see* Mean square displacement (MSD)
M. semitendinosus muscle (ST muscle), 265
MT firmness tester, *see* Magness–Taylor firmness tester (MT firmness tester)
Multiangle light scattering technique (MALS technique), 334
Multilaser scatter imaging system, 243
Multilinear regression models (MLR models), 294
Multiple regression analysis, 322–323
Multiple scattering, 344–345
Multispectral scattering imaging, 289–290
Multispectral imaging
    spectrograph, 239
    techniques, 295–296
Multivariate
    data analysis, 323
    speckle pattern analysis, 375–376
Myosin, 261

N

Near-infrared (NIR), 256, 323
    digital projector, 173
    hyperspectral imaging system, 295–296, 299
    light, 4
    region, 30
    wavelength range, 189
Near-infrared spectroscopy (NIRS), 4, 14, 33, 390
Negative "image source", 57
Nephelometric and turbidimetric detection of antibody–antigen complex, 430–431
Neural network (NN), 239
NIR, *see* Near-infrared (NIR)
NN, *see* Neural network (NN)
Nondestructive techniques, 225
Nonlinear least-squares estimates, 126
Nonparametric phase functions, 100–102
Nonradiative relaxation, 30
Normalization, 118
Normal vector space statistics, 374–375
Numerical methods, 44, 66, 113, 121, 161

O

OCT, *see* Optical coherence tomography (OCT)
OD, *see* Optical density (OD)
ODM, *see* Original diffusion model (ODM)
OLS, *see* Ordinary least squares (OLS)
1D spectral scattering profiles, 239
One radian, 45
Online NIR system, 309
OPA, *see* "Optical Property Analyzer" (OPA)
Opaque, 20
Optical
    properties, 159
    properties of medium, 80
    reflectance in whole muscle, 264–266
    scattering, 255
    sensor technologies, 324
    spectroscopy, 254
Optical-based methods, 254, 255
Optical absorption measurement and scattering properties of food, 133
    Beer's law for absorption measurement, 134–135
    direct measurement of optically thin (single scattering) tissue, 141–144
    IAD method, 149–155
    integrating sphere technique, 135–141
    K–M method, 144–149
    photometric techniques, 134
Optical characterization of beef tenderness, 271–276
Optical characterization of fiber formation
    CCD image acquisition, 271
    laser scanning system, 269–270
    in meat analog, 266
    "photon migration"-based method, 268–269
    reflectance profiles, 267
Optical coherence tomography (OCT), 143
Optical density (OD), 33
Optical diffraction by single muscle fiber, 262–263
Optical fiber probe, 165
    fiber-array probe SRS, 166–168
    MISRS system, 169
    spatially-resolved diffuse reflectance spectra of fresh apple, 168
    translation stage SRS, 165–166
Optical interference, 10
Optical methods, 429–430
Optical microrheology, 333, 352–353
Optical or radiant intensity, 9
Optical parameters, 54, 227
Optical properties of fruits and vegetables, 226, 232; *see also* Quality assessment of fruits and vegetables
    absorption and scattering, 227–233
    anisotropy factor, 233–234
    optical absorption and reduced scattering coefficients, 228–229
"Optical Property Analyzer" (OPA), 172
Optical theories, 5
    approximate timeline, 6
    electromagnetic theory, 5, 6, 10–13
    quantum theory, 5, 6, 13–15
    ray theory, 5, 7–9
    wave–photon duality characteristic, 5
    wave theory, 5–6, 9–10
Optics, 1
Optimization, 114

Ordinary differential equation, 64
Ordinary least squares (OLS), 117–118
Original diffusion model (ODM), 123
"Overlap region", 261
Overtone transition, 14

P

Pale, soft, and exudative (PSE), 297
Paper fibers, 435
Paper microfluidics, 435
Parallel-computed MC algorithms, 96–97
Parallel computing, 97
Parameter estimation methods, 112, 114–115, 125–126
　case study, 120–129
　choice of model, 121–122
　confidence intervals of parameters, 118
　data transformation and weighting method, 122–123
　forward problem, 112–113
　inverse problem, 113–115
　OLS, 117–118
　residual analysis, 118–119
　scaled sensitivity coefficients, 116–117, 123–125
　sequential estimation, 119–120
　standard statistical assumptions, 115–116
　statistical results for optical parameters and dependent variable, 126–127
　theory of, 115–120
Parameterized Fujii method, 370
Parametric phase functions, 99–100
Partial current boundary condition (PCBC), 55–56, 65
Partial differential equations (PDEs), 113
Partial least-squares regression modeling (PLSR modeling), 295–296
Particle-enhanced light scattering immunoassays, 431
　antibody-conjugated particles, 434
　antibody–antigen binding, 432
　carboxylated polystyrene particles, 434
　paper microfluidics, 435
　particles sizes, 433
　polystyrene particles, 432–433
　Smartphone application, 436
Particle-splitting technique, 97
Particle size and size distribution, 343–344, 433
Path length of light, 237
Pathogen identification using light scattering refraction pattern imaging, 438–440
PCA, *see* Principal component analysis (PCA)
PCBC, *see* Partial current boundary condition (PCBC)
PCR, *see* Polymerase chain reaction (PCR)
PDEs, *see* Partial differential equations (PDEs)

PDMS, *see* Polydimethylsiloxane (PDMS)
PDW spectroscopy, *see* Photon density wave spectroscopy (PDW spectroscopy)
Perturbation MC, 96
Phase function, 88
Phosphate buffered saline (PBS), 436
Photoluminescence, 15, 30
　forms of, 31
Photometric techniques, 134
Photon(s), 13
　absorption, 87–88
　launching, 86
　packets, 87, 93–94
　step size, 86–87
Photon correlation spectroscopy, *see* Dynamic light scattering (DLS)
Photon density wave spectroscopy (PDW spectroscopy), 326
"Photon migration"-based method, 268–269
Photon scattering, 88
　anisotropy factor, 91
　diffusion theory, 90
　directional cosines, 92
　function of deflection and azimuthal angle, 89
　HG scattering phase function, 90
　phase function, 88
pH value, 298–299
"Pink Lady®" apples, 196
Planar sinusoidal lighting pattern, 63
Plane
　parallel to realistic geometries, 103
　scanning, 289
Plant
　plant-based materials, 30
　tissues, 20
PLSR modeling, *see* Partial least-squares regression modeling (PLSR modeling)
PM muscle, *see* *M. psoas major* muscle (PM muscle)
Point-spread function (PSF), 169
Point lasers, 395
Polarization, 6, 10–11, 13
Polydimethylsiloxane (PDMS), 434
Polydisperse samples, 341–342
Polydispersity, 342
　index, 334, 343
Polymerase chain reaction (PCR), 429
Polynomial curve fitting, 407
Polynomial equation, 267
Polystyrene particles, 432–433
Portable and movable prototype devices, 309
　freshness detection device, 309
　online NIR system, 309
　TVC detection device, 309–312
Predicted reflectance profiles, 59
Principal component analysis (PCA), 375
Principle of superposition, 10

Index

Probabilistic methods, 82
Probability distribution, 36
Processed fruit, 207–208
Propagation
   of light in medium, 120
   of light waves, 9
Proteolysis, 253, 265
Proton pumps, 365
PSE, see Pale, soft, and exudative (PSE)
Pseudorandom number, 86, 100
PSF, see Point-spread function (PSF)

Q

QE, see Quantum efficiency (QE)
QELS, see Quasi-elastic light scattering (QELS)
Qualitative analysis, 409
   individual compositions, 414
   SMA for mixture of chemical adulterants, 411–413
   spectral similarity metrics, 410
Quality assessment of fruits and vegetables, 234; see also Optical properties of fruits and vegetables
   defects, 243–245
   hyperspectral imaging configuration, 241–242
   laser scattering experimental arrangement, 237
   laser scattering image acquisition system, 238
   LD function, 240
   multilaser scatter imaging system, 243
   multispectral imaging spectrograph, 239
   other applications, 245–246
   ripeness and quality, 236–243
   summary of research, 235–236
Quantitative analysis, 414–416
Quantum efficiency (QE), 401
Quantum mechanism, see Quantum theory
Quantum physics, see Quantum theory
Quantum theory, 5, 6, 13
   luminescent light, 15
   Schrödinger equation, 14
   wave-like and particle-like behavior, 13
Quartz-halogen lamps, see Halogen lamps
Quartz iodine lamps, see Halogen lamps
Quasi-elastic light scattering (QELS), 333

R

Radial dependence of diffuse reflectance, 161
Radiance, 46–48, 52, 54
Radiant
   energy, 44–45
   intensity, 45–46
   power, 44–45
Radiant flux, see Radiant—power

Radiation transport equation, see Boltzmann function
Radiative transfer equation, see Radiative transport equation (RTE)
Radiative transfer theory, 44, 49
   energy gain due to internal source, 51
   energy gain due to scattering, 50–51
   energy loss due to extinction, 50
   energy loss due to scattering out, 50
   forms of radiation, 49
   radiant energy flowing, 49
   standard form of RTE, 51–52
Radiative transport equation (RTE), 49, 80, 134, 255
Radiative transport theory (RTT), 79–80
Radio frequency (RF), 400
Radiometric quantities, 9, 44
   fluence rate and flux, 48–49
   radiance and irradiance, 46–48
   radiant energy and power, 44–45
   radiant intensity, 45–46
Radio waves, 5
Raman chemical imaging (RCI), 388, 394
   bench-top point-scan Raman imaging system, 395
   global excitations, 395
   macro-scale RCI, 396
   SORS, 396–397
Raman data analysis techniques, 405
   data preprocessing, 405–409
   qualitative analysis, 409–414
   quantitative analysis, 414–416
Raman instruments
   components of Raman systems, 397–403
   custom-designed Raman system, 403–404
   integrated bench-top Raman systems, 403
   macro-scale RCI, 404–405
   Raman data, 404
   Raman measurement systems, 403
   spectral calibrations, 405
Raman measurement techniques, 390
   backscattering Raman spectroscopy, 390
   Raman chemical imaging, 394–397
   SERS, 394
   SORS, 391–394
   transmission Raman spectroscopy, 390–391
Raman probe, 403–404
Raman scattering, 15, 17, 31, 39–40, 388, 429–430; see also Dynamic light scattering (DLS)
   applications for food quality and safety, 416–422
   effect, 39
   principles, 388–390
   Raman data analysis techniques, 405–416
   Raman instruments, 397–405
   Raman measurement techniques, 390–397

Raman shift, 389
Raman signals, 390, 397
Raman spectral fingerprint, 297
Raman spectroscopy, 40
Raman spectrum, 389
Raman systems
  components, 397
  detectors, 400–403
  excitation sources, 397–398
  wavelength separation devices, 398–400
Rayleigh scattering, 30, 34, 99, 321, 398
Ray theory, 5; *see also* Electromagnetic theory
  refractive index of medium, 7
  scattering, 7
  total internal reflection, 8, 9
Ray tracer, 83, 103
RCI, *see* Raman chemical imaging (RCI)
Reduced albedo, 64
Reduced scattering coefficient, 38, 53, 54, 191, 227
Reflectance, 28
Reflection-grating-based dispersive Raman spectrometer, 398
Reflection-grating imaging spectrograph, 287
Reflection, 5, 21
  diffuse, 24–29
  Fresnel equations for, 21–24
  at layer boundaries, 92–93
Reflectivity of coating material, 137
Refraction, 2, 5, 7
  at layer boundaries, 92–93
Refractive index, 80, 82, 133, 143
  of medium, 7
  refractive index-matched boundary, 54
  refractive index-mismatched boundary, 54
Regression models, 210–212
Relative-weighting diffusion model (RWDM), 123
Release of energy, 30
Residual analysis, 118–119
Resolution, 322
RF, *see* Radio frequency (RF)
Rheo-DWS setup, 351
Rotational transitions, 29, 30
RTE, *see* Radiative transport equation (RTE)
RTT, *see* Radiative transport theory (RTT)
RWDM, *see* Relative-weighting diffusion model (RWDM)

S

Salmonella enteritidis (SE), 440
SAM, *see* Spectral angle mapper (SAM)
SAOR, *see* Small-amplitude oscillatory rheometry (SAOR)
Sarcomeres, 251–252
SC, *see* Sugar content (SC)

Scaled sensitivity coefficients, 116–117, 123–125
Scaling MC, 95
Scanning electron microscopy (SEM), 322
Scattering, 7, 33, 133, 159, 227–233, 323
  anisotropy, 259
  coefficient, 35, 80, 82, 142, 255, 258
  event, 36
  intensity, 275
  Mie scattering, 35–39
  phase function, 5, 35–39, 259
  process, 255–256
  profile, 36
  Raman scattering, 39–40
  Rayleigh scattering, 34
  scattering phase function, 80, 82, 99, 101, 143
  spectra parameters, 195
  vector, 337
Scattering characteristics
  Boltzmann function, 293–294
  extraction and analysis, 290
  modified Gompertz functions, 291–293
  modified Lorentzian functions, 290–291
Scattering image acquisition methods, 288
  hyperspectral scattering imaging, 288–289
  multispectral scattering imaging, 289–290
Schrödinger equation, 14
SCM, *see* Spectral correlation mapper (SCM)
SE, *see* *Salmonella enteritidis* (SE)
Secondary electron detector, 322
SECV, *see* Standard errors of cross-validation (SECV)
Self-modeling mixture analysis (SMA), 392
SEM, *see* Scanning electron microscopy (SEM)
Semi-infinite scattering media, analytical solutions to diffusion equation for, 57, 58, 60, 62, 68
  cw light illumination, 57–59
  frequency-modulated light illumination, 61–63
  short-pulsed light illumination, 59–61
  spatially modulated light illumination, 63–65
Semi-infinite turbid medium, 162
Sensing modes, 28, 29
Sensitivity matrix, 116
Sequential estimation, 119–120
SERS, *see* Surface-enhanced Raman spectroscopy (SERS)
SEV, *see* Standard error for validation (SEV)
SFD, *see* Spatial frequency domain (SFD)
SFDI, *see* Spatial frequency-domain imaging (SFDI)
Shannon entropy, 368
Shelf-life estimation, 306–308
Short-pulsed light illumination, 59–61
Short-wavelength NIR spectroscopy, 303
SID, *see* Spectral information divergence (SID)
Signal-to-noise ratios (SNRs), 166, 401

Index

Silicon, 401
Simple equations, 66
Simple two-parameter phenomenological model, 237
Simulation, 111, 112
Single-integrating-sphere-based system, 152
Single muscle fiber, optical diffraction by, 262–263
Size distribution, 321–323, 342–343
Skeletal muscles, 261
Slab geometry, 149, 150
sLASCA, *see* Spatially derived laser speckle contrast analysis (sLASCA)
Slice shear force (SSF), 295
SLS, *see* Static light scattering (SLS)
SMA, *see* Self-modeling mixture analysis (SMA)
Small-amplitude oscillatory rheometry (SAOR), 325
Smartphone application, 436
Snell's law, 7, 8, 21
SNRs, *see* Signal-to-noise ratios (SNRs)
Softening modeling, 200–204
Softening process, 213
Software correction methods, 407
Solid angle, 45, 46
Soluble solids content (SSC), 236
SORS, *see* Spatially offset Raman spectroscopy (SORS)
Soy protein, 253–254
Space–time speckle patterns (STS patterns), 367–368
Spatial contrast, 366
Spatial frequency-domain imaging (SFDI), 65, 160, 173–175, 227
Spatial frequency domain (SFD), 173
Spatially-resolved reflectance spectroscopy, *see* Spatially-resolved spectroscopy (SRS)
Spatially-resolved spectroscopy (SRS), 71, 72, 160, 227; *see also* Time-resolved reflectance spectroscopy (TRS)
  hyperspectral imaging, 170–173
  instrumentation for, 165–175
  measurement configurations, 160
  measurement principle, 161
  monochromatic imaging, 169–170
  optical fiber probe, 165–169
  optical properties measurement of food products, 175–180
  relationship between optical properties and structural properties of apple fruit, 179–180
  scheme of light propagation in two-layered turbid medium, 164
  SFDI, 173–175
  steady-state light transfer in homogenous media, 161–163
  steady-state light transfer in layered media, 163–165
  theory and modeling for, 160–165
Spatially derived laser speckle contrast analysis (sLASCA), 372
Spatially modulated light illumination, 63–65
Spatially offset Raman spectroscopy (SORS), 388, 391, 392
  offset spectra, 392
  SMA, 392, 394
  for subsurface detection, 393
Spatial–temporal correlation technique, 369
Speckle, 10
  contrast, 366–367
Spectral
  flux, 45
  intensity, 46
  irradiance, 48
  measurements, 29
  radiance, 48
  spectral domain, analysis in, 375
Spectral angle mapper (SAM), 410
Spectral correlation mapper (SCM), 410
Spectral information divergence (SID), 410
Spectralon panels, 27
Spectral power, *see* Spectral flux
Spectral scattering, 226, 284; *see also* Dynamic light scattering (DLS); Raman scattering
  area detectors, 288
  extraction and analysis of scattering characteristics, 290–294
  imaging, 306–307
  light sources, 285–286
  line-scanning hyperspectral scattering imaging system, 285
  in meat, 284
  meat quality attributes assessment, 294–300
  meat safety attributes assessment, 300–308
  optical properties of fruits and vegetables, 226–234
  portable and movable prototype devices, 309–312
  quality assessment of fruits and vegetables, 234–246
  scattering image acquisition methods, 288–290
  wavelength dispersion devices, 286–287
Spectroscopic analysis, 323
Specular reflection, 23
Sphere multiplier, 138
Spherical coordinate system, 48
"Spring Belle" peaches, 194, 195, 196
SRS, *see* Spatially-resolved spectroscopy (SRS)
SSC, *see* Soluble solids content (SSC)
SSF, *see* Slice shear force (SSF)
Standard error for validation (SEV), 239

Standard errors of cross-validation (SECV), 242
Standard MC algorithms, 83
Standard statistical assumptions, 115–116
Static light scattering (SLS), 333
    DLS vs., 333–334, 336
    light scattered from solution of macromolecules, 335
Statistical analysis, 126
Steady-state diffusion model, 121
Steady-state light transfer
    in homogenous media, 161–163
    in layered media, 163–165
Stepwise discrimination, 305
Steradian (sr), 45
Sternomandibularis beef muscle, 264
ST muscle, *see M. semitendinosus* muscle (ST muscle)
Stokes–Einstein equation, 335, 340
Stokes–Einstein relationship, 354
Stokes' law, 335
Stokes Raman scattering, 321
Stokes scattering, 388–389
STS patterns, *see* Space–time speckle patterns (STS patterns)
Sugar content (SC), 236
Support vector machine (SVM), 302, 307
Suppression, 346
Surface-enhanced Raman spectroscopy (SERS), 388, 394, 422
Surface diffuse reflection, 26
Surface reflection, 7, 23
Sweetness, 236
System of linear equations, 67

T

Target analyte, 394
Taylor series expansion of first order, 56
TCSPC, *see* Time-correlated single-photon counting (TCSPC)
TE, *see* Transverse electric (TE)
Teflon slab, 392
Televisions (TVs), 5
TEM, *see* Transmission electron microscopy (TEM)
Temporal contrast, 367
Temporal difference method, 373
Temporally derived laser speckle contrast analysis (tLASCA), 372
Tenderness, 252, 295–296
Textural profile analysis, 254
Texture, 209
    classification models, 212–213
    correlations, 209–210
    regression models, 210–212
Thermoplastic expansion, 254
Thornhill sarcomere model, 264

Three-dimension (3-D)
    multiple-scattering problems, 160
    plots, 129, 130
    position of photons, 83
3D cross-correlation experiment (3D-DLS), 346–347
Threshold values, 216–217
Time-correlated single photon counting (TCSPC), 189
Time-domain laser reflectance spectroscopy, 242
Time-resolved reflectance spectroscopy (TRS), 187, 227; *see also* Spatially-resolved spectroscopy (SRS)
    absorption and scattering spectra in fruit species, 193–198
    applications of, 188
    data analysis, 191–193
    fruit maturity, 198–209
    instrumentation and data analysis, 189–193
    internal disorders, 213–218
    setups, 190
    spectral regions and wavelengths, 192–193
    texture, 209–213
Time-resolved reflectance technique, 61
Time history speckle pattern (THSP), see Space–time speckle patterns (STS patterns)
Titin, 251–252
tLASCA, *see* Temporally derived laser speckle contrast analysis (tLASCA)
TM, *see* Transverse magnetic (TM)
"Tommy Atkins" mangoes, 196
Total attenuation coefficient, *see* Extinction coefficient
Total internal reflection, 8, 9
Total reflectance of incident light, 139
Total transmittance of sample, 140
Total viable count (TVC), 301, 306
    detection device, 309–312
    in meats, 301–302
    spoilage in meat, 300–301
Total volatile basic nitrogen (TVB-N), 302–304
Tracing photons through tissue, 83
    generic flowchart of MC algorithm, 84
    launching photon, 86
    localizing photons in tissue, 83–85
    photon absorption, 87–88
    photon scattering, 88–92
    photon step size, 86–87
    reflection and refraction at layer boundaries, 92–93
    two-layered tissue, 85
    visualization of photon absorption on log10-scale, 88, 89
Translation stage SRS, 165–166
Transmission, 21, 26
    diffuse, 24–29
    Fresnel equations for, 21–24

Index

Raman spectroscopy, 390–391
transmission-grating imaging spectrograph, 287
Transmission electron microscopy (TEM), 322
Transmittance, 28
Transparent samples, 334
Transverse electric (TE), 263
Transverse magnetic (TM), 263
TRS, *see* Time-resolved reflectance spectroscopy (TRS)
TRS system for multiwavelength measurements (TRS–MW), 189
TRS system working at discrete wavelengths (TRS–DW), 189
TRS system working at single wavelengths (TRS–SW), 189
Tunable band-pass filters, 287
Tunable light sources, 286
Tungsten halogen lamps, *see* Halogen lamps
Turbidimetry, 431
TVB-N, *see* Total volatile basic nitrogen (TVB-N)
TVC, *see* Total viable count (TVC)
TVs, *see* Televisions (TVs)
Two-color DLS, 346
Two-dimension (2-D)
inverse Fourier transform, 164
multiple-scattering problems, 160
Two-layered diffusion model, 163, 164
Two-layered turbid medium, 163, 164

U

Ultraviolet (UV), 287, 388
light, 3
Unscattered transmittance, *see* Collimated transmittance
U.S. Department of Agriculture (USDA), 16, 19, 309
UV-A rays, 3
UV-B rays, 3
UV-C rays, 3

V

Variance reduction techniques, 97
Variational method, 66, 67
Vector(s), 74
in Cartesian coordinate system, 75
dot product, 76
field, 74
in spherical coordinate system, 75
waves
Vibrational bands, 14
Vibrational transitions, 29, 30
Viscoelastic materials, 348
Visible (Vis), 98, 261
light, 4
Visible and near-infrared (VIS/NIR), 285
spectral regions, 285
spectroscopy system, 299

W

Warner–Bratzler shear force (WBSF), 253, 295
Water-holding capacity (WHC), 285, 297–298
Wave
characteristics, 2
function, 9
theory, 5–6, 9–10
Wavelength dispersion devices, 286–287, 398
dispersive Raman spectrometers, 398
FT-Raman spectrometers, 400
for Raman measurement systems, 399
Wavelet entropy (WE), 368
Wavelet transforms, 375–376
Wave–photon duality property, 2
WBSF, *see* Warner–Bratzler shear force (WBSF)
WE, *see* Wavelet entropy (WE)
Weighted generalized difference algorithm (WGD algorithm), 371
Weighting method, 122–123
WGD algorithm, *see* Weighted generalized difference algorithm (WGD algorithm)
WHC, *see* Water-holding capacity (WHC)

X

X-rays, 2–3

Z

Z-average hydrodynamic radius, 334
Zero boundary condition (ZBC), 55